Michael Chernick
15 Quail Drive
Holland, PA 1896

D1156376

OXFORD STATISTICAL SCIENCE SERIES

SERIES EDITORS

A. C. ATKINSON (*London School of Economics*)

D. A. PIERCE (*Radiation Effects Research Foundation, Hiroshima*)

M. J. SCHERVISH (*Carnegie Mellon University*)

D. M. TITTERINGTON (*University of Glasgow*)

R. J. CARROLL (*Texas A & M University*)

D. J. HAND (*Imperial College, London*)

R. L. SMITH (*University of North Carolina*)

OXFORD STATISTICAL SCIENCE SERIES

Highly Structured Stochastic Systems

Edited by

PETER J. GREEN
University of Bristol

NILS LID HJORT
University of Oslo

SYLVIA RICHARDSON
Imperial College, London

OXFORD
UNIVERSITY PRESS

OXFORD
UNIVERSITY PRESS

Great Clarendon Street, Oxford OX2 6DP

Oxford University Press is a department of the University of Oxford.
It furthers the University's objective of excellence in research, scholarship,
and education by publishing worldwide in

Oxford New York

Auckland Bangkok Buenos Aires Cape Town Chennai
Dar es Salaam Delhi Hong Kong Istanbul Karachi Kolkata
Kuala Lumpur Madrid Melbourne Mexico City Mumbai Nairobi
São Paulo Shanghai Taipei Tokyo Toronto

Oxford is a registered trade mark of Oxford University Press
in the UK and in certain other countries

Published in the United States
by Oxford University Press Inc., New York

© Oxford University Press, 2003

British Library Cataloguing in Publication Data

Data available

Library of Congress Cataloging in Publication Data

ISBN 0 19 851055 1

10 9 8 7 6 5 4 3 2 1

Typeset by the authors using L^AT_EX

Printed in Great Britain
on acid-free paper by
Biddles Ltd, *www.biddles.co.uk*

List of contributors

Elja Arjas Rolf Nevanlinna Institute, FIN-00014 University of Helsinki, Finland . . . . elja.arjas@rni.helsinki.fi

Kari Auranen Rolf Nevanlinna Institute, FIN-00014 University of Helsinki, Finland . . . . kari.auranen@rni.helsinki.fi

M. J. Bayarri Department of Statistics and O.R., Av. Dr. Moliner 50, Universitat de València, 46100 Burjassot, Valncia, Spain susie.bayarri@uv.es

Niels G. Becker National Centre for Epidemiology and Population Health, Australian National University, Canberra ACT 0200, Australia
. . . . niels.becker@anu.edu.au

Carlo Berzuini Dipartimento di Informatica e Sistemistica, University of Pavia, 27100 Pavia, Italy . . . . carlo.berzuini@unipv.it

Julian Besag Department of Statistics, University of Washington, Seattle WA 98195, USA . . . . julian@stat.washington.edu

Peter Clifford Department of Statistics, University of Oxford, 1 South Parks Road, Oxford OX1 3TG, UK . . . . peter.clifford@jesus.ox.ac.uk

Rainer Dahlaus Institute for Applied Mathematics, Universität Heidelberg, D 69120 Heidelberg, Germany . . . . . dahlhaus@statlab.uni-heidelberg.de

A. Philip Dawid Statistical Science, University College London, Gower Street, London WC1E 6BT, UK . . . . dawid@stats.ucl.ac.uk

Vanessa Didelez Statistical Science, University College London, Gower Street, London WC1E 6BT, UK . . . . vanessa@stats.ucl.ac.uk

Fabio Divino Department of Geoeconomics and Statistics, Università di Roma, Via del Castro Laurenziano, 9, 00161-Roma, Italy
. . . . fabio.divino@uniromal.it

Michael Eichler Institute of Applied Mathematics, Universität Heidelberg, INF 294, D-69120 Germany . . . . eichler@statlab.uni-heidelberg.de

Arnoldo Frigessi Norwegian Computing Centre, Blindern, 0314 Oslo, Norway
. . . . Arnoldo.Frigessi@nr.no

Alan E. Gelfand Institute of Statistics and Decision Sciences, Duke University, Durham, North Carolina, USA . . . . alan@stat.duke.edu

Walter R. Gilks MRC Biostatistics Unit, Institute of Public Health, Robinson Way, Cambridge, UK . . . . wally.gilks@mrc-bsu.cam.ac.uk

Simon J. Godsill Department of Engineering, University of Cambridge, Cambridge CB2 1PZ, UK .... sjg@eng.cam.ac.uk

Peter J. Green Department of Mathematics, University of Bristol, Bristol BS8 1TW, UK .... P.J.Green@bristol.ac.uk

R. C. Griffiths Department of Statistics, University of Oxford, 1 South Parks Road, Oxford OX1 3TG, UK .... griff@stats.ox.ac.uk

Simon C. Heath Centre National de Genotypage, 2 rue Gaston Crémieux, CP 5721, 91057 Evry Cedex, France .... simon.heath@cng.fr

Juha Heikkinen Finnish Forest Research Institute, Unioninkatu 40A, FIN-00170 Helsinki, Finland .... juha.heikkinen@metla.fi

Nils Lid Hjort Department of Mathematics, University of Oslo, Blindern, N-0316 Oslo, Norway .... nils@math.uio.no

Merrilee Hurn Department of Mathematical Sciences, University of Bath, Claverton Down, Bath, BA2 7AY. .... m.a.hurn@bath.ac.uk

Oddvar Husby Department of Mathematical Sciences, Norwegian University of Science and Technology, NO-7491 Trondheim, Norway
.... okh@math.ntnu.no

Leonhard Knorr-Held Department of Statistics, University of Munich, Ludwigstrasse 33, 80539 Munich, Germany
.... knorr-held@stat.uni-muenchen.de

Anne Korhonen (formerly Riiali) Pohjola Non-Life Insurance Company Ltd, Lapinmentie 1, 00013 Pohjola, Finland ... anne.korhonen@pohjola.fi

Jan T. A. Koster Erasmus University Rotterdam, P.O.Box 1738, 3000 DR Rotterdam, The Netherlands .... koster@fsw.eur.nl

Hans R. Künsch Seminar for Statistics, ETH Zentrum, CH-8092 Zurich, Switzerland .... kuensch@stat.math.ethz.ch

Steffen L. Lauritzen Department of Mathematical Sciences, Aalborg University, Fredrik Bajers Vej 7 G, DK-9220 Aalborg, Denmark
.... steffen@math.auc.dk

M. N. M. van Lieshout Centrum voor Wiskunde en Informatica, Kruislaan 413, 1098 SJ Amsterdam, The Netherlands .... colette@cwi.nl

Jesper Møller Department of Mathematical Sciences, Aalborg University, Fredrik Bajers Vej. 7 G, DK-9220 Aalborg, Denmark ...... jm@math.auc.dk

Julia Mortera Università degli Studi Roma Tre, Dipartimento de Economia, Via Ostiense, 139, 00154 Roma, Italy .... mortera@uniroma3.it

Anthony O'Hagan Department of Probability and Statistics, University of Sheffield, Sheffield S3 7RH, UK .... a.ohagan@sheffield.ac.uk

Philip D. O'Neill School of Mathematical Sciences,University of Nottingham, University Park, Nottingham NG7 2RD, UK .... pdo@maths.nott.ac.uk

Antti Penttinen Department of Mathematics and Statistics, University of Jyväskylä, FIN 40351 Jyväskylä, Finland ....penttine@math.jyu.fi

Sonia Petrone Instituto di Metodi Quantitativi, Università Bocconi, Viale Isonzo 25, 20136 Milano, Italy .... sonia.petrone@uni-bocconi.it

Sylvia Richardson Department of Epidemiology and Public Health, Imperial College Faculty of Medicine, Norfolk Place, London W2 1PG, UK ....sylvia.richardson@ic.ac.uk

Thomas S. Richardson Department of Statistics, University of Washington, Seattle WA 98195, USA .... tsr@stat.washington.edu

Christian P. Robert CEREMADE, Université Paris Dauphine, 75775 Paris Cedex 16, France ....xian@ceremade.dauphine.fr

Gareth O. Roberts Department of Mathematics, Lancaster University, Bailrigg, Lancaster LA1 4YW, UK ........G.O.Roberts@lancaster.ac.uk

James M. Robins Department of Epidemiology, Harvard School of Public Health, 677 Huntington Avenue, Boston MA 02115, USA .... robins@hsph.harvard.edu

Håvard Rue Department of Mathematical Sciences, Norwegian University of Science and Technology, 7491 Trondheim, Norway .... Havard.Rue@math.ntnu.no

Alexandra Mello Schmidt Instituto de Matematica - Bloco C, Universidade Federal do Rio de Janeiro IM-DME, CEP 21945-970, Rio de Janeiro R.J., Brazil ....alex@im.ufrj.br

Nuala A. Sheehan Department of Epidemiology and Public Health, University of Leicester, 22-28 Princess Road West, Leicester LE1 6TP, UK ....Nas11@le.ac.uk

Daniel A. Sorensen Danish Institute of Agricultural Sciences, Department of Genetics and Breeding, PB 50, DK-8830 Tjele, Denmark .... snfds@genetics.agrsci.dk

Peter Spirtes Department of Philosophy, Carnegie–Mellon University, Pittsburgh, PA 15213, USA .... ps7z@andrew.cmu.edu

David A. Stephens Department of Mathematics, Imperial College London, 180 Queens Gate, London SW7 2AZ, UK ....d.stephens@ic.ac.uk

Geir Storvik Department of Mathematics, University of Oslo, Blindern, N-0316 Oslo, Norway .... geirs@math.uio.no

Milan Studený Institute of Theory of Information and Automation, CZ 18208, Prague, Czech Republic .... studeny@utia.cas.cz

Jeremy Tantrum Department of Statistics, University of Washington, Seattle WA 98195, USA .... tantrum@stat.washington.edu

Simon Tavaré Program in Molecular and Computational Biology, University of Southern California, Los Angeles, CA 90089-1340, USA
....stavare@usc.edu

Alain Trubuil Institut National de la Recherche Agronomique, 78320 Jouy-en-Josas, France .... Alain.Trubuil@jouy.inra.fr

Sergey Utev School of Mathematical Sciences, University of Nottingham, Nottingham NG7 2RD, UK .... sergey.utev@nottingham.ac.uk

Aad van der Vaart Vrije Universiteit Amsterdam, De Boelelaan 1081, HV Amsterdam, The Netherlands .... aad@cs.vu.nl

Gunter Weiss Max Planck Institute for Evolutionary Anthropology, Inselstr 22, D-04275 Leipzig, Germany .... weiss@eva.mpg.de

Nanny Wermuth Abt Methodenlehre und Statistik, University of Mainz, 55099 Mainz, Germany .... wermuth@psych.uni-mainz.de

Carsten Wiuf Variagenics, 60 Hampshire St., Cambridge, MA 02139, USA
....cwiuf@variagenics.com

Contents

Introducing Highly Structured Stochastic Systems

Peter J. Green
University of Bristol, UK

Nils Lid Hjort
University of Oslo, Norway

Sylvia Richardson
Imperial College, London, UK

1 How to read this book

Highly Structured Stochastic Systems (HSSS) is the name given to a modern strategy for building statistical models for challenging real-world problems, for computing with them, and for interpreting the resulting inferences. Complexity is handled by working up from simple local assumptions in a coherent way, and that is the key to modelling, computation, inference, and interpretation. The aim of this book is to make recent developments in HSSS accessible to a general statistical audience.

Readers already familiar with the scope and principles of the HSSS approach may choose to skip most of this Introduction, and proceed directly to the topics of their choice, aided by the chapter headings in the Contents and perhaps by the guide to the structure of the book in Section 4 below. Others may be interested to read our exposition in Section 2 of the basic ideas underlying HSSS, that provide strong unity to a diverse range of statistical models and methodologies. In Section 3 we describe the origins and development of the Network and Programme in HSSS funded by the European Science Foundation (ESF), whose final activity was the production of this volume. Section 4 lays out the structure of the book itself, and our motivations for including these particular topics. In Section 5, we give our personal view of the future of the field.

2 What are highly structured stochastic systems?

HSSS combine simple local relations to build stochastic models that exhibit great complexity. Such complex stochastic models have found applications in areas as diverse as signal processing, genetics, and epidemiology. The needs of these areas have in turn stimulated important theoretical developments. By emphasizing common ideas and structures, such as graphical, hierarchical, and spatial models, and techniques, such as Markov chain Monte Carlo methods and local exact computation, cross-disciplinary work in stochastic systems has been promoted.

2.1   Background

Statistical methods that were once restricted to specialist statisticians, such as generalized linear modelling and multivariate discrimination, are now widely used by individual scientists, engineers, and social scientists, aided by statistical packages. However, these powerful and flexible techniques are still restricted by necessary simplifying assumptions, such as precise measurement and independence between observations, and it long ago became clear that in many areas such assumptions can be both influential and misleading.

There are several generic sources of complexity in data that require methods beyond the commonly-understood tools provided in mainstream statistical packages. Often data exhibit complex dependence structures, having to do for example with repeated measurements on individual items, or natural grouping of individual observations due to the method of sampling, spatial or temporal association, family relationship, and so on. Other sources of complexity are connected with the measurement process, such as having covariates at both individual and group level, errors in measuring responses and covariates, multiple measuring instruments, non-random drop-out of individuals from a longitudinal study, or missing data. For many years, such complications in data-generating processes were often cavalierly ignored in statistical analysis, leading to unquantifiable biases.

In the late 1980s, much of statistical science underwent significant transformation, stimulated by both external factors – technological developments in instrumentation and computing, and internal innovation – theoretical developments in stochastic modelling, especially in conditional independence, in spatial stochastic processes, and in Bayesian methodology. Statistics began to offer both the theoretical concepts and the computational tools to handle properly some of the challenges raised by modern science and technology.

In several domains of application – notably image processing, expert systems, and telecommunications networks – statistical science started to make a contribution to basic understanding, and to the research and development of important new methodologies. Modern stochastic modelling techniques proved capable of building highly complex probabilistic structures, to handle a great variety of practical problems. What remained to be found was a general formalism embracing the techniques mentioned, generalizing the ingredients of the models, broadening the scope of applications and allowing cross-fertilization between different areas; the key idea was the framework of hierarchical modelling.

2.2   Hierarchical models

The word 'hierarchical' in 'Bayesian hierarchical models' refers to a generic model building strategy in which latent (that is, unobserved) quantities are organized into (a small number of) discrete levels with logically distinct and scientifically interpretable functions, with probabilistic relationships between them that capture inherent features of the data. It should be contrasted, for example, with the use of the word in the phrase 'hierarchical clustering', where the levels represent only continuously-varying degrees of refinement, represented for example in a dendrogram.

Hierarchical models provide a unifying framework for dealing with all the sources of complexity identified in the preceding subsection, and of course many more besides.

Many individual subject areas within HSSS have developed their own isolated methods, software, and terms for handling such complexity: for example, *population methods* are widely used in pharmacokinetics, *multilevel* models in education and geography, *latent variable models* in psychology and econometrics, *random effects* models in biostatistics and elsewhere, *frailty* models in survival analysis, econometrics and reliability, *Bayesian networks* in artificial intelligence, *pedigrees* in genetics, and so on. These are all examples of hierarchical models as we use the term here, reflecting the presence of additional structure over the standard 'flat' statistical models.

A number of themes run through work being carried out in different disciplines, although any particular subject area may not make use of all of them:

- The need for complex hierarchical models taking into account features of the type outlined above.

- Graphical models, in which conditional independence assumptions are given a compelling pictorial form which focuses attention, and can also form the basis for organizing efficient computation.

- The growing use of Bayesian methods in which complex systems are described as full probability models: that is, all quantities, whether traditionally viewed as data, parameters, latent variables, or whatever, are expressed as random variables.

- The use of Markov chain Monte Carlo (MCMC) methods for making inferences in complex models.

2.3 Conditional independence

Dawid (1979) recognized the central role played by conditional independence in a precise description of the information conveyed by the value of one variable about others – whether observables or parameters – in a statistical model. Although there have been interesting attempts to formulate conditional independence in the abstract, the conventional interpretation in terms of conditional probabilities is the most fruitful, and underpins all of the work described in this book. By analogy with the use of conditional independence assumptions in temporal Markov processes, 'Markov properties' are statements about conditional independence in general statistical models, and the interplay between various Markov concepts has been the subject of much research; see, for example, Lauritzen (1996, chapter 3).

Hierarchical models can exhibit great complexity globally, yet have relatively simple local structure. The concept of conditional independence, whereby each variable is related locally (conditionally) to only a few other variables, is the key to both the construction and analysis of such models.

This building of global structure from local specification is central to HSSS, but does require a note of warning. There is certainly accumulated experience that the influence of local assumptions about variables that are 'remote' in a suitable sense from the objectives of inference is often negligible, but general rules are lacking. Indeed, a large part of statistical physics is concerned with 'phase transition' effects where there is infinite-range dependence. Theoretical work on these issues is badly needed, and in the meantime careful sensitivity studies should be conducted when using unfamiliar models.

2.4   Graphical models

Graphical modelling provides a diagrammatic representation of the logical structure of a joint probability distribution, in the form of a network or graph depicting the local relations among variables. The graph can have directed or undirected links or edges between the nodes, which represent the individual variables. Associated with the graph are various Markov properties that specify how the graph encodes conditional independence assumptions; see Lauritzen (1996) for comprehensive coverage of the theory of graphical models.

It is the conditional independence assumptions that give graphical models their fundamental modular structure, enabling computation of globally interesting quantities from local specifications. In this way, graphical models form an essential basis for HSSS methodologies.

More complex modelling typically requires both directed and undirected edges between variables in the model specification: a graph allowing both kinds of edges but with no directed loops is called a chain graph. This form of graph describes the majority of situations encountered so far in HSSS; there is continuing research, however, on how to augment these graphs to embrace more complex settings including hidden confounders, etc. See Richardson and Spirtes (this volume) and Cox and Wermuth (1996).

2.5   The Bayesian paradigm

Graphical models provide much insight into the structure of a probability model, and many ingredients of graphical modelling prove useful in frequentist statistical methodology, especially in multivariate analysis. However, statistical inference only benefits from the full power of the graphical formalism when it is set wholly within a probability model, that is in the Bayesian paradigm.

In the Bayesian framework, all unknown quantities, whether parameters, missing observations, or latent variables, are treated symmetrically as random variables. Their prior distributions, together with the likelihoods of observed variables, define a full joint probability model; inference about the unknowns is based on their posterior distributions (sometimes, with the aid of loss functions). This unified treatment of variables allows coherent integration and propagation of all sources of uncertainty. It gives a powerful framework in which prior scientific knowledge can be expressed, and in which inferences are made using probability calculation and decision theory, fully exploiting the graphical structure of the model.

It is therefore very natural to make inference in HSSS from a Bayesian perspective, and this is evident in many chapters in this book; conversely, the existence of HSSS has stimulated advancement of the Bayesian approach into a great variety of application areas, exploiting a degree of complexity and realism in modelling hitherto unattainable.

2.6 Computational methods

2.6.1 *Markov chain Monte Carlo*

There is a price to be paid for allowing such complexity into our statistical modelling: computation with very complex models is difficult, and exact, analytical tools are seldom available. Especially in the context of stochastic systems, novel computational methods are needed.

Markov chain Monte Carlo (MCMC) is a general method for sampling and computing expectations in multivariate stochastic processes. It is particularly useful when the correlation structure is complex or when the process has a density which is only known up to a proportionality constant, and in many cases it is the only known approach that is computationally feasible.

MCMC methods are iterative stochastic algorithms converging weakly to the distribution of interest. Ergodic theory justifies their use for sampling and computing expectations. Research has focused on the design of MCMC algorithms, and the proof of their correct convergence, together with more or less precise estimates of the rate of such convergence; this is an important issue because it is closely related to the performance of the algorithm and the design of practical stopping rules. Central limit theorems quantify the uncertainty of the estimates. There has been a renewed interest in the theory of discrete and continuous state space Markov chains, leading to important progress. A further focus of research has been to on-line monitoring, aiming at the detection of approximate convergence while the chain is running. Such monitoring is unnecessary when it is possible to construct perfect sampling methods, protocols for organizing MCMC computations that detect convergence precisely.

In the course of the 1990s, MCMC algorithms have revolutionized the practice of Bayesian statistics.

2.6.2 *EM algorithms*

MCMC is not the only class of numerical methods for which HSSS models are well-tuned. The expectation-maximization (EM) algorithm is a general approach to maximization of a likelihood in a so-called missing data problem by iteratively maximizing the *expected complete-data log-likelihood* over values of the unknown parameters, conditional on the observed data and the *current* values of those parameters.

Outside simple exponential families, the maximization step here is seldom tractable analytically. However, whenever the joint model for parameters and complete data has a simple local structure, and especially where there are no awkward normalizing constants, the complete-data log-likelihood separates into a sum of terms each involving only a few of the variables, and numerical implementation of EM is greatly facilitated.

2.6.3 *Local exact computation*

In some cases the modularity of an HSSS can be exploited to the point where the computation in even very complex systems can be handled exactly. This is the case if the system can be suitably embedded into a so-called acyclic hypergraph of local relations. Fast probability propagation algorithms have been developed to perform such embedding and to exploit this for computation; these algorithms function by passing messages between local modules and can be extremely fast and efficient, especially in discrete systems with sparse graphs. Sometimes these algorithms can be advantageously combined with MCMC methods.

2.7 Cross-fertilization

A benefit of the integration of a broad range of applications into a common model-building framework has been the opportunities for cross-fertilization of techniques between different fields. There are many successful examples of such opportunities being seized, and the following list is not exhaustive.

Statistical physicists' models were borrowed by spatial statisticians, and their stochastic algorithms adapted for general statistical computation. Markov random fields arising in spatial statistics as models, for example, for ecological phenomena, became standard tools for image modelling. Covariance selection, contingency tables and spatial modelling were ancestors of graphical modelling. Graphical models provided a formalism for expressing causality in clinical trials and epidemiology. Interpreting windowing of spatial point processes as a form of censoring in survival analysis has led to new statistical tools for point patterns.

Peeling algorithms for pedigrees turned out to be equivalent to probability propagation methods devised for expert systems, and related to the backwards–forwards methodology of hidden Markov models. In Markov chains, there was synergy between probabilistic research into their behaviour and their Monte Carlo application. MCMC methods devised by Bayesian statisticians became used in complex likelihood calculations involving missing data or intractable normalizing constants. Finally, bringing perfect simulation to bear on spatial processes led to discovery of the generic class of dominated coupling-from-the-past methods.

3 The ESF-funded HSSS Programme

Up to the early 1990s, progress in the areas reviewed in the previous section was made in a fragmentary, disconnected manner, with separate development of major ingredients of new methodologies – for example, graphical modelling – occurring in different application domains. However, the common structure of various problems was eventually recognized, and this recognition was the major motivation for an application to the ESF in 1991 for a new Scientific Network in statistical science. 'Highly Structured Stochastic Systems' (HSSS) was coined as a term to unify these types of research. The application was approved, and the Network was funded for the years 1993 to 1995. Subsequently, a full Programme was established, which ran from 1997 to 2001. These two initiatives played a major role in encouraging cross-disciplinary and international collaboration, and enabled genuine cross-fertilization between fields.

The activities funded by the Network and Programme were concerned with the underlying theory, the details of implementation of HSSS, and their application to a remarkably wide range of problems in the applied sciences. Open international conferences on the whole field of HSSS were held in Rebild, Denmark, in May 1996, and Pavia, Italy, in September 1999. A Summer School for training young researchers in MCMC theory and methodology was partially funded by HSSS. A total of 18 workshops were organized; the topics provide a good indication of the breadth of HSSS:

Highly Structured Stochastic Systems, Cortona, Italy, April 1994.

Model Criticism Concerning HSSS, Wiesbaden, Germany, September 1994.

Model Building and Model Interpretation, Luminy, France, June 1995.

Model Determination and Variable Dimension MCMC, New Forest, UK, September 1997.

Structured Backcalculation, Rome, Italy, November 1997.

Multivariate Research with Latent Variables, Wiesbaden, Germany, November 1997.

Exact Simulation, Rebild, Denmark, November/December 1997.

Statistical Models and Methods for Discontinuous Phenomena, Oslo, May 1998.

Markov Chain Monte Carlo (tutorial workshop), Limburg, The Netherlands, June 1998.

Structured Learning in Graphical Models, Tirano, Italy, September 1998.

Gene Mapping, Lammi, Finland, March 1999.

HSSS and Environmental Statistics, Willersley Castle, UK, May 1999.

On-line Monte Carlo Filtering and Prediction, Pavia, Italy, June 1999.

Statistical Modelling of Spatial and Space–Time Processes, Luminy, France, July 1999.

Data Analysis with Graphical Models – Computational Aspects, Applications, and Extensions, München, Germany, March 2000.

Bias Reduction and Confidence Estimation in Complex Models, Lillesand, Norway, June 2000.

Models and Inference in HSSS: Recent Developments and Perspectives, Luminy, France, November 2000.

Structural Stochastic Systems for Individual Behaviours, Louvain-la-Neuve, Belgium, January 2001.

HSSS funding also supported six more intimate 'research kitchens' and a total of 46 individual research visits. Much of the work presented at or generated via these workshops, research kitchens, and visits has led to publications in various journals of statistical science, covering the range from probability theory, statistics methodology, and applications.

Throughout our activities we were guided by one of the original aims of the Network, namely the encouragement of new talent into this area of research. The enthusiastic involvement of younger researchers in all the HSSS activities augurs well for the future vitality of the subject.

Other objectives of the Network and Programme were to broaden the interaction with scientists from other communities, in particular neural network researchers, dynamic modellers, geneticists, etc.; to ensure rapid dissemination of research results; to stimulate cross-fertilization across the diverse areas associated with HSSS; and to disseminate and spread expertise in HSSS internationally.

Activities throughout the seven years were directed by active scientific steering committees, whose members included the present editors and those listed in the Acknowledgements. Full details of HSSS activities can be found on the web site: http://www.maths.nottingham.ac.uk/hsss.

4   Structure of the book

The philosophy and coverage of this book were chosen by the HSSS Programme scientific steering committee. It was clear that to attempt to cover all topics embraced by the Programme's activities would be far too ambitious and lead to an unfocused volume, and one in which compromise had to be made between depth and breadth in each topic. It was therefore necessary to be selective.

It was felt that graphical modelling and MCMC methodology were so central to the field, and were involved in so many of the domains of application, that they demanded coverage in some depth. Overall, we wanted the book to be balanced between such underlying methodologies and substantive areas of application. We have sought to maintain coherence and focus by including two clusters of application-driven contributions – in the areas of spatial statistics and biological applications. Finally, we included two topics that seek to challenge the parametric assumptions that otherwise underlie most HSSS models. The result, we hope, is a well-balanced coverage of the field, although it does not claim to be completely comprehensive.

Altogether there are 15 topics in the book, and for each we have a substantial article by a leading author in the field. Most of these topics have attracted intense research attention over the duration of the Programme, and so one point of view on each is too restrictive. Our strategy for including diversity of emphasis and perspective was to hold a research workshop, at Luminy near Marseille in November 2000, at which all authors were present, together with two invited discussants for each topic. The discussants each made presentations that complemented, extended or discussed the main article. Productive interaction during and after the workshop, often involving the other participants, led to the revised articles and written discussion contributions included here. The discussions are quite heterogeneous in style, and should be read in parallel with the corresponding main article; there are no rejoinders from the main authors, but this should not be taken as necessarily indicating convergence of views. For reasons of space, the following review of the contents of the chapters concentrates on the main articles. Each of the 15 topics therefore leads to a chapter consisting of a main article and two discussions.

The first four chapters concern graphical models and causality. There is no general overview of graphical modelling, for which we refer the reader to Lauritzen (1996) or Cox and Wermuth (1996). Instead, here **Lauritzen** focuses on some innovative applications. He discusses in particular the daunting computational tasks faced by applications of realistic size and proposes new strategies for decision networks. Other domains of application – to social science and forensic genetics – are introduced by the discussants. Causality is an important and difficult concept and its interplay with directed graphical models (also known as influence diagrams) has long been controversial. **Dawid** demonstrates the use of intervention directed acyclic graphs to expose patterns of causality in a classic problem, that of partial compliance. By contrast, **Richardson and Spirtes** address the difficulty of explicit modelling of all latent variables through the introduction of ancestral graphs. Other points of view on causality are argued by the discussants, giving the reader insights into different formalisms. Another challenging topic is the interplay of causality with time, and **Dahlhaus and Eichler** address different possible notions of causal graph for time series and their relations with Granger causality.

The next group of chapters concern Monte Carlo methods, the numerical engine for inference in the great majority of current Bayesian hierarchical models. The basic ideas, underlying theory, and a wealth of applications of MCMC can be found in Gilks, Richardson, and Spiegelhalter (1996). Markov chain theory and the development of MCMC methods have somewhat diverged in the recent literature, and **Roberts** seeks to close the gap by discussing the impact of various concepts of ergodicity on practical implementations. MCMC samplers that traverse varying dimension parameter spaces have proved useful in extending the scope of Bayesian computation to model choice problems and those where the parameter length is inherently variable for increased flexibility. **Green** gives an exposition of the reversible jump method, sets it in the context of various alternative proposals for the same task, and proposes some potential generic strategies for generating trans-dimensional proposals. MCMC is difficult to use when there is a time sequence in the model as well as in the simulation, especially if it is required to process long signals in real time. **Berzuini and Gilks** discuss particle filters, using importance sampling and re-sampling in dynamic models, and borrowing ideas from MCMC methodology to introduce new steps into the standard algorithms to improve performance.

Spatial modelling has enjoyed a special role among application areas for HSSS. In the seminal paper of Besag (1974), the need to express spatial structure flexibly led to early examples of modelling using conditional independence ideas, and contributed to the genesis of graphical modelling generally. At the beginning of the 1980s, several groups of statisticians began to address problems in image analysis, and there was important cross-fertilization between the fields, including the development of the Gibbs sampler by Geman and Geman (1984). Although important work in spatial statistics has used spatial stochastic processes for particular directly observed phenomena, recent emphasis has been on the use of such models for latent processes, and that is the principal focus in the three chapters here. **Richardson** gives an overview of the types of spatial stochastic model that are used in spatial epidemiology, including both areally-aggregated and point process models, and discusses how to aggregate models from the point level. She emphasizes the interpretability of the resulting inferences, in particular in relation to explanatory variables at different scales. **Penttinen** deals with ecological applications, in particular analysing data on the joint occurrence of two species through hierarchical modelling of site suitability. The role of statistical modelling in image analysis is discussed by **Hurn, Husby, and Rue**, who review the main classes of prior image model (covering both high-level and pixel-based approaches), modelling of image data, treatment of parameters, and inference using loss functions. They demonstrate many of these strands in an application to contour detection from ultrasound data.

We next include a trio of topics whose common theme is their focus on modelling of biological systems. **Becker and Utev** use structured models informed by biological understanding to adjust for heterogeneity in infectious disease transmission, where a characteristic feature of the data is that the process is only very partially observed. They describe analytical techniques for estimating parameters that provide a tractable alternative to full Bayesian analysis in this context. There is huge growth in science related to genetics, and HSSS model-based approaches are becoming important in several fields. We have chosen to focus on two of these, omitting, for example, the whole areas of model-based sequencing and gene expression analysis. Gene-mapping is represented by the article by **Heath**, who expounds on linkage analysis, and critically reviews the range of methods for pedigree analysis from peeling by exact methods to MCMC implementation for more general problems with many marker loci. This class of models is interesting for the diversity of MCMC formulation that has been exploited. In addition to pedigree-based methods, interest is now turning to population-association gene-mapping methods, which has provoked renewed interest in population genetics models, such as the coalescent model. This is the topic of the article by **Griffiths and Tavaré** who focus on the use of the coalescent model in evolutionary genetics. Griffiths and Tavaré make use of combinatorial probabilistic arguments to quantify the distribution of the age of mutations, sometimes leading to direct or MCMC simulation-based numerical calculation.

In the final group of articles, **O'Hagan** makes a start on one of the challenges of current HSSS approaches — model criticism. He formalizes the iterative process of model building, criticism, and modification, and provides diagnostic tools addressing both local and global features of hierarchical models. While non-parametric statistics has been long-established in classical inference, Bayesian non-parametrics has a much shorter history due to the inherent complexities of mathematical formulation, statistical interpretation, and computational implementation. Among its aims is to free statisticians from the sometimes too hard modelling assumptions underlying most current HSSS applications, without essential loss of inference power. **Hjort** reviews recent developments and introduces various extensions, such as random quantile processes and Bayesian local regression.

We conclude this section by mentioning some of the topics that are inherently part of HSSS, and indeed in some cases were represented among the workshops and other activities funded by the HSSS Programme, but are not included in the book. Arguably under-represented areas include expert systems, time series in economics and finance, and in signal processing, longitudinal modelling, survival analysis, environmental statistics, geostatistics, space–time modelling, pattern recognition, perfect simulation, connections with statistical physics, genomics and bioinformatics including DNA and protein sequences, and quantum probability and statistics.

5 Expectations of the future

Although funding from the ESF under the title of Highly Structured Stochastic Systems has come to an end, the field has flourished and matured, and a substantial community of researchers has been established; we certainly expect that there will be continuing research in these topics for years to come.

Some of the issues of great current interest are: developing diagnostic and analytic tools for model criticism; understanding sensitivity of models to local specifications; designing automated and adaptive MCMC algorithms; identifying limits of causal interpretation in networks representing observational studies; introducing non-parametric elements into graphical models; and extending the theory and methodology to systems that develop over space and time, including on-line monitoring and control.

Looking to the future, many of the individual authors contributing to this volume have offered their own perspectives of probable developments in their topics. Perhaps one of the most serious generic challenges is discovering modelling strategies and computational methods to deal with very large data sets, such as arise with modern data acquisition systems in genomics, telecommunications and commercial data-mining.

Acknowledgements

We are greatly indebted to the other members of the HSSS Programme scientific steering committee: Elja Arjas, Arnoldo Frigessi, Richard Gill, Steffen Lauritzen, Tony O'Hagan, and Nanny Wermuth. They played an important role in planning this endeavour. We are also happy to have had the support of Adrian Baddeley, Alan Gelfand, Amy Racine-Poon, and David Spiegelhalter as members of the steering committee of the earlier Network, and of the members of the advisory scientific committee who provided wise counsel from the background. We are grateful to Catherine Werner, our efficient and friendly contact at the European Science Foundation in Strasbourg. Nikki Carlton and David Leslie have done a marvellous job as technical editors, integrating the different LaTeX and grammatical styles of 50 authors. Tragically, Nikki died suddenly in the closing months of this work; we are sad that she will not see the final result of her efforts, and was never fully aware of the gratitude of the authors and editors.

References

Besag, J. (1974). Spatial interaction and the statistical analysis of lattice systems (with discussion). *Journal of the Royal Statistical Society*, B, **36**, 192–236.

Cox, D. R. and Wermuth, N. (1996). *Multivariate Dependencies*. Chapman & Hall, London.

Dawid, A. P. (1979). Conditional independence in statistical theory (with discussion). *Journal of the Royal Statistical Society*, B, **41**, 1–31.

Geman, S. and Geman, D. (1984). Stochastic relaxation, Gibbs distributions and the Bayesian restoration of images. *IEEE Transactions on Pattern Analysis and Machine Intelligence*, **6**, 721–41.

Gilks, W. R., Richardson, S., and Spiegelhalter, D. J. (eds) (1996). *Markov Chain Monte Carlo in Practice*, Chapman & Hall, London.

Lauritzen, S. L. (1996). *Graphical Models*. Oxford University Press, UK.

1

Some modern applications of graphical models

Steffen L. Lauritzen
Aalborg University, Denmark

1 Introduction

In recent years there are a number of areas where graphical models have served successfully in the process of understanding, formulating, and solving problems. Although the early papers on graphical models dealt with undirected graphs (Darroch *et al.* 1980), recent applications of graphical models have predominantly been based on directed graphical models, also known as recursive models (Wermuth and Lauritzen 1983) or *Bayesian networks*, a term coined by Pearl (1986). This will also be reflected in the emphasis of the present article which will describe new areas of application and methodology for Bayesian networks, not otherwise well represented in this volume. This means in particular that we will not discuss the application of Bayesian networks to problems of genetic computation and problems of causal reasoning and discovery.

The versatility of Bayesian network methodology has in particular been demonstrated in connection with Bayesian modelling of complex biomedical systems using the general software BUGS (Gilks *et al.* 1994). We also choose not to describe this work here but refer the reader to Spiegelhalter (1998) and references listed therein.

After having described the basic elements of Bayesian networks and probability propagation algorithms in Section 2, Section 3 gives examples of the application of graphical models for *decision support*, both to medical and technical problems, and to the general area of decision analysis. In Section 4 we show how Bayesian networks are applied for *digital communication*. In Section 5 we describe the method of *variational inference*, an interesting alternative to Markov chain Monte Carlo (MCMC) methods which is popular within the machine learning community. Finally, in Section 6 we briefly mention other application areas.

2 Bayesian networks and probability propagation

This section gives a brief introduction to Bayesian networks and associated probability propagation algorithms (Lauritzen and Spiegehalter 1988; Jensen *et al.* 1990). The reader is referred to Cowell *et al.* (1999) for a detailed treatment.

A *Bayesian network* is determined by a directed acyclic graph $\mathcal{D}$ with *vertices* or *nodes* V serving also as labels for random variables $X = (X_v)_{v \in V}$, and a probability distribution of X with joint density p with respect to a product measure, which is assumed to factorize as

$$p(x) = \prod_{v \in V} p(x_v \mid x_{\mathrm{pa}(v)}), \tag{2.1}$$

where $\mathrm{pa}(v)$ denotes the set of parents in $\mathcal{D}$ of v and $x_A = (x_v)_{v \in A}$ for any subset $A \subseteq V$. Computational algorithms typically exploit a slightly different factorization over an undirected graph:

$$p(x) = \prod_{a \in \mathcal{A}} \phi_a(x), \tag{2.2}$$

where $\mathcal{A}$ is the set of *complete* subsets (i.e. subsets where all pairs of vertices have edges between them) of an undirected graph $\mathcal{G}$ and ϕ_a depends on x through x_a only. The *moral graph* $\mathcal{D}^m$ is obtained from $\mathcal{D}$ by first adding undirected edges between vertices with common children (marrying parents) and subsequently ignoring directions on all edges. If p factorizes over $\mathcal{D}$ as in (2.1), then it also factorizes over $\mathcal{D}^m$ as in (2.2). This is easily seen by letting

$$\phi_{v \cup \mathrm{pa}(v)}(x) = p(x_v \mid x_{\mathrm{pa}(v)}).$$

Whereas MCMC methods such as the Gibbs sampler typically operate on $\mathcal{D}^m$, exact propagation algorithms operate on a larger graph, which has the further property that it is *chordal* or *triangulated*, meaning that it has no cyclic induced subgraphs with more than three vertices. The cliques of a chordal graph $\mathcal{G}$ can be arranged in a so-called *junction tree*. This is a tree $\mathcal{T}$ with vertices equal to the set of cliques (maximal complete subsets) $\mathcal{C}$ of $\mathcal{G}$, which satisfies the additional property that for any two cliques $c_1, c_2 \in \mathcal{C}$ and any $c \in \mathcal{C}$ which lies on the path from c_1 to c_2 in $\mathcal{T}$, it holds that $c_1 \cap c_2 \subseteq c$. In general the junction tree $\mathcal{T}$ is not uniquely determined from $\mathcal{G}$, but the set of intersections $c \cap d$ of pairs of neighbours in $\mathcal{T}$, denoted $\mathcal{S}$ and named *separators*, can be shown to be independent of the particular tree chosen.

Exact probability propagation algorithms can be described in terms of message passing in the junction tree $\mathcal{T}$. Some variants of the algorithm operate on a basic factorization

$$p(x) \propto \frac{\prod_{c \in \mathcal{C}} \psi_c(x)}{\prod_{s \in \mathcal{S}} \psi_s(x)}, \tag{2.3}$$

where $\psi_a, a \in \mathcal{C} \cup \mathcal{S}$, are functions stored in the appropriate cliques and separators which as usual depend on x through x_a only, and thus have low complexity when the cliques are of low cardinality. The factorization (2.3) is easily constructed from the factorization (2.2) as long as all complete subsets in $\mathcal{D}^m$ are subsets of cliques in $\mathcal{G}$.

Messages are now sent between neighbours in the tree according to a specific *schedule*. An efficient message-passing schedule allows a clique to send exactly one message to each of its neighbours and only after it has already received messages from all its other neighbours. Thus the leaves in the tree are first allowed to send messages, where a leaf of the junction tree is a clique that has only a single neighbour. When a message is sent from c to d via the separator $s = c \cap d$, the following operations are performed:

$$\psi_c^{\downarrow s}(x_s) \leftarrow \sum_{y_{c \setminus s}} \psi_c(x_s, y_{c \setminus s}); \quad \psi_d(x_d) \leftarrow \psi_d(x_d) \frac{\psi_c^{\downarrow s}(x_s)}{\psi_s(x_s)}; \quad \psi_s(x_s) \leftarrow \psi_c^{\downarrow s}(x_s),$$

i.e. first the *s-marginal* $\psi_c^{\downarrow s}$ of ψ_c is calculated by summing out over all variables not in s, then the clique potential ψ_d is modified by multiplication with the ratio $\psi_c^{\downarrow s}/\psi_s$, and finally the separator potential ψ_s is replaced by $\psi_c^{\downarrow s}$.

The factorization (2.3) is invariant under the message-passing operation. Using an efficient schedule as described above, it can be shown that when exactly two messages have been sent along every branch of the junction tree, it holds that

$$\psi_a(x_a) \propto p^{\downarrow a}(x_a) = \sum_{y_{V \setminus a}} p(x_a, y_{V \setminus a}) \quad \text{for all } a \in \mathcal{C} \cup \mathcal{S}.$$

Generalizations of the message-passing scheme use different definitions of the marginalization operation $\downarrow$ and the multiplication used in the basic factorization and the message computation, but work otherwise in the same basic fashion (Shenoy and Shafer 1990; Lauritzen and Jensen 1997). For example, *dynamic programming* algorithms can be described in this way by replacing products and ratios with sums and differences, and marginalizing by maximization rather than summation (Dawid 1992).

3 Decision support and expert systems

An important area of application of Bayesian networks has been that of systems for decision support and expert systems. This area has also contributed vigorously to the theoretical development of graphical models.

3.1 Medical diagnosis

Some of the first practical attempts to build probabilistic expert systems involved very large graphs (Henrion *et al.* 1991) and were made in the area of medical diagnosis. This includes the system MUNIN (Andreassen *et al.* 1989) developed at Aalborg University for electromyographic diagnosis of diseases of the muscle–nerve system, but also the probabilistic reconstruction (Shwe *et al.* 1991) of the QMR/INTERNIST system for general medical diagnosis (Miller *et al.* 1982), as well as the PATHFINDER system for diagnosis of lymph vertex pathology (Heckerman *et al.* 1991). The latter has been converted to the commercial system INTELLIPATH, but generally the medical systems have had limited success in being transformed into widely available tools for medical decision support.

In the first phase of development of MUNIN, it reached the level displayed in Fig. 1 (Andreassen *et al.* 1992) with more than 1000 nodes each having be-

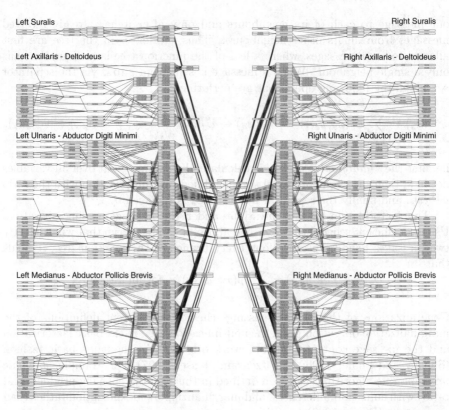

FIG. 1. Bayesian network for MUNIN. This network describes a pair of three
   muscle–nerve systems in a (micro)human and contains a little more than 1000
   nodes. Reprinted from Andreassen *et al.* (1992), p. 149, with permission from
   Elsevier Science.

tween 2 and 30 states. This network is concerned with a so-called 'microhuman',
having only three muscles on each side of the body, each group of muscles being
controlled from a single nerve with associated branches.

Based upon electromyographic recordings corresponding to observed values
at some of the 'outer' nodes of the network, conditional probabilities of states
at the 'inner' nodes are calculated by probability propagation. Initially a huge
number of conditional probabilities are specified based on neurophysiological
knowledge and some calibration exercises. Although an overwhelming number of
probabilities had to be specified, this process was aided by repeated substructures
in the network and local modelling.

Although work is still progressing towards making MUNIN into a practical diagnostic tool and the microhuman prototype has shown quite satisfactory
performance (Andreassen *et al.* 1996), it is not obvious that MUNIN will ever

develop into a usable system for diagnostic support, in particular because the complexity of a full-scale system of this type seems daunting.

A medical system of a different type is that of DIAS which is meant for planning insulin dosages for diabetes patients (Andreassen *et al.* 1994). The system is not quite so monstrous in size and the Bayesian network is essentially a discretized version of a system of differential equations regulating the metabolism of insulin, the level of discretization corresponding to 24 hourly time slices. The conditional probabilities are modelled using the published literature which measures various rates of metabolism. The Bayesian network is augmented with decision nodes that prescribe dosages of insulin of different type, and with utility nodes that measure the well-being of the patients. In addition there are nodes describing individual insulin sensitivity, which is then 'learnt' during a hospitalization phase with frequent measurements of blood glucose and controlled diet. The system reaches a dosage prescription by maximizing the expected utility using a simple gradient search. A simplified graphical representation of a part of the Bayesian network of the system is shown in Fig. 2. Although experimental results seem to indicate that DIAS has good abilities for predicting blood glucose levels (Hejlesen *et al.* 1998), there is still quite a distance to practical full-scale clinical testing and deployment as a standard tool. The difficulty is exacerbated by the absence of clinical standards to be used for comparison.

We finally mention TREAT which is a rather recent system built for assignment of antibiotics for severe bacterial infections of the urinary tract (Leibovici *et al.* 2000). This system is simpler than the previously mentioned systems and may have better chances of becoming a practical tool. We refrain from describing TREAT here in any detail.

3.2 Troubleshooting with Bayesian networks

It seems to have been somewhat easier for Bayesian networks to be used for diagnosis in an industrial context, for example where they form the basis for troubleshooting systems (Heckerman *et al.* 1995). There are probably several reasons for this: industrial systems are much simpler than the 'system' of a human being; knowledge about the construction of a technical system is more readily available; and there are not the same ethical problems involved in using machines for decision support and manipulation of the systems.

Microsoft and Hewlett Packard have developed systems for printer troubleshooting, and some of the HP systems are just about to be deployed for use in printer hotline support. The systems are built by a project group SACSO involving personnel from Aalborg University, Hewlett Packard Company, and HUGIN Expert Ltd, using rather simple and Bayesian networks for a large number (about 200) of *error scenarios*. In this context a tool has been developed for building general troubleshooting systems (Skaanning 2000). For a detailed desciption of the SACSO systems, see Skaanning *et al.* (2000).

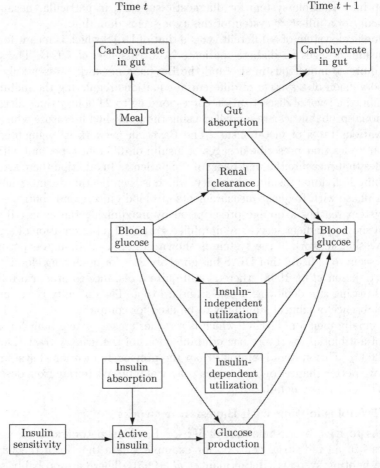

FIG. 2. Part of a one hour time slice of the Bayesian network for DIAS, showing
    how a meal and an insulin injection affects the concentration of carbohydrate
    in the gut and glucose in the blood.

3.3   Decision networks

Influence diagrams (IDs) (Howard and Matheson 1984) are compact representations of decision problems with uncertainty. They are natural extensions of
Bayesian networks and similar local computation algorithms for finding whether
optimal strategies exist (Olmsted 1983; Shachter 1986; Shenoy 1992; Jensen *et
al.* 1994). Such IDs consist of a Bayesian network augmented with decision and
utility nodes together with a protocol that identifies which variables are known
to the decision maker when decisions are to be taken.

The application of IDs to many real decision problems is hindered by a general tendency that the complexity of finding globally optimal strategies is daunting, despite these algorithms.

Lauritzen and Nilsson (2001) relax the standard assumption in an influence diagram of 'no forgetting', i.e. that values of observed variables and decisions which have previously been taken are remembered at all later times. They denote these more general diagrams by LIMIDs (LImited Memory Influence Diagrams). Such diagrams were also studied by Zhang *et al.* (1994) with a similar motivation under the general name of decision networks.

The fictitious example below is taken from Lauritzen and Nilsson (2001) and illustrates the notion of a LIMID and its relation to a traditional ID. We refer to this example as PIGS.

A pig breeder is growing pigs for a period of four months and subsequently selling them. During this period the pig may or may not develop a certain disease. If the pig has the disease at the time when it must be sold, the pig must be sold for slaughtering and its expected market price is then 300 DKK (Danish kroner). If it is disease free, its expected market price as a breeding animal is 1000 DKK.

Once a month, a veterinary doctor sees the pig and makes a test for presence of the disease. If the pig is ill, the test will indicate this with probability .80, and if the pig is healthy, the test will indicate this with probability .90. At each monthly visit, the doctor may or may not treat the pig for the disease by injecting a certain drug. The cost of an injection is 100 DKK.

A pig has the disease in the first month with probability .10. A healthy pig develops the disease in the subsequent month with probability .20 without injection, whereas a healthy and treated pig develops the disease with probability .10, so the injection has some preventive effect. An untreated pig which is unhealthy will remain so in the subsequent month with probability .90, whereas the similar probability is .50 for an unhealthy pig which is treated. Thus spontaneous cure is possible but treatment is beneficial on average.

The story now continues in two versions. In the version traditionally described by influence diagrams, the pig breeder will at all times know whether the pig has been treated earlier and also the previous test results. This story corresponds to a (finite) *partially observed Markov decision process* (POMDP). As opposed to fully observed Markov decision processes (Howard 1960), the complexity of exact POMDP algorithms grows quickly with time and they fail therefore to provide the optimal solutions desired; see White (1991) for a survey. To represent this version of PIGS by a LIMID, we let h_i, $i = 1, \ldots, 4$, denote whether the pig is *healthy* or *unhealthy* in the ith month and t_i, $i = 1, 2, 3$, represent the corresponding test results, which are said to be *positive* if they indicate presence of the disease, and otherwise *negative*. The nodes d_i, $i = 1, 2, 3$, correspond to the decisions *treat* or *leave*, the latter implying that no injection is made. The utility nodes u_1, u_2, and u_3 represent the potential injection costs, whereas u_4 is the (expected) market price of the pig as determined by its health

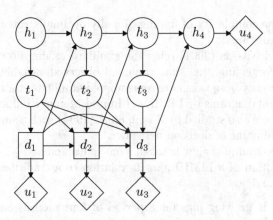

FIG. 3.   LIMID representation of the version of PIGS traditionally described
by an influence diagram. The full previous treatment and test history is
available when decisions must be taken.

at the fourth month. The corresponding diagram is displayed in Fig. 3. The assumption of 'no forgetting' is made explicit by links from predecessors of decision
nodes.

In another version of the story, the pig breeder does not keep individual
records for his pigs and has to make his decision knowing only the test result
for the given month. The memory has been limited to the extreme of only
remembering the present, thereby avoiding the complexity explosion due to remembering the past. This version of the story could not be represented by an
ID. The corresponding LIMID is displayed in Fig. 4.

The complexity of finding fully optimal strategies within LIMIDs is in general
prohibitive. Lauritzen and Nilsson (2001) describe a procedure called *Single
Policy Updating* (SPU) which leads to strategies that are locally optimal, in
the sense that no policy modification involving only a single decision node can
increase the expected utility of the strategy. A message-passing algorithm for
each SPU step is also given. General conditions are established for local optimal
strategies to be globally optimal. Algorithms for identifying such cases and the
corresponding optimal strategies are provided. In addition, algorithms are given
for reducing a given LIMID by identifying and removing redundant memory links.
Since IDs are special cases of LIMIDs, these steps lead in particular to more
efficient algorithms for solving problems which can be represented by traditional
IDs (Lauritzen and Nilsson 2001). The increase in efficiency is partly due to
complexity reduction from removing redundant memory links, and partly to the
SPU algorithm having smaller complexity than previously known algorithms.

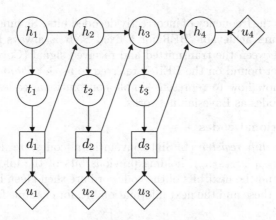

FIG. 4.   LIMID for the PIGS decision problem when only the current test result
is available at the time of decision.

4   Digital communication

The correction of errors due to electronic noise in digital communication plays
an important role in modern society. Such problems have a natural formulation
in terms of Bayesian networks (Frey 1998).

4.1   Coding and decoding noisy communication channels

Information is generally represented in binary vectors $u = (u_1, \ldots, u_K)$ of length
K. Before transmission, information vectors u are *encoded*, i.e. mapped into
binary *codewords* $x = (x_1, \ldots, x_N)$. Typically $N > K$ to enable error correction
when *decoding* noisy measurements $y = (y_1, \ldots, y_N)$ of x to obtain a reconstruction $\hat{u}$ of the original information vector u. A code as above is a (K, N)-*code*
and $R = K/N$ is the *rate* of the code.

The Additive Gaussian White Noise (AGWN) channel is a standard model
for digital communication. Here U is uniformly distributed on $\{0,1\}^K$, X is a deterministic and known function of U with values in $\{0,1\}^N$, and $Y \sim N(x, \sigma^2 I)$.

The tasks are to construct encoders with good error-correcting properties,
and reliable and fast algorithms for decoding. Some algorithms are of a purely
algebraic nature and therefore not discussed here. Typical probabilistic decoding
methods use Viterbi decoding (Viterbi 1967) or the forward–backward algorithm
– also known as the BCJR algorithm (Bahl *et al.* 1974) – to find the value of $\hat{u}$
that maximizes $P(u \mid y)$ jointly or bitwise. These algorithms can be seen as instances of the general probability propagation algorithms described in Section 2.

The *signal-to-noise ratio* (SNR) is the ratio between the power in the transmitted signal and noise. It is normally indicated in decibels (dB) and given as

$$10 \log_{10} \text{SNR} = 10 \log_{10} \frac{1}{2R\sigma^2}.$$

The quality of a coded communication channel is measured by its *bit error rate*

(BER), which is the frequency of incorrectly decoded bits. Shannon (1948) shows that the minimum achievable BER for a fixed SNR is always larger than the cross-entropy between the transmitted and received signal (Cover and Thomas 1991). This lower bound on the BER is referred to as the *Shannon floor*.

Below we show how to represent simple convolutional codes and the more complex turbocodes as Bayesian networks.

4.2   Convolutional codes

Linear feedback shift register (LFSR) convolutional codes are determined by a register $r_k = (s_{k-1}, \ldots, s_{k-p})$, holding previous bits of the information vector and codeword, and the next bits of the codeword are then given by binary linear combinations of these and the next information bit, for example for a LFSR code with $R = 1/2$ as

$$x_{2k} = a_0 u_k \oplus a_1 s_{k-1} \oplus \cdots \oplus a_p s_{k-p}, \qquad x_{2k+1} = b_0 u_k \oplus b_1 s_{k-1} \oplus \cdots \oplus b_p s_{k-p},$$

where $\oplus$ denotes addition modulo 2. A simple such code (Lin and Costello 1983) with $p = 7$ and $s_k = u_k$ is given as

$$x_{2k} = u_k, \qquad x_{2k+1} = u_k \oplus u_{k-1} \oplus u_{k-3} \oplus u_{k-5} \oplus u_{k-6} \oplus u_{k-7}.$$

A Bayesian network for this code is shown in Fig. 5. Convolutional codes are simple to decode using probability propagation algorithms, but their error-correcting properties are less impressive and the best convolutional codes perform at approximately 3.7 dB from the Shannon floor.

4.3   Iterative decoding and turbocodes

Recent discoveries have led to the construction and use of superefficient turbocodes. These codes are typically based on a pair of identical convolutional codes

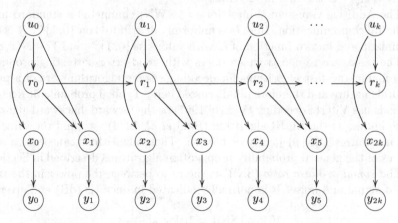

FIG. 5. Bayesian network for a simple convolutional code of rate 1/2.

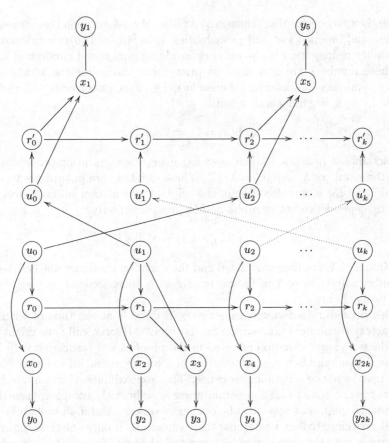

FIG. 6. Bayesian network for a rate $1/2$ turbocode with blocksize k. The sequence $(u'_1, \ldots, u'_k)$ is obtained from $(u_1, \ldots, u_k)$ by a known permutation (scrambler).

that are combined with a *scrambler*, which permutes a block of information bits u of size K to form $u' = (u'_1, \ldots, u'_K) = (u_{\pi(1)}, \ldots, u_{\pi(K)})$. The convolutional code is then applied both to u and u'. To keep the rate of the code at $1/2$, only every second bit of each codeword is transmitted. A Bayesian network for this construction is shown in Fig. 6.

The exact probabilistic decoding of these codes has a complexity that is exponential in the block size K, this being due to the scrambling of codebits. It is a remarkable and empirical fact that an iterative scheme of a type to be described below gives fast decoding with impressive results. Turbocodes have been constructed in this way with a BER only 0.5 dB from the Shannon floor (Berrou and Glavieux 1996; McEliece *et al.* 1998).

One instance of the iterative scheme runs as follows, where we let y and y'

denote the two transmitted sequences. At first, the information bits are assumed uniform and independent and probabilities $\lambda_i = p(u_i = 1 \mid y)$ are calculated by probability propagation in the network involving only u and children of u.

These numbers are now used as prior probabilities for the scrambled sequence, *ignoring dependence between bits*, i.e. it is (incorrectly) alleged that $u'_i, i = 1, \ldots, k$, are independent with

$$p(u'_i = 1) = \lambda_{\pi(i)}.$$

The scrambled network is then decoded using these as prior probabilities to yield the numbers $\lambda'_i = p'(u'_i = 1 \mid y')$. These numbers are in turn used as prior probabilities for a new decoding of the original information bit sequence, again ignoring dependence between bits and letting (incorrectly)

$$p(u_{\pi(i)} = 1) = \lambda'_i.$$

New values of λ_i are then calculated and the iteration continues until all λ-values are sufficiently close to 1 or 0, and the signal is then decoded by assigning the most probable bit.

This procedure is formally incorrect and there is at the time of writing no satisfactory theoretical explanation for its success. Heuristically one might argue that the only approximation involved in the procedure is associated with ignoring the dependence between information bits when passing information between the original and the scrambled sequence. This approximation error might be less serious for two reasons: if the permutation is sufficiently complex, then the independence might not appear to be obviously wrong. And if all probabilities are indeed very close to 0 and 1 because there essentially is only one codeword that is compatible with the sequence of bits received, then the distance to independence may be insignificant.

Since the appearance of turbocodes, MacKay (1999) has demonstrated similar performance of so-called *low density parity check codes*, also known as *Gallager codes*, with iterative probabilistic decoding. Such codes were invented by Gallager (1963) but were subsequently 'forgotten' because no efficient decoding algorithm was available. They consist in their simplest form of the information sequence u supplemented with a sequence of parity checks c, leading to a binary sequence $t = (u, c)$ satisfying $Ht = 0 \pmod 2$, where H is a very sparse matrix of zeros and ones; typically H could have dimensions $10\,000 \times 20\,000$ with three non-zero bits per column.

These codes are also decoded iteratively by probabilistic methods where messages are repeatedly sent along the edges of the graph with incidence matrix H. If this graph were a tree, then the iteration would converge after exactly two rounds, yielding correct marginal posterior probabilities for each bit. However, even though this is not the case for interesting H-matrices, the iteration still tends empirically to identify the correct bits although with incorrect probabilities associated. Again, the theoretical justification for the success of this approximative method is lacking, but heuristically the approximate correctness seems to be associated with the fact that the cycles are typically very long indeed.

We note that iterative probability propagation algorithms on graphs with cycles have also been used for the computation of approximate likelihoods for genetic pedigrees and identification of consistent genotype configurations (Lange and Goradia 1987; Janss *et al.* 1995; Wang *et al.* 1996; Heath 1998).

5 Machine learning and variational inference

Graphical models have recently been successfully applied in the general area of machine learning and pattern recognition (Neal 1996; Frey 1998). Models in this area are typically large Bayesian networks where exact inference by probability propagation is not possible. When inference need not be made with full accuracy, approximate methods may be used, for example MCMC algorithms, typically of the sequential type (Berzuini and Gilks, Chapter 7, this volume).

An interesting alternative to MCMC algorithms has become popular within the machine learning community. In its simple form, this is known as the *mean field method* and, as with MCMC methods, it originates from physics (Parisi 1988). In its modern and more general form it is known as *variational inference* (Jordan *et al.* 1998).

Variational inference addresses the problem where x is observed, y is unobserved, and $p(y \mid x)$ or $p(x)$ is desired, but exact computation of these is untractable. For simplicity we describe the method in the discrete case, although this is inessential and the method applies quite generally. The basic starting point is the following identity, where q is an arbitrary probability distribution:

$$\log p(x) = \sum_y q(y) \log p(x) = \sum_y q(y) \log \frac{p(x,y)}{p(y \mid x)}$$

$$= \sum_y q(y) \log \frac{p(x,y)}{q(y)} + \sum_y q(y) \log \frac{q(y)}{p(y \mid x)}$$

$$= \sum_y q(y) \log \frac{p(x,y)}{q(y)} + D(q \,\|\, p^*).$$

Here $p^*(y) = p(y \mid x)$ is the desired conditional distribution and $D(q \,\|\, p^*)$ is the Kullback–Leibler divergence between q and p^*. This divergence is necessarily non-negative, leading to the inequality

$$\log p(x) \geqslant \sum_y q(y) \log \frac{p(x,y)}{q(y)}, \tag{5.1}$$

the difference between the right-hand and left-hand sides being $D(q \,\|\, p^*)$. The idea is now to choose q within some class of tractable distributions so as to make this divergence small. The chosen q is then used as an approximation to p^*, and the right-hand side of (5.1) as an approximation to the log-likelihood

$\log p(x)$, which simultaneously is a lower bound. Maximizing this lower bound over a class of distributions $\mathcal{Q}$ is then equivalent to minimizing the approximation error $D(q \,\|\, p^*)$. Note that the Kullback–Leibler divergence has q and p^* reversed compared with what would be common in most contexts.

The classical mean field approximation concerns the case where $y = (y_\alpha)_{\alpha \in V}$ is a set of variables and q is chosen among product measures. The optimization of q is performed coordinatewise and iteratively. For fixed $q_\beta, \beta \neq \alpha$, the Kullback–Leibler divergence is minimized in q_α. To perform this minimization we write, for an arbitrary product measure q,

$$
\begin{aligned}
D(q \,\|\, p^*) &= \sum_y q(y) \log \frac{q(y)}{p^*(y)} \\
&= \sum_y q(y) \log \frac{q(y_\alpha) q(y_{V \setminus \alpha})}{p^*(y_\alpha \,|\, y_{V \setminus \alpha}) p^*(y_{V \setminus \alpha})} \\
&\sim \sum_y q(y) \log \frac{q(y_\alpha)}{p^*(y_\alpha \,|\, y_{V \setminus \alpha})} \\
&= \sum_{y_\alpha} q(y_\alpha) \log \frac{q(y_\alpha)}{r(y_\alpha)},
\end{aligned}
$$

where '$\sim$' indicates that the difference between the right-hand side and the left-hand side is constant in q_α for fixed $q_{V \setminus \alpha}$ and $r(y_\alpha) = \exp\left\{\mathrm{E}_q \log p^*(y_\alpha \,|\, Y_{V \setminus \alpha})\right\}$. Thus the optimal choice for q_α is given as

$$
q(y_\alpha) \propto \exp\left\{\mathrm{E}_q \log p^*(y_\alpha \,|\, Y_{V \setminus \alpha})\right\}. \tag{5.2}
$$

For α varying, this gives a system of equations known as the *mean field equations*, where the expression 'mean field' refers to the quantity $\mathrm{E}_q \log p^*(y_\alpha \,|\, Y_{V \setminus \alpha})$: the true (random) field $\log p(y_\alpha \,|\, Y_{V \setminus \alpha})$ acting on y_α is replaced by its mean.

Equation (5.2) should be contrasted with the analogous equation satisfied by the true conditional distribution $\pi = p^*$ which forms the basis for the Gibbs sampler:

$$
\pi(y_\alpha) = \mathrm{E}_\pi p^*(y_\alpha \,|\, Y_{V \setminus \alpha}).
$$

Typically, if for fixed y_α the true field depends on a large subset of variables $A \subseteq V \setminus \alpha$, then it will with high probability be close to its mean, implying that $\exp(\mathrm{E} \log p^*)$ will be close to $\mathrm{E} p^*$ and the mean field approximation method will be very accurate. Note that in this situation the Gibbs sampler is often computationally inconvenient so the methods complement each other well.

The general versions of variational inference let q vary in a more complex set of distributions, just making sure that these are tractable in the sense that expectations with respect to q can be calculated, for example, by exact probability propagation methods. For examples of recent applications of variational inference, see Murphy (1999) and Pavlovic *et al.* (1999).

It is of particular interest to consider the application of variational methods to Bayesian inference problems, where the unobserved quantities y are partitioned into incidental random variables z and parameters θ of interest. Suppose, furthermore, that a prior density π for θ has been chosen from a family $\mathcal{P}$ which is conjugate under complete observation, i.e. it holds for all (x, z) that

$$\pi(\theta) \in \mathcal{P} \implies \pi(\theta \,|\, x, z) \in \mathcal{P}$$

and that this family is log-convex in the sense that if

$$\pi_1, \pi_2 \in \mathcal{P} \text{ and } \lambda_1, \lambda_2 > 0 \implies c(\lambda_1, \lambda_2)^{-1} \pi_1^{\lambda_1} \pi_2^{\lambda_2} \in \mathcal{P},$$

where $c(\lambda_1, \lambda_2)$ is an appropriate normalization constant.

If we are now seeking distributions q that factorize as $q(z, \theta) = h(z)k(\theta)$ it can be seen that the approximating density k will always be a member of $\mathcal{P}$. This follows because the mean field equation for k becomes

$$k(\theta) \propto \exp \sum_z \log \pi(\theta \,|\, x, z) h(z) = \prod_z \pi(\theta \,|\, x, z)^{h(z)},$$

which is a log-convex combination of complete data posteriors. This fact was exploited in a special case by Attias (1999) and observed more generally by Humphreys and Titterington (2000). Note that the argument can be applied also to the other factor in the product. If the conditional densities $p(z \,|\, \theta, x)$ factorize according to a graphical model as

$$p(z \,|\, \theta, x) = \prod_{c \in \mathcal{C}} \phi_c(z \,|\, \theta, x),$$

where $\phi_c(z)$ depend on z through z_c only, then the approximating distribution $h(z)$ will factorize according to the same graphical model.

This could be exploited in a general purpose system for variational Bayesian inference in graphical models. Such a system, using a BUGS-like interface for its basic setup and named VIBES (Variational Inference in Bayesian Networks) is under attempted development by C. M. Bishop, J. Winn, and D. J. Spiegelhalter (personal communication) although currently the approximating family $\mathcal{Q}$ is simply chosen as the family of appropriately factorizing Gaussian distributions, so the essential ambition of the inference is to obtain approximate values of the first two marginal moments of the posterior distributions.

It should be noted that in many cases the variational approximation may be quite inferior to other simple approximations (Humphreys and Titterington 2001) and it is generally also rather computationally intensive.

6   Other applications

An area where graphical models seems to gain an increasing importance is that of *data mining*, involving the analysis of huge bodies of multivariate data, collected in a variety of ways. In contrast to what we have previously described, these applications seem generally more amenable to graphical models based on undirected graphs, and various algorithms for model selection or learning are then applied to identify a suitable graph for the model directly from the data collected. This type of application has been discussed by Heckerman *et al.* (2000), Giudici (2001), Giudici and Castelo (2001), and Cortes and Pregibon (2001).

We conclude this kaleidoscopic review of applications of Bayesian networks by briefly mentioning an unconventional application in music (Raphael 1999, 2001a,b). This work is concerned with making a prerecorded musical accompaniment – 'music minus one' – listen and adapt its pitch and tempo to follow a soloist. We refrain from giving the details of this beautiful application and refer the reader to listen at `http://fafner.math.umass.edu/reverie`.

References

Andreassen, S., Jensen, F. V., Andersen, S. K., Falck, B., Kjærulff, U., Woldbye, M., Sørensen, A., Rosenfalck, A., and Jensen, F. (1989). MUNIN – an expert EMG assistant. In *Computer-Aided Electromyography and Expert Systems* (ed. J. E. Desmedt), pp. 255–77. Elsevier Science, Amsterdam.

Andreassen, S., Falck, B., and Olesen, K. G. (1992). Diagnostic function of the microhuman prototype of the expert system MUNIN. *Electroencephalography and Clinical Neurophysiology*, **85**, 143–57.

Andreassen, S., Benn, J. J., Hovorka, R., Olesen, K. G., and Carson, E. R. (1994). A probabilistic approach to glucose prediction and insulin dose adjustment: Description of metabolic model and pilot evaluation study. *Computer Methods and Programs in Biomedicine*, **41**, 153–65.

Andreassen, S., Rosenfalck, A., Falck, B., Olesen, K. G., and Andersen, S. K. (1996). Evaluating the diagnostic performance of the expert EMG assistant MUNIN. *Electroencephalography and Clinical Neurophysiology*, **101**, 129–44.

Attias, H. (1999). Inferring parameters and structure of latent variable models by variational Bayes. In *Proceedings of the Fifteenth Annual Conference on Uncertainty in Artificial Intelligence (UAI-99)*, pp. 21–30. Morgan Kaufmann, San Francisco.

Bahl, L., Cocke, J., Jelinek, F., and Raviv, J. (1974). Optimal decoding of linear codes for minimizing symbol error rate. *IEEE Transactions on Information Theory*, **20**, 284–7.

Berrou, C. and Glavieux, A. (1996). Near optimum error correcting coding and decoding: Turbo-codes. *IEEE Transactions on Communication*, **44**, 1261–71.

Cortes, C. and Pregibon, D. (2001). Signature-based methods for data streams. *Data Mining and Knowledge Discovery*, **5**, 167–82.

Cover, T. M. and Thomas, J. A. (1991). *Elements of Information Theory*. Wiley, New York.

Cowell, R. G., Dawid, A. P., Lauritzen, S. L., and Spiegelhalter, D. J. (1999). *Probabilistic Networks and Expert Systems.* Springer-Verlag, New York.

Darroch, J. N., Lauritzen, S. L., and Speed, T. P. (1980). Markov fields and log-linear models for contingency tables. *Annals of Statistics*, **8**, 522–39.

Dawid, A. P. (1992). Applications of a general propagation algorithm for probabilistic expert systems. *Statistics and Computing*, **2**, 25–36.

Frey, B. J. (1998). *Graphical Models for Machine Learning and Digital Communication.* MIT Press, Cambridge, MA.

Gallager, R. G. (1963). *Low Density Parity Check Codes.* MIT Press, Cambridge, MA.

Gilks, W. R., Thomas, A., and Spiegelhalter, D. J. (1994). A language and a program for complex Bayesian modelling. *The Statistician*, **43**, 169–78.

Giudici, P. (2001). Bayesian data mining, with application to benchmarking and credit scoring. *Applied Stochastic Models in Business and Industry*, **17**, 69–81.

Giudici, P. and Castelo, R. (2001). Association models for web mining. *Data Mining and Knowledge Discovery*, **5**, 183–96.

Heath, S. C. (1998). Generating consistent genotypic configurations for multi-allelic loci and large complex pedigrees. *Human Heredity*, **48**, 1–11.

Heckerman, D., Horvitz, E. J., and Nathwani, B. (1991). Toward normative expert systems I: The pathfinder project. *Methods of Information in Medicine*, **31**, 90–105.

Heckerman, D., Breese, J. S., and Rommelse, K. (1995). Decision-theoretic troubleshooting. *Communications of the ACM*, **38**, 49–57.

Heckerman, D., Chickering, D. M., Meek, C., Rounthwaite, R., and Kadie, C. (2000). Dependency networks for inference, collaborative filtering, and data visualization. *Journal of Machine Learning Research*, **1**, 49–75.

Hejlesen, O. K., Andreassen, S., Frandsen, N. E., Sørensen, T. B., Sandø, S. H., Hovorka, R., and Cavan, D. A. (1998). Using a double blind controlled clinical trial to evaluate the function of a diabetes advisory system: A feasible approach? *Computer Methods and Programs in Biomedicine*, **56**, 165–73.

Henrion, M., Breese, J. S., and Horvitz, E. J. (1991). Decision analysis and expert systems. *AI Magazine*, **12**, 64–91.

Howard, R. A. (1960). *Dynamic Programming and Markov Processes.* MIT Press, Cambridge, MA.

Howard, R. A. and Matheson, J. E. (1984). Influence diagrams. In *Readings in the Principles and Applications of Decision Analysis* (eds R. A. Howard and J. E. Matheson). Strategic Decisions Group, Menlo Park, CA.

Humphreys, K. and Titterington, D. M. (2000). Approximate Bayesian inference for simple mixtures. In *COMPSTAT — Proceedings in Computational Statis-*

tics (eds J. G. Bethlehem and P. G. M. Heijden), pp. 331–6. Springer-Verlag, New York.

Humphreys, K. and Titterington, D. M. (2001). Some examples of recursive variational approximations for Bayesian inference. In *Advanced Mean Field Methods — Theory and Practice* (eds M. Opper and D. Saad), pp. 179–95. MIT Press, Cambridge, MA.

Janss, L. L. G., van Arendonk, J. A. M., and van der Werf, J. H. J. (1995). Computing approximate monogenic model likelihoods in large pedigrees with loops. *Genetic Selection and Evolution*, **27**, 567–79.

Jensen, F. V., Lauritzen, S. L., and Olesen, K. G. (1990). Bayesian updating in causal probabilistic networks by local computation. *Computational Statistics Quarterly*, **4**, 269–82.

Jensen, F., Jensen, F. V., and Dittmer, S. L. (1994). From influence diagrams to junction trees. In *Proceedings of the Tenth Annual Conference on Uncertainty in Artificial Intelligence* (eds R. L. de Mantaras and D. Poole), pp. 367–73. Morgan Kaufmann, San Francisco.

Jordan, M. I., Ghahmarani, Z., Jaakola, T. S., and Saul, L. K. (1998). An introduction to variational methods for graphical models. In *Learning in Graphical Models* (ed. M. I. Jordan). MIT Press, Cambridge, MA.

Lange, K. and Goradia, T. M. (1987). An algorithm for automatic genotype elimination. *American Journal of Human Genetics*, **40**, 250–6.

Lauritzen, S. L. and Spiegelhalter, D. J. (1988). Local computations with probabilities on graphical structures and their application to expert systems (with discussion). *Journal of the Royal Statistical Society*, B, **50**, 157–224.

Lauritzen, S. L. and Jensen, F. V. (1997). Local computation with valuations from a commutative semigroup. *Annals of Mathematics and Artificial Intelligence*, **21**, 51–69.

Lauritzen, S. L. and Nilsson, D. (2001). Representing and solving decision problems with limited information. *Management Science*, **47**, 1235–51.

Leibovici, L., Fishman, M., Schønheider, H. C., Riekehr, C., Kristensen, B., Shraga, I., and Andreassen, S. (2000). A causal probabilistic network for optimal treatment of bacterial infections. *IEEE Transactions on Knowledge and Data Engineering*, **12**, 517–28.

Lin, S. and Costello, D. J. (1983). *Error Control Coding: Fundamentals and Applications*. Prentice Hall, Englewood Cliffs, NJ.

MacKay, D. J. C. (1999). Good error-correcting codes based on very sparse matrices. *IEEE Transactions on Information Theory*, **45**, 399–431.

McEliece, R. J., MacKay, D. J. C., and Cheng, J.-F. (1998). Turbo decoding as an instance of Pearl's belief propagation algorithm. *IEEE Journal on Selected Areas in Communications*, **16**, 140–52.

Miller, R. A., Pople, H. E. J., and Myers, J. D. (1982). Internist-1, an experimental computer-based diagnostic consultant for general internal medicine. *New England Journal of Medicine*, **307**, 468–76.

Murphy, K. (1999). A variational approximation for Bayesian networks with discrete and continuous latent variables. In *Proceedings of the Fifteenth Annual Conference on Uncertainty in Artificial Intelligence (UAI–99)*, pp. 457–66. Morgan Kaufmann, San Francisco.

Neal, R. (1996). *Bayesian Learning for Neural Networks*. Springer-Verlag, New York.

Olmsted, S. M. (1983). *On Representing and Solving Decision Problems*. PhD Thesis, Department of Engineering–Economic Systems, Stanford University, Stanford, CA.

Parisi, G. (1988). *Statistical Field Theory*. Addison-Wesley, Redwood City, CA.

Pavlovic, V., Frey, B. J., and Huang, T. S. (1999). Variational learning in mixed-state dynamic graphical models. In *Proceedings of the Fifteenth Annual Conference on Uncertainty in Artificial Intelligence (UAI–99)*, pp. 522–30. Morgan Kaufmann, San Francisco.

Pearl, J. (1986). Fusion, propagation and structuring in belief networks. *Artificial Intelligence*, **29**, 241–88.

Raphael, C. (1999). Automatic segmentation of acoustic musical signals using hidden Markov models. *IEEE Transactions on Pattern Analysis and Machine Intelligence*, **21**, 360–70.

Raphael, C. (2001a). Can the computer learn to play music expressively? In *Artificial Intelligence and Statistics 2001* (eds T. Jaakkola and T. S. Richardson), pp. 113–20. Morgan Kaufmann, San Francisco.

Raphael, C. (2001b). A probabilistic expert system for automatic musical accompaniment. *Journal of Computational and Graphical Statistics*, **10**, 487–512.

Shachter, R. D. (1986). Evaluating influence diagrams. *Operations Research*, **34**, 871–82.

Shannon, C. E. (1948). A mathematical theory of communication. *Bell System Technical Journal*, **27**, 379–423, 623–56.

Shenoy, P. P. (1992). Valuation-based systems for Bayesian decision analysis. *Operations Research*, **40**, 463–84.

Shenoy, P. P. and Shafer, G. (1990). Axioms for probability and belief–function propagation. In *Uncertainty in Artificial Intelligence 4* (eds R. D. Shachter, T. S. Levitt, L. N. Kanal, and J. F. Lemmer), pp. 169–98. North-Holland, Amsterdam.

Shwe, M., Middleton, B., Heckerman, D., Henrion, M., Horvitz, E. J., and Lehmann, H. (1991). Probabilistic diagnosis using a reformulation of the INTERNIST-1 / QMR knowledge base I: The probabilistic model and inference algorithms. *Methods of Information in Medicine*, **30**, 241–55.

Skaanning, C. (2000). A knowledge acquisition tool for Bayesian-network troubleshooters. In *Proceedings of the Sixteenth Annual Conference on Uncertainty in Artificial Intelligence (UAI–2000)*, pp. 549–57. Morgan Kaufmann, San Francisco.

Skaanning, C., Jensen, F. V., and Kjærulff, U. (2000). Printer troubleshooting using Bayesian networks. In *Proceedings of the Thirteenth International Conference on Industrial and Engineering Applications of AI and Expert Systems*, Lecture Notes in Computer Science, No. 1821, pp. 367–79. Springer-Verlag, New York.

Spiegelhalter, D. J. (1998). Bayesian graphical modelling: A case-study in monitoring health outcomes. *Applied Statistics*, **47**, 115–33.

Viterbi, A. J. (1967). Error bounds for convolutional codes and an asymptotically optimum decoding algorithm. *IEEE Transactions on Information Theory*, **13**, 260–9.

Wang, T., Fernando, R. L., Stricker, C., and Elston, R. C. (1996). An approximation to the likelihood for a pedigree with loops. *Theoretical and Applied Genetics*, **93**, 1299–309.

Wermuth, N. and Lauritzen, S. L. (1983). Graphical and recursive models for contingency tables. *Biometrika*, **70**, 537–52.

White, C. C. (1991). A survey of solution techniques for the partially observed Markov decision process. *Annals of Operations Research*, **32**, 215–30.

Zhang, N. L., Qi, R., and Poole, D. (1994). A computational theory of decision networks. *International Journal of Approximate Reasoning*, **11**, 83–158.

1A
Analysing social science data with graphical Markov models

Nanny Wermuth
University of Mainz, Germany

1   Introduction

The term graphical Markov models has been suggested by Michael Perlman, University of Washington, for multivariate statistical models in which a joint distribution satisfies independence statements that are captured by a graph. The study and development of these models is such an active research area that some of their properties are not yet discussed in recent statistical books concentrating on them (Cox and Wermuth 1996; Lauritzen 1996; Edwards 2000). We ask here:

- How do they relate to models used more traditionally for data analysis?
- What do they offer in addition?
- Are case studies available?

In independence graphs used to summarize aspects of detailed statistical analyses, each vertex or node represents a variable feature of individuals under study. These features may be categorical; in this case they are denoted by capital letters $A, B, C, \ldots$, are modelled by discrete random variables, and are drawn as dots. Alternatively, they may have numerical values of substantive meaning. Then, they are denoted by capital letters $X, Z, U, \ldots$, are typically modelled by continuous variables, and are drawn as circles. If this distinction is not important, then the individual components of a vector random variable Y may be denoted by $Y_1, Y_2, \ldots$, or, more compactly, just by integers $1, 2, \ldots$.

Often substantive knowledge is strong enough to specify a fully ordered sequence of the variables which starts with a background variable, ends with a response of primary interest, and has single intermediate variables. These intermediate variables are both potential responses to variables of the past, and potentially explanatory to variables of the future. In this case, no variable is taken to be explanatory by itself and an independence graph fitted to the responses will be fully directed and acyclic. For these models we give examples of both research questions and theoretical results.

2   A motivating research example

Consider the two research questions:
'Who admits to being not concerned about protecting the environment?'

A=1, no concern about protecting the environ- ment 7.0%	B=1, no own political impact expected 15.4%	C=1, at risk of social exclusion 12.8%	D=1, own education at lower level 45.3%	E=1, parents' education, both at lower level : 73.9% F=1, resp.' age group, 40-65 years: 57.3% G=1, gender, fem: 49.9%
Primary response		Intermediate variables		Background variables

FIG. 1.

'How does such an attitude develop?'

We use the answers from 1228 respondents, aged between 18 and 65 years, from the General Social Survey in Germany in 1998. Figure 1 shows a first ordering of seven variables which are categorized to be binary, together with the observed percentages for the stated category. Most variables are based on answers to a single question but, for instance, risk of social exclusion is derived from several aspects such as no or incomplete vocational training and extended periods of unemployment.

After checking for interactive effects (Cox and Wermuth 1994), and using a likelihood-ratio-based model selection strategy, we concluded that each of the univariate conditional distributions here is well described by logistic regressions having two main effects. For each response the important explanatory variables are shown in the graph by arrows pointing directly from the former to the latter. The factorization of the joint density f_V can be read off the graph to be

$$f_V = f_{A|B,G} f_{B|C,D} f_{C|D,G} f_{D|E,F} f_G f_{E,F}.$$

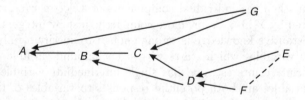

For each response, the direction and strength of the dependencies can be read off from the estimated conditional probabilities. Table 1 shows for which levels of the explanatory variables the lowest and highest percentages are observed for level 1 of each response, together with these observed percentages.

One main advantage of the graph is the possibility of tracing developments. For instance, the path from G to C to B to A describes that women are at higher risk for social exclusion than men, that those at higher risk for social

TABLE 1.

Level of response:	$A = 1$		$B = 1$		$C = 1$		$D = 1$	
Explanatory variables:	BG		CD		DG		EF	
Levels with highest percentage:	1,2	26.6	1,1	40.7	1,1	28.2	1,1	65.2
Levels with lowest percentage:	2,1	2.2	2,2	8.2	2,2	4.7	2,2	7.7

exclusion are less likely to believe in having an own political impact, and that those perceiving to have no political impact are more likely to be unconcerned about the environment.

Another important use of a directed acyclic graph is that its consequences can be derived if only subsets of the variables are considered and if subpopulations are selected. General answers have been given recently (Wermuth and Cox 2001a). For instance, the independence graph implied if the variables A, F, and E are ignored, that is the graph for the joint distribution of all remaining variables, is, in this example, again a directed acyclic graph, shown below.

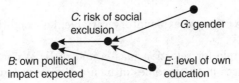

This follows by summing over the variables A, F, and E in the joint density f_V, in which each response depends only on the directly important explanatory variables.

3 Some consequences of directed acyclic graphs

3.1 Consequences in simple cases

For relations among only three variables, we now show examples of consequences that have been described in the literature as spurious dependence, spurious association, and selection bias. The graphs help to visualize the concepts.

The first example for spurious dependence concerns the question of discrimination against women and data from the German labour market for academics, whose field of qualification was either mechanical engineering or home economics.

The well-fitting independence graph is

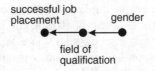

Ignoring the intermediate variable, i.e. marginalizing over B (⫫) leaves A dependent on C: the data appear to indicate discrimination, since men have a more than five times higher chance of successful job placement.

However, including the information of the field of qualification by fixing levels of the intermediate variable, i.e. conditioning on B (■), shows A independent of C.

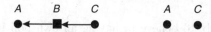

The second example for spurious association was used by Y. Yule more than 100 years ago to argue that correlation is not causation. Ignoring the common explanatory variable, i.e. marginalizing over Z, leaves Y and X associated.

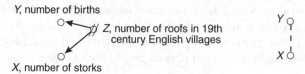

Fixing levels of the common explanatory variable, i.e. conditioning on Z, shows Y independent of X.

The third example for selection bias is due to H. Wainer. He showed how systematic differences in the incomes of men and women having the same level of formal schooling get covered up when different scales are used for income in graphs showing for both genders a systematic increase in income with higher levels of formal schooling.

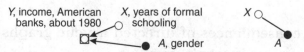

Overall, the level of formal schooling, X, is independent of gender, A. This shows after ignoring the common response variable, i.e. after marginalizing over Y. But, after selecting a fixed level of the common response variable, i.e. after conditioning on Y, the explanatory variables become associated: within a given income group women have a higher level of formal schooling.

3.2   Some consequences of large graphs

In general, for all variable pairs, simple matrix calculations can be used to derive whether a directed acyclic graph implies, for instance, marginal independence. An edge or incidence matrix is a way of storing the information in an independence graph with zeros for missing edges and ones for edges present. We let row i in an edge matrix correspond to node i in a graph and let the node set be ordered as $V = (1, \ldots, d_V)$ so that all ij-arrows for $j > i$ point from j to i. The edge matrix of a directed acyclic graph, $\mathcal{A}$, is then an upper-triangular matrix with ones along the diagonal and a one in position (i, j) if and only if there is an ij-arrow present in the graph.

In the language for such directed graphs, it has become a convention to call the node of a directly explanatory variable a parent and the node of a direct response variable a child. The node of an indirectly explanatory variable is named an ancestor and the node of an indirect response variable a descendant.

A directed acyclic graph is then often called the parent graph, G_{par}^V, with edge matrix $\mathcal{A}$. The graph obtained from it by adding a direction-preserving arrow for every ancestor–descendant relation is called the overall ancestor graph, G_{anc}^V, with edge matrix $\mathcal{B}$. An undirected graph of dashed or broken lines is induced by the parent graph G_{par}^V which has a missing ij-edge (and a missing ji-edge) if and only if Y_i is implied to be marginally independent of Y_j. It is called the induced overall covariance graph, G_{cov}^V. The name derives from joint Gaussian distributions for which marginal independencies are reflected as zeros in the covariance matrix. We shall explain here why the edge matrix of the induced overall covariance graph is the indicator matrix of $\mathcal{BB}^T$, where an indicator matrix $\text{In}(M)$ of a matrix M has a one in position (i,j) if and only if the element of M in this position is non-zero. We give first an example with 6 nodes.

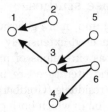 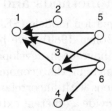 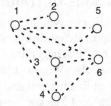

the parent graph G_{par}^V with edge matrix $\mathcal{A}$	the overall ancestor graph G_{anc}^V with edge matrix $\mathcal{B}$	the induced overall covariance graph G_{cov}^V, edge matrix $\text{In}(\mathcal{BB}^T)$

The corresponding edge matrices $\mathcal{A}$ and $\mathcal{B}$ are

$$\mathcal{A} = \begin{pmatrix} 1 & 1 & 1 & 0 & 0 & 0 \\ & 1 & 0 & 0 & 0 & 0 \\ & & 1 & 0 & 1 & 1 \\ & & & 1 & 0 & 1 \\ \mathbf{0} & & & & 1 & 0 \\ & & & & & 1 \end{pmatrix}; \quad \mathcal{B} = \begin{pmatrix} 1 & 1 & 1 & 0 & 1 & 1 \\ & 1 & 0 & 0 & 0 & 0 \\ & & 1 & 0 & 1 & 1 \\ & & & 1 & 0 & 1 \\ \mathbf{0} & & & & 1 & 0 \\ & & & & & 1 \end{pmatrix}.$$

In this small example it may be checked directly for which pairs the factorization of the joint density given by the parent graph,

$$f_{1,\ldots,6} = f_{1|2,3} f_2 f_{3|5,6} f_{4|6} f_5 f_6,$$

implies that $f_{ij} = f_i f_j$ by integrating over all remaining variables.

The edge matrix of an overall ancestor graph, $\mathcal{B}$, is the indicator matrix of

$$I + \sum_{r \geq 1} (\mathcal{A} - I)^r,$$

where I is the identity matrix and $(\mathcal{A} - I)^r$ counts for each $i < j$ the number of direction-preserving paths of length r present in the parent graph between them.

A matrix product BB^T has in position (i, j) the element $b_{ij} + \sum_{k > j} b_{ik} b_{jk}$. Therefore, if $\text{In}(BB^T)$ is the induced edge matrix, then there is an additional ij-one if and only if non-adjacent nodes i and j have a node k as a common parent in the ancestor graph. Thus, an additional ij-edge in G_{cov}^V compared with G_{par}^V arises if and only if either j is an ancestor of i or i and j have a common ancestor. This statement is equivalent to Pearl's (1988) separation criterion for directed acyclic graphs when the conditioning set is empty. The matrix result completes the search for the proper paths for all pairs at once.

By a similar simpler argument the induced overall concentration graph, G_{con}^V, can be shown to have edge matrix $\text{In}(\mathcal{A}^T \mathcal{A})$. It is an undirected graph of full lines, where each edge concerns the conditional relation of two variables given all remaining ones, $i \perp\!\!\!\perp j \mid V \setminus \{i, j\}$. The name derives from joint Gaussian distributions, for which these independencies are reflected as zeros in the concentration matrix, which is the inverse of the covariance matrix.

4   Relations to traditional methods and case studies

In the social sciences, structural equation models (SEMs) and, more generally, linear structural relation models (Bollen 1989) have been used extensively for analysing multivariate data. They have been developed as extensions of path analysis models (Wright 1934), which in the econometric literature are better known as linear recursive equations with uncorrelated residuals (Goldberger 1964). Graphical Markov models provide a different extension in which both categorical and numerical features can be modelled. In the subclass of chain graph models, joint distributions are decomposed recursively into conditional joint distributions and simplified by conditional independencies. There are no theoretical restrictions on the form of the conditional distributions; however, algorithms for computing estimates under each specified model are not yet available generally.

It has recently been shown (Koster 1999) how an independence graph can be associated with each Gaussian structural equation model, to read off the graph all independence statements implied by the model. But, while in chain graphs every missing edge corresponds to an independence statement and every edge present can be associated with a specific conditional or marginal association of the variable pair, this does not hold in general for structural equation models. An edge present in the graph relates directly to a parameter in an equation, but may be connected in a complicated way to any statement about the conditional or marginal association of the variable pair. Similarly, a variable pair with an edge missing may be associated no matter which conditioning set is chosen.

Results about fitting chain graphs approximately with the help of univariate conditional regressions, and results for deriving chain graphs induced by directed acyclic graphs (Wermuth and Cox 2001a) can be viewed as supplementing existing, useful data analysis tools. Local fitting of univariate conditional distributions permits the break up of seemingly complex structures into tractable subcomponents. These components may then be directly related to substantive knowledge available about subsets of the variables under study.

Some case studies using chain graph models are by Klein *et al.* (1995), Hardt (1995), Cox and Wermuth (1993; 1996, Chapter 6; 2001), Pigeot *et al.* (1999, 2000), Stanghellini *et al.* (1999), Wermuth and Cox (2001b), and Cheung and Andersen (2002).

Acknowledgements

Support by the HSSS programme of the European Science Foundation and the Radcliffe Institute for Advanced Study at Harvard University is gratefully acknowledged.

1B
Analysis of DNA mixtures using Bayesian networks

Julia Mortera
Università Roma Tre, Italy

Here we present an extension, rather than a discussion, of the article by S. Lauritzen, where we describe another modern application of Bayesian networks, concerning forensic identification using DNA profiles. In Dawid *et al.* (2002), Bayesian networks (Cowell *et al.* 1999) were used to analyse complex cases of forensic inference, involving missing data on one or more of the relevant individuals in paternity testing, genetic mutation, and identification on a large pedigree. Another interesting and complex problem which can readily be handled by using a Bayesian network (BN) is the interpretation of DNA profiles when the trace evidence contains a mixture of genetic material from more than one person. After reformulating the problem as a BN, existing fast general software such as HUGIN (http://www.hugin.com) can be used to perform the numerical computations.

Mixed trace evidence has been studied by, among others, Weir *et al.* (1997) and Evett and Weir (1998). For example, in the infamous O. J. Simpson case (Weir 1995) the evidence found at the scene of the crime revealed that several people must have contributed to the trace. The interpretation of DNA profiles from biological samples is complicated when the samples contain material from more than one individual. This is common in rape cases where the sample may contain biological material from the victim, the perpetrator, one or more consensual partners, or from multiple perpetrators. DNA mixtures can also occur in criminal cases where the victim and one or more suspects might produce a mixed trace as a consequence of a scuffle or brawl. The complexity of mixed traces is in part due to the very large number of combinations of genotypes that must be considered.

TABLE 1. Criminal case: mixed trace data.

Marker	Victim (vgt)		Suspect (sgt)		Mixed trace (mix)		Likelihood Ratio
D3S1358	18	18	16	16	16	18	11.35
VWA	17	17	17	18	17	18	15.43
TH01	6	7	6	7	6	7	5.48
TPOX	8	8	8	11	8	11	3.00
D5S818	12	13	12	12	12	13	14.79
D13S317	8	8	8	11	8	11	24.45
FGA	22	26	24	25	22 24	25 26	76.92
D7S820	8	10	8	11	8	10 11	4.90
Overall likelihood							3.93×10^8

A DNA profile consists of measurements on different markers, each representing a genotype consisting in two unordered bands (alleles), one inherited from the father and the other from the mother, although it is not possible to distinguish which is which. For example, columns 2 and 3 of Table 1 show the observed DNA profiles of the victim and the suspect, respectively. The observed crime scene evidence, shown in column 4 of Table 1, has four distinct bands on marker FGA and three distinct bands on marker D7S820; this clearly indicates that the trace is a mixture of two or more individuals' DNA profiles, since a single individual can at most have two distinct bands on each marker.

Let the random variable M represent the set of unknown individuals that contributed to the mixed trace. The crime scene evidence can be denoted by $\gamma_M = x$, where γ_M is the potential mixture of DNA profiles and the set of alleles, x, is its observed value. Furthermore, as in Dawid and Mortera (1996), a set α of individuals from one or more populations, or gene pools, are examined and for each individual $i \in \alpha$, the DNA profile χ_i is observed. In the mixture problem the set α of measured profiles might include the profiles of the victim v, a possible suspect s, and a consensual partner, as well as those profiles contained in the database.

We assume that, prior to observing evidence on γ_M, the information on an individual's DNA profile χ does not give any information on whether he contributed to the mixture, i.e. knowledge on χ is irrelevant to the identity of individuals in M, thus

$$M \perp\!\!\!\perp \chi. \tag{1}$$

For simplicity, suppose we are interested in comparing the following two hypotheses: $H_0 : M = \{v, s\}$ versus $H_1 : M = \{v, U\}$, where U represents an unknown individual. For cases where the victim's and the suspect's DNA profiles, χ_v and χ_s respectively, are compatible with x and $\chi_v \cup \chi_s = x$, then, given all the evidence, and from (1), the likelihood ratio is given by

$$\frac{\Pr(\gamma_M = x | \chi_\alpha, H_0)\Pr(\chi_\alpha | H_0)}{\Pr(\gamma_M = x | \chi_\alpha, H_1)\Pr(\chi_\alpha | H_1)} = \frac{1}{\sum_y \Pr(\chi_U = y | \chi_\alpha)}, \tag{2}$$

where y is ranging over the set of DNA profiles such that $\chi_v \cup y = x$, that is the set of possible profiles compatible with x that could have been left by unknown individual(s) U. Although most problems concerning mixtures can be solved algebraically (Evett and Weir 1998), their structuring via a BN avoids case by case analysis and greatly simplifies the solution of complex problems.

Example: Mixed Trace in a Criminal Case

Table 1 shows a real example from the casework of the Forensic Laboratory at the Università Cattolica del Sacro Cuore, Italy, where a mixed trace was found at the scene of the crime. DNA profiles were taken from the crime trace, the victim, v, and the suspect, s. The DNA profiles used consist of measurements on eight 'short tandem repeat' (STR) markers. Each marker has a finite number of discrete alleles. From databases of DNA profiles estimates of the frequencies of the different alleles, in different populations, can be estimated.

We assume Hardy–Weinberg and linkage equilibrium, i.e. independence of alleles within and across markers. This assumption greatly simplifies the BN, since each DNA marker in the profile is handled in a separate network and the overall likelihood ratio for the quantity of interest is simply obtained as a product of the likelihood ratios for each marker. We will furthermore assume that contributors are unrelated allowing for independence of the alleles between individuals, and that all contributors are from the same population. Here we will, for simplicity, also assume that the mixed stain is known to be from two individuals.

The case in Table 1 was analysed using the BN given in Fig. 1. The observed genotypes of the victim and the suspect plus the mixed trace are represented by (grey) nodes `vgt`, `sgt`, and `mix`, respectively. Since the mixed trace is assumed to come from two contributors, two unobserved (white) nodes, `T1gt` and `T2gt` are introduced, representing the genotypes of the unknown contributors to trace 1 and 2, respectively. In order to simplify the calculations, the network needs

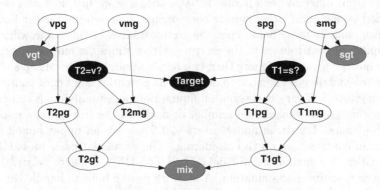

FIG. 1. Network for a simple DNA mixed stain.

to be represented at the most disaggregated level possible. For this reason it is useful to introduce the unobserved nodes, representing paternal and maternal genes for each genotype, e.g. vpg and vmg for vgt, etc. Furthermore, target nodes (black) T1=s? (and T2=v?) represent the following hypotheses or queries: is trace T1 (T2) from the suspect, *s* (victim, *v*), or is it not? The target node (Target) represents the four possible combinations of its parent nodes T1=s? and T2=v?.

For each node in the network, a table of probabilities, conditional on its parent nodes, is given as follows: population gene frequencies are used for founder gene nodes; individual's genotype nodes are given by a simple deterministic relationship between gene nodes; and the mixture node is given by the union of genotypes T1gt and T2gt. The paternal (maternal) gene of trace 1 (trace 2) is either identical to the corresponding gene of the suspect (victim) or is given by the relevant population gene frequencies, according to whether the target node T1=s? (T2=v?) is true or false. For illustration, the HUGIN file containing the network for marker D7S820 is available at http://www.maths.nott.ac.uk/hsss/. After entering the evidence at grey nodes sgt, vgt, and mix, it is propagated through the network to give the updated marginal posterior probabilities of all nodes. Uniform prior probabilities were used on the hypotheses nodes, so that after propagation the likelihood for each combination of hypotheses can be found in the target node.

The overall likelihood ratio (obtained as a product of those for each marker) is roughly 4×10^8, an extremely incriminating value. It represents the likelihood ratio for comparing $H_0 : M = \{v, s\}$ versus $H_1 : M = \{v, U\}$, as in (2).

One of the many useful features of BNs is their modular representation. Note that the network in Fig. 1 is formed by two similar components, representing the two possible contributors to the mixture. This approach is particularly suitable for solving complex cases concerning multiple contributors, for ascertaining the number of contributors to the mixture, and for handling cases where contributors are from different gene pools. In fact, the network in Fig. 1 can easily be extended by including other similar components, representing further known or unknown contributors. Sometimes the circumstances of the crime will dictate the number of contributors to the mixture. Other times the number may not be known but the presence of more than two bands at any marker indicates that the number of contributors is more than one. The potential number of contributors to the mixture will vary between a minimum number of individuals sufficient to explain the observed maximum number of bands in the trace, and a maximum plausible value. Lauritzen and Mortera (2002) derive an upper bound on the number of contributors worth considering. The particular prior probability on the number of contributors will depend on non-DNA evidence relative to the particular case under examination. A network can be built to handle the uncertainty about the number of contributors by including a node representing the plausible number of contributors, which according to its value will activate or inactivate the components of the network representing the different contributors to the mixed trace. These extensions will be reported elsewhere.

The general algorithm used in a BN (Lauritzen and Spiegelhalter 1988) is similar to the peeling algorithm (Cannings *et al.* 1978) applied in large complex pedigree studies (see also Heath, Chapter 12, this volume) but a BN has more flexible properties. For example, complex forensic problems having both mixed traces and missing information on relevant individuals in a pedigree can be solved by piecing together different BNs representing these problems. A further important feature is that both genetic and non-genetic information can be represented in the same network. As illustrated here, and in Dawid *et al.* (2002), query or hypotheses nodes can be represented in the network. This feature has important applications not only in forensic identification, but in standard analysis of complex pedigrees, where often the problem of uncertainty in the pedigree is ignored.

Acknowledgement

We are grateful to Vincenzo Pascali, Forensic Science Laboratory, Università Cattolica del Sacro Cuore, for providing the data.

Additional references in discussion

Bollen, K.A. (1989). *Structural Equations with Latent Variables.* Wiley, New York.

Cannings, C., Thomson, E. A., and Skolnick M. H. (1978). Probability functions on complex pedigrees. *Advances in Applied Probability*, **10**, 26–61.

Cheung, S. Y. and Andersen, R. (2002). Time to read: Family resources and educational inequalities. *Journal of Comparative Family Studies.* To appear.

Cox, D. R. and Wermuth, N. (1993). Linear dependencies represented by chain graphs (with discussion). *Statistical Science*, **8**, 204–18; 247–77.

Cox, D. R. and Wermuth, N. (1994). Tests of linearity, multivariate normality and adequacy of linear scores. *Journal of the Royal Statistical Society*, **C43**, 347–55.

Cox, D. R. and Wermuth, N. (1996). *Multivariate Dependencies – Models, Analysis and Interpretation.* Chapman & Hall, London.

Cox, D. R. and Wermuth, N. (2001). Some statistical aspects of causality. *European Sociological Review*, **17**, 65–74.

Dawid, A. P. and Mortera, J. (1996). Coherent analysis of forensic identification evidence. *Journal of the Royal Statistical Society*, B, **58**, 425–43.

Dawid, A. P., Mortera, J., Pascali, V., and van Boxel, D. (2002). Probabilistic expert systems for forensic inference from genetic markers. *Scandinavian Journal of Statistics*, **29**, 577–95.

Edwards, D. (2000). *Introduction to Graphical Modelling.* Springer-Verlag, New York.

Evett, I. W. and Weir, B. S. (1998). *Interpreting DNA Evidence.* Sinauer, Sunderland, MA.

Goldberger, A. S. (1964). *Econometric Theory.* Wiley, New York.

Hardt, J. (1995). *Chronifizierung und Bewältigung bei Schmerzen.* Lengerich, Pabst.

Klein, J. P., Keiding, N., and Kreiner, S. (1995). Graphical models for panel studies, illustrated on data from the Framingham heart study. *Statistics in Medicine,* **14,** 1265–90.

Koster, J. (1999). On the validity of the Markov interpretation of path diagrams of Gaussian structural equation systems of simultaneous equations. *Scandinavian Journal of Statistics,* **26,** 413–31.

Lauritzen, S. L. (1996). *Graphical Models.* Oxford University Press, UK.

Lauritzen, S. L. and Mortera, J. (2002). Bounding the number of contributors to mixed stains. Research Report R-02-2003, Department of Mathematical Sciences, Aalborg University.

Pearl, J. (1988). *Probabilistic Reasoning in Intelligent Systems.* Morgan Kaufmann, San Mateo, CA.

Pigeot, I., Caputo, A., and Heinicke, A. (1999). A graphical chain model derived from a model selection strategy for the sociologists graduates study. *Biometrical Journal,* **41,** 217–34.

Pigeot, I., Heinicke, A., Caputo, A., and Brüderl, J. (2000). The professional career of sociologists: a graphical chain model reflecting early influences and associations. *Allgemeines Statistisches Archiv,* **84,** 3–21.

Stanghellini, E., McConway, K. J., and Hand, D. J. (1999). A discrete variable chain graph for applicants for credit. *Journal of the Royal Statistical Society,* C, **48,** 239–51.

Weir, B. S. (1995). DNA statistics in the Simpson matter. *Nature Genetics,* **11,** 366–8.

Weir, B. S., Triggs, C. M., Starling, L., Stowell, L. I., Walsh, K. A. J., and Buckleton, J. (1997). Interpreting DNA mixtures. *Journal of Forensic Sciences,* **42,** 213–22.

Wermuth, N. and Cox, D. R. (2001a). Joint response graphs and separation induced by triangular systems. Research report available at http://www.radcliffe.edu/fellowships/current/2002/wermuth.html.

Wermuth, N. and Cox, D. R. (2001b). Graphical models: Overview 9. In *International Encyclopedia of the Social and Behavioural Sciences* (eds P. B. Baltes and N. J. Smelser), pp. 6379–86. Elsevier, Amsterdam.

Wright, S. (1934). The method of path coefficients. *Annals of Mathematical Statistics,* **5,** 161–215.

2

Causal inference using influence diagrams: the problem of partial compliance

A. Philip Dawid
University College, London, UK

1   Introduction

In this article I aim to illustrate the fruitfulness, for purposes of causal modelling and inference, of an elaboration of graphical modelling based on influence diagrams. These can be used to model, in a clear and unambiguous way, probabilistic relationships that are regarded as invariant features across a range of circumstances. They have simple and expressive semantics, are easy to manipulate, and support straightforward and meaningful analysis.

As a specific example, I consider in detail the problem of partial compliance, where we wish to allow for the possibility that a patient may choose to deviate from a recommended course of treatment. Although no new results are presented, the intention is to show how the use of extended influence diagrams to model this situation results in increased generality, as well as improved clarity of specification, interpretation, and analysis.

I also relate this approach to other graphical representations, particularly the functional models of Pearl (1995a). I argue that this additional restriction on the framework is unnecessary, all meaningful results that flow from it being obtainable from our more general probabilistic approach.

2   Causal models

There are many different possible understandings of what is, or should be, meant by 'causal modelling', and many ways in which one can try to put it into effect. My own philosophy (Dawid 2000, especially Section 1 of the Rejoinder to the Discussion) is that causal modelling is not, in essence, any different from any other kind of scientific or statistical modelling. All models are mental constructs, by means of which we attempt to interpret, relate, and explain empirical observations, both past and future. And any such model will have its appropriate 'ambit', describing, at an appropriate level of detail, the external circumstances under which it is intended to apply. The empirical adequacy or inadequacy of a model is to be judged by how well it performs at predicting observable quantities, within its own ambit.

Most probabilistic and statistical models have a fairly limited ambit, often restricted to repeated trials under some unchanging set of conditions. A causal model differs only in being more ambitious: specifically, in relating together, within a broader framework, cases involving a variety of sets of conditions. It typically does this by postulating the existence of certain features that are considered invariant across a range of conditions. For example, it might set out a certain kind of dependence of a response variable on other independent variables, supposing this dependence to be the same for a variety of different ways in which the values of the independent variables might have come about (e.g. perhaps they arose at the whim of Nature, or perhaps they were manipulated by an experimenter). Just like for any other model, the ambit of a causal model must be clearly understood and described, and its limitations fully appreciated: it would be simply inappropriate to try to apply it beyond the specific sets of circumstances for which it is intended. On the other side of the coin, in order for a causal model to be severely tested it must be pitted against its predictions over the full variety of circumstances encompassed by its ambit of intended applicability.

A causal model, then, is just one that is intended to apply across a broad, but clearly defined and circumscribed, range of circumstances, and which incorporates certain invariant features that appear unchanged in all of these. It is perfectly acceptable to call these invariant features 'causal mechanisms' if desired – so long as we do not attach too much deep significance to this term. In particular, if we were to investigate a wider range of circumstances, or the same circumstances at a deeper level, then we should not be surprised to find that previously invariant features are no longer so. From this point of view, notions of cause are both provisional and relative. Science advances by developing causal models that successfully explain broader and broader classes of circumstances. Science typically also attempts to find causal explanations at deeper and deeper levels of detail, but such reductionism is best viewed as providing a different kind of explanation, with a different ambit, rather than superseding causal explanations at a cruder level of description: these continue to be valid within their intended ambits. This relativistic philosophy has no truck with such concepts as 'ultimate cause' or 'true causal model', and thereby eliminates many perplexing philosophical conundra concerning the 'true nature' of cause, as simply misconceived.

A commitment to the importance of some sort of invariant feature is implicit (though less often explicit) in most of the various frameworks that have been proposed to support causal modelling within a statistical context. Thus in Rubin's 'potential response' framework (Rubin 1974) there is supposed to exist a well-determined response of a unit to each of the various treatments that might be applied to it – even though at most one of these responses can ever be observed (by applying the relevant treatment). If so, that response must, automatically, be the same, no matter how the treatment came to be chosen. Pearl (1993a,b), following Spirtes *et al.* (1993), builds probabilistic models in which certain conditional probabilities are supposed to be the same under different ways in which the conditioning variables might have arisen; while Pearl (1995a) assumes that certain functional relationships between variables are similarly invariant under different observational regimes.

The approach taken in the present article is an elaboration of that of Spirtes *et al.* (1993) and Pearl (1993a,b). The invariant features in our models will be conditional probability distributions, rather that functional relations. The novelty (such as it is) of our approach is that it attempts to be as explicit as possible, at all times, about just what invariance assumptions are being made. This is effected by the use of a special graphical model, the 'extended influence diagram', that allows transparent representation of the assumptions built in to the model.

3   Extended influence diagrams

Extended influence diagrams (EIDs) are generalizations of influence diagrams (IDs) (Howard and Matheson 1984; Shachter 1986; Oliver and Smith 1990), which are themselves a special form of directed acyclic graph (DAG) (Cowell *et al.* 1999). Such a diagram will contain nodes of various kinds, representing random variables, parameters, decisions/interventions, etc. (however, we have no need for value nodes in the current work, nor shall we be concerned with transforming or solving influence diagrams to find optimal courses of action). Some pairs of nodes are connected by arrows, in such a way that no directed cycles are formed. Each node v has a (possibly empty) set $\mathrm{pa}(v)$ of 'parents' – those nodes from which arrows lead into v. If v is a random node, then there is an associated specification of its conditional distribution given its parents – this is supposed to be unaffected by further conditioning on any other nodes that are not 'descendants' of v. If v is a decision node, then its parents represent the full extent of the information supposed available to the decision-maker at the time that that decision has to be taken.

A simple example of an ordinary influence diagram is presented in Fig. 1. This involves three random nodes, B, D, and E, and two decision nodes, A and C. In this and similar diagrams, a random variable will be represented by a circle (solid if observable, dotted if unobservable), and an external intervention by a

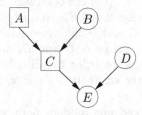

FIG. 1. Causal influence diagram.

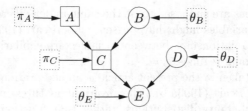

FIG. 2. Extended influence diagram.

solid square. We refer to all these five nodes, which are potentially observable, as 'domain variables'. Decision A is taken at the outset, with no knowledge of other variables. In parallel, and entirely independently, variable B is generated (from some externally specified distribution $p(B)$). Decision C is then taken in the light of the observed values at A and B, while random variable D is generated (from $p(D)$), independently of all the other variables so far. Finally, random variable E is generated from a specified distribution $p(E \mid C, D)$, using the realized values of C and D but no other variables.

A more complete representation of this problem is given by the *extended influence diagram* of Fig. 2. In addition to the domain variables, this contains additional 'parameter nodes': θ_B (holding the specification of $p(B)$), θ_D (holding $p(D)$) and θ_E (holding $p(E \mid C, D)$); as well as 'strategy nodes': π_A (specifying the way, perhaps randomized, in which A is initially chosen) and π_C (specifying the dependence, again possibly randomized, of the choice of C on the values of A and B). We represent both parameter nodes and strategy nodes by dotted squares.

Further background on the construction, properties, and applications of EIDs may be found in Lauritzen (2000) and Dawid (2002).

3.1 Conditional independence

A model represented by an ID or EID incorporates some explicit assumptions of conditional independence: that each domain variable be independent of its 'non-descendants', conditional on only its 'parents' (and, for IDs such as that of Fig. 1, on any relevant externally specified distributional or functional forms that are not explicitly represented in the diagram). For example, Fig. 2 encodes an assumption that the distribution of E, given all other variables, is completely determined by the values of C and D and the specification θ_E of the conditional distributions of E given any values of C and D; and is entirely unaffected by further knowledge of the value of B; or the distribution θ_B that it was drawn from; or the value of A; or the strategies π_A and π_C that were used to select A and C.

Note that we have allowed non-random parameter and strategy nodes as potential conditioning variables – although it is typically not appropriate (except in a Bayesian analysis) to include them on the left-hand side of a conditioning statement. These extended semantics enable us to use the EID of Fig. 2 to be more careful and specific than would be possible using only the ID of Fig. 1. As we shall see, this facility is particularly useful for representing the invariant features of causal models.

Once an ID or EID has been constructed, it is possible to read off from it further properties of conditional independence, logically implied by its explicit inputs. This may be done by means of the 'd-separation' criterion of Verma and Pearl (1990), or the equivalent, and generally more straightforward, 'moralization criterion' of Lauritzen *et al.* (1990). Thus from Fig. 2 we can deduce, for example,

$$B \perp\!\!\!\perp (E, \pi_A, \theta_D, \theta_E) \mid (A, C, \theta_B, \pi_C), \qquad (3.1)$$

which (with the interpretation and notation for conditional independence introduced by Dawid (1979)) says that the conditional distribution of B given A, C, and E does not depend on the value of E, nor on the way in which A was chosen, nor on the distributions $p(D)$ and $p(E \mid C, D)$; although it may depend on the distribution $p(B)$ and the strategy used to select C.

4 Partial compliance

The remainder of this article consists of a specific illustration of the application of the above formulation, aimed at showing how it can be used to clarify assumptions and guide analysis. The problem studied is that of making causal inferences from experiments in which patients may not comply with the treatments they have been assigned (Manski 1990; Efron and Feldman 1991; Balke 1995; Balke and Pearl 1993, 1994, 1997). A thorough account may be found in Chapter 8 of Pearl (2000).

4.1 Encouragement trial

Consider an encouragement trial, in which each patient is randomly assigned, or encouraged, to take either active treatment or placebo. We introduce an assignment indicator Z, taking value 1 for treatment assignment and 0 for placebo assignment. However, the patient may not comply with the assignment, so that the treatment actually taken, $D = 0$ or 1, may differ from Z, and this in a way which may vary from patient to patient. We finally observe a binary response Y, whose probabilistic dependence on the treatment D actually taken may also vary from patient to patient. We model the problem conditional on any observed covariates, which can thus be regarded as fixed and need not be mentioned further. Even so, the possible common dependence of D and Y on further unobserved patient characteristics will typically imply that Y cannot be assumed independent of Z given D. Let the parameter ϕ describe the full joint distribution of (Y, D) given Z, which is clearly all that is estimable from this encouragement trial. We shall denote $\Pr(Y = y, D = d \mid Z = z, \phi)$ by $\phi_{yd.z}$ $(y, d, z = 0, 1)$.

In such a case, it might *sometimes* be reasonable to accept the following:

Assumption A_1 There exists an unobserved 'latent' variable U for which the encouragement trial can be modelled by the EID of Fig. 3.

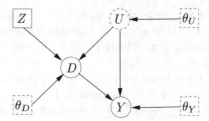

FIG. 3. Encouragement experiment: all variables.

We shall call a problem thus represented by Fig. 3 a *partial compliance model* (PCM). Typically U will be a multivariate description of some characteristic or characteristics of the patient.

Assumption A_1 is equivalent to assuming certain (conditional) independencies between the variables in the problem. Specifically, these are:

(i) $Z \perp\!\!\!\perp U$, represented in Fig. 3 by the absence of any arrow between Z and U. If U is a pre-existing patient characteristic, then this requires that U be not taken into account in any way when deciding which treatment Z to assign (this might be assured by randomization, for example); otherwise, it requires that the distribution of U should not be affected, either directly or indirectly, by (knowledge of) treatment assignment (for example, there must be no common cause of Z and U).

(ii) $Y \perp\!\!\!\perp Z \mid (D, U, \theta_Y)$, represented by the absence of any arrow from Z to Y. This expresses the fact that, once the treatment D actually taken and the characteristic U are specified, the resulting uncertainty about Y, as described by the distribution θ_Y for Y given (D, U), is entirely unaffected by further knowledge of the treatment Z that was originally assigned. We might call such a patient characteristic U a *sufficient concomitant* with respect to encouragement.

Under Assumption A_1 we can regard U as encompassing characteristics of the patient that might be relevant, either to the patient's decision as to what treatment D he or she should take (after learning the assigned treatment Z), or to mediating the effect of the treatment taken, D, on the response Y. We are thus explicitly allowing both treatment application and response to treatment to depend on hidden patient characteristics. Clearly, this is liable to introduce biases into any simple interpretation of the data of the encouragement trial.

The joint distribution θ of (Y, D, U) given Z is described in Fig. 3 by means of: θ_U, specifying the distribution of U; θ_D, specifying that of D given (Z, U); and θ_Y, specifying that of Y given (D, U). Since U is not observed, θ is not fully estimable from the encouragement trial, even when U is clearly defined. The most that can be estimated is the distribution ϕ, for (Y, D) conditional on Z. This may be derived from θ by suitably marginalizing out over U. Specifically, we have:

$$\phi_{yd.z} = \mathrm{E}\{\Pr(D = d | Z = z, U, \theta_D) \Pr(Y = y | D = d, U, \theta_Y)\}, \qquad (4.1)$$

where the expectation is over the distribution θ_U of U.

Not every conditional distribution ϕ can arise in such a way. For example, if $\phi_{00.0} = 1$, then (4.1) implies $\Pr(Y = 0 | D = 0, U, \theta_Y) = 1$ almost surely, whence $\Pr(Y = 1 | D = 0, U, \theta_Y) = 0$ almost surely, which in turn implies $\phi_{10.1} = 0$. The collection of values of ϕ which are consistent with Assumption A_1 is a convex set Φ, which is further characterized in Lemma 4.1 below.

Ideally, the latent variable U represented in Fig. 3 would be a clearly identified real-world quantity, and the validity of (i) and (ii) of Assumption A_1 would be based on firm scientific understanding and/or evidence. More usually we will not be in a position to identify such a variable U. We caution that in general, even when $\phi \in \Phi$, there is no compelling reason to posit the existence of an unobserved variable for which Assumption A_1 holds. To make this assumption is to say something non-trivial about how the world is. And even when it can be made, there is no reason why the variable U should be essentially unique. However, if the structure of Fig. 3 does apply for two different definitions of U (and thus of θ), applying (4.1) in each case must yield the same distribution ϕ for the observables.

Balke and Pearl (1994) use a representation similar to that of Fig. 3, but with the additional feature that D is specified to be a function of (Z, U), and Y a function of (D, U). We call such a model a *functional partial compliance model* (FPCM), and we indicate functional relationships on the diagram by means of hollow arrowheads. A PCM will be a FPCM if and only if the conditional distributions described by θ_D and θ_Y are degenerate (given $Z = 0$ or 1, and any value of U), i.e. put all their mass on a single outcome (0 or 1).

A notable feature of the present analysis is that we are able to reproduce all the results that have previously been based on FPCM representations with the weaker assumptions embodied in the general probabilistic PCM of Fig. 3. This is in keeping with the general philosophy of Dawid (2000), which is extremely sceptical of 'counterfactual' approaches, i.e. those that invoke or (as is the case for FPCMs – see Section 4.2 below) imply the simultaneous existence of all the potential responses to all the alternative interventions that could in principle be made on the same unit.

In what follows we shall proceed under Assumption A_1. For clarity, and to concentrate on essentials, we assume that a large encouragement experiment has allowed us accurately to identify $\phi \in \Phi$. By (4.1), this supplies partial information about θ. We shall suppose further that this is the full extent of our knowledge of θ.

4.2   Minimal representation

Define the following random variables (functions of the distribution θ as well as the random quantity U):

$$
\begin{aligned}
\eta_0 &:= \Pr(Y = 1 \mid D = 0, U, \theta), \\
\eta_1 &:= \Pr(Y = 1 \mid D = 1, U, \theta), \\
\delta_0 &:= \Pr(D = 1 \mid Z = 0, U, \theta), \\
\delta_1 &:= \Pr(D = 1 \mid Z = 1, U, \theta).
\end{aligned}
\tag{4.2}
$$

We note that the conditioning in η_0 and η_1 could be extended to include Z, without effect, since $Y \perp\!\!\!\perp Z \mid (D, U, \theta)$. Moreover, η_0 and η_1 depend on θ only through θ_Y, and δ_0 and δ_1 only through θ_D. The vector $\tau := (\eta_0, \eta_1, \delta_0, \delta_1)$, a function of U and (θ_Y, θ_D) that takes values in the unit 4-cube $\mathbf{B} := [0,1]^4$, determines, and essentially describes, the joint distribution of (Y, D) given (Z, U).

We can always, without any essential loss, replace U in Fig. 3 by its functional reduction τ (and θ_U by the induced marginal distribution θ_τ for τ over $\mathbf{B}$ when $U \sim \theta_U$). Then θ_τ, which can in principle be an arbitrary distribution over $\mathbf{B}$, fully captures the relevant structure of the original problem. We may term this equivalent form of the problem its *minimal representation*, τ the *minimal concomitant*, and θ_τ the *minimal distribution*.

Note that the values of η_0 and η_1 depend on (U and) θ_Y; likewise, the values of δ_0 and δ_1 depend on (U and) θ_D. Once these random variables are constructed, their joint distribution θ_τ over $\mathbf{B}$ will further depend on the distribution θ_U of U. So θ_τ will depend (in a fairly complicated way) on the whole of θ. If there are differing specifications for the latent variable U in Fig. 3, all describing the same real-world problem, then these may lead to differing specifications of τ, and hence different distributions θ_τ. In particular, a given probabilistic structure for the observables in the problem will typically have more than one minimal representation.

It is easy to see that the original model is a FPCM if and only if the distribution θ_τ is concentrated on the 16-point set $\hat{\mathbf{B}} := \{0,1\}^4$ of extreme points of $\mathbf{B}$, so that each of η_0, η_1, δ_0, and δ_1 is a binary $(0,1)$ variable. We can then (but only then) interpret η_d as Y_d, the 'potential value' of Y were we to have $D = d$; and similarly δ_z as the potential value of D were we to have $Z = z$. In this special case, all these potential responses can be considered as having a well-defined, simultaneous existence, even though they cannot be observed simultaneously. That is, a FPCM is a special kind of counterfactual model, as described and criticized by Dawid (2000).

4.3   Constraints on ϕ

Define random variables (functions of U and θ, with values in $[0,1]$): $\xi_{yd.z} := \Pr(Y = y, D = d \mid Z = z, U, \theta)$; and let $\xi \in [0,1]^8$ denote the random vector with these components. Using the definitions of the δs and ηs, we find:

$$\begin{aligned}
\xi_{00.0} &= (1-\eta_0)(1-\delta_0), \\
\xi_{01.0} &= (1-\eta_1)\delta_0, \\
\xi_{10.0} &= \eta_0(1-\delta_0), \\
\xi_{11.0} &= \eta_1\delta_0, \\
\xi_{00.1} &= (1-\eta_0)(1-\delta_1), \\
\xi_{01.1} &= (1-\eta_1)\delta_1, \\
\xi_{10.1} &= \eta_0(1-\delta_1), \\
\xi_{11.1} &= \eta_1\delta_1.
\end{aligned} \tag{4.3}$$

Note that there are only six mathematically independent equations here, since

$$\sum_{y=0}^{1}\sum_{d=0}^{1} \xi_{yd.z} \equiv 1, \qquad z = 0, 1.$$

Since $U \perp\!\!\!\perp (Z, \theta) \mid \theta_U$, we have

$$\mathrm{E}(\xi_{yd.z}) = \phi_{yd.z}, \qquad y, d, z = 0, 1, \tag{4.4}$$

the expectation being under the distribution θ_τ. Again, only six of these eight equations are mathematically independent. From extensive encouragement data (but without observation of U), we can only learn the six mathematically independent quantities $(\phi_{yd.z})$, the expectations under θ_τ of the six mathematically independent functions $(\xi_{yd.z})$ of τ for $y, d, z = 0, 1$, but nothing else about θ_τ.

Let $\Xi : \mathbf{B} \to [0,1]^8$ denote the mapping defined by (4.3), and let its range $\Xi(\mathbf{B})$, a subset of (a six-dimensional subspace in) $[0,1]^8$, be $\mathbf{T}$. We also introduce $\hat{\mathbf{T}} := \Xi(\hat{\mathbf{B}})$. Explicitly, the mapping Ξ takes the following form on $\hat{\mathbf{B}}$:

τ					ξ							
(η_0	η_1	δ_0	δ_1)		($\xi_{00.0}$	$\xi_{01.0}$	$\xi_{10.0}$	$\xi_{11.0}$	$\xi_{00.1}$	$\xi_{01.1}$	$\xi_{10.1}$	$\xi_{11.1}$)
(0	0	0	0)		(1	0	0	0	1	0	0	0)
(0	0	0	1)		(1	0	0	0	0	1	0	0)
(0	0	1	0)		(0	1	0	0	1	0	0	0)
(0	0	1	1)		(0	1	0	0	0	1	0	0)
(0	1	0	0)		(1	0	0	0	1	0	0	0)
(0	1	0	1)		(1	0	0	0	0	0	0	1)
(0	1	1	0)		(0	0	0	1	1	0	0	0)
(0	1	1	1) $\to$			(0	0	0	1	0	0	0	1)
(1	0	0	0)		(0	0	1	0	0	0	1	0)
(1	0	0	1)		(0	0	1	0	0	1	0	0)
(1	0	1	0)		(0	1	0	0	0	0	1	0)
(1	0	1	1)		(0	1	0	0	0	1	0	0)
(1	1	0	0)		(0	0	1	0	0	0	1	0)
(1	1	0	1)		(0	0	1	0	0	0	0	1)
(1	1	1	0)		(0	0	0	1	0	0	1	0)
(1	1	1	1)		(0	0	0	1	0	0	0	1)

$$\tag{4.5}$$

Note that there are only 12 distinct points in $\hat{\mathbf{T}}$, because $\Xi(\tau)$ takes on the same value for $\tau = (0,0,0,0)$ and $(0,1,0,0)$; for $(1,0,0,0)$ and $(1,1,0,0)$; for $(0,0,1,1)$ and $(1,0,1,1)$; and for $(0,1,1,1)$ and $(1,1,1,1)$. Let $\mathbf{H}$ denote the convex hull of the 12-point set $\hat{\mathbf{T}}$. It can be shown (for example, using the computer program PORTA (Christof and Loebel 1998), which transforms between dual representations of a convex polytope) that $\phi = (\phi_{yd.z}) \in \mathbf{H}$ if and only if $\phi_{yd.z} \geqslant 0$ $(y,d,z = 0,1)$, $\sum_{y=0}^{1} \sum_{d=0}^{1} \phi_{yd.z} = 1$ $(z = 0,1)$, and the following additional inequalities are satisfied:

$$\begin{aligned} \phi_{00.0} + \phi_{10.1} &\leqslant 1, \\ \phi_{10.0} + \phi_{00.1} &\leqslant 1, \\ \phi_{11.0} + \phi_{01.1} &\leqslant 1, \\ \phi_{01.0} + \phi_{11.1} &\leqslant 1. \end{aligned} \tag{4.6}$$

These are exactly the non-trivial inequalities implied by the 'instrumental variable' inequality (Pearl 1995b)

$$\phi_{0d.z_0} + \phi_{1d.z_1} \leqslant 1$$

$(d, z_0, z_1 = 0,1)$.

Clearly the set Φ of possible values for ϕ that can satisfy (4.4) is the convex hull of $\mathbf{T}$, and so $\mathbf{H} \subseteq \Phi$.

Lemma 4.1 $\Phi = \mathbf{H}$.

Proof It is enough to show that $\mathbf{T} \subseteq \mathbf{H}$. Since $\mathbf{H}$ is the intersection of all closed half-spaces containing $\hat{\mathbf{B}}$, we thus need to show that any such half-space contains $\mathbf{T}$. Let then $c(\xi)$ be an affine function of ξ such that $c(\xi) \geqslant 0$, all $\xi \in \hat{\mathbf{T}}$; that is, $c\{\Xi(\tau)\} \geqslant 0$, all $\tau \in \hat{\mathbf{B}}$. But $c\{\Xi(\tau)\}$ is easily seen to be an affine, and hence monotonic, function of each component of τ when the other three are fixed. It readily follows that the minimum of $c\{\Xi(\tau)\}$, as τ ranges over $\mathbf{B}$, is attained on $\hat{\mathbf{B}}$, and hence that $c\{\Xi(\tau)\} \geqslant 0$, all $\tau \in \mathbf{B}$; i.e. $c(\xi) \geqslant 0$, all $\xi \in \mathbf{T}$. □

5 Causal inference

We have not yet addressed the purpose of all the above modelling and analysis, which is 'to make inferences about the causal effect of the treatment actually taken, D, on outcome, Y'. Before we can do so, we need to clarify the terms 'causal inference' and 'causal effect'.

5.1 Intervention experiment

We take a wholeheartedly decision-theoretic approach, as described in Dawid (2000, Section 6). We consider a (perhaps hypothetical) *intervention experiment*, in which the value of D can be definitively 'set', rather than merely encouraged. If we knew (or could estimate) the consequent distributions of Y in such an experiment, in the two cases in which D is set either to 0 or to 1, we would have all the information we would need to solve any decision problem involving the choice of intervention for a new patient (so long as the utility depends only on the eventual outcome Y).

Since there are many ways in which such an intervention might be performed, it is important to be very clear in our understanding of the physical nature of the intervention being considered, before attempting to describe it by mathematical models; and the validity of any analysis must be relative to such a specific understanding.

We introduce the *intervention response probabilities* $\omega_d := \Pr(Y = 1 | F_d = d, \omega)$ $(d = 0, 1)$. In the intervention setting these probabilities are all we need to know to settle any decision problem as to which treatment to give (Dawid 2000, Section 6). The object of causal inference can thus be taken as $\omega := (\omega_0, \omega_1)$, or some suitable summary, such as

$$\alpha := \omega_1 - \omega_0. \tag{5.1}$$

In particular, comparison of α with the indifference value 0 is sufficient to solve any treatment decision problem where the utility depends only on the value of the eventual binary outcome Y.

5.2 Combination

If we had actually performed an intervention experiment, with many patients in each treatment group, then there would clearly be no problem in estimating the desired response probabilities (ω_0, ω_1), or α. As it is, we only have the data from the encouragement experiment. Clearly we can proceed no further unless we can make acceptable assumptions linking the two experiments together. The specific assumption that we shall make is a strengthening of Assumption A_1:

Assumption A_2 There exists an unobserved variable U for which the conditional independencies embodied in Fig. 4 apply.

Figure 4 is to be interpreted in the following way. The intervention node F_D has three possible states: 0, 1, and $\emptyset$. When $F_D = \emptyset$, the diagram represents the encouragement trial. Then the interpretation, structure, and distribution of all

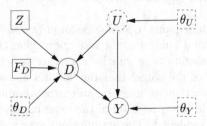

FIG. 4. Partial compliance.

the other variables are exactly as in Fig. 3. In particular, we have, for example,

$$\eta_0 = \Pr(Y = 1 \mid F_D = \emptyset, D = 0, U, \theta_Y).$$

When $F_D = 0$ or 1, this is intended to describe the intervention experiment, with the appropriate setting of D. (In this intervention case, the variable Z becomes irrelevant.) Also, $D = F_D$ whenever $F_D \neq \emptyset$ – this being the only property of the EID of Fig. 4 we shall need that is not explicitly represented on the graph by the moralization criterion.

Under Assumption A_2, the link between the two experimental situations is embodied in the following further properties, expressed by missing edges in the diagram, which relate the intervention to some sufficient concomitant with respect to encouragement, U:

(iii) $F_D \perp\!\!\!\perp U$. This is a two-fold condition, both aspects of which are important. First, it requires that the distribution of the latent variable U should be the same in both experiments, observational and interventional. Secondly, in the intervention experiment, U should not affect, nor be directly or indirectly affected by, the value at which the treatment is set. (If the model is interpreted as applying conditionally on certain observed covariates, then so too should these requirements be imposed at that level – which may serve to make them more acceptable.)

(iv) $Y \perp\!\!\!\perp (Z, F_D) \mid (D, U)$. This strengthens (ii) above to require that, given the treatment D applied and the value of the latent variable U, the conditional distribution of the response Y is completely unaffected by the way in which D came to take on its value and, in particular, by whether that was by encouragement or by intervention.

When (i)–(iv) hold, we shall term U a *totally sufficient* concomitant. Note particularly that we are not requiring that these properties hold for all unobserved concomitants, merely for (at least) one such.

A combined encouragement–intervention problem represented by Fig. 4 may be termed a *causal partial compliance model* (CPCM). Note how such a model clearly and explicitly describes what (probabilistic) features are being considered invariant across the two different regimes, encouragement and intervention – a major advantage of this approach over others that have been applied to this problem.

As we continually take pains to stress, assumptions such as A_2 are assertions about the real world, and analyses that rest upon them can only be of interest when the underlying assumptions can be regarded as acceptable properties of those specific aspects of the world that our models are intended to represent. Acceptability of Assumption A_2 must be an entirely empirical question, having regard to the real-world interpretation of the experiments (both encouragement or intervention) being modelled. The assumptions may well be reasonable in one specific interpretation, but not in another. Causal inference cannot proceed in in a vacuum, but must be intimately linked to real decision problems.

Even if we were able to perform intervention, as well as encouragement, experiments, without observation of U the most that we could estimate of the probabilistic structure of the problem would be the six mathematically independent quantities $\phi_{yd.z}$ (from encouragement data) and the further two quantities ω_d (from intervention data) – a total of eight mathematically independent probabilities. Clearly this is very sparse information about θ_τ, which can be a general joint distribution for four continuous variables.

Because Fig. 4 is supposed to apply, in particular, to the intervention experiment, we must now have

$$\omega_d = \Pr(Y = 1 \mid F_D = d, \theta) \tag{5.2}$$

as derived from Fig. 4, and thus, in particular,

$$\alpha = \Pr(Y = 1 \mid F_D = 1, \theta) - \Pr(Y = 1 \mid F_D = 0, \theta). \tag{5.3}$$

5.3 Inference

In focusing on the quantities in (5.2) or (5.3) we are assuming that (as represented in Fig. 4 by the absence of an arrow from U to F_D) information on patient characteristics U will not be available when we consider setting D. If, hypothetically, we could observe this information, obtaining say $U = u$, the relevant quantities would then be $\Pr(Y = 1 \mid F_D = d, U = u, \theta)$, rather than ω_d ($d = 0, 1$); and thus, rather than base treatment decisions on α, we would use

$$\mathrm{ICE}(u, \theta) := \Pr(Y = 1 \mid F_D = 1, U = u, \theta) - \Pr(Y = 1 \mid F_D = 0, U = u, \theta), \tag{5.4}$$

the *individual causal effect* for a patient having $U = u$.

Now

$$\Pr(Y = 1 \mid F_D = d, U, \theta) = \Pr(Y = 1 \mid F_D = d, D = d, U, \theta) \tag{5.5}$$
$$= \Pr(Y = 1 \mid D = d, U, \theta_Y) \tag{5.6}$$
$$= \Pr(Y = 1 \mid F_D = \emptyset, D = d, U, \theta_Y) \tag{5.7}$$
$$= \eta_d. \tag{5.8}$$

In the above (5.5) follows since $F_D = d$ implies $D = d$, and (5.6) and (5.7) from $Y \perp\!\!\!\perp (F_D, \theta) \mid (D, U, \theta_Y)$. In particular, $\Pr(Y = 1 \mid F_D = d, U = u, \theta)$, as a function of u, could be estimated from the encouragement data if U were also observed, so that we could estimate θ_Y.

It now follows that

$$\mathrm{ICE}(U, \theta) = \zeta := \eta_1 - \eta_0. \tag{5.9}$$

We further note that, for $d = 0, 1$,

$$\Pr(Y = 1 \mid F_D = d, \theta) = \mathrm{E}\left\{ \Pr(Y = 1 \mid F_D = d, U, \theta) \mid F_D = d, \theta \right\}$$
$$= \mathrm{E}\left\{ \Pr(Y = 1 \mid F_D = d, U, \theta_Y) \mid \theta_U \right\}, \tag{5.10}$$

since $U \perp\!\!\!\perp (F_D, \theta) \mid \theta_U$. This implies that

$$\omega_d = \mathrm{E}(\eta_d \mid \theta_U) \tag{5.11}$$

and thus that the causal comparison $\alpha = \omega_1 - \omega_0$ can also be regarded as the *average causal effect*,

$$\alpha = \mathrm{ACE} := \mathrm{E}\{\mathrm{ICE}(U, \theta_Y) \mid \theta_U\} \tag{5.12}$$
$$= \mathrm{E}(\zeta \mid \theta_U).$$

In particular (and unsurprisingly), the causal parameters ω and α depend on θ only through (θ_Y, θ_U), the distribution θ_D of D given (Z, U) in the experiment being irrelevant to their definition.

It is worth remarking that, for any fixed real-world interpretation of the variables Z, D, and Y and the intervention F_D, (5.12) holds for any latent variable U for which the assumptions embodied in Fig. 4 are valid. Changing U will typically change θ_U and θ_Y, and η_0 and η_1, but cannot affect the distribution ω of Y given F_D in the intervention experiment, nor therefore α. Equation (5.12) may then be regarded as a justification for the common, but usually unmotivated, practice of focusing attention on ACE, even when the variable U is left unspecified.

From $U \perp\!\!\!\perp F_D \mid \theta_U$ we further have $\alpha = \mathrm{E}\{\mathrm{ICE}(U, \theta_Y) \mid F_D = \emptyset, \theta_U\}$, so that α could also be be estimated from encouragement data, if observation on U, and hence estimation of θ_U, were possible. However, α (and *a fortiori* ω) is typically not estimable from an encouragement trial in which U is unobserved, which is what we suppose here.

5.4   Bounding the intervention response probabilities

Our encouragement data arise conditionally on $F_D = \emptyset$ and on Z. For any patient we observe Z, D, and Y, but not U. From these data we can estimate any aspects of the conditional distribution ϕ of (Y, D) given Z and $F_D = \emptyset$. Thus, if the experiment is large, we shall, essentially, know the values of

$$\phi_{yd.z} = \mathrm{Pr}(Y = y, D = d \mid F_D = \emptyset, Z = z, \theta) \tag{5.13}$$

$(y, d, z = 0, 1)$. Subject to the constraints of (4.4) on θ_τ, we wish to know what can be said about the intervention response probabilities ω_d.

An argument essentially identical to that of Lemma 4.1 shows that the set of feasible combinations for $(\phi, \omega_0, \omega_1)$ is just the convex hull of the 16 (now distinct) points obtained by extending each of the vectors displayed on the right-hand side of (4.5) with its associated values for η_0 and η_1. Again applying the program PORTA to represent this convex polytope in dual form, we obtain the following defining inequalities:

$$\omega_0 \leqslant \min \left\{ \begin{array}{c} 1 - \phi_{00.0} \\ 1 - \phi_{00.1} \\ \phi_{01.0} + \phi_{10.0} + \phi_{10.1} + \phi_{11.1} \\ \phi_{10.0} + \phi_{11.0} + \phi_{01.1} + \phi_{10.1} \end{array} \right\}, \qquad (5.14)$$

$$\omega_0 \geqslant \max \left\{ \begin{array}{c} \phi_{10.1} \\ \phi_{10.0} \\ \phi_{10.0} + \phi_{11.0} - \phi_{00.1} - \phi_{11.1} \\ -\phi_{00.0} - \phi_{11.0} + \phi_{10.1} + \phi_{11.1} \end{array} \right\}, \qquad (5.15)$$

$$\omega_1 \leqslant \min \left\{ \begin{array}{c} 1 - \phi_{01.1} \\ 1 - \phi_{01.0} \\ \phi_{10.0} + \phi_{11.0} + \phi_{00.1} + \phi_{11.1} \\ \phi_{00.0} + \phi_{11.0} + \phi_{10.1} + \phi_{11.1} \end{array} \right\}, \qquad (5.16)$$

$$\omega_1 \geqslant \max \left\{ \begin{array}{c} \phi_{11.1} \\ \phi_{11.0} \\ -\phi_{01.0} - \phi_{10.0} + \phi_{10.1} + \phi_{11.1} \\ \phi_{10.0} + \phi_{11.0} - \phi_{01.1} - \phi_{10.1} \end{array} \right\}. \qquad (5.17)$$

We further find that, for given ϕ, ω_0 and ω_1 can vary independently, subject only to the above inequalities.

Once they have been suggested, it is straightforward to verify the above inequalities. For example, using (4.4) and (5.11) with (4.3), we see that the inequality

$$\omega_0 \leqslant \phi_{01.0} + \phi_{10.0} + \phi_{10.1} + \phi_{11.1} \qquad (5.18)$$

is equivalent to

$$E(\gamma) \geqslant 0, \qquad (5.19)$$

where

$$\gamma := (1 - \eta_1)\delta_0 + \eta_0(1 - \delta_0) + \eta_0(1 - \delta_1) + \eta_1\delta_1 - \eta_0$$
$$= (1 - \eta_1)\delta_0 + \eta_0(1 - \delta_0) + (\eta_1 - \eta_0)\delta_1.$$

Arguing as for Lemma 4.1, we see that the minimum value of γ, as τ ranges over **B**, is attained at a point in $\hat{\mathbf{B}}$. And, on checking all 16 points of $\hat{\mathbf{B}}$, it can be verified that the minimum value of γ on $\hat{\mathbf{B}}$ is 0. Since thus $\gamma \geqslant 0$ on **B**, (5.19) follows. All the other inequalities can be verified similarly.

It is of some interest that the only additional inequalities (not involving ω_0 and ω_1) that PORTA finds are the trivial ones: that the $(\phi_{yd.z})$ should be non-negative, and sum to 1 for fixed z. This is because (as is easily confirmed) the non-trivial constraints of (4.6) are already implied by (5.14)–(5.17). In particular, if we attempt to use (5.14)–(5.17) to set bounds on ω_0 and ω_1, the resulting intervals will both be non-empty, and so we will get a solution, if and only if

$\phi \in \Phi$. Thus we do not need to check $\phi \in \Phi$ in advance, but will automatically discover whether or not this holds as we conduct the above analysis for ω.

Once we have the inequalities of (5.14)–(5.17), we can easily bound any function of ω should we so wish – for example, α, or ω_1/ω_0.

5.5   Bounds for α

The bounds thus obtained directly for α contain some that are redundant. In order to characterize the minimal set of defining inequalities, we use PORTA yet again, this time after extending the right-hand sides of (4.5) with the values for ζ: in order, these values are

$$0 \quad 0 \quad 0 \quad 0 \quad 1 \quad 1 \quad 1 \quad 1 \quad -1 \quad -1 \quad -1 \quad -1 \quad 0 \quad 0 \quad 0 \quad 0.$$

The output now yields the following necessary and sufficient inequalities involving α:

$$\alpha \leqslant \alpha^* := \min \left\{ \begin{array}{c} 1 - \phi_{10.0} - \phi_{01.1} \\ 1 - \phi_{01.0} - \phi_{10.1} \\ \phi_{00.0} - \phi_{01.0} + \phi_{11.0} + \phi_{00.1} + \phi_{01.1} \\ \phi_{00.0} + \phi_{01.0} + \phi_{00.1} - \phi_{01.1} + \phi_{11.1} \\ \phi_{00.1} + \phi_{11.1} \\ \phi_{00.0} + \phi_{11.0} \\ \phi_{10.0} + \phi_{11.0} + \phi_{00.1} - \phi_{10.1} + \phi_{11.1} \\ \phi_{00.0} - \phi_{10.0} + \phi_{11.0} + \phi_{10.1} + \phi_{11.1} \end{array} \right\}, \tag{5.20}$$

$$\alpha \geqslant \alpha_* := \max \left\{ \begin{array}{c} \phi_{00.0} + \phi_{11.1} - 1 \\ \phi_{11.0} + \phi_{00.1} - 1 \\ -\phi_{01.0} - \phi_{10.0} + \phi_{11.0} - \phi_{10.1} - \phi_{11.1} \\ -\phi_{10.0} - \phi_{11.0} - \phi_{01.1} - \phi_{10.1} + \phi_{11.1} \\ -\phi_{01.1} - \phi_{10.1} \\ -\phi_{01.0} - \phi_{10.0} \\ -\phi_{00.0} - \phi_{01.0} + \phi_{00.1} - \phi_{01.1} - \phi_{10.1} \\ \phi_{00.0} - \phi_{01.0} - \phi_{10.0} - \phi_{00.1} - \phi_{01.1} \end{array} \right\}. \tag{5.21}$$

This time we also obtain all the inequalities of (4.6) for ϕ, which thus have to be checked separately, ahead of asserting the above bounds for α.

Given specific values for $\phi \in \Phi$, we deduce that the causal parameter α must lie within the interval $[\alpha_*, \alpha^*]$ – but, without any further external information, can assert nothing else about α.

6   Functional representation

Balke and Pearl (1994) derived the inequalities of (5.20) and (5.21) by formulating and symbolically solving a pair of linear programs; they similarly derived (5.14)–(5.17) in Balke and Pearl (1997) (without, however, explicitly noting the 'variation independence' of ω_0 and ω_1). However, their analyses are based on the seemingly restrictive additional assumption that the model is a FPCM. Note that, whenever we have a FPCM, on replacing U with the minimal concomitant τ we obtain another FPCM, which we may call the *canonical form* of the original.

We stress that our own treatment in Section 5.4 did not make use of any assumptions beyond A_2; in particular, we did not, either in the formulation or in the analysis, require that the model be a FPCM. However, our analysis did make use of the property (an extension of Lemma 4.1) that any consistent set of values for (ϕ, α) is attainable by a distribution θ_τ concentrated on $\hat{\mathbf{B}}$, which thus corresponds to a FPCM. This immediately implies that, for the purely mathematical purpose of setting bounds on α (or on ω) given ϕ, nothing is lost by imposing this restriction.

Another way of seeing this is through the following analysis, which shows how, given any CPCM, we can construct a FPCM that is observationally equivalent to it, i.e. produces exactly the same values for ϕ and ω, the only quantities estimable from experiments (be they encouragement or observational) that do not involve observation of U.

As explained in Dawid (2002), we can introduce purely artificial 'error variables' ε_D and ε_Y, independent of each other and of (U, Z, F_D), to manipulate Fig. 4 into an equivalent form with purely functional relationships, as in Fig. 5. All the randomness in the problem is thus isolated into the error terms. (It is important to note, however, that this extension can be done in many, functionally inequivalent, ways: there is no unique way of constructing a functional representation of a given, or inferred, CPCM.) If we now define $U^* := (U, \varepsilon_D, \varepsilon_Y)$, the picture may be drawn as in Fig. 6. This looks just like the original, Fig. 4, except that all the relationships in it are deterministic, and the latent variable U has been redefined. Our construction ensures that, if we were to marginalize over ε_D and ε_Y in Fig. 5, then the probabilistic structure of the remaining variables would still be as described by Fig. 4. In particular, nothing we have done can have affected ϕ and ω. Consequently, we do not lose any generality by assuming that the CPCM we are dealing with is in fact a FPCM. And, having done so, we can further replace U^* in Fig. 6 with its corresponding minimal concomitant τ^*, whose distribution $\theta_{\tau^*}^*$ is now confined to $\hat{\mathbf{B}}$.

Note that the above replacement of U by U^*, and consequently τ by τ^* and θ_τ by $\theta_{\tau^*}^*$, will affect the definition and value of the individual causal effect

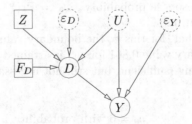

FIG. 5. Initial functional model.

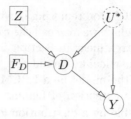

FIG. 6. Transformed functional model.

ICE, which will now always be 0, −1, or 1. However, by (5.12) we still have ACE = α, and thus, because of the clear decision-theoretic interpretation of α, independent of the specification of U, the average causal effect ACE will be unchanged when U is replaced by U^*. This can also be seen directly as follows. We have $\Pr(Y = 1 | D = 1, U) = \mathrm{E}\{\Pr(Y = 1 | D = 1, U, \varepsilon_Y) | D = 1, U\}$. Since $\varepsilon_Y \perp\!\!\!\perp D | U$, this becomes $\mathrm{E}\{\Pr(Y = 1 | D = 1, U, \varepsilon_Y) | U\}$. But $\Pr(Y = 1 | D = 1, U^*) = \Pr(Y = 1 | D = 1, U, \varepsilon_Y)$. Consequently, $\mathrm{E}\{\Pr(Y = 1 | D = 1, U)\} = \mathrm{E}\{\Pr(Y = 1 | D = 1, U^*)\}$, and similarly when $D = 0$, confirming that ACE is the same, whether calculated for U or for U^*. We can therefore conduct our inference for ACE = α equally well in the original CPCM or in the constructed FPCM. The same holds if we replace the FPCM by its canonical version.

In general there are many possible distributions θ_τ over $\mathbf{B}$ consistent with given 'observable' quantities $(\phi_{yd.z})$ and α. We have shown that it is always possible to confine such a distribution to the vertex set $\hat{\mathbf{B}}$, but even then this is not generally unique. Nor is it necessary to confine the distribution in this way, even in cases where one of the bounds in (5.20) or (5.21) is attained. For example (Balke and Pearl 1994), for $(\phi_{yd.z}) = (0.919, 0, 0.081, 0, 0.315, 0.139, 0.073, 0.473)$, the bounds on α are $\alpha_* = 0.392$ and $\alpha^* = 0.780$. In fact, from (5.14)–(5.17) we obtain the stronger conclusion:

$$0.081 \leqslant \omega_0 \leqslant 0.081,$$
$$0.473 \leqslant \omega_1 \leqslant 0.861,$$

showing that (in this special case) we have achieved exact identification of the 'placebo intervention response probability', $\omega_0 := \Pr(Y = 1 | F_D = 0) = 0.081$, from the encouragement data.

It may be verified that, for this ϕ, the boundary value $\alpha^* = 0.780$ for α (or equivalently the boundary value 0.861 for ω_1) is attained for any distribution θ_τ over $\mathbf{B}$ with the following (sufficient, but still not necessary) properties:

Marginally:

$$\delta_0 = 0 \text{ with probability } 1,$$

$$\delta_1 = \begin{cases} 0 & \text{with probability } 0.388, \\ 1 & \text{with probability } 0.612. \end{cases}$$

Given $\delta_1 = 0$:

$$E(\eta_0 \mid \delta_1 = 0) = \frac{73}{388},$$

$$E(\eta_1 \mid \delta_1 = 0) = 1$$

$$\text{(so that } \delta_1 = 0 \Rightarrow \eta_1 = 1\text{)}.$$

Given $\delta_1 = 1$:

$$E(\eta_0 \mid \delta_1 = 1) = \frac{8}{612},$$

$$E(\eta_1 \mid \delta_1 = 1) = \frac{473}{612}.$$

In particular, conditionally on $\delta_1 = 1$ the above requirements only involve the marginal expectations of η_0 and η_1, and otherwise say nothing about their joint distribution. Thus there remains a great deal of arbitrariness in choosing a distribution θ_τ consistent with the given values for (ϕ, α), and this remains true even if we confine it to $\hat{\mathbf{B}}$ (when we in fact obtain a one-parameter family of possible θ_τ).

7 Conclusions

Extended influence diagrams provide a clear and transparent representation of probabilistic causal models, helping to guide one along the path towards sensible causal inferences. Such inferences are not assisted by more complex or more restrictive representations, such as functional models, which are based on a totally unnecessary deterministic view of the world.

Acknowledgements

I am grateful to Jamie Robins for first setting me the challenge of dealing with the partial compliance problem within a fully probabilistic framework; and to him and Judea Pearl for valuable discussions on some of the issues involved.

References

Balke, A. A. (1995). *Probabilistic Counterfactuals: Semantics, Computation, and Applications*. PhD thesis, Department of Computer Science, University of California, Los Angeles.

Balke, A. A. and Pearl, J. (1993). Nonparametric bounds on causal effects from partial compliance data. Technical Report R–199, Cognitive Systems Laboratory, Computer Science Department, University of California, Los Angeles.

Balke, A. A. and Pearl, J. (1994). Counterfactual probabilities: Computational methods, bounds and applications. In *Proceedings of the Tenth Annual Conference on Uncertainty in Artificial Intelligence* (eds R. L. de Mantaras and D. Poole), pp. 46–54.

Balke, A. A. and Pearl, J. (1997). Bounds on treatment effects from studies with imperfect compliance. *Journal of the American Statistical Association*, **92**, 1172–6.

Christof, T. and Loebel, A. (1998). *PORTA Version 1.3.2.* Available online at URL:
http://www.iwr.uni-heidelberg.de/groups/comopt/software/PORTA.

Cowell, R. G., Dawid, A. P., Lauritzen, S. L., and Spiegelhalter, D. J. (1999). *Probabilistic Networks and Expert Systems*. Springer-Verlag, New York.

Dawid, A. P. (1979). Conditional independence in statistical theory (with discussion). *Journal of the Royal Statistical Society*, B, **41**, 1–31.

Dawid, A. P. (2000). Causal inference without counterfactuals (with discussion). *Journal of the American Statistical Association*, **95**, 407–48.

Dawid, A. P. (2002). Influence diagrams for causal modelling and inference. *International Statistical Review*, **70**, 161–89.

Efron, B. and Feldman, D. (1991). Compliance as an explanatory variable in clinical trials. *Journal of the American Statistical Association*, **86**, 9–26.

Howard, R. A. and Matheson, J. E. (1984). Influence diagrams. In *Readings in the Principles and Applications of Decision Analysis* (eds R. A. Howard and J. E. Matheson). Strategic Decisions Group, Menlo Park, CA.

Lauritzen, S. L. (2000). Causal inference from graphical models. In *Complex Stochastic Systems* (eds O. E. Barndorff-Nielsen, D. R. Cox, and C. Klüppelberg), pp. 63–107. CRC Press, London.

Lauritzen, S. L., Dawid, A. P., Larsen, B. N., and Leimer, H.-G. (1990). Independence properties of directed Markov fields. *Networks*, **20**, 491–505.

Manski, C. F. (1990). Nonparametric bounds on treatment effects. *American Economic Review, Papers and Proceedings*, **80**, 319–23.

Oliver, R. M. and Smith, J. Q. (1990). *Influence Diagrams, Belief Nets and Decision Analysis*. Wiley, Chichester.

Pearl, J. (1993a). Aspects of graphical models connected with causality. In *Proceedings of the Forty-ninth Session of the International Statistical Institute*, pp. 391–401.

Pearl, J. (1993b). Comment: Graphical models, causality and intervention. *Statistical Science*, **8**, 266–9.

Pearl, J. (1995a). Causal diagrams for empirical research (with discussion). *Biometrika*, **82**, 669–710.

Pearl, J. (1995b). Causal inference from indirect experiments. *Artificial Intelligence in Medicine*, **7**, 561–82.

Pearl, J. (2000). *Causality*. Cambridge University Press, UK.

Rubin, D. B. (1974). Estimating causal effects of treatments in randomized and nonrandomized studies. *Journal of Educational Psychology*, **66**, 688–701.

Shachter, R. D. (1986). Evaluating influence diagrams. *Operations Research*, **34**, 871–82.

Spirtes, P., Glymour, C., and Scheines, R. (1993). *Causation, Prediction and Search*. Springer-Verlag, New York.

Verma, T. and Pearl, J. (1990). Causal networks: semantics and expressiveness. In *Uncertainty in Artificial Intelligence 4* (eds R. D. Shachter, T. S. Levitt, L. N. Kanal, and J. F. Lemmer), pp. 69–76. North-Holland, Amsterdam.

2A
Commentary: causality and statistics

Elja Arjas
Rolf Nevanlinna Institute, University of Helsinki, Finland

Causality is a challenging topic for anyone to consider in a formal way. Already the concept itself is problematic, and often people have sharply different opinions of its foundations. But causality is perhaps a particularly challenging topic for a statistician. Rather than just trying to formulate views on some underlying philosophical issues, a statistician is often faced with the concrete problem of how to find empirical support in favour, or against, or even prove or disprove, a causal claim made in some substantive concrete context. A common solution is to altogether avoid using terminology that would refer to causality. Most textbooks on elementary statistics first give the warning that causality and correlation are not the same, and then they go on to discuss only correlation. Yet one cannot dispute the fact that nearly always, if one tests the hypothesis that a correlation or regression coefficient is zero against the alternative, then a small p-value is interpreted as evidence, or even as proof, that one of the considered variables is in some sense a cause of the other. From a practical point of view, the important question is: Has statistics something better to offer than this?

As mentioned in Dawid's article, a popular model for an explicit consideration of causal effects uses the idea of so called *potential outcomes* (or *responses*). For example, in the context of clinical trials the model postulates that, for each treatment to which a patient could be assigned, there corresponds a pre-existing, and in that sense deterministic, value of the considered outcome or response variable. Denoting generically a considered individual by i, a treatment by a, and the corresponding response by $Y_{i,a}$, the contrast $Y_{i,a} - Y_{i,a'}$ between treatments a and a' on i is then called the *individual causal effect*. Of course, even if such potential outcomes existed in some sense, at most one of the values $Y_{i,a}$ and $Y_{i,a'}$ can ever be realized. The outcome that is not realized is sometimes called *counterfactual*.

It is very much a matter of taste whether the simultaneous existence of potential outcomes, or their joint distribution, should be postulated in a statistical model for dealing with causality. Neyman (1923), Rubin (e.g. 1974, 1978) and Robins (e.g. 1986, 1998) do so explicitly, whereas Dawid clearly rejects the whole idea of potential outcomes as unnecessary. I generally agree with his criticism, thinking that Dawid's 'relativistic' position towards causality makes a great deal of sense, and particularly so in the context of concrete statistical inference where individual causal effects, even if they could be said to exist in some meaningful way, can never be estimated. A technical aspect favouring Dawid's preference of conditional distributions over 'conditional random variables' such as $Y_{i,a}$ is that the latter, when indexed by all combinations of individuals i and treatment options a, leads to an excessive number of variables in the model, with no real benefit to the causal conclusions that can be drawn.

Having rejected the potential outcomes formulation, Dawid then reserves the term 'causal effect' for a difference between two conditional probabilities, instead of a difference between two random variables. (If the response variables were not binary, as is the case here, then it would be straightforward to replace them by conditional expectations of the response, or of some function of it.) In Dawid's terminology, causal effect gets the epithet *individual* if the probabilities (or expectations) are conditioned on an unobserved individual characteristic U, and *average* if U is again removed from the conditioning by taking an expected value with respect to its distribution. The same idea of considering contrasts between probabilities or expectations, instead of the outcomes themselves, has been used in many other works on probabilistic causality, e.g. in Suppes (1970), Arjas and Eerola (1993), and Eerola (1994).

Dawid's criticism of the *functional partial compliance model* (FPCM) of Balke and Pearl is based on a similar argument. He is obviously right, in that when the final conclusion is expressed as a contrast between two conditional probabilities, then there cannot be mathematical reasons to assume that the model variables are, in some specific way, functionally dependent on each other. I think that the reason why some people prefer formulations based on random variables and their functions, rather than probabilities, is that they find the former to be somehow closer to their intuitive idea about causality.

Although Dawid does not mention the word 'confounding' in his article, one might well say that it is a central theme in it. In Fig. 4 there is an arrow from the 'individual characteristic' node U to both the decision (treatment) node D and the outcome node Y, reflecting the possibility that, in the real world, some unobserved characteristic may be influencing both. This is why Dawid, corresponding to Pearl's 'do' or 'set' conditioning, then introduces the *forced treatment* node F_D into the model, in order to guarantee that D is a genuine control factor for Y, and not merely some indicator of what value the unobserved potential confounder variable U might have taken. The crucial assumption (iv) in Section 5.2 means essentially that, while U may be a confounder, it is also a *sufficient concomitant* in the sense that, if one considers the conditional distribution of Y given both U and D, it would be irrelevant whether the choice of D was indeed 'forced' or not.

I find Dawid's analysis and solution of this problem very elegant, and I have no real quarrel with what he has done. But some of his wordings I find unnecessarily strong, and perhaps occasionally slightly misleading. Thus I have some difficulty in following Dawid's thoughts when he writes '... assumptions such as A_2 are assertions about the real world', and continues with the claim that '... acceptability of Assumption A_2 must be an entirely empirical question, having regard to the real-world interpretation of the experiments ... being modelled'. I cannot see that modelling and interpretation of models could ever be 'entirely empirical questions'. Of course, empirical observations are guiding the model-building and can be used as a yardstick of how well a model fits to reality, for example, by being able to predict the values of future observables. But I doubt that there would be many non-trivial concrete instances in which the analyst would actually be able to name exactly what characteristic or variable, if it could be measured, would constitute a totally sufficient concomitant in the sense that Dawid's model postulates (i)–(iv) would be satisfied. But if one cannot even name such a variable, how can one say that the acceptability of some postulate concerning it is an empirical question? This point is made only stronger by the fact that forced treatments, while their presence in the model appears to be necessary for a genuine causal interpretation, have an element of the same difficulty as was encountered in the potential outcomes: in the considered encouragement trial, forced interventions are only a counterfactual mental construct.

I agree with Dawid that the use of graphical models is a nice way to communicate information concerning model structure, often involving subtle conditional independence hypotheses, in a causal context. This is an important aspect, since – as I am convinced – all causal reasoning requires some amount of 'mental constructs' to which there are no corresponding observables. But I find somewhat troubling that nothing in the whole article is said about time, in spite of the fact that it is generally thought to be axiomatic to any notion of causality that 'cause must precede the effect'. Of course, one can argue that the direction of the arrows in the graphs indicates in which direction time evolves. Moreover, in examples such as clinical trials there may not be any real randomness in 'when things happen', as the event times are then typically specified beforehand in a study protocol. Nevertheless, there are important examples in which the progression of time is intrinsic to the problem at hand and in which, when dependence on time is made explicit, the corresponding graphical model becomes very large and clumsy. For example, one can consider a situation in which a patient's condition is being monitored over time, and decisions of what and when medication should be given are based on information that has accumulated from such monitoring. The situation will be complicated further if the patient's compliance is incomplete, perhaps depending on his present condition, the medication prescribed, and possibly also on side-effects encountered earlier. In such a case, a causal model, albeit a very complicated one, is more naturally formulated in terms of stochastic processes in which the process evolution depends progressively on the past, rather than in terms of 'ordinary' random variables and their joint distributions (e.g. Parner and Arjas 2000).

A final point that I would like to make concerns the concept of probability in causal inference. While stressing the role of probabilistic model formulations and the use of probabilities in assessing causal effects, Dawid is remarkably silent about how such probabilities should be interpreted. In a way, he puts this issue aside by writing 'Thus, if the experiment is large, we shall, essentially, know the values of $\phi_{yd.z} = \Pr(Y = y, D = d | F_D = \emptyset, Z = z, \theta)$ $(y, d, z = 0, 1)$.' For a frequentist statistician, these known numbers are numerical values for relative frequencies in an idealized infinite population. For a Bayesian, on the other hand, $\phi_{yd.z}$ would represent a subjective degree of belief in that a generic individual assigned to treatment z would actually take treatment d and then produce response y. The difference may seem slight, and indeed, the numerical value evaluated from large amounts of data will turn out 'essentially' the same regardless of the inferential paradigm applied. The differences become perhaps more apparent if one considers the probabilities η_0, η_1, δ_0, and δ_1, which are all conditional on a value of the unobserved individual characteristic U. This being so, a frequentist statistician needs to think again about a population of individuals sharing the same value of U. From a Bayesian perspective the corresponding conditional probabilities readily make sense on an individual level, without having to think about infinite populations for the purpose of interpretation. This is particularly relevant if one is dealing with an actual decision problem (as when a physician decides how to treat an individual patient). The Bayesian approach to inference has the additional advantage that if (as is the case always in practice) there are limited amounts of data and the causal analysis requires actual statistical estimation of model parameters, then the final result from such analysis can always be expressed in terms of predictive probabilities, distributions, or expectations.

Many people seem to reject this emphasis on predictive probabilities in a causal context, essentially on the grounds that they think of causality and causal effects as having an objective existence, and which therefore should not be described by measures that are openly subjective. For me, computing and using predictive distributions is the best tool that statistics can offer for a causal analysis. But, having this disagreement between people with a different adherence to causality, perhaps it would help communication if the terminology that Dawid uses were softened a little. One could well reserve the term 'individual causal effect' for those who want to include 'potential outcomes' in their model world. But when it comes to comparing probabilities or expected values that are evaluated on the basis of empirical data, why not call these 'predicted causal effects', or simply 'predicted effects'?

2B
Semantics of causal DAG models and the identification of direct and indirect effects

James M. Robins
Harvard School of Public Health, USA

Directed acyclic graphs (DAGs) are commonly used to represent causal models. The article by Dawid posits a causal model that is closely related to the model of Spirtes *et al.* (1993) and the model of Pearl (1993b). In this discussion I will compare and contrast the semantics of DAGs representing the Spirtes *et al.* model with that of DAGs representing the non-parametric structural equation (NPSE) model of Pearl (1995a) and the finest fully randomized causally interpreted structured tree graph (FRCISTG) model of Robins (1986). This discussion will be more philosophical than other contributions to this volume for the following reason: the major controversies in this field are often focused upon the causal rather than the statistical interpretation of various analytic procedures. For example, the finest FRCISTG and NPSE models both assume the existence of counterfactual variables; Dawid denies their existence, and the Spirtes *et al.* (1993) model is 'agnostic'. To give the flavour of the issues involved I will review in detail one controversy. The controversy concerns the question of whether, when, and how the direct effects of a treatment on an outcome can be separated by means of statistical analysis from the treatment's indirect effects. The discussion is organized as follows. In Section 1, I define the causal models that are to be compared. In Section 2, I collect mathematical results on the identification of direct and indirect effects. In Section 3, I discuss the substantive implications of these results.

1 Causal models and their DAG representation

We are given a DAG G with a vertex set of random variables $V = (V_1, \ldots, V_M)$ with density $f_V(v)$ ordered so that V_j is not a descendant of V_m for $m > j$. We shall use the following notational conventions. For any random variable Z, we let a calligraphic $\mathcal{Z}$ denote the support (i.e. the set of possible realizations z) of Z. For any $z_1, \ldots, z_m$, define $\bar{z}_m = (z_1, \ldots, z_m)$. By convention $\bar{z}_0 \equiv z_0 \equiv 0$. Let X denote any subset of V and let x be a realization of X. Both the finest FRCISTG and NPSE causal models assume the existence of the counterfactual random variable $V_m(x)$ encoding the value the variable V_m would have if, possibly contrary to fact, X were set to x, where $V_m(x)$ is assumed to be well defined in the sense that there is reasonable agreement as to the hypothetical intervention (i.e. closest possible world) that sets X to x (Robins and Greenland 2000).

Finest FRCISTG causal model A finest FRCISTG model assumes (i) all one-step-ahead counterfactuals $V_m(\bar{v}_{m-1})$ exist; (ii) $V_m(\bar{v}_{m-1}) \equiv V_m(pa_m)$ is a function of $\bar{v}_{m-1}$ only through the values pa_m of V_m's parents on G; (iii) both the observed variables V_m and the counterfactuals $V_m(x)$ for any $X \subset V$ are obtained recursively from the $V_m(\bar{v}_{m-1})$, e.g. $V_3 = V_3\{V_1, V_2(V_1)\}$ and $V_3(v_1) = V_3\{v_1, V_2(v_1)\}$; and (iv)

$$\{V_{m+1}(\bar{v}_m), \ldots, V_M(\bar{v}_{M-1})\} \perp\!\!\!\perp V_m \mid \overline{V}_{m-1} = \bar{v}_{m-1},$$

$$\text{for all } m \text{ and all } \bar{v}_{M-1} \in \overline{\mathcal{V}}_{M-1}, \qquad (1.1)$$

where $\bar{v}_k$ is a subvector of $\bar{v}_{M-1}$ for $k < M - 1$.

NPSE causal model A NPSE model assumes that there exists mutually independent random variables U_m and deterministic unknown functions f_m such that the counterfactual $V_m(\bar{v}_{m-1}) \equiv V_m(pa_m)$ is given by $f_m(pa_m, U_m)$ and both the observed variables V_m and the counterfactuals $V_m(x)$ for any $X \subset V$ are obtained recursively from the $V_m(\bar{v}_{m-1})$ as above.

Under a NPSE causal model

$$\{V_{m+1}(\bar{v}_m), \ldots, V_M(\bar{v}_{M-1})\} \perp\!\!\!\perp V_m(\bar{v}_{m-1}^{**}) \mid \overline{V}_{m-1} = \bar{v}_{m-1}^*$$

$$\text{for all } m, \text{ all } \bar{v}_{M-1} \in \overline{\mathcal{V}}_{M-1}, \text{ and all } \bar{v}_{m-1}^{**}, \bar{v}_{m-1}^* \in \overline{\mathcal{V}}_{m-1}. \qquad (1.2)$$

Thus a NPSE model is a finest FRCISTG but the converse is false, because a FRCISTG assumes independence of $\{V_{m+1}(\bar{v}_m), \ldots, V_M(\bar{v}_{M-1})\}$ and $V_m(\bar{v}_{m-1}^{**})$ given $V_{m-1} = \bar{v}_{m-1}^*$ only when $\bar{v}_{m-1}^{**} = \bar{v}_{m-1}^* = \bar{v}_{m-1}$.

In my 1995 *Biometrika* comment on Pearl (1995a), I proved the following:

Lemma 1.1 *If a DAG G represents a FRCISTG, then the density $f_V(V)$ of the observables V satisfies the Markov factorization*

$$f_V(v) = \prod_{j=1}^{M} f(v_j \mid pa_j). \qquad (1.3)$$

Remark In my 1995 *Biometrika* comment, I incorrectly claimed in my Lemma 1 that a NPSE model and a finest FRCISTG were equivalent. Butch Tsiatis pointed out to me that I had failed to note that an NPSE model satisfied the stronger assumption of (1.2).

Before defining the agnostic causal model of Spirtes *et al.* (1993) we need to discuss intervention distributions and the g-computation algorithm functional.

Intervention distributions on FRCISTGs Suppose we are given a set of variables $X = \{X_1, \ldots, X_k\} \subset V$ and an intervention DAG $G^{\tilde{}}$ that agrees with DAG G except the parents $PA_m^{\tilde{}}$ of $X_m \in X$ may differ from the parents of X_m on

G. A non-random G˜-specific treatment regime g˜ is a collection of functions g˜ $=$ $\{g_{\bar{1}}, \ldots, g_{\bar{k}} \, ; \, g_m^- : \mathcal{PA}_m^- \to \mathcal{X}_m\}$ that gives the value $g_m^-(pa_m^-)$ that we will set X_m to when PA_m^- takes the value pa_m^-. When for each m, X_m has no parents on G˜, so that $g_m^-(pa_m^-)$ is a constant, say x_m^*, we say regime g˜ is non-dynamic and write g˜ $= x^* = \{x_1^*, \ldots, x_k^*\}$. Otherwise, g˜ is dynamic. The counterfactual random variable $V_j(g$˜$)$ associated with regime g˜ is recursively defined as follows: (i) when $V_j \in V \backslash X$, $V_j(g$˜$)$ is the one-step-ahead counterfactual $V_j(\bar{v}_{j-1})$ evaluated at $\bar{v}_{j-1} = \overline{V}_{j-1}(g$˜$)$ and (ii) when $V_j = X_m \in X$, $V_j(g$˜$)$ is $g_m^-(pa_m^-)$ with pa_m^- equal to the counterfactual $PA_m^-(g$˜$)$.

Lemma 1.2 (Robins 1986) *If DAG G represents a FRCISTG, then for any set of variables $X \subset V$, and associated intervention DAG G˜, and any treatment regime g˜, the (intervention) density $f_{V(g$˜$)}(v)$ of the counterfactual $V(g$˜$)$ is a functional of the density $f_V(v)$ of V and thus is non-parametrically identified from data V. This functional, which I have referred to as the g-computation algorithm functional or density (hereafter g-functional or density), is the density f_{g˜$}(v)$ obtained by modifying the product on the right-hand side of (1.3) as follows: if $V_j = X_m \in X$, remove the term $f(v_j \mid pa_j)$ from the product and set v_j to the value $g_m^-(pa_m^-)$ elsewhere in (1.3).*

Agnostic causal model The agnostic causal model of Spirtes *et al.* (1993) effectively assumes that the joint distribution of V factors as in (1.3) and that the joint density of V under the regime g˜ on a graph G˜ is given by the g-functional f_{g˜$}(v)$. Although this agnostic model assumes that the density of V under the intervention g˜ is well defined, the model makes no reference to counterfactual variables and is agnostic as to their existence. In his article, Phil Dawid embraces a restricted version of the agnostic model in which only a subset of the variables in V can be manipulated (i.e. set) which he refers to as the decision variables. This model is closely related to the causal model discussed by Heckerman and Shachter (1995). The randomized causally-interpreted structured tree graph (RCISTG) of Robins (1987a,b) likewise restricts the set of variables in V that can be manipulated. The relationship of a RCISTG model to an FRCISTG model is analogous to that of Dawid's restricted agnostic model to the agnostic model. In his article Dawid was interested in the marginal intervention distribution f_{g˜$}(y) = \int \ldots \int f_{g$˜$}(v) \, d\mu(y^c)$ of a subset Y of the variables in $V = (Y, Y^c)$, say, when data are obtained only on some subset V^* of the variables in V. In this case one wishes to know whether the intervention distribution f_{g˜$}(y)$ of Y is identified from (i.e. is a functional of) the marginal distribution $f_{V^*}(v^*)$ of V^* and, if not, to set bounds on f_{g˜$}(y)$. Sufficient conditions for identification have been derived by Galles and Pearl (1995) for univariate (i.e. time-independent) interventions, Pearl and Robins (1995) for non-dynamic regimes, and Robins (1997) for dynamic regimes. We refer the reader to the above references for additional discussion.

2   Direct and indirect effects

Define a causal DAG model to be a manipulative causal DAG model if the only causal effects that are non-parametrically identified from the joint distribution of the variables on the DAG are those that could in principle be checked by manipulation of (equivalently, experimental intervention on or setting of) the DAG variables. That is, a manipulative causal model is one in which all the causal predictions of the model can in principle be checked (i.e. tested) by experimental intervention. The finest FRCISTG model is a manipulative model because the causal parameters that are non-parametrically identified from data on V are all functions of the counterfactual intervention densities $f_{V(g^-)}(v)$. Specifically, suppose we measure data on all the variables V on G in a large study population so that we can regard $f_V(v)$ as known. Then to check our finest FRCISTG causal model, we could take an as-yet-untreated population exchangeable with the study population and intervene by forcing them to follow some regime g^-, allowing us to empirically estimate the intervention distribution $f_{V(g^-)}(v)$. If, for some regime g^- associated with a graph $G^{\sim}$, the g-functional $f_{g^-}(v)$ differs from the intervention distribution $f_{V(g^-)}(v)$, then we can conclude that our causal model is false, as would occur if there were a common cause of two variables in V that was not itself included in V. This argument also implies that the agnostic causal model is a manipulative model. Of course in practice such intervention tests may be impossible to carry out for logistical reasons (e.g. some variables V_m cannot be measured or there is no untreated population that one regards as exchangeable with the study population) or for ethical reasons.

If, however, a causal model is non-manipulative and thus non-parametrically identifies causal effects that do not correspond to the effect of an experimental intervention, then there is no way, even in principle, that one could check the correctness of all the model predictions. Robins (1986) imposed the independence assumption (1.1) precisely because (1.1) is the independence assumption that identifies all manipulative effects $f_{V(g^-)}(v)$ without identifying any non-manipulative effects. In the course of the following discussion of direct and indirect effects, we show the NPSE model is a non-manipulative model, because it implies the stronger independence assumption (1.2).

DAG 1

Robins and Greenland (1992) (hereafter R&G) define the pure direct effect (PDE) of a (dichotomous) exposure X on Y not acting through the intermediate variable Z to be the mean of Y under exposure to X had, contrary to fact, X's effect on the intermediate Z been blocked (that is, had Z remained at its value under non-exposure thereby eliminating all indirect effects) minus the mean of

Y under non-exposure to X. That is, under a NPSE or FRCISTG model,

$$\text{PDE} = \text{E}\left[Y\left\{x = 1, Z\left(x = 0\right)\right\}\right] - \text{E}\left[Y\left(x = 0\right)\right]$$
$$= \text{E}\left[Y\left\{x = 1, Z\left(x = 0\right)\right\}\right] - \text{E}\left[Y\left(x = 0, Z\left(x = 0\right)\right)\right],$$

since $\text{E}\left[Y\left(x = 0\right)\right] = \text{E}\left[Y\left(x = 0, Z\left(x = 0\right)\right)\right]$. Here $Y\left(x, z\right)$ is the counterfactual value of Y with (X, Z) set to (x, z) and $Z\left(x\right)$ is the counterfactual value of Z when X is set to x. The total indirect effect (TIE) of a (dichotomous) exposure X on Y is the total effect of X on Y minus the PDE. The motivation underlying this definition is that any effect of X on Y that is not purely direct must have an indirect contribution. Thus,

$$\text{TIE} = \text{E}\left[Y\left(x = 1\right) - Y\left(x = 1, Z\left(x = 0\right)\right)\right]$$
$$= \text{E}\left[Y\left(x = 1, Z\left(x = 1\right)\right)\right] - \text{E}\left[Y\left\{x = 1, Z\left(x = 0\right)\right\}\right],$$

since $\text{E}\left[Y\left(x = 1\right)\right] - \text{E}\left[Y\left(x = 0\right)\right]$ is by definition the total (equivalently, net or overall) exposure effect.

Similarly, the pure indirect effect (PIE) of X on Y through an intermediate variable Z is defined to be the mean of Y under non-exposure to X but with Z set to its exposed value minus the mean of Y under non-exposure to X. That is, under a NPSE or FRCISTG model,

$$\text{PIE} = \text{E}\left[Y\left(x = 0, Z\left(x = 1\right)\right)\right] - \text{E}\left[Y\left(x = 0\right)\right]. \tag{2.1}$$

In this contrast the only effect of X on Y is indirect in that the effect is relayed through X's effect on Z. The total direct effect (TDE) of X on Y not through an intermediate variable Z is the total effect of X on Y minus the PIE. Thus,

$$\text{TDE} = \text{E}\left[Y\left(x = 1\right)\right] - \text{E}\left[Y\left(x = 0, Z\left(x = 1\right)\right)\right].$$

Pearl (2001) adopted our definitions but changed nomenclature. He refers to pure direct and indirect effects as natural direct and indirect effects. Under the agnostic causal model, the concept of the total and pure, indirect and direct effects is not defined since the counterfactuals $\text{E}\left[Y\left\{x = 1, Z\left(x = 0\right)\right\}\right]$ and $\text{E}\left[Y\left(x = 0, Z\left(x = 1\right)\right)\right]$ are not assumed to exist.

The direct effect of X when Z is set to z (i.e. $\text{E}\left[Y\left(1, z\right)\right] - \text{E}\left[Y\left(0, z\right)\right]$) is identified from $f_V\left(v\right)$ by the g-formula under all three models. But, in general, the contrast $\text{E}\left[Y\left(1, z\right)\right] - \text{E}\left[Y\left(0, z\right)\right]$ differs from both the PDE and TDE contrasts and differs depending on whether z is set to 1 or to 0. Indeed, since the intervention mean $\text{E}\left[Y\left(x\right)\right]$ is identified under any of the three causal models, determining whether the TIE, PIE, TDE, and PDE are identified is equivalent to determining whether $\text{E}\left[Y\left(x, Z\left(x^*\right)\right)\right]$, $x \neq x^*$, is identified. Below we will prove that $\text{E}\left[Y\left(x, Z\left(x^*\right)\right)\right]$ is not a manipulative effect and thus is not identified under a finest FRCISTG model. However, Pearl (2001) proved that under a NPSE model $\text{E}\left[Y\left(x, Z\left(x^*\right)\right)\right]$ is identified for certain DAGs. As an example consider a DAG, such as DAG 1 or DAG 2, where X has no parents and Z is a non-descendant of Y. Pearl showed that if

$$Y(x, z) \perp\!\!\!\perp Z(x^*) \mid X \quad \text{for all } z, \tag{2.2}$$

then under either a FRCISTG or NPSE model

$$E[Y(x, Z(x^*))] = \int E[Y(x, z)] \, dF_{Z(x^*)}(z). \tag{2.3}$$

Now (2.3) is non-parametrically identified from $f_V(v)$ under all three causal models since the intervention parameters $E[Y(x, z)]$ and $f_{Z(x^*)}(z)$ are identified by the g-formula. However, no finest FRCISTG model implies (2.2). On the other hand, it follows from (1.2) that (2.2) holds for the NPSE model represented by DAG 1. Note that (2.2) will not hold for the NPSE model represented by DAG 2. Indeed $E[Y(x, Z(x^*))]$ will never be identified from $f_V(v)$ for an NPSE model represented by any DAG which contains a descendant C of X that is an ancestor of both Z and Y. In summary, under a NPSE causal model, PIE and PDE would be non-parametrically identified based on DAG 1 but not on DAG 2.

DAG 2

A non-manipulative model We now turn to the question of whether $E[Y(x, Z(x^*))]$ can be identified by manipulating (i.e. setting) the variables on G. Now, as noted by R&G, we could identify $E[Y(x, Z(x^*))]$ if we could manipulate X to x^*, observe $Z(x^*)$, then 'return each subject to their pre-intervention state', manipulate X to x and Z to $Z(x^*)$, and finally observe $Y(x, Z(x^*))$. However, such an intervention strategy usually will not exist because we cannot 'return each subject to their pre-intervention state' by any conceivable real-world intervention (as, for example, if the outcome Y were death). As a result, because we cannot observe the same subject under both $X = x$ and $X = x^*$, we are unable to directly observe the joint distribution of $Z(x)$ and $Z(x^*)$. It follows that we cannot identify $E[Y(x, Z(x^*))]$ by any manipulation of the variables on G owing to the impossibility of differentiating exposed ($x = 1$) subjects, whose value of Z is attributable to X (i.e. $Z(x) \neq Z(x^*)$) from those whose value of Z is not ($Z(x) = Z(x^*)$). We will thus refer to $E[Y(x, Z(x^*))]$ and the pure and total, direct and indirect effect contrasts as non-manipulative parameters.

From the above, we conclude that the NPSE causal model is a non-manipulative model. One could argue that manipulative models are preferable because the predictions of non-manipulative models are not, even in principle, testable by experiment. Here are some counter-arguments defending the use of non-manipulative models.

First, the assumptions required to identify $E[Y(x, Z(x^*))]$ are analogous to the assumptions necessary to identify manipulative causal effects such as the total effect $E[Y(x)] - E[Y(x^*)]$ from observational (i.e. non-randomized) data, and we do not wish to argue against using observational data to estimate effects of exposures that cannot be tested experimentally for ethical or logistical reasons. The following sentence makes the analogy. Because we cannot observe the same subject under both $X = x$ and $X = x^*$, we can generally only (i) identify $E[Y(x)] - E[Y(x^*)]$ if we assume that within levels of baseline variables $Y(j) \perp\!\!\!\perp X$, $j = 0, 1$, and (ii) identify $E[Y(x, Z(x^*))]$ if we assume (2.2) holds within levels of baseline variables. Secondly, the fact that our observational study estimate of $E[Y(x)] - E[Y(x^*)]$ could, in principle, be checked by experiment is of no import when in fact the check cannot be carried out for either ethical or logistical reasons. Third, even when an experiment can be conducted, if one is uncertain that the available experimental subjects are exchangeable with the observational study subjects, an observational estimate of $E[Y(x)] - E[Y(x^*)]$ cannot be checked, as it is possible that any difference between observational and experimental estimates is wholly due to a lack of exchangeability. Fourth, suppose one would assume DAG 1 is an FRCISTG so that

$$Y(x, z) \perp\!\!\!\perp Z(x) \mid X = x \quad \text{for all } z, x. \tag{2.4}$$

It is hard to construct realistic (as opposed to mathematical) scenarios in which one would accept (2.4) as true but not (2.2), as it is unlikely one would accept either (2.4) or (2.2) as true unless one believed that $Z = Z(X)$ was effectively randomly assigned by nature within levels of X, in which case both (2.2) and (2.4) would be true.

An alternative identifying assumption Even if one is convinced by the above four arguments for using an NPSE model, I suspect that, in practice, this model could rarely be used to identify the non-manipulative parameters corresponding to pure and total, direct and indirect effects as it would be unusual to have sufficient prior causal knowledge to impose the identifying assumption that there is no variable, say U, that is both affected by X and is a common cause of Z and Y (i.e. there is no descendant U of X that is an ancestor of both Z and Y). (Note that if such a U exists, then it must be included on the DAG G in order for G to represent any of our three causal models.) Thus, one might wish to consider alternative identifying assumptions such as the following no-interaction assumption.

No-interaction assumption $Y(x, z) - Y(x^*, z)$ is a random function $B(x, x^*)$ of x and x^* that does not depend on z. We write $E[B(x, x^*)]$ as $b(x, x^*)$.

This assumption states that, at the individual level, the magnitude of the direct effect of x compared with x^* on the outcome Y is the same on an additive scale for all z. A detailed mechanistic discussion of this assumption is given in R&G (1992). As noted by Pearl (2001), this assumption is satisfied in the usual linear SEM model and has been used to identify direct and indirect effects in the structural equation literature. Indeed, in the linear SEM model, $Y(x, z) - Y(x^*, z)$ is usually assumed to be a deterministic function of x and x^*. The following theorem shows that direct and indirect effects are identified by a FRCISTG model under the no-interaction assumption. As in Pearl (2001), we generalize our definitions to non-dichotomous treatments by defining the effects of x compared with x^* as follows:

$$\text{PDE}(x, x^*) = \text{E}[Y\{x, Z(x^*)\}] - \text{E}[Y(x^*)],$$
$$\text{TIE}(x, x^*) = \text{E}[Y(x)] - \text{E}[Y\{x, Z(x^*)\}],$$
$$\text{PIE}(x, x^*) = \text{E}[Y(x^*, Z(x))] - \text{E}[Y(x^*)],$$
$$\text{TDE}(x, x^*) = \text{E}[Y(x)] - \text{E}[Y(x^*, Z(x))].$$

These new definitions reduce to the old on choosing $x = 1$ and $x^* = 0$.

Theorem 2.1 *Under the no-interaction assumption,*

$$\text{PDE}(x, x^*) = \text{TDE}(x, x^*) = b(x, x^*),$$
$$\text{PIE}(x, x^*) = \text{TIE}(x, x^*), \text{ and}$$
$$\text{TDE}(x, x^*) + \text{TIE}(x, x^*) = \text{E}[Y(x)] - \text{E}[Y(x^*)].$$

Given data on V, all these quantities are identified in a FRCISTG causal model.

Proof It follows immediately from the no-interaction assumption that

$$\text{PDE}(x, x^*) = \text{TDE}(x, x^*) = \text{E}[B(x, x^*)] = b(x, x^*).$$

The equality $\text{PDE}(x, x^*) = \text{TDE}(x, x^*)$ immediately implies both $\text{PIE}(x, x^*) = \text{TIE}(x, x^*)$ and $\text{TDE}(x, x^*) + \text{TIE}(x, x^*) = \text{E}[Y(x)] - \text{E}[Y(x^*)]$. Finally, $b(x, x^*) = \text{E}[Y(x, z)] - \text{E}[Y(x^*, z)]$, $\text{E}[Y(x)]$, and $\text{E}[Y(x^*)]$ are identified in an FRCISTG model. $\square$

It seems biologically rather unlikely that the no-interaction assumption will hold when Z affects Y. The no-interaction assumption can be tested in an FRCISTG model since it implies the testable restriction that $\text{E}[Y(x, z) - Y(x^*, z)]$ does not depend on z. More realistic assumptions with weaker consequences are considered in the following theorem. We say that X and Z never interact negatively if $x > x^*$ implies $Y\{x, z\} - Y\{x^*, z\}$ is non-decreasing in z. We say that X is non-preventive for Z if $Z(x) \geqslant Z(x^*)$ when $x > x^*$.

Theorem 2.2 *If X and Z never interact negatively and X is non-preventive for Z, then for $x > x^*$ $\text{TDE}(x, x^*) \geqslant \text{PDE}(x, x^*)$, $\text{TIE}(x, x^*) \geqslant \text{PIE}(x, x^*)$, $\text{PIE}(x, x^*) + \text{PDE}(x, x^*) \leqslant \text{E}[Y(x)] - \text{E}[Y(x^*)] \leqslant \text{TDE}(x, x^*) + \text{TIE}(x, x^*)$.*

Proof If we can show that $\text{TDE}(x, x^*) - \text{PDE}(x, x^*) \geqslant 0$, then the remainder of the theorem follows at once from the basic definitions of the quantities involved. Now $\mathrm{TDE}\,(x,x^*) - \mathrm{PDE}\,(x,x^*) \;=\; \mathrm{E}\,\{Y\,(x,Z\,(x)) - Y\,(x^*,Z\,(x))\} -$ $\mathrm{E}\,\{Y\,(x,Z(x^*)) - Y\,(x^*,Z(x^*))\} \;\geqslant\; 0$ because $Y\,(x,Z\,(x)) - Y\,(x^*,Z\,(x)) \;\geqslant\;$ $Y\,(x,Z(x^*)) - Y\,(x^*,Z(x^*))$ under the suppositions of the theorem.   □

3   Substantive considerations

In this section we investigate through a particular example whether and when scientific interest might lie in the estimation of pure and total direct and indirect effects, regardless of whether they are identifiable. R&G (1992) consider a setting in which X is smoking, Z is hypercholesterolemia, and Y is cardiovascular disease, and data are available from a randomized trial of smoking cessation. For simplicity, we take all variables to be dichotomous $(0,1)$ variables and assume any non-compliance to be completely at random. We let $Y = 1$, $Z = 1$, and $X = 1$ denote the presence of cardiovascular disease, hypercholesterolemia, and continued smoking. We assume the no-interaction assumption is false because in some subjects hypercholesterolemia produced coronary artery stenosis (narrowing), the narrowed artery was blocked by a blood clot caused by smoking-induced platelet aggregation, and the blocked artery resulted in a heart attack. Thus the smoking effect $\mathrm{E}\,[Y\,(x=1)] - \mathrm{E}\,[Y\,(x=1,z=0)]$ that could be eliminated by controlling all hypercholesterolemia (i.e. setting z to 0) differs from the TIE of smoking $\mathrm{E}\,[Y\,(x=1)] - \mathrm{E}\,[Y\,\{x=1,Z\,(x=0)\}]$. We are interested in providing adjuvant treatments for smokers who either cannot or will not stop smoking. R&G state that $\mathrm{E}\,[Y\,(x=1)] - \mathrm{E}\,[Y\,(x=1,z=0)]$ would be the parameter of public health policy interest whenever (i) there exists an adjuvant therapy that effectively controls hypercholesterolemia (i.e. a cholesterol lowering drug) but (ii) there is no intervention that specifically blocks smoking's effect on cholesterol. R&G go on to say that if a cholesterol lowering drug was unavailable but there was a drug that completely blocked smoking's ability to elevate cholesterol, then it would be the TIE that would be of public health interest.

We now argue that R&G's latter statement is correct only as an approximation. We will assume that a NPSE model represented by DAG 1 is the true state of nature so that (2.2) holds and thus pure and total, direct and indirect effects are identified via (2.3). Furthermore, we assume there is a drug A that completely blocks the effect of smoking X on cholesterol Z but does not affect the direct effect of smoking on Y. If R&G's final statement were precisely correct then, were all continuing smokers in the trial given drug A, the mean of Y would be $\mathrm{E}\,[Y\,\{x=1,Z\,(x=0)\}]$. We now show this need not be the case.

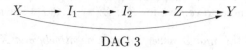

DAG 3

Essentially, any causal pathway $X \to Z$ can be further elaborated by adding to the DAG variables that mediate the effect of X on Z. For example, on DAG 3 the effect $X \to Z$ is shown to be through the intermediates I_1 and

I_2. For expositional simplicity, we shall assume that on any elaborated graph there is only one path from X to Z as no qualitatively new issues arise when there are multiple paths. Now if DAG 3 is a NPSE model, then so is DAG 1 since the variables (I_1, I_2) being marginalized over are not a common cause (i.e. parent) of any two variables on DAG 1. It is for this reason that when estimating the effect of setting any variable on DAG 1, we neither require data on (I_1, I_2) nor need to include them on DAG 1. Now from DAG 3 we observe that a drug A will succeed in blocking all of X's effect on Z by blocking the effect of X on I_1, the effect of I_1 on I_2, or the effect of I_2 on Z. However, the counterfactual mean of Y were all continuing smokers given drug A can differ in each case and will, as shown in the following paragraph, in general, equal $E[Y\{x = 1, Z(x = 0)\}]$ only if drug A blocks the effect of X on I_1 and I_1 is the unique child of X on the 'maximally elaborated path' from X to Z. We say that a path $X = I_0 \rightarrow I_1 \rightarrow \cdots \rightarrow I_{J-1} \rightarrow I_J = Z$, denoted by P, is 'maximally elaborated' if all variables on the causal chain from X to Z are included on P. For the sake of the following argument, assume such a maximally elaborated path exists.

Remark Mathematically we define P to be maximally elaborated if for all $j, 0 \leqslant j \leqslant J$, and all variables I^* such that neither (I_j, I^*) nor (I^*, I_{j+1}) have degenerate joint distributions, the DAG that replaces P by the path $X = I_0 \rightarrow I_1 \rightarrow \cdots \rightarrow I_j \rightarrow I^* \rightarrow I_{j+1} \cdots \rightarrow I_{J-1} \rightarrow I_J = Z$ is not a causal DAG.

Suppose, without loss of generality, that DAG 3 contains the maximally elaborated path from X to Z and drug A blocked the effect of I_1 on I_2 by, as one possibility, forcing I_2 to be $I_2(i_1 = 0)$, say, regardless of the value of I_1. Then among smokers given A, Z will equal the counterfactual $Z(i_1 = 0)$ and thus the mean of Y is $E[Y(x = 1, i_1 = 0)] = \int E[Y(x = 1, z)] \, dF_{Z(i_1=0)}(z)$ while, by (2.3), $E[Y\{x = 1, Z(x = 0)\}] = \int E[Y(x = 1, z)] \, dF_{Z(x=0)}(z)$. These quantities differ because the drug blocks not only the effect of X on Z but also the effect of I_1 on Z and many subjects have non-zero values of I_1 due to non-smoking-related causes that are the source of the NPSE model error term U_{I_1} associated with I_1. Indeed, if these integrals did not differ, we would have succeeded in identifying $E[Y\{x = 1, Z(x = 0)\}]$ by experimental manipulation of X to 1 and I_1 to 0 on DAG 3, contradicting the fact that $E[Y\{x = 1, Z(x = 0)\}]$ is a non-manipulable parameter. However, if drug A blocked the effect of X on I_1 by making I_1 equal to $I_1(x = 0)$ even when we set X to 1, then the mean of Y among smokers given intervention A is indeed $\int E[Y(x = 1, z)] \, dF_{Z(x=0)}(z)$. In this case, intervention with A would not correspond to setting or manipulating a variable on causal DAGs 1 or 3. Because it is unlikely that drug A acts on the first link of the maximally elaborated path, we conclude that $E[Y\{x = 1, Z(x = 0)\}]$ and thus the TIE, although possibly of mechanistic interest, will rarely be of direct public health interest, except as an approximation.

What if, following a suggestion of Pearl, we wished to predict the effect on Y of a chemically modified cigarette that totally blocks smoking's ability to elevate cholesterol. Even in this case, the effect may not be $E[Y\{x = 1, Z(x = 0)\}]$, because the modified cigarette need not necessarily block the first link on the maximally elaborated path from X to Z. However, if we could assume that a modified cigarette is more likely than our drug A to act at an early link on the causal pathway from X to Z, then $E[Y\{x = 1, Z(x = 0)\}]$ would probably better approximate the effect of a modified cigarette on Y than the effect of drug A.

It is not possible to reach agreement on a hypothetical intervention (closest possible world) under which $Y\{x = 1, Z(x = 0)\}$ could be observed, even if we allow for interventions other than the setting of variables, since a maximally elaborated path from X to Z may not exist and is at best ill-defined (in the sense that it is unclear what criteria are to be used in judging a path to be maximally elaborated).

Additional references in discussion

Arjas, E. and Eerola, M. (1993). On predictive causality in longitudinal studies. *Journal of Statistical Planning and Inference*, **34**, 361–86.

Eerola, M. (1994). *Probabilitic Causality in Longitudinal Studies*, Lecture Notes in Statistics, No. 92. Springer-Verlag, Berlin.

Galles, D. and Pearl, J. (1995). Testing identifiability of causal effects. In *Uncertainty and Artificial Intelligence 11* (eds T. Besnard and S. Hanks), pp. 185–95. Morgan Kaufmann, San Francisco.

Heckerman, D. and Shachter, R. (1995). Decision-theoretic foundations for causal reasoning. *Journal of Artificial Intelligence Research*, **3**, 405–30.

Neyman, J. (1923). On the application of probability theory to agricultural experiments. Essay on principles, Section 9. Translated in *Statistical Science*, **5**, 465–80, 1990.

Parner, J. and Arjas, E. (2000). Causal reasoning from longitudinal data. Unpublished manuscript.

Pearl, J. (2001). Direct and indirect effects. Technical Report R-273, Cognitive Systems Laboratory, Computer Science Department, University of California, Los Angeles.

Pearl, J. and Robins, J. M. (1995). Probabilistic evaluation of sequential plans from causal models with hidden variables. In *Uncertainty and Artificial Intelligence 11* (eds T. Besnard and S. Hanks), pp. 185–95. Morgan Kaufmann, San Francisco.

Robins, J. M. (1986). A new approach to causal inference in mortality studies with sustained exposure periods – application to control of the healthy worker survivor effect. *Mathematical Modelling*, **7**, 1393–512.

Robins, J. M. (1987a). Errata to 'A new approach to causal inference in mortality studies with sustained exposure periods – application to control of the healthy worker survivor effect.' *Computers and Mathematics with Applications*, **14**, 917–21.

Robins, J. M. (1987b). Addendum to 'A new approach to causal inference in mortality studies with sustained exposure periods – application to control of the healthy worker survivor effect.' *Computers and Mathematics with Applications*, **14**, 923–45.

Robins, J. M. (1997). Causal inference from complex longitudinal data. In *Latent Variable Modeling and Applications to Causality*, Lecture Notes in Statistics, No. 120 (ed. M. Berkane), pp. 69–117. Springer-Verlag, New York.

Robins, J. M. (1998). Structural nested failure time models. In *Encyclopedia of Biostatistics* (eds P. Armitage and T. Colton), pp. 4372–89. Wiley, Chichester.

Robins, J. M. and Greenland, S. (1992). Identifiability and exchangeability for direct and indirect effects. *Epidemiology*, **3**, 143–55.

Robins, J. M. and Greenland, S. (2000). Comment on 'Causal inference without counterfactuals' by A. P. Dawid. *Journal of the American Statistical Society*, **95**, 477–82.

Rubin, D. B. (1978). Bayesian inference for causal effects: the role of randomization. *Annals of Statistics*, **6**, 34–58.

Suppes, P. (1970). *A Probabilistic Theory of Causality*. North-Holland, Amsterdam.

3

Causal inference via ancestral graph models

Thomas S. Richardson
University of Washington, Seattle, USA

Peter Spirtes
Carnegie–Mellon University, Pittsburgh, USA

1 Introduction

In this article we introduce a new class of graphical models, called ancestral graph Markov models. These models are designed to provide a framework for making inferences about causal structure, and in special cases treatment effects, from background knowledge and non-randomized samples. We first provide some informal background and motivation for the models we subsequently introduce; the concepts used here will be defined more precisely in later sections.

1.1 Graphical models

A graphical model is a set of distributions associated with a graph consisting of vertices and edges. The graph encodes relations of marginal and conditional independence via a 'Markov property'. A 'directed acyclic graph' (DAG) is a graph in which every edge is directed ($\rightarrow$) and there are no directed cycles. If there is an edge $x \rightarrow y$, then x is said to be a 'parent' of y, and y a 'child' of x.

1.2 Directed acyclic graphs and data-generating processes

We associate with a DAG those distributions that can be factored into a product of conditional densities for each variable given its parents in the graph. Consequently a sample from any distribution associated with a DAG may be generated in a step-wise fashion. For the DAG in Fig. 1(a), we have:

$$p(x_1, x_2, x_3, x_4, t_1, t_2) = p(x_3)\, p(x_4)\, p(t_1)\, p(t_2 \mid t_1)\, p(x_1 \mid x_3, t_1, t_2)\, p(x_2 \mid x_4, t_2).$$

To draw a sample, a value is first assigned to x_3, x_4 and t_1 by drawing from $p(x_3)$, $p(x_4)$ and $p(t_1)$ respectively. A value is then assigned to t_2 by drawing from $p(t_2 \mid t_1)$. Finally, values may be assigned to x_1 and x_2 by drawing from $p(x_1 \mid x_3, t_1, t_2)$ and $p(x_2 \mid x_4, t_2)$, respectively. Different DAGs may be associated with the same set of distributions. For instance, reversing the direction of the edge between t_1 and t_2 in Fig. 1(a) would lead to a different factorization, but a distribution will factor according to the new graph if and only if it factors according to the

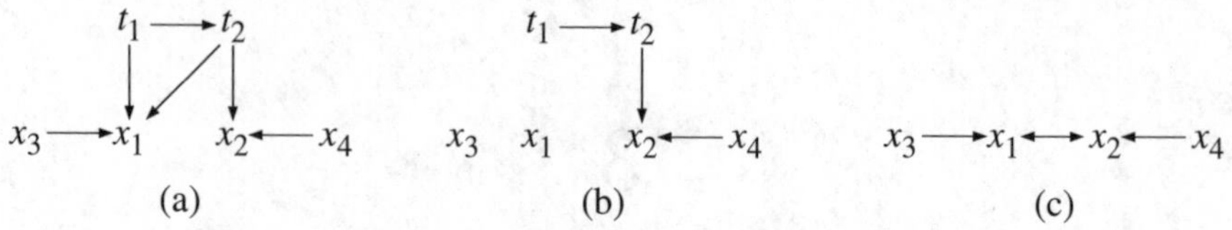

FIG. 1. (a) A directed acyclic graph $\mathcal{D}$; (b) the DAG resulting from intervening on x_1 in $\mathcal{D}$; (c) an ancestral graph representing the independence relations induced by $\mathcal{D}$ on the (x_1, x_2, x_3, x_4) margin when t_1 and t_2 are not observed.

original. DAGs that are associated with the same set of distributions are said to be 'Markov equivalent'.

1.3 Causal models

The purpose of many studies is to guide policy. Effective guidance requires making predictions about the effects of a given policy if it were to be adopted. Predicting the effect of a policy that fixes a given variable requires modifying the model since the policy changes the mechanism by which that variable is assigned its value. If we were to implement a policy that fixes the value of x_1 for the DAG in Fig. 1(a), then the value assigned to x_1 would no longer depend on the values assigned to t_1, x_3, or t_2. This can be represented graphically by removing those edges that point into x_1, as shown in Fig. 1(b). Together, a distribution and a DAG are said to be 'causal' if the distribution factors according to the DAG *and* the DAG describes the method by which the distribution should be changed in response to external interventions.

1.4 Inferring causal structure

If, for some reason, we know which variables occur in the true causal DAG, and know the population distribution when no policy is in effect, but do not know the structure of the causal DAG, then we could exploit the fact that the population distribution must factor according to this unknown DAG in order to make inferences about the DAG's structure. Consider the DAG $\mathcal{D}$ in Fig. 1(a). It can be shown that in any DAG that is Markov equivalent to $\mathcal{D}$, intervening on x_1 will have no effect on the distribution of t_1, since in all such DAGs, t_1 is a parent of x_1. Hence we may make this inference without knowing which DAG in the Markov equivalence class is the true causal DAG. However, we cannot calculate the effect that fixing t_1 will have on t_2 since although t_1 is a parent of t_2 in $\mathcal{D}$, there are DAGs Markov equivalent to $\mathcal{D}$ in which t_2 is a parent of t_1. Inference about the structure of the true causal DAG from a sample, rather than the population, is simply a problem of statistical model selection.

1.5 The problem of unobserved variables

The method outlined above was predicated on the assumption that we knew the set of variables that occurs in the true causal DAG, and had a sample from the distribution over those variables. In many contexts these assumptions are highly unrealistic. It is often the case that some causes relevant to a study have not been measured, or cannot be measured directly. Such unmeasured variables are often termed 'latent'. One could expand the class of models considered to include DAG models that explicitly include these latent variables. However, this leads to a host of statistical and inferential difficulties: the parameters may no longer be identifiable; the best known algorithms for determining whether a parameter is identifiable are exponential in the number of variables; for those parameters that are identifiable it is not known whether maximum likelihood estimates exist or are unique; the usual model selection scores, e.g. the Bayesian Information Criterion (BIC), that are asymptotically consistent for DAG models without latent variables, are not known to be asymptotically consistent when latent variables are present; the number of different models is infinite, because there is no upper limit to the number of latent variables.

1.6 Ancestral graphs

An alternative approach, taken here, simply focuses on the independence structure over the observed variables that results from the presence of latent variables, without explicitly including latent variables in the model. This may be represented graphically by permitting bi-directed ($\leftrightarrow$) edges in the graph. If t_1 and t_2 are unobserved in the DAG $\mathcal{D}$ in Fig. 1(a), then in this particular case we simply remove t_1 and t_2, and introduce an edge $x_1 \leftrightarrow x_2$ as shown in Fig. 1(c). The bi-directed edge simply allows for the existence of an unmeasured common cause of x_1 and x_2, but does not explicitly model how many latent variables there are, or their exact relationships with other latent and observed variables. We also introduce undirected edges ($-$) in order to accommodate the possibility of hidden variables that have been conditioned on rather than marginalized. For representing DAGs with hidden variables, we do not need to consider arbitrary graphs containing these three edges; a special sub-class, that we term 'ancestral graphs', is sufficient. In the Gaussian case, the associated model may be parameterized via a covariance selection model and a recursive set of linear equations with correlated errors. The resulting model class retains many of the desirable properties of DAG models without latent variables: all parameters are identified; there are a finite number of models; the BIC is asymptotically consistent. However, we still cannot distinguish between different Markov equivalent models. To address this issue we introduce another type of graph, called a 'partial ancestral graph' (PAG), which can represent structural features that are common to a Markov equivalence class of ancestral graphs. PAGs can be used to determine the effects of manipulations which are identifiable, and the function of the joint distribution over the observables which identifies them.

1.7 Overview

The remainder of the article is organized as follows: in Section 2 we define basic graphical concepts and notation; in Section 3 we define undirected and directed acyclic graphical Markov models; Section 4 introduces ancestral graph Markov models; Section 5 reviews the causal interpretation of DAG models; Section 6 describes how ancestral graphs may be used to make causal inferences when the structure is unknown; Section 7 relates this approach to other work; Section 8 contains the discussion. Many of the results relating to ancestral graphs are proved in Richardson and Spirtes (2002) (hereafter [RS02]). However, causal inference and partial ancestral graphs are not discussed there.

2 Basic graphical terms and concepts

We will consider graphs containing three types of edge: *undirected* ($-$), *directed* ($\rightarrow$), and *bi-directed* ($\leftrightarrow$). We use the following terminology to describe relations between variables in such a graph:

$$
\text{If } \left\{ \begin{array}{l} \alpha - \beta \\ \alpha \leftrightarrow \beta \\ \alpha \rightarrow \beta \\ \alpha \leftarrow \beta \end{array} \right\} \text{ in } \mathcal{G}, \text{ then } \alpha \text{ is a } \left\{ \begin{array}{l} \text{neighbour} \\ \text{spouse} \\ \text{parent} \\ \text{child} \end{array} \right\} \text{ of } \beta, \text{ and } \left\{ \begin{array}{l} \alpha \in \mathrm{ne}_{\mathcal{G}}(\beta) \\ \alpha \in \mathrm{sp}_{\mathcal{G}}(\beta) \\ \alpha \in \mathrm{pa}_{\mathcal{G}}(\beta) \\ \alpha \in \mathrm{ch}_{\mathcal{G}}(\beta) \end{array} \right\}.
$$

Note that the three edge types should be considered as distinct symbols. In particular,

$$
\alpha - \beta \;\neq\; \alpha \rightleftarrows \beta \;\neq\; \alpha \leftrightarrow \beta.
$$

All graphs considered in this article will have at most one edge between each pair of vertices. If there is an edge $\alpha \rightarrow \beta$, or $\alpha \leftrightarrow \beta$, then there is said to be *an arrowhead at β on this edge*. Conversely, if there is an edge $\alpha \rightarrow \beta$, or $\alpha - \beta$, then there is said to be a *tail at α*. If there is an edge between a pair of vertices, then these vertices are *adjacent*. We do not allow a vertex to be adjacent to itself. A *path* is a sequence of distinct vertices that are adjacent.

A vertex α is said to be an *ancestor* of a vertex β if *either* there is a directed path $\alpha \rightarrow \cdots \rightarrow \beta$ from α to β, *or* $\alpha = \beta$. A vertex α is said to be *anterior* to a vertex β if there is a path μ between α and β on which every edge is either of the form $\gamma - \delta$, or $\gamma \rightarrow \delta$ with δ between γ and β, or $\alpha = \beta$. Such a path is said to be an *anterior path* from α to β. We apply these definitions disjunctively to sets:

$$
\mathrm{an}(X) = \{\alpha \mid \alpha \text{ is an ancestor of } \beta \text{ for some } \beta \in X\};
$$

$$
\mathrm{ant}(X) = \{\alpha \mid \alpha \text{ is anterior to } \beta \text{ for some } \beta \in X\}.
$$

Our usage of the terms 'ancestor' and 'anterior' differs from Lauritzen (1996), but follows Frydenberg (1990).

3 Graphical Markov models

An *independence model* $\mathfrak{I}$ over vertex set V is a set of triples, $\langle X, Y \mid Z \rangle$, where X, Y, and Z are disjoint subsets of V, and Z may be empty. The set V also indexes a collection of random variables. If $\langle X, Y \mid Z \rangle \in \mathfrak{I}$, then we say that X is conditionally independent of Y given Z in $\mathfrak{I}$. A *graphical Markov model* consists of a graph $\mathcal{G}$, and a (global) Markov property C, which associates with the graph an independence model $\mathfrak{I}_C(\mathcal{G})$. Note that we will use the usual shorthand whereby X denotes both a set of vertices and the associated set of random variables; following the notation of Dawid (1979) we write $X \perp\!\!\!\perp Y \mid Z \ [P]$ to indicate that X is independent of Y given Z in the distribution P.

We may associate with an independence model $\mathfrak{I}$, a set of distributions

$$\mathcal{P}(\mathfrak{I}) = \{ P \mid \langle X, Y \mid Z \rangle \in \mathfrak{I} \Rightarrow X \perp\!\!\!\perp Y \mid Z \ [P] \}.$$

Note that if $P \in \mathcal{P}(\mathfrak{I})$, then the set of independence relations which hold in P may be a strict superset of those in $\mathfrak{I}$.

A distribution P is said to be *faithful* or *Markov perfect* with respect to an independence model $\mathfrak{I}$ if

$$\langle A, B \mid Z \rangle \in \mathfrak{I} \quad \text{if and only if} \quad A \perp\!\!\!\perp B \mid Z \ [P].$$

An independence model $\mathfrak{I}$ is said to be *probabilistic* if there is a distribution P that is faithful to $\mathfrak{I}$. Note that any independence model that is probabilistic will obey the semi-graphoid axioms given by Dawid (1979); see also Pearl (1988).

3.1 Example: undirected graphical Markov models

An *undirected graph (UG)*, $\mathcal{U}$, is a graph containing undirected edges $x - y$. The independence model $\mathfrak{I}_S(\mathcal{U})$ is given by the usual separation criterion, S:

$$\langle X, Y \mid Z \rangle \in \mathfrak{I}_S(\mathcal{U}) \quad \text{if } X \text{ and } Y \text{ are separated by } Z \text{ in } \mathcal{U}.$$

The set of distributions $\mathcal{P}(\mathfrak{I}_S(\mathcal{U}))$ has been studied in considerable detail. In particular, when the variables are discrete or continuous and Gaussian, the associated models are always identified, have a well-defined dimension, and procedures exist for finding the maximum likelihood estimate, which is unique (Lauritzen 1996).

3.2 Example: directed acyclic graphical Markov models

A DAG $\mathcal{D}$ is a graph containing only directed edges $x \to y$, subject to the restriction that there are no directed cycles. There are two equivalent graphical criteria that define the global Markov property for DAGs: Pearl's d-separation criterion, and the moralization criterion (Lauritzen *et al.* 1990). The d-separation criterion is a special case of the m-separation criterion defined below for ancestral graphs. Any distribution that satisfies the global Markov property with respect to a DAG $\mathcal{D}$ may be factorized as follows:

$$p(x_V) = \prod_{\alpha \in V} p(x_\alpha \mid x_{\mathrm{pa}(\alpha)}) \tag{3.1}$$

assuming that the relevant densities exist. This factorization is central to the causal interpretation of DAG models described in Section 5.

4   Ancestral graph Markov models

We now introduce the class of ancestral graph Markov models, which extends the classes of undirected graphs and DAGs.

4.1   Definition of an ancestral graph

An *ancestral* graph $\mathcal{G}$ is a graph in which the following conditions hold for all vertices α in $\mathcal{G}$:

(i) $\alpha \notin \text{ant}(\text{pa}(\alpha) \cup \text{sp}(\alpha))$;

(ii) if $\text{ne}(\alpha) \neq \emptyset$, then $\text{pa}(\alpha) \cup \text{sp}(\alpha) = \emptyset$.

In words, condition (i) requires that if α and β are joined by an edge with an arrowhead at α, then α is not anterior to β. Condition (ii) requires that there are no arrowheads present at a vertex which is an endpoint of an undirected edge. Condition (i) implies that if α and β are joined by an edge with an arrowhead at α, then α is not an ancestor of β. This is the motivation for terming such graphs 'ancestral' (see also Proposition 4.1 below). Note that if $\mathcal{G}$ is an undirected graph, or a directed acyclic graph, then $\mathcal{G}$ is an ancestral graph.

Let $\text{un}_{\mathcal{G}} \equiv \{\alpha \mid \text{pa}_{\mathcal{G}}(\alpha) \cup \text{sp}_{\mathcal{G}}(\alpha) = \emptyset\}$ be the set of vertices at which no arrowheads are present in $\mathcal{G}$. Any ancestral graph can be split into an undirected graph $\mathcal{G}_{\text{un}_{\mathcal{G}}}$, and an ancestral graph containing no undirected edges $\mathcal{G}_{V \setminus \text{un}_{\mathcal{G}}}$; any edge between a vertex $\alpha \in \text{un}_{\mathcal{G}}$ and a vertex $\beta \in V \setminus \text{un}_{\mathcal{G}}$ takes the form $\alpha \to \beta$. This decomposition is shown schematically in Fig. 2. (See [RS02], Lemma 3.7.)

An ancestral graph is uniquely determined by its adjacencies (or 'skeleton') and anterior relations (see Corollary 3.10 in [RS02]):

Proposition 4.1 *If $\mathcal{G}_1$ and $\mathcal{G}_2$ are two ancestral graphs with the same vertex set V and the same adjacencies, and if for all pairs of adjacent vertices $\alpha, \beta \in V$, $\alpha \in \text{ant}_{\mathcal{G}_1}(\beta) \Leftrightarrow \alpha \in \text{ant}_{\mathcal{G}_2}(\beta)$, then $\mathcal{G}_1 = \mathcal{G}_2$.*

4.2   The global Markov property for ancestral graphs

We now extend Pearl's d-separation criterion (Pearl 1988), defined originally for DAGs, to ancestral graphs. A non-endpoint vertex ζ on a path is a *collider on the path* if the edges preceding and succeeding ζ on the path have an arrowhead

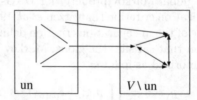

FIG. 2. Schematic showing the decomposition of an ancestral graph into an undirected graph and a graph containing no undirected edges.

at ζ, i.e. $\rightarrow \zeta \leftarrow$, $\leftrightarrow \zeta \leftrightarrow$, $\leftrightarrow \zeta \leftarrow$ or $\rightarrow \zeta \leftrightarrow$. A non-endpoint vertex ζ on a path that is not a collider is a *non-collider on the path*. A path between vertices α and β in an ancestral graph $\mathcal{G}$ is said to be *m-connecting given a set Z* (possibly empty) if

(i) every non-collider on the path is not in Z, and

(ii) every collider on the path is in $\text{an}_{\mathcal{G}}(Z)$.

If there is no path m-connecting α and β given Z, then α and β are said to be *m-separated given Z*. For disjoint sets X, Y, and Z, where Z may be empty, X and Y are *m-separated given Z*, if for every pair α and β, with $\alpha \in X$ and $\beta \in Y$, α and β are m-separated given Z. We denote the independence model resulting from applying the m-separation criterion to $\mathcal{G}$ by $\mathfrak{I}_m(\mathcal{G})$. This criterion is an extension of Pearl's d-separation criterion to ancestral graphs in that in a DAG, $\mathcal{D}$, a path is d-connecting if and only if it is m-connecting. See Fig. 3 for two examples of m-connecting paths.

The global Markov property for ancestral graphs may also be formulated as a generalization of the moralization criterion for DAGs. (See [RS02], Section 3.5.) In addition we have:

Theorem 4.2 *If $\mathcal{G}$ is an ancestral graph, then $\mathfrak{I}_m(\mathcal{G})$ is probabilistic, in particular there is a normal distribution $N \in \mathcal{P}(\mathfrak{I}_m(\mathcal{G}))$, which is faithful to $\mathfrak{I}_m(\mathcal{G})$.*

See Theorem 7.5 [RS02]. It follows that any independence model $\mathfrak{I}_m(\mathcal{G})$ given by an ancestral graph will obey the semi-graphoid axioms. This may also be proved more directly by using the definition of m-separation.

4.3 Pairwise Markov property: maximal ancestral graphs

Independence models described by DAGs and undirected graphs satisfy pairwise Markov properties with respect to these graphs, hence every missing edge corre-

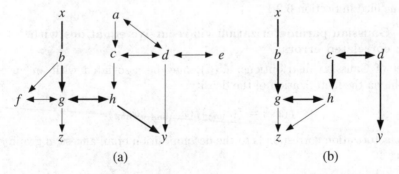

(a)   (b)

FIG. 3. Examples of the global Markov property for ancestral graphs. In each graph thicker edges form a path m-connecting x and y given $\{z\}$.

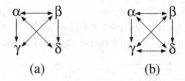

(a)   (b)

FIG. 4. (a) The simplest example of a non-maximal ancestral graph: γ and δ are not adjacent, but are m-connected given every subset of $\{\alpha, \beta\}$, hence $\mathfrak{I}_m(\mathcal{G}) = \emptyset$; (b) the unique ancestral extension of the graph in (a) with the same (trivial) independence model.

sponds to a conditional independence (Lauritzen 1996, p. 32). This is not true in general for an arbitrary ancestral graph, as shown by the graph in Fig. 4(a).

This motivates the following definition: an ancestral graph $\mathcal{G}$ is said to be *maximal* if, for every pair of vertices α and β, if α and β are not adjacent in $\mathcal{G}$, then there is a set Z $(\alpha, \beta \notin Z)$, such that $\langle \{\alpha\}, \{\beta\} \mid Z \rangle \in \mathfrak{I}_m(\mathcal{G})$. Thus a graph is maximal if every missing edge corresponds to at least one independence in the corresponding independence model.

The use of the term 'maximal' is motivated by the following:

Proposition 4.3 *If $\mathcal{G}$ is a maximal ancestral graph, and $\mathcal{G}$ is a subgraph of $\mathcal{G}^*$ with the same vertex set as $\mathcal{G}^*$, then $\mathfrak{I}_m(\mathcal{G}) = \mathfrak{I}_m(\mathcal{G}^*)$ implies $\mathcal{G} = \mathcal{G}^*$.*

Hence maximal ancestral graphs are maximal in the sense that no additional edge may be added to the graph without changing the independence model. The following theorem gives the converse.

Theorem 4.4 *If $\mathcal{G}$ is an ancestral graph, then there exists a unique maximal ancestral graph $\bar{\mathcal{G}}$ formed by adding $\leftrightarrow$ edges to $\mathcal{G}$ such that $\mathfrak{I}_m(\mathcal{G}) = \mathfrak{I}_m(\bar{\mathcal{G}})$.*

The graph $\bar{\mathcal{G}}$ may be easily constructed from $\mathcal{G}$ via the transformation $\mathcal{G} \mapsto \mathcal{G}[_\emptyset^\emptyset$, defined in Section 6.2.1.

4.4 Gaussian parameterization via recursive equations with correlated errors

A set of Gaussian distributions, $\mathcal{N}(\mathcal{G})$, may be associated with an ancestral graph, via the factorization of the density:

$$f(x_V) = f(x_{\text{un}_{\mathcal{G}}})f(x_{V \setminus \text{un}_{\mathcal{G}}} \mid x_{\text{un}_{\mathcal{G}}}). \tag{4.1}$$

This factorization corresponds to the decomposition of an ancestral graph given in Fig. 2.

- The **undirected component** $f(x_{\text{un}_{\mathcal{G}}})$ may be parameterized via an undirected graphical Gaussian model, also known as a covariance selection model (Dempster 1972; Lauritzen 1996).

- The **directed component**, $f(x_{V \setminus un_\mathcal{G}} \mid x_{un_\mathcal{G}})$, may be parameterized via a set of recursive equations with correlated errors as follows:

  (i) Associate with each ν in $V \setminus un_\mathcal{G}$ a linear equation, expressing X_ν as a linear function of the variables for the parents of ν plus an error term:

$$X_\nu = \mu_\nu + \sum_{\pi \in \mathrm{pa}(\nu)} b^*_{\nu\pi} \cdot X_\pi + \epsilon_\nu.$$

  (ii) Specify a non-singular multivariate normal distribution over the error variables $(\epsilon_\nu)_{\nu \in V \setminus un_\mathcal{G}}$ (with mean zero) satisfying the condition that

$$\text{if there is no edge } \alpha \leftrightarrow \beta \text{ in } \mathcal{G}, \text{ then } \mathrm{cov}(\epsilon_\alpha, \epsilon_\beta) = 0,$$

  but otherwise unrestricted.

It is a direct consequence of conditions (i) and (ii) in the definition of an ancestral graph, given in Section 4.1, that any variable that is a parent of ν is uncorrelated with ϵ_ν. It follows that the coefficients b^* have a population interpretation as regression coefficients; see Theorem 8.7 in [RS02]. Furthermore, all the parameters of the model are identified and the model forms a curved exponential family; see Corollary 8.8 and Theorem 8.23 in [RS02].

This parameterization is important for the purposes of statistical inference because software packages exist for estimating the component models: MIM (Edwards 1995) fits undirected Gaussian models via the IPS algorithm; AMOS (Arbuckle 1997), EQS (Bentler 1986), Proc CALIS (SAS Publishing 1995) and LISREL (Jöreskog and Sörbom 1995) are packages that fit structural equation models via numerical optimization.

4.5 Gaussian independence models

We define the Gaussian independence model associated with an ancestral graph $\mathcal{G}$ to be the set of non-singular Gaussian distributions that satisfy the global Markov property for $\mathcal{G}$:

$$\mathcal{N}(\mathfrak{I}_m(\mathcal{G})) = \{N \mid N \in \mathcal{N}_{|V|}, \text{ s.t. if } \langle X, Y \mid Z \rangle \in \mathfrak{I}_m(\mathcal{G}), \text{ then } X \perp\!\!\!\perp Y \mid Z \, [N]\},$$

where $\mathcal{N}_{|V|}$ is the set of non-singular Gaussian distributions of dimension $|V|$.

It is an important property of maximal ancestral graphs that the set of Gaussian distributions given via recursive equations, as described in the previous section, is exactly the set given implicitly by the independence model:

Theorem 4.5 *If $\mathcal{G}$ is a maximal ancestral graph, then*

$$\mathcal{N}(\mathcal{G}) = \mathcal{N}(\mathfrak{I}_m(\mathcal{G})).$$

Thus any Gaussian distribution obtained from the parameterization above will satisfy the global Markov property for $\mathcal{G}$, and conversely, any Gaussian distribution obeying the Markov property may be parameterized in this way. If $\mathcal{G}$

is not maximal, then the first implication holds, but the converse does not; see Theorem 8.14 in [RS02].

Two ancestral graphs $\mathcal{G}_1$ and $\mathcal{G}_2$ are said to be *Markov equivalent* if $\mathfrak{I}_m(\mathcal{G}_1) = \mathfrak{I}_m(\mathcal{G}_2)$. An immediate corollary of Theorem 4.5 is that the Gaussian models associated with two maximal ancestral graphs will be Markov equivalent if and only if the resulting models are equivalent; see Corolloary 8.19 in [RS02].

4.6 Future work: the discrete case

At present there is no general method for estimating or parameterizing ancestral graph Markov models in the discrete case. This is the focus of ongoing research between the authors and James Robins, since ancestral graph Markov models are closely related to structural nested models; see Robins (1997).

5 Causal interpretation of DAG models

There is now a well-established causal interpretation of directed acyclic graphs; see Spirtes *et al.* (1993), Pearl (1995), Pearl (2000), Lauritzen (2000). Central to this interpretation is the distinction between conditioning by *intervention* and conditioning via *observation*. The conditional distribution of Y given that X takes value x *via intervention* is the value that Y would take on if X *were forced to take on the value x*. Following Lauritzen (2000) we denote this by

$$p(y \,\|\, x) \;=\; P(Y = y \mid X \leftarrow x)$$

where $X \leftarrow x$ has the meaning 'X is forced to take on value x'. It is not possible to compute the quantity $p(y \,\|\, x)$ from the joint distribution alone; in general $p(y \,\|\, x) \neq p(y \mid x)$. It should be noted that even in contexts such as randomized experiments where we can identify the population distribution $p(y \,\|\, x)$, resulting from a given intervention, we will still be unable to identify the individual-level causal effects, which can be quite different from the average effect. This may occur if the population is made up of sub-populations which respond very differently to treatment. Cox and Wermuth (2002) discuss an example of this, due to Zivin and Choi (1991), in the context of stroke treatment.

5.1 Causal DAGs

We define a directed acyclic graph $\mathcal{D}$ and distribution P to be *causal* if P obeys the global Markov property for $\mathcal{D}$, and further, for any $\beta \in V$,

$$
\begin{aligned}
p(x_{V \setminus \{\beta\}} \,\|\, x_\beta^*) &= \left. \prod_{\alpha \in V \setminus \{\beta\}} p(x_\alpha \mid x_{\mathrm{pa}(\alpha)}) \right|_{x_\beta = x_\beta^*} \\
&= \left. \frac{p(x_V)}{p(x_\beta \mid x_{\mathrm{pa}(\beta)})} \right|_{x_\beta = x_\beta^*}.
\end{aligned}
$$

Thus if a graph and distribution are causal, then the distribution that results from intervening on a particular variable x_ν, forcing it to take on the value x_ν^*, is given by simply removing the term $p(x_\nu \mid x_{\mathrm{pa}(\nu)})$ from the joint factorization (3.1) and substituting the value x_ν^* for x_ν in any conditional density in which it appears, i.e. in $p(x_\alpha \mid x_{\mathrm{pa}(\alpha)})$ for any variable α for which $\nu \in \mathrm{pa}(\alpha)$. Note that 'causality' is thus a property of the pair $(P, \mathcal{D})$. Though we do not do so here, this definition may easily be generalized to include interventions on sets of variables. An immediate consequence of this definition is that if $\mathcal{D}$ and P are causal, then the conditional density $p(x_\alpha \mid x_{\mathrm{pa}(\alpha)})$ remains stable under intervention on any other variable x_β, with $\beta \neq \alpha$. The formula for the intervention distribution (5.1) may be contrasted with that obtained by the usual observational conditioning:

$$p(x_{V \setminus \{\beta\}} \mid x_\beta^*) = \left. \frac{p(x_V)}{p(x_\beta)} \right|_{x_\beta = x_\beta^*}. \tag{5.1}$$

5.2 Examples

Consider the DAG shown in Fig. 5(a). Associated with this graph is the following factorization:

$$p(w, x, y, z) = p(w)p(x)p(z)p(y \mid x, z, w).$$

It then follows that

$$p(w, y, z \parallel x^*) = p(w)p(z)p(y \mid x^*, z, w),$$

where we are again using the independence of w and z. Hence for this graph,

$$p(y \parallel x^*) = \sum_{z,w} p(y \mid x^*, z, w)p(z)p(w) \tag{5.2}$$

$$= p(y \mid x^*) \tag{5.3}$$

since $x \perp\!\!\!\perp \{z, w\}$. For the DAG shown in Fig. 5(b), eqn (5.2) holds, but not (5.3), since the global Markov property does not imply $x \perp\!\!\!\perp z$. Consequently for this graph we have:

$$p(y \parallel x^*) = \sum_z p(y \mid x^*, z)p(z).$$

This is an instance of the back-door formula; see Theorem 7.1 of Spirtes *et al.* (1993), Pearl (1995), and Section 8.4 of Lauritzen (2000).

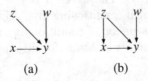

(a)   (b)

FIG. 5. Two simple DAGs leading to different expressions for $p(x \parallel y)$.

6   Identification of causal effects when the causal DAG is unknown

For the remainder of this article we consider the context in which an unknown causal DAG, possibly containing hidden variables, generated the data. We then outline a two-stage approach to inferring causal effects:

(1) Inferences are made concerning the structure of the underlying DAG by performing model determination within the class of maximal ancestral graph models.

(2) Distributions resulting from interventions are calculated if the inferences concerning structure that have been performed under (1) imply that such distributions are identifiable.

We first show that the independence structure implied by the causal DAG among the observed variables in a selected sub-population can always be represented by a maximal ancestral graph. This justifies the approach in step (1). We do not discuss here specific methods for performing model determination. We then illustrate step (2) via simple examples.

6.1   Marginalizing and conditioning independence models ($\mathfrak{I}[^S_L$)

In this section we define formal operations of marginalizing and conditioning for independence models.

An independence model $\mathfrak{I}$ *after marginalizing out a subset* L is simply the subset of triples that do not involve any vertices in L. More formally we define:

$$\mathfrak{I}[_L \equiv \Big\{\langle X, Y \mid Z\rangle \,\Big|\, \langle X, Y \mid Z\rangle \in \mathfrak{I};\; (X \cup Y \cup Z) \cap L = \emptyset\Big\}.$$

If $\mathfrak{I}$ contains the independence relations present in a distribution P, then $\mathfrak{I}[_L$ contains the subset of independence relations remaining after marginalizing out the 'Latent' variables in L; see Theorem 7.1 in [RS02].

An independence model $\mathfrak{I}$ *after conditioning on a subset* S is the set of triples defined as follows:

$$\mathfrak{I}[^S \equiv \Big\{\langle X, Y \mid Z\rangle \,\Big|\, \langle X, Y \mid Z \cup S\rangle \in \mathfrak{I};\; (X \cup Y \cup Z) \cap S = \emptyset\Big\}.$$

Thus if $\mathfrak{I}$ contains the independence relations present in a distribution P, then $\mathfrak{I}[^S$ constitutes the subset of independencies holding among the remaining variables after conditioning on S. The letter S is used because *S*election effects represent one context in which conditioning may occur. See Cox and Wermuth (1996, p. 44), for further discussion of conditioning in this context.

Combining these definitions we obtain, for disjoint sets S and L:

$$\mathfrak{I}[^S_L \equiv \Big\{\langle X, Y \mid Z\rangle \,\Big|\, \langle X, Y \mid Z \cup S\rangle \in \mathfrak{I};\; (X \cup Y \cup Z) \cap (S \cup L) = \emptyset\Big\}.$$

Note that, by definition, $\mathfrak{I}[^\emptyset_\emptyset = \mathfrak{I}$.

6.1.1 *Example*

Consider the following independence model:

$$\mathfrak{I}^* = \Big\{ \langle \{a,x\}, \{b,y\} \mid \{t\} \rangle, \langle \{a,x\}, \{b\} \mid \emptyset \rangle, \langle \{b,y\}, \{a\} \mid \emptyset \rangle, \langle \{a,b\}, \{t\} \mid \emptyset \rangle \Big\}.$$

In fact, $\mathfrak{I}^* \subset \mathfrak{I}_m(\mathcal{D})$, where $\mathcal{D}$ is the DAG in Fig. 6(i). ($\mathfrak{I}^*$ is a subset of $\mathfrak{I}_m(\mathcal{D})$ since, for example, $\langle \{a\}, \{b\} \mid \emptyset \rangle \in \mathfrak{I}_m(\mathcal{D})$.) In this case:

$$\mathfrak{I}^*\big[^{\emptyset}_{\{t\}} = \Big\{ \langle \{a,x\}, \{b\} \mid \emptyset \rangle, \langle \{b,y\}, \{a\} \mid \emptyset \rangle \Big\}, \ \mathfrak{I}^*\big[^{\{t\}}_{\emptyset} = \Big\{ \langle \{a,x\}, \{b,y\} \mid \emptyset \rangle \Big\}.$$

6.2 Marginalizing and conditioning for ancestral graphs

Given an ancestral graph $\mathcal{G}$ with vertex set V, for arbitrary disjoint sets S and L (both possibly empty) we now define a graphical transformation

$$\mathcal{G} \mapsto \mathcal{G}\big[^{S}_{L}$$

in such a way that the independence model corresponding to the transformed graph will be the independence model obtained by marginalizing and conditioning the independence model of the original graph.

Though we define this transformation for any ancestral graph $\mathcal{G}$, our primary motivation is the case in which $\mathcal{G}$ is the underlying (causal) DAG that is partially observed.

6.2.1 *Definition of $\mathcal{G}\big[^{S}_{L}$*

For disjoint sets S and L, the graph $\mathcal{G}\big[^{S}_{L}$ has vertex set $V \setminus (S \cup L)$, and edges specified as follows:

If α and β, are s.t. $\forall Z$, with $Z \subseteq V \setminus (S \cup L \cup \{\alpha, \beta\})$,

$$\langle \{\alpha\}, \{\beta\} \mid Z \cup S \rangle \notin \mathfrak{I}_m(\mathcal{G}),$$

and
$$
\left.\begin{cases}
\alpha \in \mathrm{ant}_{\mathcal{G}}(\{\beta\} \cup S); \ \beta \in \mathrm{ant}_{\mathcal{G}}(\{\alpha\} \cup S) \\
\alpha \notin \mathrm{ant}_{\mathcal{G}}(\{\beta\} \cup S); \ \beta \in \mathrm{ant}_{\mathcal{G}}(\{\alpha\} \cup S) \\
\alpha \in \mathrm{ant}_{\mathcal{G}}(\{\beta\} \cup S); \ \beta \notin \mathrm{ant}_{\mathcal{G}}(\{\alpha\} \cup S) \\
\alpha \notin \mathrm{ant}_{\mathcal{G}}(\{\beta\} \cup S); \ \beta \notin \mathrm{ant}_{\mathcal{G}}(\{\alpha\} \cup S)
\end{cases}\right\}, \text{ then }
\left.\begin{cases}
\alpha - \beta \\
\alpha \leftarrow \beta \\
\alpha \rightarrow \beta \\
\alpha \leftrightarrow \beta
\end{cases}\right\} \text{ in } \mathcal{G}\big[^{S}_{L}.
$$

In words, $\mathcal{G}\big[^{S}_{L}$ is a graph containing the vertices that are not in S or L. Two vertices α and β are adjacent in $\mathcal{G}\big[^{S}_{L}$ if α and β are m-connected in $\mathcal{G}$ given every subset that contains all vertices in S and no vertices in L. If α and β are adjacent in $\mathcal{G}\big[^{S}_{L}$, then there is an arrowhead at α if and only if α is anterior to neither β nor any vertex in S, and a tail otherwise.

It may be shown that $\mathcal{G}\big[^{S}_{L}$ is always maximal. It follows that $\mathcal{G}$ is maximal if and only if $\mathcal{G}\big[^{\emptyset}_{\emptyset} = \mathcal{G}$; see Corollaries 4.19 and 5.2 in [RS02].

The rationale for introducing this transformation is the following result (see [RS02], Theorem 4.18):

Theorem 6.1 *If $\mathcal{G}$ is an ancestral graph over V, and S and L are disjoint subsets of V, then*

$$\Im_m(\mathcal{G})[^S_L = \Im_m(\mathcal{G}[^S_L).$$

Note that the stated equality holds between two independence models, *not* two sets of distributions. The transformation described above provides a method for constructing a graph that represents the independence restrictions that a given DAG model imposes on a subset of variables after marginalizing and conditioning. However, perhaps more significantly, it guarantees that the class of ancestral graph Markov models is sufficiently large that the independence relations imposed by an unknown DAG on some margin, after conditioning, can always be represented. Consequently, when representing the independence structure implied by the causal DAG among the observed variables in a selected sub-population, we are guaranteed that no 'extra' edges will be included, solely owing to the class of Markov models that was used. Were we to perform a model search in the class of chain graphs, DAGs, or undirected graphs, over the observed variables, no such guarantee would exist.

6.2.2 *Examples*

Consider the DAG, $\mathcal{D}$, shown in Fig. 6(a). Suppose that we set $L = \{t\}$, $S = \emptyset$. First consider the adjacencies that will be present in the transformed graph $\mathcal{D}[^{\emptyset}_{\{t\}}$. It follows directly from the definition that vertices that are adjacent in the original graph will also be adjacent in the transformed graph, if they are present in the new graph, since adjacent vertices are m-connected given any subset of the remaining vertices. Hence the pairs (a, x) and (b, y) will be adjacent in $\mathcal{D}[^{\emptyset}_{\{t\}}$. In addition, x and y will be adjacent since any set m-separating x and y in $\mathcal{D}$ contains t, hence there is no set $Z \subseteq \{a, b\}$ such that $\langle\{x\}, \{y\} \mid Z\rangle \in \Im_m(\mathcal{D})$. In addition, $\langle\{a\}, \{b, y\} \mid \emptyset\rangle, \langle\{b\}, \{a, x\} \mid \emptyset\rangle \in \Im_m(\mathcal{D})$, hence there are no other adjacencies. It remains to determine the type of these three edges in $\mathcal{D}[^{\emptyset}_{\{t\}}$. Since $x \notin \mathrm{ant}_{\mathcal{D}}(y)$ and $y \notin \mathrm{ant}_{\mathcal{D}}(x)$, the edge between x and y is of the form $x \leftrightarrow y$. Similarly the other edges are $a \to x$ and $b \to y$. Thus the graph $\mathcal{D}[^{\emptyset}_{\{t\}}$ is as shown in Fig. 6(b). There is no DAG or UG with vertex set $\{a, b, x, y\}$ that represents these independence relations. Consequently any DAG or UG representing this distribution would contain 'extra' adjacencies.

Now suppose that $L = \emptyset$ and $S = \{t\}$. Since $\langle\{a, x\}, \{b, y\} \mid \{t\}\rangle \in \Im_m(\mathcal{D})$,

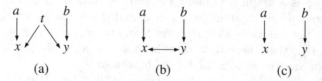

(a)   (b)   (c)

FIG. 6. (a) A simple DAG model, $\mathcal{D}$; (b) the graph $\mathcal{D}[^{\emptyset}_{\{t\}}$; (c) the graph $\mathcal{D}[^{\{t\}}_{\emptyset}$.
    See text for further explanation.

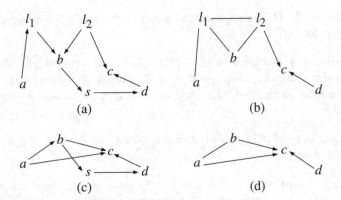

FIG. 7. (a) Another DAG, $\mathcal{D}'$; (b) the graph $\mathcal{D}'[_{\emptyset}^{\{s\}}$; (c) the graph $\mathcal{D}'[_{\{l_1,l_2\}}^{\emptyset}$; (d) the graph $\mathcal{D}'[_{\{l_1,l_2\}}^{\{s\}}$.

it follows that (a, x) and (b, y) are the only pairs of adjacent vertices present in the transformed graph $\mathcal{D}[_{\emptyset}^{\{t\}}$, and hence this graph takes the form shown in Fig. 6(c). Note that $\mathfrak{I}^*[_{\emptyset}^{\{t\}} \subset \mathfrak{I}_m(\mathcal{D}[_{\emptyset}^{\{t\}})$, where $\mathfrak{I}^*$ was defined in Section 6.1.1.

Another example of this transformation is given in Fig. 7, with a more complex DAG $\mathcal{D}'$. Note the edge between a and c that is present in $\mathcal{D}'[_{\{l_1,l_2\}}^{\{s\}}$.

6.3   Relating $\mathcal{G}$ and $\mathcal{G}[_L^S$

The transformation $\mathcal{G} \mapsto \mathcal{G}[_L^S$ destroys structural information, but $\mathcal{G}[_L^S$ reflects the structure of $\mathcal{G}$ in the following way:

Theorem 6.2 *For an ancestral graph $\mathcal{G}$ with vertex set $V = O \dot{\cup} S \dot{\cup} L$, if $\alpha \in O$ then the following hold:*

$$\text{ant}_\mathcal{G}(\alpha) \setminus \left(\text{ant}_\mathcal{G}(S) \cup L\right) \subseteq \text{ant}_{\mathcal{G}[_L^S}(\alpha) \subseteq \text{ant}_\mathcal{G}(\{\alpha\} \cup S) \setminus \left(S \cup L\right), \quad (6.1)$$

$$\left(\text{un}_\mathcal{G} \cup \text{ant}_\mathcal{G}(S)\right) \setminus (S \cup L) \subseteq \text{un}_{\mathcal{G}[_L^S}. \quad (6.2)$$

In words, (6.1) states that if α and β are in $\mathcal{G}[_L^S$ and β is anterior to α but not S in $\mathcal{G}$, then β is also anterior to α in $\mathcal{G}[_L^S$, while conversely, if β is anterior to α in $\mathcal{G}[_L^S$, then β is anterior to either α or S in $\mathcal{G}$. Equation (6.2) states that the vertices in $\text{un}_{\mathcal{G}[_L^S}$ are either in $\text{un}_\mathcal{G}$ or anterior to S in $\mathcal{G}$. Consequently, if in $\mathcal{G}[_L^S$ β is anterior to α and there is some edge with an arrowhead at β, then β is an ancestor of α in $\mathcal{G}$: by (6.1) β is either anterior to α or S in $\mathcal{G}$; by (6.2) β is not anterior to S in $\mathcal{G}$, since by definition $\beta \notin \text{un}_{\mathcal{G}[_L^S}$. See also Sections 4.2.5 and 4.2.6 in [RS02].

The following results show that the introduction of undirected edges is naturally associated with conditioning, while bi-directed edges are associated with marginalizing.

Proposition 6.3 *If $\mathcal{G}$ is an ancestral graph that contains no undirected edges, then neither does $\mathcal{G}[_L^{\emptyset}$.*

In particular, if we begin with a DAG then undirected edges will only be present in the transformed graph if $S \neq \emptyset$; likewise it follows from the next Proposition that bi-directed edges will only be present if $L \neq \emptyset$. (See [RS02], Section 4.2.8.)

Proposition 6.4 *If $\mathcal{G}$ is an ancestral graph that contains no bi-directed edges, then neither does $\mathcal{G}[_\emptyset^S$.*

Although the transformed graph still retains information about ancestral and anterior relations, information is lost concerning which vertices lie on which paths. For example, marginalization of t from either of the DAGs $\mathcal{D}_1$ or $\mathcal{D}_2$ in Fig. 8(a) will result in the graph $\mathcal{G}_1$ shown in Fig. 8(b). The presence of the edge $b \rightarrow y$ in the transformed graph reflects the fact that in both of the DAGs containing t, b is an ancestor of y, and further that the global Markov property applied to these graphs does not imply $\langle b, y \mid W \rangle$ for any $W \subseteq \{c, x\}$. The remarks following Theorem 6.2 further imply that b is an ancestor of y in any DAG which may be transformed into the graph in Fig. 8(b) (since b is anterior to y, and there is an arrowhead at b in $\mathcal{G}_1$). Note, however, that we are unable to tell whether or not the directed path from b to y in the DAG prior to transformation includes x. (See also Spirtes *et al.* (1993, Section 6.8).) For these reasons, a directed edge $x \rightarrow y$ in an ancestral graph should *not* be interpreted as indicating a direct causal relation between x and y. The interpretation of such an edge is given directly by the definition of the transformation $\mathcal{D} \mapsto \mathcal{D}[_L^S$: it indicates that in the DAG $\mathcal{D}$, x is anterior to y or to a vertex in S, but y is neither anterior to x nor to any vertex in S, and in addition there is no set W such that $\langle x, y \mid W \rangle \in \mathfrak{I}_m(\mathcal{D})[_L^S$. In particular, in some cases the transformed graph $\mathcal{D}[_L^S$ may still be a DAG, say $\mathcal{D}'$. However, if the original DAG $\mathcal{D}$ is causal this does *not* imply that $\mathcal{D}'$ is causal; edges in $\mathcal{D}'$ should only be interpreted with respect to the original DAG $\mathcal{D}$. (See $\mathcal{D}_1$ and $\mathcal{G}_1$ in Fig. 8 for an example.)

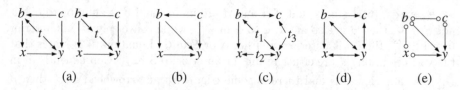

(a)   (b)   (c)   (d)   (e)

FIG. 8. (a) Two DAGs, $\mathcal{D}_1$ and $\mathcal{D}_2$; (b) the ancestral graph $\mathcal{G}_1 = \mathcal{D}_1[_{\{t\}} = \mathcal{D}_2[_{\{t\}}$; (c) another DAG, $\mathcal{D}_3$; (d) the ancestral graph $\mathcal{G}_2 = \mathcal{D}_3[_{\{t_1, t_2, t_3\}}$; (e) the partial ancestral graph representing the equivalence class containing $\mathcal{G}_1$ and $\mathcal{G}_2$.

6.4 Markov equivalence classes of maximal ancestral graphs

It follows directly from maximality that if two maximal ancestral graphs (MAGs) are Markov equivalent, then they have the same adjacencies. However, they may have different arrowheads or tails; see Fig. 9(a) for an example. In order to represent the arrowheads and tails that are present in every graph in an equivalence class we introduce the *partial ancestral graph (PAG)* Ψ for a Markov equivalence class $\mathfrak{G}$ of MAGs, which is a graph defined as follows:

(i) Ψ has the same adjacencies as every graph $\mathcal{G} \in \mathfrak{G}$.

(ii) There is an arrowhead ($\blacktriangleright$) at a vertex v on an edge ϵ in Ψ if and only if in every graph $\mathcal{G} \in \mathfrak{G}$ there is also an arrowhead present at v on ϵ.

(iii) There is a tail ($-$) at a vertex v on an edge ϵ in Ψ if and only if in every graph $\mathcal{G} \in \mathfrak{G}$ there is a tail present at v on ϵ.

(iv) There is a circle ($\circ$) at v on ϵ in Ψ if and only if there exist graphs $\mathcal{G}, \mathcal{G}' \in \mathfrak{G}$ and there is an arrowhead at v on ϵ in $\mathcal{G}$, and a tail at v on ϵ in $\mathcal{G}'$.

Figure 9(ii) shows the PAG corresponding to the Markov equivalence class shown in (i). The presence of the $x \longrightarrow y$ edge in the PAG reflects the presence of this edge in all four of the graphs in the equivalence class, similarly with the arrowheads at x on the $z \circ\!\!\longrightarrow x$ and $w \circ\!\!\longrightarrow x$ edges. As noted before, the presence of a directed edge $x \longrightarrow y$ and an arrowhead at x in an ancestral graph $\mathcal{G}$ indicates that x is an ancestor of y in any causal DAG that is transformed into $\mathcal{G}$ by marginalizing and conditioning. Consequently, if the true causal DAG is unknown, but we have been able to find the partial ancestral graph shown in Fig. 9(ii) through a model search, then we may infer that x is an ancestor of y in the true causal DAG. See Fig. 8 for another example: (v) shows the PAG for the Markov equivalence class containing the ancestral graphs shown in (ii) and (iv).

In general, model search techniques will only be able to find the correct PAG asymptotically if all of the independence relations holding among the observed variables in the selected sub-population result from the structure of the true causal DAG, and not from 'parametric cancellation' (Cox and Wermuth 1996), also referred to as a 'violation of faithfulness' (Spirtes *et al.* 1993) or lack of

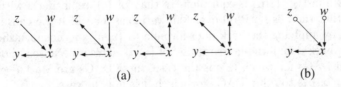

(a) (b)

FIG. 9. (a) A Markov equivalence class of maximal ancestral graphs; (b) the corresponding PAG.

'stability' (Pearl 2000).

Furthermore, it can be shown that for any causal DAG that is transformed into one of the graphs shown in Fig. 9(i), it holds that:

$$p(y \parallel x) = p(y \mid x)$$

since in each of these graphs x is an ancestor of y, and $y \perp\!\!\!\perp \{z, w\} \mid x$. Consequently, *if* we have been able to find the PAG in Fig. 9(ii), then we are able to calculate the distribution of y after intervening on x, since this distribution is invariant for any true causal DAG corresponding to a maximal ancestral graph represented by this PAG. General methods for identifying some causal effects from $\mathfrak{I}_m(\mathcal{D})[^{\emptyset}_L$ are given in Spirtes *et al.* (1993, Chapter 7).

It will not always be possible to calculate $p(y \parallel x)$ even if there is an edge $y \longleftarrow x$ in the PAG. For instance, if there is no arrowhead present at x, then it is possible that x is an ancestor of a selection variable in the true causal DAG, and not y. Even when selection variables are ruled out, there are situations in which it is possible to infer that x is an ancestor of y in the true causal DAG, but it is not possible to compute $p(y \parallel x)$.

6.5 PAGs over four variables

In Fig. 10 we show all the PAGs that represent Markov equivalence classes of MAGs with four vertices. The number of distinct classes given by permutation of the vertices is shown in parentheses. There are 251 classes in total: 185 contain at least one MAG which has no bi-directed or undirected edges and hence is a DAG (left-hand column); three do not contain any DAGs, but do contain a single UG (centre column); 63 do not contain any DAGs or UGs (right-hand column).

There are 11 PAGs which contain a specific directed edge $x \longrightarrow y$, which are shown in Fig. 11. Of these, the six in Fig. 11(a) do not have arrowheads at x, and so may correspond to DAGs in which x is an ancestor of a selection variable but not y, hence the intervention distribution $p(y \parallel x)$ cannot be computed. For the PAGs shown in Fig. 11(b), we have $p(y \parallel x) = p(y \mid x)$. For those in Fig. 11(c), we have

$$p(y \parallel x) = \sum_u p(y \mid x, u) p(u),$$

which is the back-door formula applied to u. The formula for $p(y \parallel x)$ for the PAG shown in Fig. 11(d) is equivalent to that for (c), replacing u with v.

Note that there is no PAG over four variables for which the causal effect is identified by applying the back-door formula to $\{u, v\}$ and not identified by applying the back-door formula to a subset of $\{u, v\}$. This is because although there are causal DAGs for which this is the case, these DAGs are Markov equivalent (over $\{u, v, x, y\}$) to other DAGs for which it is not the case.

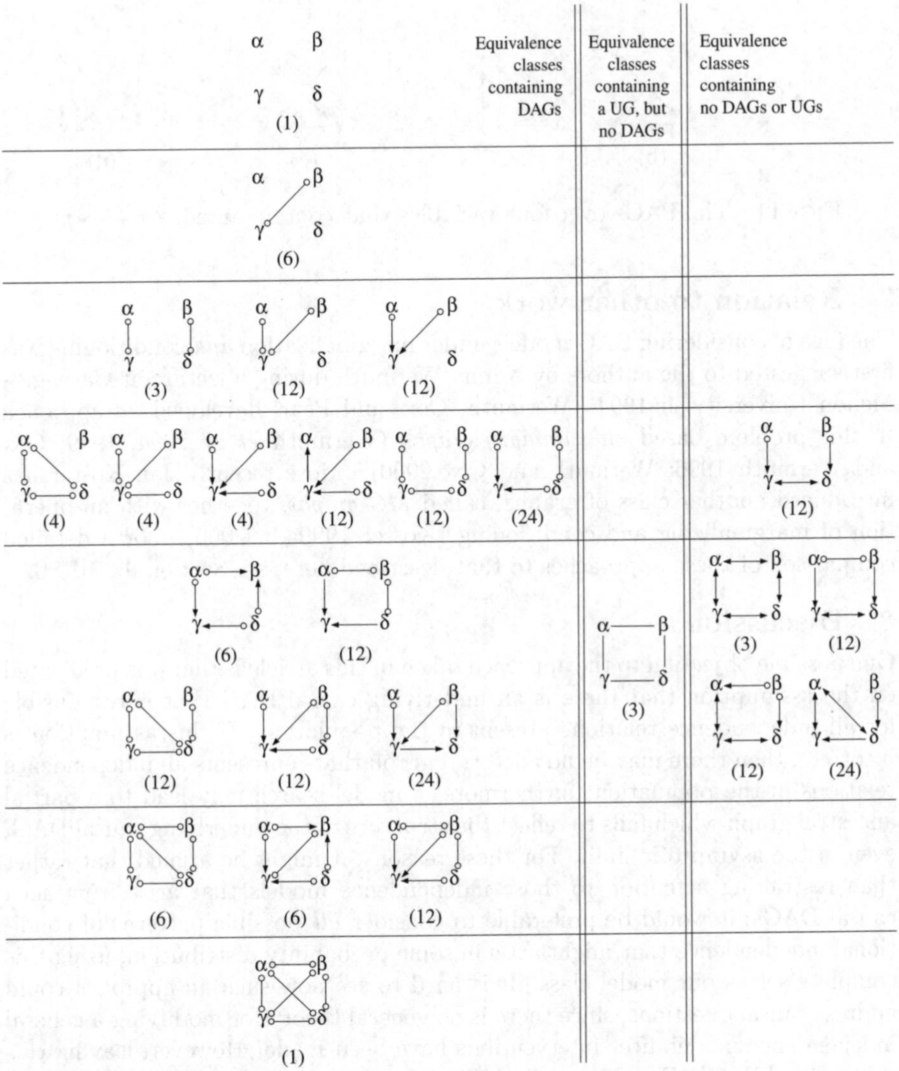

FIG. 10. Partial ancestral graphs representing all Markov equivalence classes of maximal ancestral graphs on four variables. See text for further explanation.

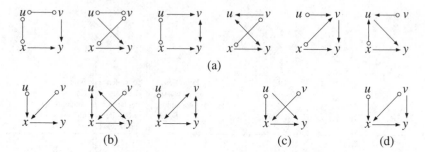

(a)

(b)   (c)   (d)

FIG. 11. The PAGs over four variables that contain an edge $x \longrightarrow y$.

7 Relation to other work

The idea of considering DAG models under marginalization *and* conditioning was
first suggested to the authors by Nanny Wermuth during a lecture at Carnegie–
Mellon University in 1994. Wermuth, Cox, and Pearl developed an approach
to this problem based on *summary graphs* (Wermuth *et al.* 1994, 1999; Cox
and Wermuth 1996; Wermuth and Cox 2000). More recently Jan Koster has
introduced another class of graphs, called *MC graphs*, together with an operation of marginalizing and conditioning (Koster 1999a,b, 2000). For a detailed
comparison of these approaches to that described here, see Section 9 of [RS02].

8 Discussion

One possible objection to the approach taken in this article is that it is predicated
on the assumption that there is an underlying causal DAG that is responsible
for all independence relations present in the population. If this assumption is
incorrect, then there may be no ancestral graph that represents all independence
relations in the population; furthermore, a model search may lead to a partial
ancestral graph which fails to reflect the structure of the underlying causal DAG,
even in the asymptotic limit. For these reasons, it might be argued that rather
than restricting attention to those independence models that arise from such
causal DAGs, it would be preferable to consider *all* possible patterns of conditional independence that might arise in some probability distribution, using this
complete set as our model class. It is hard to see how such an approach could
address causal questions, since there is no general theory for modifying a general
independence model after interventions have been made. However, leaving this
aside, Pavel Boček, Fero Matúš, and Milan Studený have shown that when there
are just four variables there are already 18 300 possible independence models for
discrete distributions; see Matúš (1999) and references therein. Consequently,
we believe that from the point of view of statistical modelling it is infeasible
to model conditional independence structures without restricting attention to a
more tractable sub-class.

Another possible objection to the approach taken here is that often information, such as time order, is available. However, when hidden variables may be present, knowledge about ordering may not yield any information which is relevant for restricting the class of possible independence models. Though it would clearly be desirable to incorporate information about ordering into the process of model determination, it is not straightforward to see how this may be implemented in practice; see Section 5 of Lauritzen and Richardson (2002).

In this article we assumed that the observed variables were part of a larger causal DAG. If feedback is present, then this assumption may be too restrictive. When feedback of certain types is present it may be more reasonable to assume that the variables are part of a chain graph or a directed cyclic graph. Lauritzen and Richardson (2002) provide a formal causal interpretation of chain graphs; see Richardson (1996) and Spirtes (1995) for a discussion of cyclic graphs.

Acknowledgements

This research was supported by the U.S. Environmental Protection Agency, and the U.S. National Science Foundation through grants DMS-9972008 and DMS-9873442. The work presented here benefited greatly from discussions that the authors had with Jan Koster, Steffen Lauritzen, Milan Studený, and Nanny Wermuth at meetings held as part of the Highly Structured Stochastic Systems Programme of the European Science Foundation.

References

Arbuckle, J. L. (1997). *AMOS User's Guide Version 9.6*. SPSS, Chicago.

Bentler, P. M. (1986). *Theory and Implementation of EQS: A Structural Equations Program*. BMDP Statistical Software, Los Angeles.

Cox, D. R. and Wermuth, N. (1996). *Multivariate Dependencies: Models, Analysis and Interpretation*. Chapman & Hall, London.

Cox, D. R. and Wermuth, N. (2002). Causal inference and statistical fallacies. In *Encyclopedia of Behavioral and Social Sciences*. Wiley, Chichester.

Dawid, A. (1979). Conditional independence in statistical theory (with discussion). *Journal of the Royal Statistical Society*, B, **41**, 1–31.

Dempster, A. P. (1972). Covariance selection. *Biometrics*, **28**, 157–75.

Edwards, D. M. (1995). *Introduction to Graphical Modelling*. Springer-Verlag, New York.

Frydenberg, M. (1990). The chain graph Markov property. *Scandinavian Journal of Statistics*, **17**, 333–53.

Jöreskog, K. and Sörbom, D. (1995). *LISREL 8: User's Reference Guide*. Scientific Software International, Chicago.

Koster, J. T. A. (1999a). Linear structural equations and graphical models. Lecture notes, The Fields Institute, Toronto.

Koster, J. T. A. (1999b). On the validity of the Markov interpretation of path diagrams of Gaussian structural equation systems with correlated errors. *Scandinavian Journal of Statistics*, **26**, 413–31.

Koster, J. T. A. (2000). Marginalizing and conditioning in graphical models. Technical Report, Erasmus University, Rotterdam.

Lauritzen, S. L. (1996). *Graphical Models*. Clarendon Press, Oxford.

Lauritzen, S. L. (2000). Causal inference from graphical models. In *Complex Stochastic Systems* (eds O. E. Barndorff-Nielsen, D. R. Cox, and C. Klüppelberg). Chapman & Hall, London.

Lauritzen, S. L. and Richardson, T. S. (2002). Chain graph models and their causal interpretation (with discussion). *Journal of the Royal Statistical Society*, B, **64**, 321–61.

Lauritzen, S. L., Dawid, A., Larsen, B., and Leimer, H.-G. (1990). Independence properties of directed Markov fields. *Networks*, **20**, 491–505.

Matúš, F. (1999). Conditional independences among four random variables iii: final conclusion. *Combinatorics, Probability and Computing*, **8**, 269–76.

Pearl, J. (1988). *Probabilistic Reasoning in Intelligent Systems*. Morgan Kaufmann, San Mateo, CA.

Pearl, J. (1995). Causal diagrams for empirical research (with discussion). *Biometrika*, **82**, 669–90.

Pearl, J. (2000). *Causality: Models, Reasoning and Inference*. Cambridge University Press, UK.

Richardson, T. S. (1996). *Models of feedback: interpretation and discovery*. PhD thesis, Carnegie–Mellon University, Pittsburgh.

Richardson, T. S. and Spirtes, P. (2002). Ancestral graph Markov models. *Annals of Statistics*, **30**, 962–1030.

Robins, J. (1997). Causal inference from complex longitudinal data. In *Latent Variable Modelling and Applications to Causality*, Lecture Notes in Statistics, No. 120 (ed. M. Berkane). Springer-Verlag, New York.

SAS Publishing (1995). *SAS/STAT User's Guide, Version 6* (4th edn). SAS Publishing, Cary, NC.

Spirtes, P. (1995). Directed cyclic graphical representations of feedback models. In *Proceedings of the Eleventh Annual Conference on Uncertainty in Artificial Intelligence* (eds P. Besnard and S. Hanks). Morgan Kaufmann, San Francisco.

Spirtes, P., Glymour, C., and Scheines, R. (1993). *Causation, Prediction and Search*, Lecture Notes in Statistics, No. 81. Springer-Verlag, New York.

Wermuth, N. and Cox, D. R. (2000). A sweep operator for triangular matrices and its statistical applications. Technical Report 00-04, ZUMA Institute, Mannheim, Germany.

Wermuth, N., Cox, D. R., and Pearl, J. (1994). Explanations for multivariate structures derived from univariate recursive regressions. Technical Report 94–1, ZUMA Institute, Mannheim, Germany.

Wermuth, N., Cox, D. R., and Pearl, J. (1999). Explanations for multivariate structures derived from univariate recursive regressions. Technical Report Revision 94–1, ZUMA Institute, Mannheim, Germany.

Zivin, J. A. and Choi, D. W. (1991). Stroke therapy. *Scientific American*, **265**, 59–63.

3A
Other approaches to the description of conditional independence structures

Milan Studený
Academy of Sciences, Prague, Czech Republic

1 Introduction

This contribution concerns some aspects complementary to the article by Richardson and Spirtes. In my opinion, the article relates three basic approaches to interpretation of graphs whose nodes correspond to random variables.

(1) The first possible interpretation of a graph of this type is that it serves as a synoptic pictorial representation of a *data-generating process*. This phrase is occasionally used to name a whole collection of assumptions about specific functional relationships between the random variables considered, e.g. recursive linear structural equation models represented by acyclic directed graphs (Andersson *et al.* 1999) or LISREL models represented by reciprocal graphs (Koster 1996).

(2) The second interpretation is that a graph encodes a certain *conditional independence structure* (CI structure). The respective statistical model consists of a class of distributions within a considered framework (e.g. the framework of non-degenerate Gaussian distributions) which is globally *Markovian* with respect to the graph; that is, it satisfies CI statements which are determined by the graph through a graphical criterion, e.g. the moralization criterion (Lauritzen 1996) or the c-separation criterion (Studený and Bouckaert 1998) for chain graphs.

(3) The third possible interpretation is that a graph describes a specific *parameterization* of a certain sub-class of the class of distributions within a considered framework, namely those parameterizations in which real parameters correspond to the edges of a graph, e.g. the parameterization induced by ancestral graphs (Richardson and Spirtes 2002) used within the framework of non-degenerate Gaussian distributions.

The charm of the article is that these three approaches are equivalent for (maximal) ancestral graphs within the Gaussian framework. This contribution is limited to the CI interpretation but not to the Gaussian framework. Some cited results are valid within the framework of discrete probability distributions. On the other hand, other graphical approaches and one non-graphical approach are mentioned. However, *directed graphs with cycles* (Spirtes 1995), *covariance graphs* (Kauermann 1996), and *MC graphs* (Koster 2000) are omitted here.

2 Chain graphs

A *chain graph* (CG) is a graph that has both directed edges, called *arrows*, and undirected edges, called *lines*, but without semi-directed cycles – for a formal definition see Lauritzen (1996). Three basic classes of CGs were introduced with the purpose of describing CI structures.

(1) *Classic CGs* were introduced by Lauritzen and Wermuth (1989) and studied in more detail by Frydenberg (1990). They are also named LWF CGs (after Lauritzen, Wermuth, and Frydenberg).

(2) *Alternative CGs,* or more precisely alternatively interpreted CGs, were proposed by Andersson *et al.* (2001). They are also named AMP CGs (after Andersson, Madigan, and Perlman).

(3) *Joint-response CGs,* introduced by Cox and Wermuth (1996), allow additional 'dashed' lines and arrows as well as the usual 'solid' lines and arrows (with different interpretations).

Note that joint-response CGs can be viewed as a generalization of both classic and alternative CGs: solid lines and arrows yield the classic interpretation while solid lines and dashed arrows lead to the alternative interpretation. However, the theory of joint-response CGs needs to be developed to include the concept of a global Markov property; the general definition of an induced CI structure is expected as a result of future research. As far as I know, only the pairwise Markov property, which describes a graph with a CI structure only partially, has been introduced and the respective global Markov property has been established in three special cases.

There are two graphical criteria which determined the class of globally Markovian distributions for classic CGs. The original *moralization criterion* used by Frydenberg (1990), which is based on stepwise transformation of the original CG into an undirected graph, was shown to be equivalent to the *c-separation criterion* (Studený and Bouckaert 1998) which directly tests whether trails in the original CG are blocked. The point is that the c-separation criterion allows proof of the consistency result saying that every classic CG describes perfectly the CI structure induced by a discrete probability distribution. *Markov equivalence,* i.e. the situation when two graphs have the same class of globally Markovian distributions, was characterized in graphical terms in Frydenberg (1990). It was shown there that every Markov equivalence class contains a distinguished representative with the greatest number of lines, called the *largest chain graph* (LCG). A recovery algorithm for LCGs on the basis of an induced CI structure was described in Studený (1997); graphical characterization of LCGs was described in Volf and Studený (1999).

Results on alternative chain graphs are analogous. Andersson *et al.* (2001) proposed an *augmentation criterion* which is an analogue of the stepwise moralization criterion for classic CGs. The respective *p-separation criterion* and the corresponding consistency result for alternative CGs within the Gaussian framework is in Levitz *et al.* (2001). Markov equivalence of alternative CGs was characterized by Andersson *et al.* (2001). There is no direct analogue of the LCG but the concept of *essential graph* was proposed there instead.

3 Annotated graphs

A regular *annotated graph* is an undirected graph whose edges are annotated by sets of other nodes in a certain way – for a formal definition see Paz *et al.* (2000). The corresponding *membership algorithm* then ascribes a formal CI structure to every annotated graph of this type. Note that Paz (2001) explains that this algorithm can be viewed as a generalization of the moralization criterion mentioned above. Regular annotated graphs thus generalize undirected graphs, acyclic directed graphs, and classic CGs in the sense that every CI structure induced by these graphs can be described by an annotated graph as well.

4 Structural imsets

However, graphical methods can hardly describe all probabilistic CI structures. For example, a long-term effort to characterize all probabilistic CI structures over four variables (Matúš 1999) resulted in the conclusion that that there exist 18 300 structures of this type – for an overview see Studený and Boček (1994). On the other hand, the number of classic CG models over four variables is only 200.

This motivated an effort to describe probabilistic CI structures by more general objects of discrete mathematics. The objects that seem suitable for this purpose are certain integer-valued functions on the power set of the set of variables, called *structural imsets*; for a formal definition see Studený (2002). A detailed exposition of this approach is in Studený (2001).

The main result is that every CI structure induced by either a discrete distribution or by a non-degenerate Gaussian distribution or by a conditional Gaussian distribution from Lauritzen (1996) can be described perfectly by a structural imset. Moreover, the lattice of CI structures induced by structural imsets has pleasant mathematical properties which give a chance that structural imsets can be implemented on a computer.

Acknowledgements

Research supported by grant GAČR 201/01/1482.

3B
On ancestral graph Markov models

Jan T. A. Koster
Erasmus University, Rotterdam, The Netherlands

Let me first congratulate the authors on their interesting article on ancestral graph Markov models. Readers of the article are strongly advised to go through the details of the longer 'parent' paper as well (Richardson and Spirtes 2002) to appreciate even further the elegant way in which the properties of this class of Markov models is developed. In my comment I will focus on the relationship between parameterized Gaussian ancestral graph models and Gaussian linear structural equation modelling (LSEM) as it is my conviction that any extension to the edifice of graphical models should also have something to offer to those who work in the time-honoured LSEM tradition. Let me begin by bringing to mind the basic structure of an ancestral graph $\mathcal{G} = (V, E)$. Let $u := \{\alpha \in V \mid \text{pa}(\alpha) = \text{sp}(\alpha) = \emptyset\}$, and $d := V \setminus u$. As follows from the definition of an ancestral graph, and is indeed noted in Section 4.1, the basic structure of $\mathcal{G}$ is as shown in Fig. 1.

In a Gaussian parameterization of an ancestral graph $\mathcal{G}$ each vertex $\nu \in V$ represents a random variable X_ν such that the vector $X := (X_\nu)_{\nu \in V} = \begin{bmatrix} X_u \\ X_d \end{bmatrix}$ satisfies the linear equation $X_d = BX_d + CX_u + E$, where $\begin{bmatrix} X_u \\ E \end{bmatrix} \sim N_{|V|}(\mu, \Gamma)$,

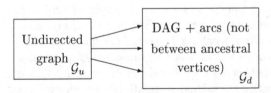

FIG. 1. Schematic structure of an ancestral graph. The subgraph $\mathcal{G}_u$ is a simple undirected graph; the subgraph $\mathcal{G}_d$ is a directed acyclic graph (DAG) to which arcs (bi-directed edges) may be added between vertices which are not ancestral to each other; arrows may also connect vertices in u and vertices in d, though only in the indicated direction.

$\Gamma = \begin{bmatrix} \Lambda^{-1} & 0 \\ 0 & \Omega \end{bmatrix}$.[1] It is stipulated that the entries of the matrices B, C, $\Lambda = \mathrm{cov}(X_u)^{-1}$ and $\Omega = \mathrm{cov}(E)$ either contain a structural zero or a free parameter. Essentially, whenever there is an arrow in $\mathcal{G}$ there is a free parameter in the corresponding entry of either B or C; a line induces a free parameter in Λ, whereas an arc corresponds to a free parameter in Ω. Note that this is all very similar to 'classical' LSEM (cf. Jöreskog and Sörbom 1995), except for the fact that it is now Λ, the *inverse* covariance matrix of the exogenous variables X_u which is being parameterized instead of the covariance matrix itself. This means that a covariance selection model (Dempster 1972) is specified for the distribution of X_u. Of course, in the case of the path diagram of a classical LSEM being an ancestral graph we in fact have $u = \emptyset$, so parameterized (Gaussian) ancestral graph models can be viewed as a generalization of this subclass of classical LSEMs. The point I wish to make, however, is that I am not sure if this extension will be very appealing to researchers who apply linear structural equation models, particularly since modelling the distribution of the exogenous variables is often considered not very interesting. Indeed, in many cases the LSEM does not constrain the distribution of the exogenous variables at all; that is, it is a model for the conditional distribution of X_d given X_u rather than for the joint distribution. A convincing example might remove some of my scepticism in this area.

My main concern regards the general restrictions imposed by ancestral graphs on the type of model they can represent. One can have at most a single edge between any pair of distinct vertices. In ancestral graphs a strong emphasis is put on anterior relationships. In fact, if two distinct vertices α and β are adjacent, then there will be a tail end at α (as in $\alpha - \beta$, or $\alpha \to \beta$) iff α is anterior to β. Equivalently, there will be an arrow head at α (as in $\beta \to \alpha$, or $\alpha \leftrightarrow \beta$) iff α is not anterior to β. This entails that reciprocal relationships and, more generally, (semi-)directed cycles are forbidden. Clearly, these characteristics are direct extensions to the class of ancestral graphs of similar properties of DAGs. However, in a linear structural equation model of, say, a supply–demand market equilibrium system feedback loops appear quite naturally (Strotz and Wold 1960). Another constraint imposed by ancestral graphs is that there cannot be a bi-directed edge between a vertex and any of its ancestors. Again, I think this is restrictive in practice as residual correlation between two variables, one of which is an ancestor of the other, is encountered regularly in panel studies and repeated measurement designs. The limited scope of ancestral graph Markov models is also apparent from the result stated in Section 4.4 that the free parameters in B and C are regression coefficients. The positive side of this is that the parameters associated with arrows in the graph are always identified and can be consistently estimated by ordinary least squares estimation. However, it also follows that

---

[1] Our notation here differs slightly from the authors' notation: their matrix B equals the present block matrix $\begin{bmatrix} I & 0 \\ -C & I - B \end{bmatrix}$.

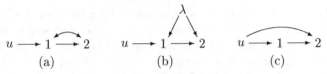

FIG. 2. Linear structural equation model: three repeated measurements with correlated residual variables (left); latent variable model (middle); ancestral graph model after marginalization over λ (right).

a model in which not all these parameters are regression coefficients cannot be represented by an ancestral graph. I think one of the reasons why the LISREL software became so extremely popular after its introduction in the 1970s is that it allows social science researchers to estimate models outside the 'regression domain'. For this reason too it is not be expected that Gaussian ancestral graph modelling will be seen as a full alternative to classical LSEM.

Since the operations of marginalizing and conditioning may introduce new edges to a graph and even change some of its existing edges, the emphasis on anterior relationships sometimes has consequences which some researchers who apply LSEMs may find surprising. Consider the following Gaussian linear system of three repeated measurements, X_u, X_1, and X_2, and non-zero covariance between the residual variables E_1 and E_2:

$$\left[\begin{array}{c} X_1 \\ X_2 \end{array} \right] = \left[\begin{array}{cc} 0 & 0 \\ b & 0 \end{array} \right] \left[\begin{array}{c} X_1 \\ X_2 \end{array} \right] + \left[\begin{array}{c} c \\ 0 \end{array} \right] X_u + \left[\begin{array}{c} E_1 \\ E_2 \end{array} \right],$$

where $\mathrm{cov}(E_1, E_2) \neq 0$. The path diagram of this system is given by Fig. 2 (left), and a latent variable model explaining the residual covariance by postulating a latent common factor λ would have Fig. 2 (middle) as its path diagram.[2] Since this is a DAG, and hence an ancestral graph, one can now marginalize again over λ (following the procedure of Section 6.2 in Chapter 3) and thus a graph is obtained in which the original arc $1 \leftrightarrow 2$ is replaced by the arrow $u \rightarrow 2$ (Fig. 2, right). This clearly shows that the usual interpretation of an arrow as indicating a causal effect is invalid in situations where the graph is not a priori known to be the path diagram of a structural equations system. To be fair, the authors warn explicitly against interpreting the presence of an arrow in ancestral graphs as indicating a causal effect, but my point is that, if Fig. 2 (middle) is the true situation, from a structural point of view marginalizing over the latent variable λ should render Fig. 2 (left), *not* (right).

I conclude with a critical remark concerning the way the authors introduce 'conditioning by intervention' in Section 5. They say, 'The conditional distribution of Y given that X takes value x *via intervention* is the value Y would take on if X *were forced to take on the value x*' (emphasis in original). I can

---

[2]Note that the residual variables are not represented by a vertex in the path diagram. A non-zero covariance between two residual variables is represented by an arc between the vertices representing the associated endogenous variables.

accept this as a description of the intuition behind the notion of 'conditioning by intervention', not as a definition. As it is, the expressions $p(y \parallel x)$ and $P(Y = y \mid X \leftarrow x)$ both remain undefined. Consequently, the definition in Section 5.1 of the directed acyclic graph $\mathcal{D}$ and distribution P being causal is not rigorous either. In Pearl (1995, 2000) conditioning by intervention is defined starting from a set of non-linear structural equations representing a causal theory based on subject-matter knowledge. The expression for $p(y \parallel x)$ then follows as a theorem.

Additional references in discussion

Andersson, S. A., Madigan, D., Perlman, M. D., and Richardson, T. S. (1999). Graphical Markov models in multivariate analysis. In *Multivariate Analysis, Design of Experiments, and Survey Sampling* (ed. S. Ghosh), pp. 187–229. Dekker, New York.

Andersson, S. A., Madigan, D., and Perlman M. D. (2001). Alternative Markov properties for chain graphs. *Scandinavian Journal of Statistics*, **28**, 33–85.

Kauermann, G. (1996). On a dualization of graphical Gaussian models. *Scandinavian Journal of Statistics*, **23**, 105–16.

Koster, J. T. A. (1996). Markov properties on nonrecursive causal models. *Annals of Statistics*, **24**, 2148–78.

Lauritzen, S. L. and Wermuth, N. (1989). Graphical models for asociation between variables, some of which are qualitative and some quantitative. *Annals of Statistics*, **17**, 31–57.

Levitz, M., Perlman, M. D., and Madigan, D. (2001). Separation and completeness properties for AMP chain graph Markov models. *Annals of Statistics*, **29**, 1751–84.

Paz, A. (2001). A fast version of Lauritzen's algorithm for checking representation of independencies. In *Proceedings of WUPES 2000*, to appear in *Soft Computing*.

Paz, A., Geva, R. Y., and Studený, M. (2000). Representation of irrelevance relations by annotated graphs. *Fundamenta Informaticae*, **42**, 149–99.

Strotz, R. H. and Wold, H. O. A. (1960). Recursive versus nonrecursive systems: an attempt at synthesis. *Econometrica*, **28**, 417–27.

Studený, M. (1997). A recovery algorithm for chain graphs. *International Journal of Approximate Reasoning*, **17**, 265–93.

Studený, M. (2001). *On mathematical description of probabilistic conditional independence structures*. Dr. Sc. thesis, Institute of Information Theory and Automation, Prague.

Studený, M. (2002). On stochastic conditional independence: the problem of characterization and description. *Annals of Mathematics and Artificial Intelligence*, **35**, 323–41.

Studený, M. and Boček, P. (1994). CI-models arising among 4 random variables. In *Proceedings of WUPES 1994*, pp. 268–82.

Studený, M. and Bouckaert, R. R. (1998). On chain graph models for description of conditional independence structures. *Annals of Statistics*, **26**, 1434–95.

Volf, M. and Studený, M. (1999). A graphical characterization of the largest chain graphs. *International Journal of Approximate Reasoning*, **20**, 209–36.

4
Causality and graphical models in time series analysis

Rainer Dahlhaus and Michael Eichler
Universität Heidelberg, Germany

1 Introduction

Over the last few years there has been growing interest in graphical models and in particular in those based on directed acyclic graphs as a general framework to describe and infer causal relations (Pearl 1995, 2000; Dawid 2000; Lauritzen 2001). This new graphical approach is related to other approaches to formalize the concept of causality, such as Neyman and Rubin's potential-response model (Neyman 1935; Rubin 1974; Robins 1986) and path analysis or structural equation models (Wright 1921; Haavelmo 1943). The latter concept has been applied in particular by economists to describe the equilibrium distributions of systems that typically evolve over time.

In this article we take up the idea behind the dynamic interpretation of structural equations and discuss (among others) graphical models in which the flow of time is exploited for causal inference. Instead of the intervention-based causality concept used by Pearl (1995) and others we apply a simpler notion of causality which is based on the obvious fact that an effect cannot precede its cause in time. Assuming that the variables are observed at different time points the non-causality relations in the system can be derived by examining the (partial) correlation between the variables at different time lags. In the same way we are able to identify the direction of directed edges in graphical models thus avoiding the assumption of an a priori given ordering of the variables as, for example, in directed acyclic graphs for multivariate data. Since we have repeated measurements of the variables over time we are in a time series setting. More restrictively we focus in this article on the situation where we observe only one long multivariate, stationary time series (in contrast, for example, to the panel situation where we have multiple observations of possibly short series).

The concept of causality we use is the concept of Granger causality (Granger 1969) which exploits the natural time ordering to achieve a causal ordering of the variables. More precisely, one time series is said to be Granger causal for another series if we are better able to predict the latter series using all available information than if the information apart from the former series had been used. This concept of causality has been discussed extensively in the econometrics literature, with several extensions and modifications. Eichler (1999, 2000) has used the definition of Granger causality to define causality graphs for time series. We discuss these graphs in Section 2 together with time series chain graphs and partial correlation graphs for time series (Dahlhaus 2000). In Section 3 we discuss Markov properties and in Section 4 statistical inference for these graphs. Section 5 contains two data examples. Finally, Section 6 offers some concluding remarks. In particular we discuss the use of the presented graphs for causal inference and possible sources for wrong identification of causal effects.

2   Graphical models for multivariate time series

Let $X = \{X_a(t), t \in \mathbb{Z}, a = 1, \ldots, d\}$ be a d-variate stationary process. Throughout the article we assume that X has positive definite spectral matrix $f(\lambda)$ with eigenvalues bounded and bounded away from zero uniformly for all $\lambda \in [-\pi, \pi]$. Let $V = \{1, \ldots, d\}$ be the set of indices. For any $A \subseteq V$ we define $X_A = \{X_A(t)\}$ as the multivariate sub-process given by the indices in A. Furthermore, $\overline{X}_A(t) = \{X_A(s), s < t\}$ denotes the past of the sub-process X_A at time t.

There exist several possibilities for defining graphical models for a time series X. We can distinguish between two classes of graphical models. In the first class the variable $X_a(t)$ at a specific time t is represented by a separate vertex in the graph. This leads to generalizations of classical graphical models such as the time series chain graph introduced in Definition 2.1. In the second class of graphical models the vertex set only consists of the components X_a of the series, which leads to a coarser modelling of the dependence structure of the series. As we will see below, this leads to mixed graphs in which directed edges reflect Granger causality whereas the contemporaneous dependence structure is represented by undirected edges. These graphs are termed Granger causality graphs (Definition 2.4). In addition we have also partial correlation graphs for time series (Definition 2.6) which generalize classical concentration graphs to the time series situation.

Here we restrict ourselves to linear association and linear Granger non-causality, which formally can be expressed in terms of conditional orthogonality of closed linear subspaces in a Hilbert space of random variables (e.g. Eichler 2000, Appendix A.1). For random vectors X, Y, and Z, X and Y are conditionally orthogonal given Z, denoted by $X \perp Y \mid Z$, if X and Y are uncorrelated after the linear effects of Z have been removed. It is clear that this linear approach cannot capture the full causal relationships of processes that are partly non-linear in nature. The need for this restriction results from problems in statistical inference. For example, there exist non-parametric methods for estimating linear dependencies whereas in the general case non-parametric inference seems hardly possible, particularly in view of the curse of dimensionality. Even if parametric models are used, inference for non-linear models is often impractical due to a large number of parameters, and many applications therefore are restricted to linear models. We note that the graphical modelling approach can be extended to the non-linear case by replacing conditional orthogonality by conditional independence, which leads to the notion of strong Granger causality (Eichler 2000, Section 5; 2001). For Gaussian processes the two meanings of the graphs are identical of course. Large parts of the results in this article also hold for these general graphs. For details we refer to the discussion in Section 6.

2.1 Time series chain graphs

The first approach for defining graphical time series models naturally leads to chain graphs. Lynggaard and Walther (1993) introduced dynamic interaction models for time series based on the classical LWF Markov property for chain graphs (Lauritzen and Wermuth 1989; Frydenberg 1990). Here we discuss an alternative approach which defines the graph according to the AMP Markov property of Andersson *et al.* (2001). One reason for this is that there exists an intimate relation between the use of the AMP Markov property in time series chain graphs and the recursive structure of a large number of time series models which then immediately characterize the graph. Another reason is that the AMP Markov property is related to the notion of Granger causality which is used in the definition of Granger causality graphs. In particular, the AMP Markov property allows one to obtain the Granger causality graph from the time series chain graph by simple aggregation.

Definition 2.1 (Time series chain graph) The time series chain graph (TSC graph) of a stationary process X is the chain graph $G_{\mathrm{TS}} = (V_{\mathrm{TS}}, E_{\mathrm{TS}})$ with $V_{\mathrm{TS}} = V \times \mathbb{Z}$ and edge set E_{TS} such that

$$(a, t - u) \longrightarrow (b, t) \notin E_{\mathrm{TS}} \Leftrightarrow u \leqslant 0 \text{ or } X_a(t - u) \perp X_b(t) \,|\, \overline{X}_V(t) \backslash \{X_a(t - u)\},$$

$$(a, t - u) \longrightarrow (b, t) \notin E_{\mathrm{TS}} \Leftrightarrow u \neq 0 \text{ or } X_a(t) \perp X_b(t) \,|\, \overline{X}_V(t) \cup \{X_{V \backslash \{a, b\}}(t)\}.$$

Since the process is stationary we have $(a, t) \longrightarrow (b, t) \notin E_{\mathrm{TS}}$ if and only if $(a, s) \longrightarrow (b, s) \notin E_{\mathrm{TS}}$ for all $s \in \mathbb{Z}$. The same shift invariance also holds for the directed edges. We further note that the above conditions guarantee that the process satisfies the pairwise AMP Markov property for G_{TS} (Andersson *et al.* 2001).

Example 2.2 (Vector autoregressive processes) Suppose X is a linear vector autoregressive (VAR) process

$$X(t) = A(1)X(t - 1) + \cdots + A(p)X(t - p) + \varepsilon(t), \qquad (2.1)$$

where the errors $\varepsilon(t)$ are independent and identically distributed with mean 0 and covariance matrix Σ. If $G_{\mathrm{TS}} = (V_{\mathrm{TS}}, E_{\mathrm{TS}})$ denotes the TSC graph, then it can be shown that

$$(a, t - u) \longrightarrow (b, t) \in E_{\mathrm{TS}} \Leftrightarrow u \in \{1, \ldots, p\} \text{ and } A_{ba}(u) \neq 0, \qquad (2.2)$$

i.e. the directed edges in the graph reflect the recursive structure of the time series. Furthermore, the undirected edges specify a covariance selection model

R. Dahlhaus and M. Eichler

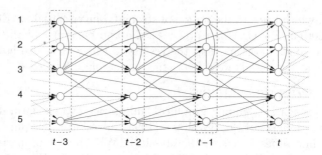

FIG. 1. TSC graph G_{TS} for the VAR process in Example 2.2.

(e.g. Dempster 1972) for the errors $\varepsilon(t)$ and we have with $K = \Sigma^{-1}$:

$$(a,t) \text{ ---- } (b,t) \in E_{\text{TS}} \Leftrightarrow \varepsilon_a(t) \perp \varepsilon_b(t) \,|\, \varepsilon_{V\setminus\{a,b\}}(t) \Leftrightarrow K_{ab} \neq 0.$$

As an example we consider the five-dimensional VAR(2) process with parameters

$$A(1) = \begin{pmatrix} \frac{3}{5} & 0 & \frac{1}{5} & 0 & 0 \\ 0 & \frac{3}{5} & 0 & -\frac{1}{5} & 0 \\ \frac{2}{5} & \frac{1}{3} & \frac{3}{5} & 0 & 0 \\ 0 & 0 & 0 & -\frac{1}{2} & \frac{1}{5} \\ 0 & 0 & \frac{1}{5} & 0 & \frac{2}{5} \end{pmatrix}, \quad A(2) = \begin{pmatrix} 0 & 0 & -\frac{1}{5} & 0 & 0 \\ 0 & 0 & 0 & 0 & 0 \\ 0 & 0 & 0 & 0 & 0 \\ 0 & 0 & \frac{1}{5} & 0 & \frac{1}{3} \\ 0 & 0 & 0 & 0 & -\frac{1}{5} \end{pmatrix},$$

$$\Sigma = \begin{pmatrix} 1 & \frac{1}{2} & \frac{1}{3} & 0 & 0 \\ \frac{1}{2} & 1 & -\frac{1}{3} & 0 & 0 \\ \frac{1}{3} & -\frac{1}{3} & 1 & 0 & 0 \\ 0 & 0 & 0 & 1 & 0 \\ 0 & 0 & 0 & 0 & 1 \end{pmatrix}.$$

The above conditions for the edges in E_{TS} lead to the TSC graph in Fig. 1.

2.2 Granger causality graphs

We now define non-causality as we will use it throughout this article, and then Granger causality graphs (Eichler 2000). These are mixed graphs whose vertex set consists only of the components of the series. For directed edges we use the notion of Granger causality (Granger 1969) and for undirected edges we use the same definition as for the TSC graphs. We mention that there has been a lively discussion in the econometrics literature on various definitions of causality; for overviews we refer to Geweke (1984) and Aigner and Zellner (1988).

Definition 2.3 (Noncausality) X_a is noncausal for X_b relative to the process X_V, denoted by $X_a \nrightarrow X_b \,[X_V]$, if

$$X_b(t) \perp \overline{X}_a(t) \,|\, \overline{X}_{V\setminus\{a\}}(t).$$

Furthermore, X_a and X_b are partially contemporaneously uncorrelated relative to the process X_V, denoted by $X_a \nsim X_b \, [X_V]$ if

$$X_a(t) \perp X_b(t) \,|\, \overline{X}(t), X_{V\setminus\{a,b\}}(t).$$

Definition 2.3 can be retained if X_a and X_b are replaced by multivariate sub-processes X_A and X_B, respectively.

In a multivariate setting there are different possible causality patterns such as direct and indirect causality, feedback, or spurious causality (Hsiao 1982). These patterns can be formally described by the use of graphs.

Definition 2.4 (Causality graph) The Granger causality graph of a stationary process X is the mixed graph $G_C = (V, E_C)$ such that for all $a, b \in V$ with $a \neq b$

(i) $a \longrightarrow b \notin E_C \Leftrightarrow X_a \nrightarrow X_b \, [X_V]$,

(ii) $a \longrightarrow b \notin E_C \Leftrightarrow X_a \nsim X_b \, [X_V]$.

For simplicity we will speak only of causality graphs instead of Granger causality graphs. We implicitly assume that each component also depends on its own past. This could be expressed by directed self-loops. Since the insertion of these loops does not change the separation properties for the graph we omit these loops for the sake of simplicity.

The relation between the TSC graph $G_{TS} = (V_{TS}, E_{TS})$ and the causality graph $G_C = (V, E_C)$ is simple:

Proposition 2.5 *(Aggregation) Let G_C and G_{TS} be the causality graph and the TSC graph, respectively, of a stationary process X. Then we have*

(i) $a \longrightarrow b \notin E_C \Leftrightarrow (a, t - u) \longrightarrow (b, t) \notin E_{TS} \quad \forall u > 0 \quad \forall t \in \mathbb{Z}$,

(ii) $a \longrightarrow b \notin E_C \Leftrightarrow (a, t) \longrightarrow (b, t) \notin E_{TS} \quad \forall t \in \mathbb{Z}$.

Proof (i) follows from a replicated application of the intersection property (C5) in Lauritzen (1996). Details are omitted. (ii) is straightforward.   □

Example 2.2 (contd) From Proposition 2.5 and (2.2) it follows that $a \longrightarrow b \notin E_C$ if and only if $A_{ba}(1) = A_{ba}(2) = 0$. The resulting causality graph of $X = \{X(t)\}$ is shown in Fig. 2. From this graph we can see, for example, that

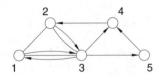

FIG. 2. Causality graph G_C for the VAR process in Example 2.2.

X_1 is non-causal for X_4 relative to the full process. More intuitively, the directed paths from 1 to 4 suggest that X_1 causes X_4 indirectly. This indirect cause seems to be mediated by X_3 since all paths from 1 to 4 intersect 3. In the next section we will see that such causality relations can indeed be derived formally from the graph.

2.3 Partial correlation graphs

We now introduce the partial correlation graph for a time series (Dahlhaus 2000). This undirected graph is the counterpart of a concentration graph for ordinary variables. The graph has a simple separation concept and allows additional conclusions about the dependence structure of the series.

Definition 2.6 (Partial correlation graph) The partial correlation graph (PC graph) $G_{PC} = (V, E_{PC})$ of a stationary process X is given by

$$a - b \notin E_{PC} \Leftrightarrow X_a \perp X_b \,|\, X_{V \setminus \{a,b\}}.$$

There exists a nice characterization of PC graphs in terms of the inverse of the spectral matrix of the process. Let $f_{ab}(\lambda)$ be the cross-spectrum between X_a and X_b, $f(\lambda) = (f_{ab}(\lambda))_{a,b \in V}$ be the spectral matrix, and $g(\lambda) = f(\lambda)^{-1}$ its inverse. Then it can be shown (see Proposition 2.7) that

$$a - b \notin E_{PC} \Leftrightarrow g_{ab}(\lambda) = 0 \quad \forall \lambda \in [-\pi, \pi].$$

That is, we have a similar characterization as for concentration graphs. More important, the absolute rescaled inverse $|d_{ab}(\lambda)| = |g_{ab}(\lambda)|/(g_{aa}(\lambda)g_{bb}(\lambda))^{1/2}$ is a measure for the strength of the 'connection' between X_a and X_b. To see this let $Y_{a|V \setminus \{a,b\}}$, $Y_{b|V \setminus \{a,b\}}$ be the residual series of X_a, respectively X_b, after the linear effects of the other components $X_{V \setminus \{a,b\}}$ have been removed. Then the partial cross spectrum $f_{ab|V \setminus \{a,b\}}(\lambda)$ is defined as the cross spectrum between the residual series $Y_{a|V \setminus \{a,b\}}$ and $Y_{b|V \setminus \{a,b\}}$, and

$$R_{ab|V \setminus \{a,b\}}(\lambda) = f_{ab|V \setminus \{a,b\}}(\lambda)/\left(f_{aa|V \setminus \{a,b\}}(\lambda)f_{bb|V \setminus \{a,b\}}(\lambda)\right)^{1/2} \quad (2.3)$$

is the partial spectral coherence (cf. Brillinger 1981, Chapters 7 and 8), which is a kind of partial correlation between X_a and X_b 'at frequency λ'. Then we have

Proposition 2.7 *Under the regularity assumptions on X we have for $a \neq b$*

$$g_{aa}(\lambda) = 1/f_{aa|V \setminus \{a\}}(\lambda) \quad and \quad d_{ab}(\lambda) = -R_{ab|V \setminus \{a,b\}}(\lambda).$$

In particular, $d_{ab}(\lambda) = 0$ for all $\lambda \in [-\pi, \pi]$ if and only if $a - b \notin E_{PC}$.

For a proof see Dahlhaus (2000, Theorem 2.4).

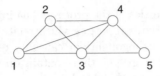

FIG. 3. Partial correlation graph G_{PC} for the VAR process in Example 2.2.

Example 2.2 (contd) For general VAR processes of the form (2.1) we have $f(\lambda) = (2\pi)^{-1}A^{-1}(e^{-i\lambda})\Sigma A^{-1}(e^{i\lambda})'$ with $A(z) = I - A(1)z - \cdots - A(p)z^p$. Consequently $g(\lambda) = f(\lambda)^{-1} = 2\pi A(e^{i\lambda})'\Sigma^{-1}A(e^{-i\lambda})$ and the restriction $g_{ab} \equiv 0$ thus leads to the following parameter constraints:

$$\sum_{u=0\vee h}^{p\wedge p+h} \sum_{j,k=1}^{d} K_{jk}A_{ja}(u)A_{kb}(u+h) = 0, \qquad h = -p,\ldots,p, \qquad (2.4)$$

with $A(0) = I$ and $K = \Sigma^{-1}$. From these constraints we can derive the PC graph for the VAR(2) model from Example 2.2. Assuming that the terms in (2.4) are non-zero whenever one summand is non-zero we obtain an edge $a - b \in E_{\text{PC}}$ whenever a and b are connected by a directed or an undirected edge in G_C. Additionally we find that for $a = 1$ and $b = 4$ the equation for $h = 1$ yields $K_{12}A_{24}(1) \neq 0$, which implies that 1 and 4 are also adjacent in G_{PC}. This leads to the PC graph in Fig. 3. In general further conditional orthogonalities are possible under additional restrictions on the parameters. However, for this particular model an elaborate examination shows that this is not the case.

The condition $K_{12}A_{24}(1) \neq 0$ in the example implies that $1 - 2 \leftarrow 4$ is a subgraph of G_C. After insertion of the edge $1 - 4$ this subgraph becomes complete, that is, any two vertices in the subgraph are adjacent. Thus the PC graph of the process has been obtained from the causality graph by completing this subgraph and then converting all directed edges to undirected edges. This operation is called moralization. In the next section we will see that this relationship between the causality graph and the PC graph holds for all processes satisfying the regularity assumptions in Section 2.

Summarizing: the three graphs G_{TS}, G_C, and G_{PC} are related by two mappings which correspond to the operations of aggregation and moralization:

$$G_{\text{TS}} \xmapsto{\text{Proposition 2.5}} G_C \xmapsto{\text{Remark 3.5}} G_{\text{PC}}.$$

Thus in a certain sense the graphs visualize the dependence structure of the process at different resolution levels, the TSC graph having the finest resolution.

3 Markov properties

The time series graphs introduced in the previous section visualize the pairwise interaction structure between the components of the process, i.e. they reflect the pairwise Markov properties of the process. The interpretation of the graphs is enhanced by global Markov properties which relate the separation properties of a graph to conditional orthogonality or causality relations between the components of the process. For graphs that contain directed edges there are two main approaches to defining global Markov properties. One approach utilizes separation in undirected graphs by applying the operation of moralization to appropriate subgraphs while the other approach is based on path-oriented criteria like d-separation (Pearl 1988) for directed acyclic graphs.

3.1   Global Markov properties for PC graphs

Partial correlation graphs generalize concentration graphs for multivariate variables. Thus the corresponding global Markov property is naturally based on the separation in undirected graphs. For an undirected graph $G = (V, E)$ and disjoint subsets $A, B, S \subseteq V$ we say that S separates A and B, denoted by $A \bowtie B \,|\, S$ $[G]$, if every path between A and B in G necessarily intersects S. Then the following result was shown by Dahlhaus (2000).

Theorem 3.1 *Let* $G_{\mathrm{PC}} = (V, E_{\mathrm{PC}})$ *be the PC graph of* X. *Then for all disjoint subsets* A, B, S *of* V

$$A \bowtie B \,|\, S \ [G_{\mathrm{PC}}] \Rightarrow X_A \perp X_B \,|\, X_S.$$

We say that X *satisfies the global Markov property for* G_{PC}.

Example 2.2 (contd)   In the graph in Fig. 3 the subset $\{3, 4\}$ separates $\{1\}$ and $\{5\}$. Thus it follows by Theorem 3.1 that $X_1 \perp X_5 \,|\, X_{\{3,4\}}$.

3.2   Global Markov properties for TSC graphs

Separation in undirected graphs can also be used to define global Markov properties for directed graphs, chain graphs, or mixed graphs. This approach is based on 'moralization' of subgraphs induced by certain subsets of vertices.

Let $G = (V, E)$ be a mixed graph. The undirected subgraph $G^{\mathrm{u}} = (V, E^{\mathrm{u}})$ is obtained from G by removing all directed edges from G. Furthermore, a subset A of V induces the subgraph $G_A = (A, E_A)$, where E_A is obtained from E by keeping edges with both endpoints in A. As motivated in the previous section, a mixed graph can be moralized by completing certain subgraphs. Following Andersson et al. (2001) we call a triple (a, b, c) an immorality (respectively, a flag) if $a \rightarrow c \leftarrow b$ $(a \rightarrow c - b)$ is a subgraph of G but a and b are not adjacent in G. Furthermore, the subgraph induced by (a, b, c, d) is called a 2-biflag if it contains the subgraph $a \rightarrow c - d \leftarrow b$ and a and b are not adjacent. The possible forms of immoralities, flags, and 2-biflags in mixed graphs are depicted in Fig. 4, where dotted lines between vertices indicate possible further edges. Most of these edges, however, are excluded in the case of chain graphs. We note that in mixed graphs with multiple edges a subgraph can be both an immorality and a flag.

FIG. 4. Configurations in mixed graphs which lead to an additional edge between vertices a and b in the moral graph: (a) flag, (b) immorality, and (c) 2-biflag.

Definition 3.2 Let $G = (V, E)$ be a mixed graph. The moral graph $G^m = (V, E^m)$ derived from G is defined to be the undirected graph obtained by completing all immoralities, flags, and 2-biflags in G and then converting all directed edges in G to undirected edges.

In TSC graphs the orthogonality relations of the process can be described by the global AMP Markov property for chain graphs (Andersson *et al.* 2001). Here we give a slightly different definition of the global AMP Markov property which is based on marginal ancestral subgraphs instead of extended subgraphs. We directly give the general definitions for mixed graphs which include chain graphs as a special case.

As in Frydenberg (1990) we call a vertex a an ancestor of u if either $a = u$ or there exists a directed path $a \longrightarrow \cdots \longrightarrow u$ in G. For $U \subseteq V$ let $\mathrm{an}(U)$ be the set of ancestors of elements in U. The marginal ancestral graph $G\langle A\rangle = (\mathrm{an}(A), E\langle A\rangle)$ induced by A is given by the subgraph $G_{\mathrm{an}(A)}$ with additional undirected edges $a - b$ whenever a and b are not separated by $\mathrm{an}(A)\backslash\{a, b\}$ in G^u. Owing to these additional edges the sub-process $X_{\mathrm{an}(A)}$ satisfies the pairwise Markov properties for the marginal ancestral graph $G\langle A\rangle$.

Theorem 3.3 *Let $G_{\mathrm{TS}} = (V_{\mathrm{TS}}, E_{\mathrm{TS}})$ be the TSC graph of X and A, B, and S be disjoint subsets of V_{TS}. Then if $A \bowtie B \mid S \ [G_{\mathrm{TS}}\langle A \cup B \cup S\rangle^m]$ we have*

$$\{X_a(t), (a, t) \in A\} \perp \{X_b(t), (b, t) \in B\} \mid \{X_s(t), (s, t) \in S\}.$$

We say that X satisfies the global AMP Markov property for G_{TS}.

Proof Since for random variables X, Y_1, Y_2, and Z the orthogonality relations $X \perp Y_i \mid Z$, $i = 1, 2$, always imply $X \perp (Y_1, Y_2) \mid Z$, the pairwise and the global AMP Markov properties for G_{TS} are equivalent (Andersson *et al.* 2001).   ☐

Example 2.2 (contd) Since X is a VAR(2) process the directed edges in the graph in Fig. 1 have at most lag 2. The same holds for the edges in $G_{\mathrm{TS}}\langle A \cup B \cup S\rangle^m$ for all sets A, B, and S. Thus it follows that $X(t-3)$ and $X(t)$ are conditionally orthogonal given $X(t-1)$ and $X(t-2)$.

The global AMP Markov property can be used to derive general causality relations from the graph involving sub-processes of X. More precisely, for $A, B \subseteq S$

X_A is non-causal for X_B relative to X_S whenever $\bar{A}_t \bowtie B_t \,|\, \bar{S}_t \backslash \bar{A}_t \;\; [G_{\mathrm{TS}} \langle \bar{S}_t \cup B_t \rangle^{\mathrm{m}}]$, where $\bar{A}_t = \{(a,s), a \in A, s \leqslant t\}$ and $A_t = \{(a,t), a \in A\}$. In the special case where $A = \{a\}$, $B = \{b\}$ and $S = V \backslash \{a,b\}$ this gives the pairwise causality relations relative to the full process X.

3.3 Global Markov properties for causality graphs

In causality graphs each component $X_a = \{X_a(t)\}$ of the process is represented by just a single vertex a in the graph. Thus each vertex a has to be interpreted either as $\overline{X}_a(t)$ or as $X_a(t)$ depending on the type and the direction of the adjacent edges. All information on the pairwise causality relations is encoded only in the type and the direction of the edges. Since the directions of the edges are removed by moralization we cannot hope to retrieve unidirectional causalities by this method. Nevertheless, the moralization criterion can be used to derive global conditional orthogonality relations between the components of the series, i.e. in this case the vertex a stands for X_a as in PC graphs.

Theorem 3.4 X satisfies the global AMP Markov property for G_C, i.e.

$$A \bowtie B \,|\, S \;\; [G_C \langle A \cup B \cup S \rangle^{\mathrm{m}}] \;\Rightarrow\; X_A \perp X_B \,|\, X_S.$$

For a proof see Eichler (2001).

Remark 3.5 If we set $A = \{a\}$, $B = \{b\}$, and $S = V \backslash \{a,b\}$, then the corresponding marginal ancestral graph is the moral graph G_C^{m} of G_C itself irrespective of a and b. Thus we have for all $a, b \in V$

$$\{a\} \bowtie \{b\} \,|\, V \backslash \{a,b\} \;\; [G_C^{\mathrm{m}}] \;\Leftrightarrow\; a - b \notin E_C^{\mathrm{m}} \;\Rightarrow\; X_a \perp X_b \,|\, X_{V \backslash \{a,b\}},$$

i.e. the process X satisfies the pairwise Markov property for the moral graph G_C^{m}. In particular, G_{PC} is a subgraph of G_C^{m} which together with Theorem 3.1 implies that X satisfies also the global Markov property for the moral graph G_C^{m}.

The global AMP Markov property for G_C is stronger than the global Markov property for G_C^{m} or G_{PC}. To illustrate this we consider the simple causality graph $1 \rightarrow 3 \leftarrow 2$. With $A = \{1\}$, $B = \{2\}$, and $S = \emptyset$ the global AMP Markov property yields $X_1 \perp X_2$. On the other hand, the moral graph G_C^{m} is complete and consequently no conditional orthogonality relation can be read off the graph.

In TSC graphs the derivation of causal relations by the moralization criterion is based on the fact that the past and the present of a component at time t are represented by different subsets of vertices in the TSC graph. To preserve the causal ordering when moralizing causality graphs Eichler (2000) introduced a special separation concept for mixed graphs which is based on the idea of splitting the past and the present of certain variables and considering them together in a chain graph. For any subset B of $U \subseteq V$ this splitting of the past and the present for the variables in B can be accomplished by augmenting the moral graph $G_C \langle U \rangle^{\mathrm{m}}$ with new vertices b^* for all $b \in B$, which then represent the present of these variables, that is, b^* corresponds to $X_b(t)$, whereas all other vertices stand for the past at time t, for example a for $\overline{X}_a(t)$. Let $B^* = \{b^*|b \in B\}$ be the set of augmented vertices. The new vertices in B^* are joined by edges such that we obtain a chain graph with two chain components, $\mathrm{an}(U)$ and B^*, which reflect the AMP Markov properties of the process. More precisely we define the augmentation chain graph $G\langle U\rangle_{B^*}^{\mathrm{aug}}$ as the mixed graph $(\mathrm{an}(U) \cup B^*, E^{\mathrm{aug}})$ with edge set E^{aug} such that for all $a_1, a_2 \in U$ and $b_1, b_2 \in B$:

$$a_1 - a_2 \notin E^{\mathrm{aug}} \Leftrightarrow a_1 - a_2 \notin E\langle U\rangle^{\mathrm{m}},$$
$$a_1 \rightarrow b^* \notin E^{\mathrm{aug}} \Leftrightarrow a_1 \rightarrow b_1 \notin E,$$
$$b_1^* - b_2^* \notin E^{\mathrm{aug}} \Leftrightarrow b_1 \bowtie b_2 \,|\, B\backslash\{b_1, b_2\} \;[G^{\mathrm{u}}].$$

Let $Y = (Y_v)_{v\in\mathrm{an}(U)}$ and $Y^* = (Y_b^*)_{b\in B}$ be random vectors with components $Y_v = \overline{X}_v(t)$ and $Y_b^* = X_b(t)$. Then (Y, Y^*) satisfies the pairwise AMP Markov property with respect to $G\langle U\rangle_{B^*}^{\mathrm{aug}}$. This suggests defining the global Markov property for causality graphs by the global AMP Markov property in augmentation chain graphs.

Definition 3.6 X satisfies the global causal Markov property for the mixed graph $G = (V, E)$ if for all $U \subseteq V$ and all disjoint partitions A, B, C of U:

$$A \bowtie B^* \,|\, U\backslash A \;[(G\langle U\rangle_{B^*}^{\mathrm{aug}})^{\mathrm{m}}] \Rightarrow X_A \nrightarrow X_B \;[X_U],$$
$$A^* \bowtie B^* \,|\, S \cup C^* \;[(G\langle U\rangle_{U^*}^{\mathrm{aug}})^{\mathrm{m}}] \Rightarrow X_A \nsim X_B \;[X_U].$$

The assumptions on the spectral matrix in Section 2 now guarantee the equivalence of pairwise and global causal Markov properties.

Theorem 3.7 *Let $G_C = (V, E_C)$ be the causality graph of X. Then X satisfies the global causal Markov property with respect to G_C.*

For a proof see Eichler (2000, Theorem 3.8).

Example 2.2 (contd) As already mentioned, an intuitive interpretation of the causality graph in Fig. 2 suggests that X_1 has only an indirect effect on X_4, which is mediated by X_3. That this interpretation is indeed correct can now be shown by deriving the relation $X_1 \nrightarrow X_4 \;[X_{\{1,3,4\}}]$ from the corresponding augmentation chain graph $G\langle\{1,3,4\}\rangle_{\{4^*\}}^{\mathrm{aug}}$.

Since the ancestral set generated by $\{1,3,4\}$ is equal to the full set V we start from the moral graph G^{m} in Fig. 3. Augmenting the graph with a new vertex 4^* and joining this with vertex 4 and its parents 3 and 5 by arrows pointing towards 4^* we obtain the augmentation chain graph $G\langle\{1,3,4\}\rangle_{\{4^*\}}^{\mathrm{aug}}$ in Fig. 5(a). As the graph does not contain any flag or 2-biflag the removal of directions yields the moral graph $(G\langle\{1,3,4\}\rangle_{\{4^*\}}^{\mathrm{aug}})^{\mathrm{m}}$ in Fig. 5(b). In this graph the vertices 1 and 4^* are separated by the set $\{3,4\}$ and hence the desired non-causality relation follows from Theorem 3.7. Similarly, it can be shown that

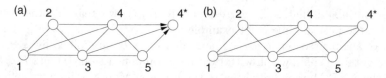

FIG. 5. Illustration of the global causal Markov property for the VAR in Example 2.2: (a) Augmentation chain graph $G\langle\{1,3,4\}\rangle_{\{4^*\}}^{\text{aug}}$ and (b) its moral graph $(G\langle\{1,3,4\}\rangle_{\{4^*\}}^{\text{aug}})^{\text{m}}$.

X_1 and X_4 are contemporaneously partially uncorrelated relative to the same sub-process $X_{\{1,3,4\}}$ (Eichler 2000, Example 3.9).

3.4 The p-separation criterion

The moralization criterion in the previous sections is not a separation criterion in G_{TS} of G_{C} itself because the ancestral graphs and augmented moral graphs appearing in the definitions of the global AMP and global causal Markov properties vary with the subsets A, B, and S. For AMP chain graph models Levitz *et al.* (2001) presented a pathwise separation criterion, called p-separation, which is similar to the d-separation criterion for directed acyclic graphs. In Eichler (2001) it is shown that the global AMP Markov property and the global causal Markov property for causality graphs equivalently can be formulated by an adapted version of the p-separation criterion.

4 Inference for graphical time series models

4.1 Conditional likelihood for graphical autoregressive models

We first discuss the case of fitting VAR(p) models restricted with respect to a time series chain graph. In practice the true TSC graph is unknown and one has to apply model selection strategies to find the best approximation for the true graph. Depending on the model discrepancy this leads to different model selection criteria such as AIC or BIC, the Akaike or Bayesian Information Criteria. More precisely, let $G = (V_{\text{TS}}, E)$ be a graph from the class $\mathcal{G}_{\text{TS}}(p)$ of all time series chain graphs whose edges have at most lag p and which are invariant under translation in the sense that for all $a, b \in V$ and all $t, s, u \in \mathbb{Z}$ we have $(a, t-u) \longrightarrow (b, t) \notin E$ if and only if $(a, s-u) \longrightarrow (b, s) \notin E$ and $(a, t) \!-\! (b, t) \notin E$ if and only if $(a, s) \!-\! (b, s) \notin E$. We consider Gaussian VAR(p) models of the form

$$X(t) = A(1)X(t-1) + \cdots + A(p)X(t-p) + \varepsilon(t), \qquad \varepsilon(t) \overset{\text{iid}}{\sim} \mathcal{N}(0, \Sigma_G),$$

where the parameters $A_G = (A(1), \ldots, A(p))$ and $K_G = \Sigma_G^{-1}$ satisfy

$$A_{ba}(u) = 0 \quad \text{if } (a, t) \longrightarrow (b, t+u) \notin E \qquad \text{and} \qquad K_{ab} = 0 \quad \text{if } (a, t) \!-\! (b, t) \notin E.$$

We call a VAR model with these constraints on the parameters a VAR(p, G) model.

Given observations $X(1), \ldots, X(T)$ let $T_0 = T - p$. The conditional log-likelihood function of a Gaussian VAR(p, G) model for the data can be written as (Lütkepohl 1993)

$$\mathcal{L}_T\left(A_G, K_G\right) = \frac{d}{2}\log(2\pi) - \frac{1}{2}\log|K_G| + \frac{1}{2T_0}\sum_{t=p+1}^{T} e(t)' K_G e(t),$$

where $e(t) = X(t) - \sum_{u=1}^{p} A(u)X(t - u)$. Differentiating the log-likelihood we obtain equations for the maximum likelihood estimate which can be solved numerically by an iterative algorithm (Dahlhaus and Eichler 2001). Furthermore, we can use model selection criteria to select the optimal VAR(p, G) model among all models with $1 \leqslant p \leqslant P$ and $G \in \mathcal{G}_{\text{TS}}(p)$. Since in practice the number of possible models is too large one has to use special model selection strategies.

Alternatively we can fit VAR(p) under the restrictions of a mixed graph G from the class of all causality graphs $\mathcal{G}_C$. Such models are a special case of the VAR(p, G) models since by Proposition 2.5 for each $G \in \mathcal{G}_C$ there exists a chain graph $G' = (V_{\text{TS}}, E') \in \mathcal{G}_{\text{TS}}(p)$ which implies exactly the same constraints on the parameter as G, i.e. we have for all $t \in \mathbb{Z}$ $(a, t) \relbar (b, t) \notin E'$ if and only if $a \relbar b \notin E$ and $(a, t) \rightarrow (b, t + u) \notin E'$ if and only if $u > p$ or $a \rightarrow b \notin E$.

4.2   Partial directed and partial contemporaneous correlation

We now present a non-parametric approach for estimating TSC graphs that is based on the definition of the TSC graph itself. For random variables X, Y, and Z let $\text{corr}_{\text{L}}(X, Y|Z)$ denote the correlation between X and Y after the linear effects of Z have been removed. Thus we have $\text{corr}_{\text{L}}(X, Y|Z) = 0$ if and only if $X \perp Y \mid Z$. We define the partial directed correlation at lag $u > 0$ by

$$\pi_{ba}(u) = \text{corr}_{\text{L}}\left(X_b(t), X_a(t - u)\big|\overline{X}_V(t)\backslash\{X_a(t - u)\}\right)$$

and the partial contemporaneous correlation by

$$\pi_{ba}^{\circ} = \text{corr}_{\text{L}}\left(X_b(t), X_a(t)\big|\overline{X}_V(t) \cup \{X_{V\backslash\{a,b\}}(t)\}\right).$$

Then the time series chain graph G_{TS} of a process $X = \{X(t)\}$ can be characterized in terms of the partial directed and partial contemporaneous correlations,

$$(a, t - u) \rightarrow (b, t) \notin E_{\text{TS}} \Leftrightarrow \pi_{ba}(u) \neq 0,$$
$$(a, t) \relbar (b, t) \notin E_{\text{TS}} \Leftrightarrow \pi_{ba}^{\circ} \neq 0.$$

When estimating the partial directed and partial contemporaneous correlations from observations $X(1), \ldots, X(T)$ it is clear that only the observed part of the past can be taken into account. Therefore we define

$$\widehat{\pi}_{ba}(u) = \widehat{\text{corr}}_{\text{L}}\left(X_b(p), X_a(p - u)\big|\{X(1), \ldots, X(p-1)\}\backslash\{X_a(p - u)\}\right)$$

for some fixed p, with a similar definition for $\widehat{\pi}_{ba}^{0}$. In Dahlhaus and Eichler (2001) an iterative scheme for the computation of these estimators has been suggested.

The estimators $\widehat{\pi}_{ba}(u)$ and $\widehat{\pi}_{ba}^{0}$ can be used to test for the existence of an edge in the TSC graph G_{TS}. More precisely, if the edge $(a, t - u) \rightarrow (b, t)$ is absent in the true TSC graph, then $T\widehat{\pi}_{ba}^{2}(u)$ is asymptotically χ_{1}^{2}-distributed. Similarly, we can test for the absence of undirected edges in the TSC graph using the partial contemporaneous correlation. For model selection these tests have to be applied repeatedly for various edges, which raises the usual problems associated with multiple testing.

In Proposition 2.5 we have seen that aggregation of the edges in the TSC graph leads to the causality graph of the process. The idea of aggregation suggests using the sum of $\widehat{\pi}_{ba}^{2}(u)$ to test for the absence of a directed edge in the causality graph. Let $\widehat{R}_{p}$ be the empirical covariance matrix of X and $\widehat{H}$ a $p \times p$ matrix with entries $\widehat{H}_{uv} = (\widehat{R}_{p}^{-1})_{bb}(u - v)/(\widehat{R}_{p}^{-1})_{bb}(0)$. Correcting for the dependence between the $\widehat{\pi}_{ba}(u)$ at different lags we then obtain as a test statistic:

$$S_{ab} = T\, \widehat{\pi}_{ba}'\, \widehat{H}^{-1}\widehat{\pi}_{ba},$$

where $\widehat{\pi}_{ba}$ is the vector of partial directed correlations. Under the null hypothesis that X_{a} is non-causal for X_{b} the test statistic is asymptotically χ_{p}^{2}-distributed. An equivalent test statistic has been presented by Tjøstheim (1981) for testing for causality in multivariate time series, without considering causality relations between the variables at the different time points themselves.

4.3   Partial correlation graphs

Inference for PC graphs is in a certain sense more complicated than for TSC graphs or causality graphs. Suppose, for example, that one wants to test by a likelihood ratio test the hypothesis that the edge $a - b$ is missing in the PC graph of the VAR model in Example 2.2. This requires calculation of the maximum likelihood estimator under the restrictions (2.4). There does not exist an analytic solution for this problem, whereas numerical solutions seem to be difficult due to the particular non-linear structure of the restrictions.

Instead we suggest a non-parametric method based on the characterization of a missing edge by a vanishing partial spectral coherence in Proposition 2.7. By this proposition the partial spectral coherence can be obtained by inversion and rescaling of the spectral density matrix. We therefore estimate the spectral matrix and invert and rescale this estimate. As an estimator for $f_{ab}(\lambda)$ we take a kernel estimator based on the periodogram (cf. Dahlhaus 2000).

We then test the hypothesis $a - b \notin E_{PC}$ by using an approximation of the distribution of $\sup_{\lambda \in [-\pi,\pi]} |\widehat{R}_{ab|V\backslash\{a,b\}}(\lambda)|^{2}$ under the hypothesis, i.e. under $R_{ab|V\backslash\{a,b\}}(\lambda) = 0$ (for more details see Dahlhaus 2000, Section 5).

Alternatively the test can be based in the time domain on the partial correlations of X_{a} and X_{b} after removing the linear effects of $X_{V\backslash\{a,b\}}$ (see Eichler *et al.* 2000).

5   Applications

In order to illustrate the methods presented in this article we discuss two real data examples from chemistry and neurology. For further examples we refer to Gather *et al.* (2002), who analysed time series from intensive care monitoring with the help of partial correlation graphs, and Eichler *et al.* (2000), who discussed the identification of neural connectivities from spike train data.

5.1   Application to air pollution data

The method was used to analyse a five-dimensional time series of length 35 088 of air pollutants recorded half-hourly from January 1991 to December 1992 in Heidelberg. The recorded variables were CO and NO (mainly emitted from cars, house-heating, and industry), NO_2 and O_3 (created in different reactions in the atmosphere), and the global radiation intensity I_{rad}, which plays a major role in the generation of ozone. Details on these reaction can be found, for example, in Seinfeld (1986, Chapter 4). Figure 6 shows the daily course of the five variables averaged over 61 successive days in summer. CO and NO increase early in the morning due to traffic and, as a consequence, NO_2 also increases. O_3 increases later due to the higher level of NO_2 and the increase in the global radiation. Figure 6 indicates that all variables are correlated at different lags.

The following analysis is based on the residual series after subtracting the (local) average course as shown in Fig. 6. The original series contained a few missing values (less than 2%) which were completed by interpolation of the residual series with splines.

Figure 7 shows above diagonal estimates for the correlations between the variables at different lags and below diagonal estimates for the partial directed and partial contemporaneous correlations as discussed in Section 4.2. Since $\pi_{ab}(u)$ is defined only for $u > 0$ we have plotted $\pi_{ba}(-u)$ for $u < 0$ and π_{ab}^0 for $u = 0$. Furthermore, horizontal dashed lines indicate simultaneous 95% test bounds for the test $\pi_{ab}(u) = 0$ ($\pi_{ab}^0 = 0$). These dashed lines are close to zero and hardly visible. The test bounds lead to the TSC graph (not plotted) and

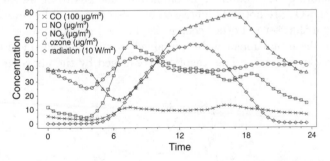

FIG. 6. Average of daily measurements over 61 days in summer.

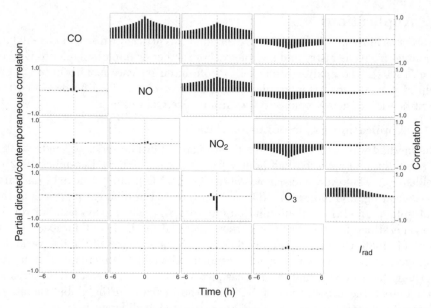

FIG. 7. Correlations (above diagonal) and partial directed and partial contemporaneous correlations (below diagonal) for air pollution data.

the causality graph in Fig. 8.

The graph reflects the creation of O_3 from NO_2 and the decrease in O_3 leading to an increase in NO_2 (indicated by the very strong negative partial contemporaneous correlation between O_3 and NO_2 in Fig. 7). The graph also confirms that global radiation plays a major role in the process of O_3 generation. Furthermore, there is a very strong partial contemporaneous correlation between CO and NO (both are emitted from cars, etc.). The meaning of the other edges, and of some of the missing edges, is less obvious. Chemical reactions between air pollutants are very complex. In particular, one has to be aware of the fact that NO_2 and O_3 are not only increased but also decreased by several chemical reactions and that several other chemicals play an important role.

Some of these reactions can be explained by the photochemical theory (Seinfeld 1986, Section 4.2). This theory is confirmed by the above graph. First,

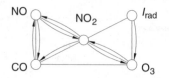

FIG. 8. Causality graph for the air pollution data.

the edge between I_{rad} and NO_2 represents the photolysis of NO_2. Second, the edges between NO and NO_2 are present but Fig. 7 show that the partial directed and partial contemporaneous correlations are less strong than the correlation between CO and NO_2. This indicates that mainly the concentration of CO (and not of NO) is responsible for the generation of NO_2, which means that NO_2 is mainly generated via a radical reaction (where CO is involved) and not in a direct reaction (where CO is not involved). This also is in accordance with the photochemical theory for ozone generation.

The example shows the effect of the discretization interval (30 minutes) on the graph. Obviously this interval is too large to resolve the direction in the chemical reactions from the time flow. Thus most of the causality is detected as undirected partial contemporaneous correlation instead of directed Granger causality. There is one exception: the strong partial contemporaneous correlation between CO and NO is due to the fact that both are emitted at the same time from cars, i.e. we have a confounder which is causal for both CO and NO.

The example also reveals the need for a more quantitative analysis in that one wants to discriminate between major effects (such as $NO_2 — O_3 — I_{rad}$) and minor effects (such as $CO — O_3$) leading to the idea of edges with different grey levels or colours reflecting the 'strength' of a connection. This has to be pursued, in particular for applications with large data sets.

An analysis of the same data set discretized at a 4 hour interval with PC graphs can be found in Dahlhaus (2000).

5.2 Application to human tremor data

The second example in this section is concerned with the identification of neurological signal transmission pathways in the investigation of human tremor. Tremor is defined as the involuntary, oscillatory movement of parts of the body, mainly the upper limbs. It has been shown that patients suffering from Parkinson's disease show a tremor-related cortical activity which can be detected in the EEG time series by cross-spectral analysis (Timmer *et al.* 2000).

In one experiment the EEG and the surface electromyogram (EMG) of the left-hand wrist extensor muscle in a healthy subject have been measured during an externally enforced oscillation of the hand with 1.9 Hz. The EMG signal was band-pass filtered to avoid aliasing effects and undesired slow drifts. Additionally the signal was digitally full wave rectified. The resulting time series reflects the muscle activity encoded in the envelope of the originally measured signal. The EEG recordings were performed using a 64-channel EEG system. The potential field measured over the scalp was transformed into the reference free current distribution which reflects the underlying cortical activity by calculating second spatial derivatives. All data were simultaneously sampled at 1000 Hz. For the analysis a subseries of length $T = 40\,000$ has been used.

In a preliminary analysis significant coherence has been found between the EMG and the EEG of channels P1 and C2P. Figure 9(a) displays estimates for the spectra on the diagonal and estimates for the ordinary and the partial spectral

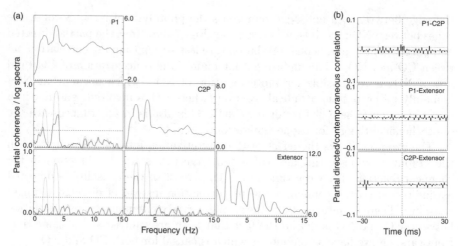

FIG. 9. Human tremor data. (a) On diagonal: logarithm of auto-spectra; below
    diagonal: spectral coherences (dotted) and partial spectral coherences (solid).
    (b) Partial directed and partial contemporaneous correlations.

coherence between the respective time series below the diagonal. The higher
harmonics in the spectra indicate that the process has a non-linear dynamic.

A significant coherence at the 5% level is found at the tremor frequency
and its higher harmonics between all three channels. The partial coherence
between the extensor EMG and the EEG of channel P1 falls below the 5%
threshold. Consequently there is no edge between P1 and the extensor in the
corresponding partial correlation graph, which is given in Fig. 10(a). The graph
suggests that the coherence between P1 and the EMG signal does not represent
a direct connection, but that the signal transfer is mediated via C2P.

For a further investigation of the interrelations and in particular the directions of signal transmission, we also estimated the partial directed and partial
contemporaneous correlations given in Fig. 9 (b). The processes seem to be only
weakly correlated and the interrelations that were clearly discernible in the frequency domain are almost undetectable in the time domain. Using the threshold
for the identification of the edges we obtain the causality graph in Fig. 10(b).
The graph reveals that the cortical activity at location C2P is directly caused by
the oscillation of the hand, whereas the cortical activity at P1 is only indirectly

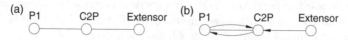

FIG. 10. (a) Causality graph and (b) partial correlation graph for human tremor
    data.

influenced via C2P. The edges between P1 and C2P indicate that signals between the two regions in the cortex are transmitted in both directions with very short delay. On the other hand, the delay of about 25 ms in the transmission of signals from the extensor to C2P corresponds quite well to the length of the pathway.

The example shows that due to the periodic nature of tremor, oscillation frequency domain methods are much better suited for identifying the corresponding neurological connectivities. Therefore it seems to be advantageous to have a frequency domain method to estimate the causality graphs. A first step in that direction is partial directed coherence (Sameshima and Baccalá 1999), but so far there does not exist any test based on this statistic. In the context of neurophysiological point process data Dahlhaus *et al.* (1997) suggested an indirect method: they first identified the PC graph and then in addition used the partial phase information to infer the directions in the interrelation between the point processes.

6   Discussion

The recent development of causal Markov models has shown that graphical models provide a general framework for describing causal relations. In this article our aim was to contribute to the discussion of causality concepts and their relation to graphical models. Focusing on the temporal aspect of causality – an effect cannot precede its cause in time – we based our approach to causal modelling on the concept of Granger causality. Causal Markov models are then defined by merging Granger causality with mixed graphs.

In applications the causality graph describing the causal relations between the variables is typically unknown. Owing to the time series setting, the directions of the edges in the graph and thus of the causal effects can be identified from the data. However, there are also various problems which may lead to the wrong identification of causal effects.

First, Granger causality measures only the association between variables at different times and therefore cannot distinguish between direct causal effects and an association that is due to an unmeasured confounder. This problem was discussed briefly by Granger (1969) when introducing his notion of causality. Granger assumed in his definition that one conditions on 'all the available information in the universe' and he was of course aware of the fact that in practice inference is only based on the available information contained in the process under study. If the effects of a confounder on two variables are lagged in time, then the resulting association between lagged variables falsely leads to a directed edge between the corresponding vertices in the causality graph. However, in certain situations the global causal Markov properties can be used to distinguish between a direct causal and a confounding effect. As an example we consider the human tremor data in Section 5.2. The causality graph in Fig. 10 shows a direct feedback between the two EEG channels P1 and C2P. The two channels might also be influenced by a third region in the brain. On the other hand, a bivariate analysis of P1 and the extensor shows that the latter is still causal for the former. This contradicts the assumption of a confounder which would imply that the hand movements are non-causal for the cortical activity in channel P1. Therefore C2P must have a direct causal effect on P1, but we cannot exclude that part of the association is due to a confounder. For more details see Section 4 of Eichler (2000).

Second, the identification of non-linear causal effects is hampered by the restriction to linear methods. Here, graphs defined in terms of conditional independence or strong Granger causality provide an alternative that captures the complete dependence structure of the process. The results in Sections 2 and 3 also hold for these graphs with one notable exception: the pairwise and the global Markov properties for the TSC and the causality graphs in general are no longer equivalent, only under an additional assumption on the process. Alternatively the graphs can be defined in terms of the so-called block-recursive Markov properties, which are still equivalent to the global ones. For details we refer the reader to Section 5 of Eichler (2000) and Eichler (2001). However, we mention that it is hardly possible to make inference for arbitrary non-linear causal relationships since strictly speaking we have only one observation of the multivariate process. This problem could only be overcome by assuming specific non-linear time series models.

A third limitation arises from the fact that the measured variables typically evolve continuously over time but are observed only at discrete times. If the time between a cause and its effect is shorter than the sampling interval, then the corresponding discretized component series are only contemporaneously correlated and a directed edge is detected falsely as an undirected one. (We note that contemporaneous correlation may also be due to confounding by an unmeasured process.) For the same reasons an indirect causality mediated by measured variables may lead to a directed edge in the causality graph.

The third problem can be avoided by replacing the discrete version of Granger causality by local Granger causality and local instantaneous causality, which were introduced by Comte and Renault (1996) for the investigation of causal relations between time-continuous processes. We note that local Granger causality is related to the notion of local independence (Schweder 1970; Aalen 1987), which has been used by Didelez (2001) for the definition of directed graphs for marked point processes.

Finally, we note that the concept of Granger causality is not restricted to stationary time series but also applies to more general situations such as non-stationary time series, panels of time series, or point processes. We conjecture that the graphical modelling approach can be applied in these cases in a similar way.

References

Aalen, O. O. (1987). Dynamic modelling and causality. *Scandinavian Actuarial Journal*, 177–90.

Aigner, D. J. and Zellner, A. (eds.) (1988). Causality. *Journal of Econometrics*, **39**, 1–234.

Andersson, S. A., Madigan, D., and Perlman, M. D. (2001). Alternative Markov properties for chain graphs. *Scandinavian Journal of Statistics*, **28**, 33–85.

Brillinger, D. R. (1981). *Time Series: Data Analysis and Theory*, McGraw Hill, New York.

Comte, F. and Renault, E. (1996). Noncausality in continuous time models. *Econometric Theory*, **12**, 215–56.

Dahlhaus, R. (2000). Graphical interaction models for multivariate time series. *Metrika*, **51**, 157–72.

Dahlhaus, R., Eichler, M., and Sandkühler, J. (1997). Identification of synaptic connections in neural ensembles by graphical models. *Journal of Neuroscience Methods*, **77**, 93–107.

Dahlhaus, R. and Eichler, M. (2001). Statistical inference for time series chain graphs. In preparation.

Dawid, A. P. (2000). Causal inference without counterfactuals. *Journal of the American Statistical Association*, **86**, 9–26.

Dempster, A. P. (1972). Covariance selection. *Biometrics*, **28**, 157–75.

Didelez, V. (2001). *Graphical Models for Event History Analysis Based on Local Independence*. PhD thesis. Logos, Berlin.

Eichler, M. (1999). *Graphical Models in Time Series Analysis*. PhD thesis, University of Heidelberg, Germany.

Eichler, M. (2000). Granger-causality graphs for multivariate time series. Technical Report, University of Heidelberg, Germany.

Eichler, M. (2001). Markov properties for graphical time series models. Technical Report, University of Heidelberg, Germany.

Eichler, M., Dahlhaus, R., and Sandkühler, J. (2000). Partial correlation analysis for the identification of synaptic connections. Technical Report, University of Heidelberg, Germany.

Frydenberg, M. (1990). The chain graph Markov property. *Scandinavian Journal of Statistics*, **17**, 333–53.

Gather, U., Imhoff, M., and Fried, R. (2002). Graphical models for multivariate time series from intensive care monitoring. *Statistics in Medicine*, **21**, 2685–701.

Geweke, J. (1984). Inference and causality in economic time series. In *Handbook of Econometrics 2* (eds. Z. Griliches and M. D. Intriligator), pp. 1101–44. North-Holland, Amsterdam.

Granger, C. W. J. (1969). Investigating causal relations by econometric models and cross-spectral methods. *Econometrica*, **37**, 424–38.

Haavelmo, T. (1943). The statistical implications of a system of simultaneous equations. *Econometrica*, **11**, 1–12.

Hsiao, C. (1982). Autoregressive modeling and causal ordering of econometric variables. *Journal of Economic Dynamics and Control*, **4**, 243–59.

Lauritzen, S. L. (1996). *Graphical Models*. Oxford University Press, UK.

Lauritzen, S. L. (2001). Causal inference from graphical models. In *Complex Stochastic Systems* (eds E. Barndorff-Nielsen, D. R. Cox, and C. Klüppelberg). CRC Press, London.

Lauritzen, S. L. and Wermuth, N. (1989). Graphical models for association between variables, some of which are qualitative and some are quantitative. *Annals of Statistics*, **17**, 31–57.

Levitz, M., Perlman, M. D., and Madigan, D. (2001). Separation and completeness properties for AMP chain graph Markov models. *Annals of Statistics*, **29**, 1751–84.

Lütkepohl, H. (1993). *Introduction to Multiple Time Series Analysis*. Springer-Verlag, Berlin.

Lynggaard, H. and Walther, K. H. (1993). *Dynamic Modelling with Mixed Graphical Association Models*. Masters Thesis, Aalborg University.

Neyman, J. (1935). Statistical problems in agricultural experimentation (with discussion). *Journal of the Royal Statistical Society Supplement*, **2**, 107–80.

Pearl, J. (1988). *Probabilistic Inference in Intelligent Systems*. Morgan Kaufmann, San Mateo, CA.

Pearl, J. (1995). Causal diagrams for empirical research (with discussion). *Biometrika*, **82**, 669–710.

Pearl, J. (2000). *Causality*. Cambridge University Press, UK.

Robins, J. (1986). A new approach to causal inference in mortality studies with sustained exposure periods – application to control of the healthy worker survivor effect. *Mathematical Modelling*, **7**, 1393–512.

Rubin, D. B. (1974). Estimating causal effects of treatments in randomized and nonrandomized studies. *Journal of Educational Psychology*, **66**, 688–701.

Sameshima, K. and Baccalá, L. A. (1999). Using partial directed coherence to describe neuronal ensemble interactions. *Journal of Neuroscience Methods*, **94**, 93–103.

Schweder, T. (1970). Composable Markov processes. *Journal of Applied Probability*, **7**, 400–10.

Seinfeld, J. H. (1986). *Atmospheric Chemistry and Physics of Air Pollution*. Wiley, Chichester.

Timmer, J., Lauk, M., Köster, B., Hellwig, B., Häußler, S., Guschlbauer, B., Radt, V., Eichler, M., Deuschl, G., and Lücking, C. H. (2000). Cross-spectral analysis of tremor time series. *International Journal of Bifurcation and Chaos*, **10**, 2595–610.

Tjøstheim, D. (1981). Granger-causality in multiple time series. *Journal of Econometrics*, **17**, 157–76.

Wright, S. (1921). Correlation and causation. *Journal of Agricultural Research*, **20**, 557–85.

4A
Graphical models for stochastic processes

Vanessa Didelez
University College London, UK

1   Introduction

Graphical models have proven to be a valuable concept in various fields of multivariate data analysis, as underpinned by several contributions to this book. However, the pure definition of conditional independence graphs is not satisfactory for the representation of two closely related dependence concepts: causal and dynamic dependence – both sharing for instance the property of being asymmetric, as opposed to conditional dependence. In graphical models directed edges symbolize specific sets of conditional independence hypotheses. Thus, without additional assumptions they neither imply nor assume causal relations or a temporal ordering of the variables. We may even construct examples where due to latent variables or to a selection process the directions of the edges do not coincide with the flow of time or causality (Lauritzen and Richardson 2002). Therefore, extensions and modifications of conditional independence graphs have been developed. The main approaches using graphical models for an adequate representation of causal relations are due to Spirtes *et al.* (1993) and Pearl (1995, 2000). Recently, graphical models have also been modified to cope with dynamic dependencies among stochastic processes (Dahlhaus 2000; Eichler 2000; Didelez 2001).

   In what follows I want to focus on two aspects. First, the different notions of causality occurring in the indicated literature are compared. Second, the specific problems arising in the continuous-time situation are addressed.

2   Granger causality versus intervention causality

The causality graphs introduced by Eichler (2000) are based on dynamic dependencies among the components of multivariate time series. These dependencies are defined in terms of Granger causality (Granger 1969):

Granger causality   Let $\{\Omega_t\}$ be all the relevant information in the universe. Then, a sub-process $\{X_t\} \subset \{\Omega_t\}$ is causal for another process $\{Y_t\}$ if the optimal prediction w.r.t. $\{Y_t\}$ is less precise when based on $\{\Omega_t\}\backslash\{X_t\}$ instead of $\{\Omega_t\}$.

   In other words, only a process that is *directly* relevant for an optimal prediction is regarded as a cause. Obviously, the assumption of no unmeasured confounders, and thus the identifiability of Granger causes, is highly questionable in any real data situation (cf. the example by Dahlhaus and Eichler in

Chapter 4 where CO and NO are correlated because both are emitted by cars). In any given setting with finite information and without background knowledge on the causal structure, the analysis is instead mostly restricted to associations between lagged variables. Granger causality nevertheless provides a very intuitive concept of dynamic association which can be generalized to a large class of stochastic processes, as addressed below.

In contrast, the causality concept prevailing in the literature on graphical models or Bayesian networks is usually concerned with the effects of *interventions*, as opposed to the pure observation of a condition. For stochastic processes an intervention could mean that the values of a process at each point in time are set to values that have been fixed in advance. We could for instance think of persons suffering from a chronic disease who have to follow a specific scheme when taking the prescribed drugs over an indefinite period of time. However, the definition of a dynamic intervention should also allow for interventions taking place only at selected points in time instead of all the time, and for interventions to be conditional on the past (cf. Pearl 1994). An approach to the analysis of such sequential plans for longitudinal data using graphs has been proposed by Pearl and Robins (1995). Without going into the technical details, let me formulate an alternative definition of dynamic causality as follows.

Intervention causality A process $\{X_t\}$ is causal for $\{Y_t\}$ if an intervention w.r.t. the former affects the prediction of the latter.

Although not explicit in this definition, the identifiability of some causal effect might require that specific covariate processes have been measured, as shown for special cases by Pearl and Robins (1995) or Parner and Arjas (1999). But in contrast to Granger causality these do not necessarily include 'all the information in the universe'. Additionally, in the above definition it does not matter whether the effect of an intervention is due to a direct influence from $\{X_t\}$ to $\{Y_t\}$ or due to an indirect pathway. Thus (assuming no unobserved confounding process), any Granger cause is also a cause w.r.t. intervention but not vice versa.

Following the tradition of intervention graphs or influence diagrams, dynamic graphical models could be extended to take interventions into account, for instance by adding intervention nodes (Dawid, Chapter 2). As for the static situation, this should simplify the derivation of appropriate formulae for causal effects of interventions as well as the deduction of conditions for their identifiability. Note that the effect of an intervention can be regarded as a causal parameter (Pearl 2000, p. 39) which is to be estimated from observational data. This does not necessarily require the actual feasibility of an intervention.

3   The continuous-time situation

As noted by Dahlhaus and Eichler in Chapter 4, a drawback of the different graphical models for time series is their sensitivity to the choice of intervals between the measurements. Larger intervals correspond to marginalizing over the time in between. On the one hand, this might hide genuine short-term correlations. On the other hand, marginalization will typically create additional correlations due to common causes or mediating events occurring in the meantime. One is even led to suppose that the undirected edges in the TSC or causality graphs indicating instantaneous causality (being less obvious but also relevant in the PC graphs) are only due either to such meantime effects or to whole unobserved processes. Not only does this challenge the interpretation in terms of causality – in the sense of Granger or of interventions – but additionally, the alternative Markov properties that apply to TSC graphs do not agree with the foregoing explanation of instantaneous causality. It has been shown that general AMP chain graphs cannot be generated by marginalizing directed acyclic graphs representing the underlying data-generating process (Richardson 1998). Yet, it is not clear how else instantaneous causality could arise.

In addition to the foregoing problems regarding the interpretation of graphical models for time series, it is difficult to follow the advice of choosing the intervals small enough in order to prevent spurious correlation when the underlying process is continuous in time. In this situation one should instead consider a generalization of the idea of dynamic dependencies inherent in Granger causality to the continuous-time situation. The basic approach goes back to Schweder (1970) who introduced the notion of *local independence* for Markov processes. Aalen (1987) extended this to stochastic processes that admit a Doob–Meyer decomposition and Florens and Fougère (1996) showed the analogy to Granger causality. Local independence reads as follows.

Local independence A process $\{Y_t\}$ is said to be locally independent of $\{X_t\}$ given some further information $\{Z_t\}$ if the predictable part of $\{Y_t\}$ (technically, its compensator Λ_t^Y) remains the same regardless of whether it is conditional on the past $\mathcal{F}_{t-}^{XYZ}$ of all three processes or only on the past $\mathcal{F}_{t-}^{YZ}$ of $\{Y_t\}$ and $\{Z_t\}$.

Aalen (1987) additionally assumed that the innovations of the involved processes are uncorrelated because a notion of independence between processes would make no sense if they were fed by the same innovations. This assumption should in practice be carefully checked since it may again be violated by some unmeasured common causal process. Note that the above definition does not refer to 'all information in the universe'. Consequently, it does not claim to define causal relations.

A graphical representation of local independence structures is straightforward by representing the components of a multivariate process as vertices and drawing directed edges for any local dependence whilst omitting the edge when there is a local independence. The resulting local independence graphs have been analysed for the special case of counting processes in Didelez (2001). The global Markov properties in these graphs can be obtained by the application of a newly developed separation criterion called *δ-separation* which takes into account that local independence is asymmetric and that the graphs may be cyclic. The augmentation of the graphs with intervention nodes for the computation and identification of causal effects seems to be a promising approach.

While it is often assumed that point processes can be observed in continuous time by simply registering the time of each event, this rarely seems possible for other types of stochastic processes that vary steadily in time. As an example one might consider the medical time series typically arising in intensive care online monitoring which are often recorded in intervals of a few seconds. A graphical representation of the dependence structure would have to take into account that the time between consecutive measurements is always unobserved. Thus, a class of graphical models generated by marginalizing suitable continuous-time models over the intervals is called for. This should be subject to future research.

4B
Conditional association in time series

Hans R. Künsch
ETH Zurich, Switzerland

1 Concepts of conditional association in time series

Because of the natural order of time, different definitions of conditional associations (and lack thereof) between variables exist naturally, as has been discussed in this article. In the partial correlation graph, one is looking at an association between two variables at all times conditional on other variables also at all times. The other concepts introduced in this article take the time aspect into account. The directed edges in a causality graph describe the existence of an association between the current value of a variable X_b and all past values of another variable X_a, conditional on all past values of all variables except X_a. The undirected edges in the same graph describe the existence of an association between the current values of two variables X_a and X_b conditional on the past values of all variables and the current values of all variables other than X_a and X_b. Finally the directed edges in a time series chain graph describe the existence of an association between the current value of a variable X_b and the value of a variable X_a at a fixed lag in the past conditional on all past values of all variables except X_a at this fixed lag. Undirected edges in a times series chain graph describe the same associations as the undirected edges in a causality graph. The passage from the time series chain graph to the corresponding causality graph is straightforward by aggregation, see Proposition 2.5, at least if one restricts oneself to *linear* conditional orthogonality. Since pairwise independence does not imply joint independence, more care is needed in the case of conditional independence, but the authors have shown that the result remains valid.

The passage from the causality graph to the partial correlation graph is more subtle, as is shown in Section 3. Example 2.2 shows that a naive interpretation of a causal graph as an indication of the flow of information can be wrong: in the causal graph of Fig. 2 there is no direct flow between 1 and 4, but the corresponding partial correlation graph of Fig. 3 has an edge between 1 and 4. The flow of information between variables 4 and 1 is indirect in the sense that it goes via the *innovation* of variable 2, which is determined by the past of variable 4 and the past and present of variable 2. This follows also from the calculation using (2.4): the edge between 1 and 4 arises because K_{12} and $A_{24}(1)$ are both non-zero.

As the authors have shown, the partial correlation graph can be inferred from the spectrum. In the spectral domain, one could in addition consider associations restricted to a particular frequency band, thus enlarging the types of association graphs further. For instance, in EEG series it is common practice to look at different frequency bands separately. In the human tremor data of Section 5.2, the association between the extensor and C2P occurs mainly at the lowest harmonic, whereas the association between C2P and P1 occurs mainly at the second harmonic. I am not sure whether it is legitimate to conclude from this that there must be some association between the different harmonics in C2P. I also wonder which other associations can be deduced in the spectral domain. For instance, the phase spectrum is linear if there is a lagged relation between two variables with the slope indicating the delay, and according to the last sentence in Section 5.2 this can be used to identify directions of association.

2   Discovery and interpretation of structure

The first main use of graphical models is the discovery and interpretation of structure in data. Most time series are observational studies which lead to well-known difficulties. First, the cases where simple chain models give a satisfactory description of the data are comparatively rare. This is of course related to the fact that confounders are omnipresent. Even if we know which other variables are potentially important, we often have no means to measure them. For instance, in the air pollution example, data are considered from only one measuring station and thus no transport phenomena are taken into account. A related problem is the difference between conditioning by observation and conditioning by intervention. This is for instance the basic problem in the analyisis of economic time series.

State space and hidden Markov models are an attempt to include unobserved variables explicitly in the analysis. As long as one has a rather precise idea which variables should be included and what their time evolution looks like, they provide a powerful tool, in particular with the recent progress in handling non-linear and non-Gaussian situations; see, for example, Berzuini and Gilks (Chapter 7 in this volume). However, using these models in a non-parametric way to find out about the number and structure of variables that describe a system under study suffers from severe lack-of-identifiability problems.

3 Reducing the number of parameters

Most multivariate time series models suffer from a large number of parameters. For instance, an unrestricted vector autoregressive process of order p with d variables has pd^2 autoregressive parameters, $d(d-1)/2$ parameters for the variance–covariance matrix of the innovations, and d mean values. Similarly a Markov chain of order p with m states has $m^p(m-1)$ parameters if the transition probabilities are arbitrary. Clearly, in these examples, increasing p by one means a large increase in parameters. One therefore would like to have intermediate models. Time series chain graphs provide this by allowing some of these parameters to be zero. Other approaches are the inclusion of a moving average part, the use of canonical correlations, and reduced rank autoregressions; see, for instance, Akaike (1976), Velu *et al.* (1986), and Ahn and Reinsel (1988). In contrast to the approach here, they look for a small number of arbitrary linear combinations of variables in the past that are sufficient to predict the present. Yet another approach to obtain parsimonious models for multivariate time series has been discussed in Tiao and Tsay (1989). For Markov chains with discrete states, a related approach to reduce the number of unknown parameters is given by variable length Markov chains, see Rissanen (1986) and Bühlmann and Wyner (1999).

4 Model fitting

One crucial issue in model fitting is stationarity. In this article, the strategy is to reduce the data first to stationarity and then to investigate associations between variables. The extensive work on cointegration makes clear that this is not always the best strategy. Also, in the air pollution example, similar questions arise: How much information about associations between variables is already contained in the average curves of Fig. 6? How good is the reduction to stationarity? It is likely that the reduced data have at least a weekly pattern reflecting the different emission patterns over the weekends.

The non-parametric approach of this article is based on testing whether the partial correlations $\pi_{ab}(u)$ can be considered to be zero. I feel a little uneasy about the power of this procedure in view of the strong collinearity effects that are present in many time series. Repeated errors of the second kind could lead to a model with some important edges missing. Consider for instance a bivariate AR(1) model with $K_{11} = K_{22} = 1$ and $K_{12} = K_{21} = \gamma$. Then all $\pi_{ab}(u)$ are zero for $u > 1$ and by an easy calculation:

$$\pi_{ab}(1)^2 = \frac{A_{ab}(1)^2(1-\gamma^2)}{A_{ab}(1)^2(1-\gamma^2)+1}.$$

If γ is close to one, then this is small even when all $|A_{ab}(1)|$ are large. In such a case the procedure might lead to the wrong conclusion that only undirected edges are needed.

An alternative way for approximate likelihood estimation would be to select first a small number of most promising candidates G in the subclass of models with all undirected edges present. This means that we consider all possible restrictions for the matrices A, but no restrictions on K. The search in this subclass could be done in a fast way by a sophisticated variable selection algorithm from regression. In a second step, one would then compute the likelihood for all models that have the structure for A determined in the first step combined with all possible structures for K.

Additional references in discussion

Ahn, S. K. and Reinsel, G. C. (1988). Nested reduced-rank autoregressive models for multiple time series. *Journal of the American Statistical Association*, **83**, 849–56.

Akaike, H. (1976). Canonical correlation analysis of time series and the use of an information criterion. In *Systems Identification: Advances and Case Studies* (eds R. K. Mehra and D. G. Lainiotis), pp. 27–96. Academic Press, New York.

Bühlmann, P. and Wyner, A. J. (1999). Variable length Markov chains. *Annals of Statistics*, **27**, 480–513.

Florens, J. P. and Fougère, D. (1996). Noncausality in continuous time. *Econometrica*, **64**, 1195–212.

Lauritzen, S. L. and Richardson, T. S. (2002). Chain graph models and their causal interpretation. *Journal of the Royal Statistical Society*, B, **64**, 1–28.

Parner, J. and Arjas, E. (1999). Causal reasoning from longitudinal data. Technical Report, University of Helsinki, Finland.

Pearl, J. (1994). A probabilistic calculus of actions. In *Uncertainty in Artificial Intelligence 10* (eds R. Lopez de Mantaras and D. Poole), pp. 454–62. Morgan Kaufmann, San Mateo, CA.

Pearl, J. and Robins, J. (1995). Probabilistic evaluation of sequential plans from causal models with hidden variables. In *Uncertainty in Artificial Intelligence 11* (eds P. Besnard and S. Hanks), pp. 444–53. Morgan Kaufmann, San Francisco.

Richardson, T. (1998). Chain graphs and symmetric associations. In *Learning in Graphical Models* (ed. M. Jordan), pp. 231–60. Kluwer Academic, Dordrecht, The Netherlands.

Rissanen, J. (1986). Complexity of strings in the class of Markov sources. *IEEE Transactions on Information Theory*, **32**, 526–32.

Spirtes, P., Glymour, C., and Scheines, R. (1993). *Causation, Prediction and Search*, Lecture Notes in Statistics, No. 81. Springer-Verlag, New York.

Tiao, G. C. and Tsay, R. S. (1989). Model specification in multivariate time series (with discussion). *Journal of the Royal Statistical Society*, B, **51**, 157–213.

Velu, R. P., Reinsel, G. C., and Wichern, D. W. (1986). Reduced rank models for multiple time series. *Biometrika*, **73**, 109–18.

5

Linking theory and practice of MCMC

Gareth O. Roberts
Lancaster University, UK

1 Introduction

Markov chain Monte Carlo (MCMC) methodology is a mature and well-established technique in physics and computer science. However, it is still only some ten years since it became well known in the statistical community and acquired its reputation as the numerical method of choice for a wide range of problems. Its success has even inspired new and developing areas of statistical modelling, due to the existence of algorithms to perform inference on these new models. As such, MCMC has been at the heart of Highly Structured Stochastic Systems (HSSS) activity during the past 7 years, and the HSSS programme has been the catalyst for much of the leading work on the theory and methodology of MCMC methods during this time.

The simplicity of MCMC belies the scope of its applications. Suppose that we wish to simulate from a *target* distribution, but that direct simulation is impossible (perhaps because of the complexity or high dimensionality of the target). MCMC involves simulating a Markov chain with stationary distribution equal to the target density. Simulation thus proceeds until the process reaches its stationary regime from which samples of observations from the target density can be taken.

By now, a number of useful review articles and books are available to access the area of MCMC research. At the time of its publication, Gilks *et al.* (1996) was a comprehensive survey of the area for practitioners, while the more tutorial Gammerman (1997) also offers an excellent introduction. A detailed review which also contains theoretical background is given in Robert and Casella (1999). For a full theoretical survey, the reader should refer to Roberts and Tweedie (2003), which contains considerably more detail on all the concepts touched upon here. The web site http://www.statslab.cam.ac.uk/~MCMC is an excellent source for research work on all aspects of MCMC.

This is not a review of either theory or methodology of MCMC. Instead, an attempt has been made to describe some important concepts from Markov chain theory, and how they relate to MCMC algorithms. In particular, emphasis will be given to the relationship between these ideas and the properties of observed MCMC simulation output, and in turn how these properties affect the estimation of statistical quantities.

The Markov chain theory that will be described in some detail uses fundamentally the notion of a *drift condition*, which describes excursions away from the 'central sets' of a target density. A description will be given of how the properties of these excursions relate to the stability of Monte Carlo estimators from the algorithm. The ideas will all be illustrated using simple examples. The only algorithm considered in detail is the random walk Metropolis algorithm, although much of the theory described here can be equally applied to the analysis of other commonly used algorithms such as the Gibbs sampler, and various variations and hybrids between Gibbs and Metropolis.

We concentrate here on the use of drift and so-called *minorization* conditions (which will be introduced in Section 4) to study convergence. However, these techniques are not the only mathematical tools available to study MCMC algorithms and their convergence properties. We mention two different collections of techniques that have been successfully applied in an MCMC context. The use of geometric techniques such as Poincaré inequalities, Cheeger bounds, and other related inequalities for bounding algorithm convergence rates can be applied effectively in some situations (usually requiring extremely symmetric target density properties). The review by Saloff-Coste (1997) is an excellent introduction to this area. Some applications to MCMC are given in, for example, Frigessi *et al.* (1996) and Frieze *et al.* (1994). For high-dimensional algorithms, a useful technique is to study the properties of a weak limiting infinite dimensional process. Such a procedure can be used to study random walk Metropolis algorithms and Langevin algorithms (Roberts *et al.* 1997; Breyer and Roberts 1998; Roberts and Rosenthal 1998).

Whilst this article will unashamedly proceed to concentrate on a small but important area of MCMC research, there are two areas not considered here which deserve special mention. Reversible jump MCMC (Green 1995) has greatly facilitated model choice and model averaging procedures, and merits its own chapter, Chapter 6, in this volume. Perfect simulation (Propp and Wilson 1996) offers alternative implementation methods for MCMC algorithms which, when applicable, allow the difficult issue of assessing convergence to be completely circumvented. Although perfect simulation has had notable success in certain classes of statistical models, its general implementation remains problematical in problems without special structure. This fast emerging area is served by an excellent annotated bibliography, which can be found at `http://dimacs.rutgers.edu/~dbwilson/exact.html`.

The article is arranged as follows. Section 2 introduces an extremely simple one-dimensional example which turns out to be surprisingly difficult for the random walk Metropolis algorithm. Section 3 and Section 4 then introduce the necessary theoretical concepts in order to be able to understand the problems encountered in this motivating example. Detailed results describing how these results can be applied to the random walk Metropolis algorithm are described in Section 5, and in Section 6 these results are directed towards an understanding of the motivating example from Section 2. Section 7 briefly discusses the construction of computable bounds using drift conditions.

2   Motivating example

We begin with an example which we will describe and investigate more closely later on. The term 'mixing' is frequently used as an informal term to describe the idea that a Markov chain adequately explores the relevant state space. It is commonly believed that the so-called random walk Metropolis algorithm with suitably tuned proposal variance will always mix 'adequately' in small dimensional problems, at least for unimodal target densities. However, this is not always the case.

Suppose that we wish to simulate from a standard Cauchy distribution conditioned to be positive, using the random walk Metropolis algorithm with Gaussian proposals. Thus we take the target density $\pi(x) = (1 + x^2)^{-1}$ on the region $(0, \infty)$, and 0 otherwise, and propose jumps from x to $Y = x + \sigma Z$, where $Z \sim N(0, 1)$, accepting the new value with probability $\min\{1, \pi(Y)/\pi(x)\}$ or else remaining at x. Here σ was tuned so that the overall empirical acceptance rate of the algorithm (that is, the proportion of accepted moves) was around 35%.

Figure 1 shows a trace plot of 10 000 iterations of the algorithm started close

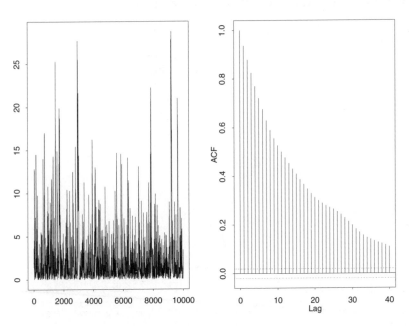

FIG. 1. A realization of a Metropolis algorithm with a light tailed proposal distribution on a Cauchy target density, together with the autocorrelation function of the logarithms of the realized values.

to its mode at 0, together with a plot of the empirical autocorrelation function. Visual inspection suggests adequate if not good mixing. This particular output satisfies standard convergence diagnostic tests such as those in CODA. (See Robert (1995), Cowles and Carlin (1996), and Brooks and Roberts (1998) for general background on convergence diagnostics.) However, note that since we have not yet introduced precise statements of what we mean by convergence, we are not in a position to comment on the relevance of the various CODA tests, many of which assess different convergence concepts.

However, in this case of course we know the true stationary probabilities, so that according to the Cauchy target density we would expect over 2% of observations (200 observations in this case) to exceed 30. However, this particular run does not manage to exceed 30 at all. How can we explain this? In particular, how can the target density be preserved by an algorithm that shows such a reluctance to visit an area of non-negligible probability?

Figure 2 offers the beginnings of an explanation of this problem. It shows two traces of the same chain started out in the right-hand tail of the target density. They illustrate very different behaviour compared with that of runs started centrally. Run (a) meanders like a Brownian motion sample path (since virtually all the moves are accepted in the tail) while run (b) returns to the area around the mode of the distribution after meandering for a while. A feature of

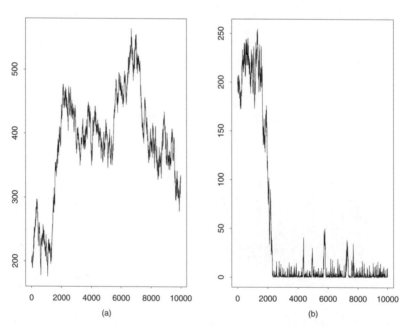

FIG. 2. The behaviour of a Metropolis algorithm with a light tailed proposal on a Cauchy target once the tail is reached.

this behaviour is the relatively high probability of an extremely long excursion away from the target density mode.

The algorithm therefore compensates for its reluctance to visit far enough into the tails of the target density, by the possibility of spending extremely long excursions out in the tails once the tail has been reached.

What are the implications of this issue for MCMC estimation? In estimating a tail probability such as $\Pr(X > 100)$ (which has value around 0.006 in this example), the distribution of the empirical estimator $\#\{X_i > 100\}/\{$number of iterations$\}$, is highly skewed – taking the value 0 with a high probability (at least for runs of length 100 000) but with a very small probability, taking an extremely large value. However, it would not really help to take larger numbers of algorithm iterations. This would only increase the probability of even longer excursions away from the distribution mode.

We have already started to discuss the properties of excursions of the algorithm away from a central set. In the next section we shall begin to formalize these ideas into a theory of convergence rates of Markov chains. It is important to discover what it is about this particular target density that causes the problems encountered in this example, in order to draw conclusions about the behaviour of algorithms more generally. In this example, it seems that the problems are related to the heaviness of the tails of the Cauchy distribution.

One plausible remedy to this is to use heavy tailed proposal distributions using, for example, a scaled Cauchy proposal. Figure 3(b) illustrates how the sample paths of this algorithm are very different from those of the light tailed proposal (in Fig. 3(a)), resembling more closely a jump process with noise than Brownian motion. Furthermore, the heavy tailed nature of the proposals allows the algorithm to return to the high density region more easily. Both of these simulation runs begin at level 1000.

A further improved algorithm can be produced by considering a logarithmic random walk algorithm, where the proposed move is to a random multiple of the current state. Figure 3(c) illustrates how such an algorithm has frequent short excursions into the tail of the target density.

All three traces in Fig. 3 are plotted on the same scale. For the logarithmic random walk algorithm, 19 observations exceed the level 3000, and are therefore out of bounds for this plotting range. According to the Cauchy target distribution, around 21 observations should exceed this level. In fact, neither of the other algorithms exceeds 3000 at all. We have focused our discussion on the estimation of tail probabilities. However, instability will also be present in the estimation of other expectations.

From these simulations it is easy to see an improvement in the mixing of the algorithms illustrated in Fig. 3 proceeding from (a) to (b) to (c). In the next section we will describe, in a non-technical fashion, the mathematical theory behind these differences.

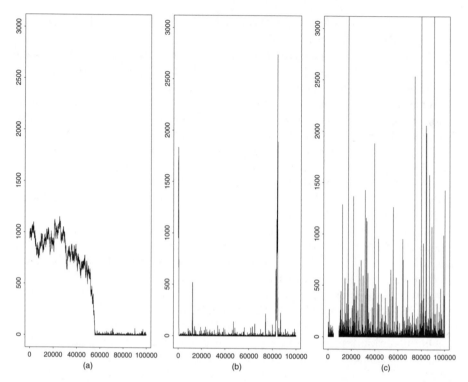

FIG. 3. A comparison of algorithms on the Cauchy target density: (a) a Metropolis algorithm with a light tailed proposal density; (b) a Metropolis algorithm with a heavy tailed proposal density; and (c) a multiplicative random walk Metropolis algorithm.

3 Mathematical framework

We begin with a description of the Metropolis–Hastings algorithm. Let us start with a *target density* π known at least to constant multiples, and assume we have also a *proposal transition kernel* $Q(\mathbf{x}, A), \mathbf{x} \in \mathbb{R}^d, A \in \mathcal{B}$ (sometimes called the candidate kernel), satisfying

$$Q(\mathbf{x}, A) \geqslant 0, \qquad Q(\mathbf{x}, \mathbb{R}^d) = 1,$$

which generates potential transitions for a discrete time Markov chain evolving on $\mathcal{X} = \mathbb{R}^d$ equipped with the Borel σ-algebra $\mathcal{B}$. In this article we will assume that $Q(\mathbf{x}, \cdot)$ is absolutely continuous, with Lebesgue density $q(\mathbf{x}, \mathbf{y})$ (in $\mathbf{y}$ for each $\mathbf{x}$), which covers many MCMC applications, although the results discussed here do extend to other cases in an obvious way.

In the Hastings algorithm, introduced in Hastings (1970), a *proposal transition* generated according to the law Q is then accepted with probability $\alpha(\mathbf{x}, \mathbf{y})$

given by

$$
\alpha(\mathbf{x}, \mathbf{y}) = \begin{cases} \min\left\{ \dfrac{\pi(\mathbf{y})}{\pi(\mathbf{x})} \dfrac{q(\mathbf{y}, \mathbf{x})}{q(\mathbf{x}, \mathbf{y})}, 1 \right\}, & \pi(\mathbf{x})q(\mathbf{x}, \mathbf{y}) > 0, \\ 1, & \pi(\mathbf{x})q(\mathbf{x}, \mathbf{y}) = 0. \end{cases} \tag{3.1}
$$

Otherwise, if a jump is rejected, then the chain remains at its original position. Thus actual transitions of the Hastings chain, which we shall denote by $\mathbf{X} = \{\mathbf{X_n}, \mathbf{n} \in \mathbf{Z^+}\}$, take place according to a law P with transition density

$$
p(\mathbf{x}, \mathbf{y}) = q(\mathbf{x}, \mathbf{y})\alpha(\mathbf{x}, \mathbf{y}), \qquad \mathbf{y} \neq \mathbf{x}, \tag{3.2}
$$

and with probability of remaining at the same point given by

$$
\Pr(\mathbf{x}, \{\mathbf{x}\}) = \int q(\mathbf{x}, \mathbf{y})[1 - \alpha(\mathbf{x}, \mathbf{y})]\,\mathrm{d}\mathbf{y}. \tag{3.3}
$$

With this choice of α we have that π satisfies

$$
\pi(A) = \int \pi(\mathbf{x})\Pr(\mathbf{x}, A)\,\mathrm{d}\mathbf{x} \qquad \text{for all } A \in \mathcal{B}. \tag{3.4}
$$

The first two examples of the three looked at in Section 2 consider the *symmetric random walk Metropolis algorithm*, which dates back to Metropolis *et al.* (1953). The Metropolis algorithm is a special case of Hastings, utilizing a symmetric candidate transition Q: that is, one for which $q(\mathbf{x}, \mathbf{y}) = q(\mathbf{y}, \mathbf{x})$. It is called 'random walk' Metropolis if further

$$
q(\mathbf{x}, \mathbf{y}) = q(\mathbf{y} - \mathbf{x}), \tag{3.5}
$$

where q is a density on $\mathbb{R}^d$.

Note that the symmetry condition ensures $q(\mathbf{x}) = q(-\mathbf{x})$. In the symmetric Metropolis algorithm, we have that the acceptance probabilities have the simpler form

$$
\alpha(\mathbf{x}, \mathbf{y}) = \begin{cases} \min\left\{ \dfrac{\pi(\mathbf{y})}{\pi(\mathbf{x})}, 1 \right\}, & \pi(\mathbf{x}) > 0, \\ 1, & \pi(\mathbf{x}) = 0. \end{cases} \tag{3.6}
$$

Thus in the main example of Section 2, $q(x, y) = (2\pi\sigma^2)^{-1/2}\exp\{-(y - x)^2/(2\sigma^2)\}$, $\pi(x) = (1 + x^2)^{-1}$, and

$$
\alpha(x, y) = \begin{cases} \min\left\{ \dfrac{1 + x^2}{1 + y^2}, 1 \right\}, & y > 0, \\ 0, & \text{otherwise.} \end{cases}
$$

In this article, one of the most important questions we shall ask is whether such chains are *geometrically ergodic*: that is, whether the n-step transition probabilities of $\mathbf{X}$, defined by

$$
P^n(\mathbf{x}, A) = \Pr(\mathbf{X}_n \in A | \mathbf{X}_0 = \mathbf{x}), \qquad n \in \mathbf{Z^+}, \mathbf{x} \in \mathcal{X}, A \in \mathcal{B},
$$

converge to π at a geometric rate. To make precise statements we need to talk about norms in which we can describe the convergence.

Convergence to stationarity of Markov chains can be studied in a variety of different norms. Often convenient (though not always appropriate) is the *total variation distance*, given for two probability measures ν_1 and ν_2 by

$$\|\nu_1 - \nu_2\| = \frac{1}{2} \int_{\mathcal{X}} |\nu_1(\mathrm{d}x) - \nu_2(\mathrm{d}x)| = \sup_{A \in B} (\nu_1(A) - \nu_2(A)). \qquad (3.7)$$

Where ν_1 and ν_2 admit densities with respect to Lebesgue measures, f_1 and f_2 say, then $\|\nu_1 - \nu_2\| = \frac{1}{2} \int_{\mathcal{X}} |f_1(x) - f_2(x)| \mathrm{d}x$.

A minimal requirement for any sensible algorithm with transition probabilities P is that of ergodicity: that is for all $x \in \mathcal{X}$,

$$\|P^n(\mathbf{x}, \cdot) - \pi(\cdot)\| := r(\mathbf{x}, n) \downarrow 0. \qquad (3.8)$$

In fact, it is easy to demonstrate that all three of the examples considered in Section 2 satisfy (3.8). It is clearly necessary to consider more refined conditions on $r(\cdot, \cdot)$ in order to compare these methods.

3.1 Rates of convergence of Markov chains

(a) A Markov chain is uniformly ergodic if $r(\mathbf{x}, n) \downarrow 0$ uniformly in $\mathbf{x}$ as $n \to \infty$.

(b) A Markov chain is geometrically ergodic if there exists a real-valued function $V(\mathbf{x})$, $\mathbf{x} \in \mathcal{X}$, and a positive constant $\rho < 1$ such that $r(\mathbf{x}, n) \leqslant V(\mathbf{x})\rho^n$ for all $\mathbf{x} \in \mathcal{X}$, and all $b \in \mathbb{Z}^+$. The infimum of all values ρ for which we can do this is called the (geometric) rate of convergence of the Markov chain.

(c) A Markov chain is polynomially ergodic at rate p for some $p > 0$ if there exists a real-valued function $h(\mathbf{x}) \geqslant 1$, $\mathbf{x} \in \mathcal{X}$, such that $r(\mathbf{x}, n) \leqslant h(\mathbf{x})n^{-r}$ for all $\mathbf{x} \in \mathcal{X}$, $r < p$, and $n \in \mathbb{Z}^+$.

Note that uniform ergodicity is a stronger condition than geometric ergodicity, although this is not immediately obvious from the definition. Under uniform ergodicity, it turns out that $r(\mathbf{x}, n) \downarrow 0$ uniformly in $\mathbf{x}$ and geometrically quickly in n.

Most MCMC algorithms are easily provably ergodic. However, very many are not uniformly ergodic, and quite surprisingly, many are not even geometrically ergodic. I shall attempt to relate these concepts to the observed behaviour of the algorithms.

4 Drift and minorization conditions

In this section, two main points will be demonstrated: we can describe excursions from small sets in terms of 'drift', and then we consider the impact of the excursion distributions on the existence of the central limit theorems important for assessing the accuracy of ergodic estimates.

From Section 2, it seems that the convergence properties of the chain are linked to properties of its excursions away from a 'central set'. We define a *small set* to be a subset of $C \subset \mathcal{X}$ with the property that there exists some $n_0 \in \mathbb{Z}^+$ such that for some positive constant ϵ and a probability measure ν,

$$P^{n_0}(\mathbf{x}, \mathrm{d}\mathbf{y}) \geqslant \epsilon \nu(\mathrm{d}\mathbf{y}). \tag{4.1}$$

Thus in n_0 steps, chains starting anywhere in C have probability ϵ of being generated from a distribution ν, which is independent of where in C the chain started from. In practice, we can often identify small sets as more familiar objects, commonly as bounded or compact sets.

To describe excursions away from the set C, we introduce the concept of a Foster–Lyapunov *drift* condition. Suppose that $V : \mathcal{X} \to [1, \infty)$ is such that P satisfies

$$\mathrm{E}[V(\mathbf{X}_1)|\mathbf{X}_0 = \mathbf{x}] \leqslant V(\mathbf{x}) - (1 - \lambda)V(\mathbf{x})^\alpha + b\mathbf{1}_C(\mathbf{x}) \tag{4.2}$$

for constants $b > 0$, $0 < \lambda < 1$, and $0 \leqslant \alpha \leqslant 1$. The parameter α is the important one here, measuring the extent of the 'drift' back towards the small set C.

The following result informally summarizes some of the important implications of the existence of drift conditions of this kind. It is not the intention to state precise results here, but the interested reader is referred to Meyn and Tweedie (1993) for details of definitions of Markov chain concepts such as aperiodicity, irreducibility, recurrence, and positivity. Let τ denote the first time the chain enters the set C.

Theorem 4.1 *We suppose throughout that C is a small set and the Markov chain is aperiodic and satisfies (4.2).*

(a) *For all $\mathbf{x} \in \mathcal{X}$, $\Pr(\tau < \infty|\mathbf{X}_0 = \mathbf{x}) = 1$ and $\mathrm{E}(\tau|\mathbf{X}_0 = \mathbf{x}) < \infty$. Furthermore, the Markov chain is positive recurrent and irreducible.*

(b) *If $\alpha = 1$, then there is a positive number s_0 such that $\mathrm{E}(e^{s\tau}|\mathbf{X}_0 = \mathbf{x}) < \infty$ for all $s < s_0, \mathbf{x} \in \mathcal{X}$; the Markov chain is geometrically ergodic.*

(c) *If $\alpha = 1$ and V is bounded, then the chain is uniformly ergodic.*

(d) *If $0 < \alpha < 1$, then $\mathrm{E}(\tau^p|\mathbf{X}_0 = \mathbf{x}) < \infty$ for all $\mathbf{x} \in \mathcal{X}$ and for all $p \leqslant 1/(1-\alpha)$. Moreover, in this case the chain is polynomially ergodic with rate $\alpha/(1 - \alpha)$.*

Thus Theorem 4.1(a) says that Metropolis–Hastings chains are automatically positive recurrent as long as return times to a small set are certain. Theorem 4.1(b) tells us that that if the geometric drift condition holds, then return times are sufficiently light-tailed to allow the existence of some return time exponential moments. This in turn guarantees geometric ergodicity. Theorem 4.1(d) says that under the weaker drift condition, polynomial moments of return times exist and polynomial ergodicity results.

This result omits a number of occasionally useful statements, for instance the four statements (a), (b), (c), and (d) actually imply the existence of a drift function V for which the appropriate drift condition (4.2) holds. Furthermore, these bounds on the moments of return times to the small set can often be strengthened to statements that hold uniformly over certain sets. However, Theorem 4.1 summarizes the main theory we require later in this article.

The proof of Theorem 4.1(a), (b), and (c) can be found in the comprehensive treatment of Meyn and Tweedie (1993), whereas (d) can be found as Theorem 3.6 of Jarner and Roberts (2002).

4.1 Central limit theorems

We have seen that the algorithm's convergence properties are closely linked to those of its excursions away from small sets. This in turn is very closely linked to the existence of central limit theorems (CLTs) for the Markov chain, which are important for any Monte Carlo implementation. We say that a $\sqrt{n}$-CLT exists for a function f if

$$n^{1/2}\left(\frac{\sum_{i=1}^{n} f(\mathbf{X}_i)}{n} - \mathrm{E}_\pi(f(X))\right) \Rightarrow \mathrm{N}(0, \tau_f \, \mathrm{var}_\pi(f(X))) \qquad (4.3)$$

for some $\tau_f \in (0, \infty)$, where τ_f denotes the *integrated autocorrelation time* for estimating the function f using the Markov chain P. It is of course important to try to construct Markov chains that mix rapidly and have a small value for τ_f. However, even more fundamentally, it is crucial for dependable estimation that τ_f is finite.

Roughly speaking, geometric ergodicity is enough to provide CLTs wherever we might expect, that is for all functions that possess finite second moments with respect to the target density π. However, in the polynomially convergent case, the situation is much more complicated. More precisely, this is given by

Theorem 4.2 *Central limit theorem for Markov chains.*

(a) *If P is geometrically ergodic and reversible, then for all functions f for which $\mathrm{var}_\pi(f(X))$ is finite, a $\sqrt{n}$-CLT exists.*

(b) *Suppose that (4.2) holds for some $\alpha < 1$, and there exists a positive constant η with $1 - \alpha \leqslant \eta \leqslant 1$ and such that $\mathrm{E}_\pi(V^{2\eta}) < \infty$. Then a $\sqrt{n}$-CLT exists for all functions f with $|f| \leqslant V^{\alpha+\eta-1}$.*

Part (a) is proved in Roberts and Rosenthal (1997) using spectral theoretic methods.

For the polynomially ergodic case, CLTs can be shown to exist only for functions that do not grow too rapidly. As an example, CLTs hold for all bounded functions if $\mathrm{E}_\pi(V^{2(1-\alpha)}) < \infty$, or $\alpha \geqslant 1/2$. Part (b) is proved (and in fact slightly generalized) in Jarner and Roberts (2002) which builds on fundamental work from Meyn and Tweedie (1993). Applications to MCMC are given in Jarner and Roberts (2000, 2002), the latter paper supporting the results with an extensive simulation study.

5 MCMC and drift conditions

The general theory described at the end of Section 4 requires the identification of natural classes of drift conditions. Fortunately, for MCMC algorithms, natural drift conditions exist in a number of important cases. Since drift conditions intuitively describe 'drift' towards small sets, it is natural to use regions to which the algorithm gravitates naturally as candidates for small sets. Since bounded sets are often small, it follows that we can usually choose a bounded region surrounding the target density mode as a small set. Taking this idea further, natural drift conditions are those that tend to quantify movement towards regions of larger target density value. Thus it is natural to use a drift function that possesses similar shaped contours to those of the target density. This motivates the use of drift functions of the form $V(\mathbf{x}) = \pi(\mathbf{x})^{-u}$ for positive constants u. This approach is successful in a wide variety of MCMC applications.

Example Suppose that π is a bounded target density on $\mathcal{X} = \mathbb{R}^d$ and we propose to run the random walk Metropolis algorithm with symmetric proposal density q, then a natural drift function turns out to be $V(x) \propto \pi(x)^{-u}$ for some $0 < u < 1/2$. We shall consider three special cases.

Proposition 5.1

(a) *If $\pi(\mathbf{x}) \propto \mathrm{e}^{-s|\mathbf{x}|}$ at least outside some compact set C, then the random walk Metropolis algorithm satisfies (4.2) with $\alpha = 1$ and any value $\lambda \in (\lambda_0, 1)$ where*

$$1 - \lambda_0 = \frac{1}{2} \int_{\mathcal{X}} (1 - \mathrm{e}^{-usz})(1 - \mathrm{e}^{-(1-u)sz}) q(\mathbf{z}) \, \mathrm{d}\mathbf{z}. \qquad (5.1)$$

Therefore, since any compact set is small, the algorithm is geometrically ergodic.

(b) *Consider the case where the tails of q are bounded by some multiple of the function $|\mathbf{x}|^{-(d+2)}$. If $\pi(\mathbf{x}) \propto |\mathbf{x}|^{-(r+d)}$ at least outside some compact set C, then the random walk Metropolis algorithm satisfies (4.2) for any $0 < \alpha < \alpha_0$ with*

$$\alpha_0 = \frac{r}{r+2}. \qquad (5.2)$$

Thus the algorithm is polynomially ergodic with rate $r/2$.

(c) *If $\pi(\mathbf{x}) \propto |\mathbf{x}|^{-(r+d)}$ at least for large enough $|\mathbf{x}|$, and q has a heavy tailed distribution with $q(\mathbf{x}) \propto |\mathbf{x}|^{-(d+\eta)}$ for all large enough $|\mathbf{x}|$, where $0 < \eta < 2$, then (4.2) is satisfied for any $0 \leqslant \alpha < \alpha_0$ with $\alpha_0 = r/(r+\eta)$. Thus the algorithm is polynomially ergodic at rate r/η.*

The conditions in Proposition 5.1 can be considerably weakened in fact, and the papers that prove these results give some directions for these generalizations. Part (a) comes from Roberts and Tweedie (1996b), though see also Jarner and Hansen (2000) and Fort and Moulines (2000a) for extensions to this result. Part

(b) is essentially from Fort and Moulines (2000b), whereas part (c) comes from Jarner and Roberts (2000). Broadly speaking, the conclusions of these extensions can be summarized informally as follows.

Behaviour of random walk Metropolis

(a) For target densities with tails uniformly bounded by exponentials in all directions, random walk Metropolis is geometrically ergodic as long as the tails are sufficiently regular. One condition that is sufficient for this requires the contours of the target density to have a curvature that converges to 0 as they move farther away from the target density mode.

(b) If π has tails that are strictly heavier than exponential, then no random walk Metropolis algorithm can be geometrically ergodic.

(c) If π resembles $|\mathbf{x}|^{-(d+r)}$ in the tails, then light tailed Metropolis proposals lead to algorithms which converge at the polynomial rate $r/2$. However, heavy tailed versions of the algorithms can increase the polynomial rate arbitrarily.

5.1   Practical implications

According to Theorem 4.1, geometric ergodicity is related to exponential moments of return times to a small set C. There are two obvious ways in which exponential moments of excursion lengths can fail to exist for a random walk Metropolis algorithm. Both of these ways can be described by the notion of excursions away from small sets being heavy tailed, as in Theorem 4.1.

(a) First, the algorithm might have 'sticky' patches where the chain spends a long time in particular sets. Suppose we can have arbitrarily bad sticky patches, i.e.

$$\Pr(\mathbf{x}_i, \{\mathbf{x}_i\}) \to 1$$

along a suitable sequence of points $\{\mathbf{x}_i\}$. Then it is easy to show that return times to small sets cannot have geometric moments. The regularity conditions we require to ensure geometric ergodicity with exponential tails is there, essentially, to preclude this kind of behaviour.

It is interesting to note that it is very common, even for standard statistical examples, for sticky regions to exist. As a simple example consider the normal-gamma density prevalent in Bayesian analysis of linear models:

$$\pi(\mu, \tau) \propto \tau^{n/2} \exp\{-\tau(S + n(\mu - \bar{x})^2)/2\}$$

for suitable constants n, S, and $\bar{x}$. Figure 4 illustrates a contour plot of the density for particular values of the constants. Note the ridges of high probability for large $|\mu|$ and the small value of the precision, τ. Along those ridges, $\Pr((\mu, \tau), \{(\mu, \tau)\}) \to 1$ and so two-dimensional random walk Metropolis fails to be geometrically ergodic. Note that since the algorithm struggles to leave these ridges, to preserve stationarity of the target density it must also have

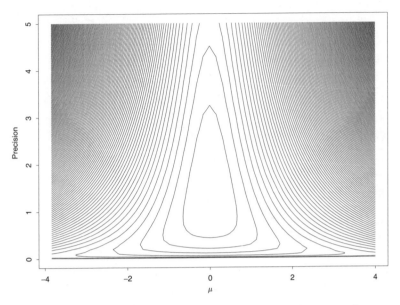

FIG. 4. A contour plot of the posterior distribution in the normal-gamma example.

difficulty in reaching out into the tails in the first place. This manifests itself in a tendency to miss the tails of μ on most MCMC sweeps of the Metropolis sampler; see Fig. 5. In the rare cases where the tail is reached, the chain will spend disproportionately long in the tails, and overestimate tail probabilities massively. Thus the overall effect is one of a highly skewed distribution of empirical tail probabilities.

(b) The second way in which random walk algorithms can fail to be geometrically ergodic is where the effect of the accept/reject mechanism becomes negligible in the tails, so that the algorithm approximates a null-recurrent random walk. Thus, because of the target density tails, the drift that the algorithm induces towards the high density regions is insufficient for a condition like (4.2) to ever hold, for any eligible function V. This is essentially why random walk Metropolis fails on heavy tailed target densities such as the Cauchy distribution seen in Section 2. In such cases, we often have the property that $\lim_{|\mathbf{x}|\to\infty} \pi(\mathbf{x}+\mathbf{y})/\pi(\mathbf{x}) = 1$, for all $\mathbf{y} \in \mathbb{R}^d$, so that by (3.6), the probability of any move being rejected recedes to zero out in the tails of the distribution.

Both types of qualitative behaviour described above can often be observed empirically, even where analytic calculations on π are not feasible.

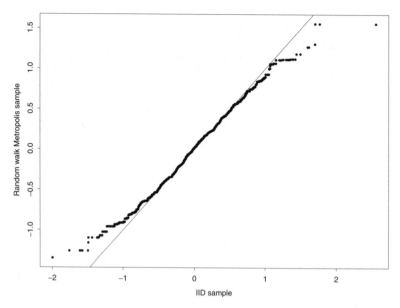

FIG. 5. A q-q plot illustrating that the random walk Metropolis algorithm fails to adequately explore the tails of the normal-gamma distribution.

5.2   Other algorithms

The discussion in Section 5.1 focused on the random walk Metropolis algorithm. However, many of its conclusions translate to many similar Metropolis–Hastings algorithms, for instance Langevin algorithms (see, for example, Roberts and Tweedie (1996a)), the simplest example of which uses the proposal density

$$q(\mathbf{x}, \mathbf{y}) = \frac{1}{(2\pi\sigma)^{d/2}} \exp\left\{ \frac{-\|\mathbf{y} - \mathbf{x} - \sigma^2 \nabla \log \pi(\mathbf{x})/2\|^2}{2\sigma^2} \right\},$$

for some suitable scaling parameter σ. Langevin methods bear some resemblance to Metropolis algorithms, with the only difference being that in the Langevin case the proposal density mean is offset by a term that pushes the algorithm towards areas of higher density. Therefore many of the results obtainable in this case are similar to those in Proposition 5.1, and can be found in Roberts and Tweedie (1996a), Stramer and Tweedie (1999), and Jarner and Roberts (2000).

Drift function methodology can also be applied successfully to many other MCMC algorithms, for example the Gibbs sampler (Chan 1993), the independence sampler (Jarner and Roberts 2002) and the so-called slice sampler (Roberts and Rosenthal 2000).

Moreover, many of the conclusions of Proposition 5.1 also carry over to the case of 'Metropolis-within-Gibbs' where the Metropolis step updates in a smaller dimensional space than the full algorithm. For instance, we can easily prove the following negative result, for example by using the capacitance ideas of Jerrum and Sinclair (1996).

Theorem 5.2 *Suppose that π is a k-dimensional density, and for each i, P_i is a reversible Markov chain which updates only the ith coordinate. Consider running a random scan of the P_is, that is a chain P with*

$$P = \frac{P_1 + P_2 + \cdots + P_k}{k}.$$

Suppose that for some component i say, P_i is a random walk Metropolis algorithm with fixed increment proposal density q, and that

$$\lim_{K \to \infty} \frac{\log \pi(X_i \in (K, \infty))}{K} = 0, \tag{5.3}$$

then P fails to be geometrically ergodic.

The peculiar requirement in (5.3) can be interpreted as the condition that the marginal distribution of the ith component has a heavier than exponential tail.

6 Examples

We return here to the simple Cauchy example introduced in Section 2. Cauchy tails satisfy

$$\pi(x) \sim cx^{-2}$$

for some constant c. By Theorem 5.1(b) the simple random walk Metropolis algorithm with a light tailed proposal satisfies (4.2) for values of α less than $\alpha_0 = 1/3$. In this case, we do not even obtain CLTs for bounded functions from Theorem 4.2.

On the other hand, the Cauchy proposal algorithm illustrated in Fig. 3(b) satisfies (4.2) for all $\alpha < \alpha_0 = 1/2$. This is not quite enough to ensure that CLTs hold for all bounded functions. (Any slightly heavier tailed proposal would achieve this.) However, it is considerably more stable than the light tailed proposal case.

The multiplicative random walk Metropolis algorithm of Fig. 3(c) is geometrically ergodic, since the tails of the density of the logarithm of a Cauchy random variable are bounded by an exponential. Thus, the algorithm is considerably more stable than either of the first two algorithms. It satisfies (4.2) for $\alpha = 1$ and therefore by Proposition 5.1(a) it is geometrically ergodic.

7 Computable bounds

This article has mainly focused on qualitative convergence results. However, it is important to recognize that the drift condition appearing in (4.2) can be used to give quantitative convergence results of the following type.

Theorem 7.1 *For any Markov chain satisfying (4.2) and (4.1) we can write*

$$\|P^n(\mathbf{x},\cdot) - \pi\| \leqslant r(\epsilon, n_0, \lambda, \alpha, b), \qquad (7.1)$$

for a function r that is explicitly specified in Roberts and Tweedie (1996b).

There is a substantial body of work on trying to find the best functions r, particularly in the geometrically ergodic case ($\alpha = 1$). See, for example, Meyn and Tweedie (1994), Rosenthal (1995), Lund *et al.* (1996), Lund and Tweedie (1996), Roberts and Tweedie (1999, 2000), and Fort and Moulines (2000a), with Roberts and Tweedie (1999) currently giving the best available results.

Most of this work assumes little or no structure to $\mathcal{X}$ and P, and is thus very generally applicable. However, this generality comes with a price – bounds are often quite conservative, and are too large to be of practical value in high-dimensional situations. The bounds rely on a comparison with a related Markov chain which is constructed directly from (4.2). Where the target density is unimodal with light tails, the random walk Metropolis algorithm can be fairly well approximated in this way and thus bounds are reasonable (Roberts and Tweedie 1996b). However, where the algorithm's qualitative behaviour is not well described by this simple model of drifting towards a small set, bounds can be very poor.

Computation of r in specific applications depends on successfully writing down explicit drift (4.2) and minorization (4.1) conditions. It is usually easy enough to do this for simple algorithms like the random walk Metropolis algorithm, but it is less often possible to find good enough bounds for the constants appearing in (4.2) and (4.1) to allow reasonable bounds to be obtained.

Improved bounds are possible in more specialized situations where extra structure is available; see, for example, Roberts and Tweedie (2000) in the case of stochastic monotonicity. The following example illustrates the potential application of these bounds.

Example The polar slice sampler (Roberts and Rosenthal 2000).

Let π be a d-dimensional log-concave target density, and write

$$f_1(\mathbf{x}) = |\mathbf{x}|^{d-1}\pi(\mathbf{x}).$$

The polar slice sampler proceeds by iterating the following two steps:

(i) Given $\mathbf{X}_n$, simulate an auxiliary variable U_{n+1} from the uniform distribution on $(0, f_1(\mathbf{X}_n))$.

(ii) Given U_{n+1}, simulate $\mathbf{X}_{n+1}$ from the probability density proportional to $|\mathbf{x}|^{-(d-1)}$ on $\{\mathbf{z}: f_1(\mathbf{z}) \geqslant U_{n+1}\}$ and 0 otherwise.

In practice this algorithm's application is limited by the difficulty of implementing (ii). However, various techniques do exist that can be applied to this problem in particular cases; see, for example, Neal (2002) and Møller *et al.* (2001).

To simplify the statement of the result, we shall define T as follows:

$$T = \inf\{n \colon \|P^n(x,\cdot) - \pi\| \leqslant 0.01\}. \tag{7.2}$$

We shall also need to define

$$A = \frac{\inf_{\boldsymbol{\theta}} M(f_1, \boldsymbol{\theta})}{\sup_{\boldsymbol{\theta}} M(f_1, \boldsymbol{\theta})},$$

where

$$M(f_1, \boldsymbol{\theta}) = \sup\{f_1(t\boldsymbol{\theta}) \, ; \; t \geqslant 0\},$$

and $\boldsymbol{\theta}$ ranges over all unit vectors in $\mathbb{R}^d$. Thus A is a measure of departure from radial symmetry.

Theorem 7.2 *Assume that*

$$f_1(\mathbf{X}_0) \geqslant 0.01 \times f_1(\mathbf{0}). \tag{7.3}$$

For various values of A, $T \leqslant n_*(A)$, where $n_*(A)$ is described in Table 1.

Thus for target densities that are reasonably close to being spherically symmetric, this algorithm converges provably quickly in a reasonably small number of iterations, so long as we do not start too far (according to (7.3)) from the mode.

There are many other ways of trying to compute or at least approximate rates of convergence of Markov chains and distances from stationarity. Already mentioned is the notion of capacitance, which has been successfully applied to some MCMC contexts (Frieze *et al.* 1994; Polson 1995; Diaconis and Saloff-Coste 1998). Other geometric and analytic methods for estimating mixing times are described in Diaconis and Saloff-Coste (1998). The difficulty with many of these ingenious methods is that the bounds are often expressed in terms of quantities that are difficult to evaluate, often involving operations such as taking infima

TABLE 1. Bounds for convergence times, $n_*(A)$ as a function of A.

A	$n_*(A)$
1.0	525
0.9	615
0.8	728
0.5	1 400
0.25	3 475
0.1	10 850
0.01	160 000

over complex sets, so are rarely applicable in statistical examples where there is insufficient symmetry to be able to avoid such calculations.

Although it is often possible to estimate drift parameters adequately, the drift condition methodology described here certainly possesses its limitations in terms of applicability in general statistical applications. For instance, it is rarely possible to find adequate bounds for minorization conditions for multiple iterations, that is where $n_0 > 1$ in (4.1), although for single iterations this is usually feasible. However, restricting to the case where $n_0 = 1$ leads to bounds that are commonly affected badly by the curse of dimensionality, where bounds increase exponentially with the dimension of the problem.

In conclusion, the drift function methodology is very natural for target distributions that give rise to unimodal densities, so that there are natural candidates for small sets, namely regions surrounding the density mode. On the other hand, it is inconceivable that the calculations of computable bounds by this method will ever develop to the extent of being routinely applicable to statistical examples.

8 Conclusions

In this article an attempt has been made to try to bring some of the main ideas behind research into MCMC theory to a broad statistical audience. It has been shown that even apparently very simple algorithms possess pitfalls which can be quite subtle and difficult to detect empirically. However, in many cases it is easy to detect problems such as the lack of geometric ergodicity, and propose alternative algorithms with more robust estimation properties.

In practice, algorithms that can be shown to be geometrically ergodic usually produce stable ergodic estimates of expectations. Central limit theorems always hold in this case, Theorem 4.2(a), and it is possible to empirically estimate its error variance by elementary techniques such as using 'batch means'. If geometric ergodicity fails, then the situation is far more sensitive. In this case, CLTs will definitely fail for *some* functions that have finite variance under the target distribution. Since it is usually very difficult to calculate polynomial rates explicitly, we cannot easily be sure about *which* functions this will apply to. Therefore, if the algorithm can be shown to converge slower than geometrically, it makes sense to try to modify or improve the algorithm so that geometric ergodicity holds.

For the random walk Metropolis algorithm, the following practical implications of the theory described here can be made.

1. Sample paths of algorithms from regions of high probability density should be scrutinized for any evidence of excursions having heavy tailed distributions which would then adversely affect the properties of ergodic estimates obtained.

2. Simple one-dimensional transformations can dramatically improve convergence properties (as in the third algorithm in Fig. 3).

3. The use of heavy tailed proposals is a sensible precaution that ought to be adopted in all problems in which the parameters have heavy tailed distribu-

tions (or the tail behaviour is unknown). There is essentially no cost to this, and in heavy tailed examples this will lead to improved stability properties of estimators.

4. It makes sense to monitor acceptance rates in different parts of the Markov chain state space in order to look out for 'sticky patches'.

It is too much to ask that we shall ever be able to determine rigorously the run lengths necessary for convergence and adequate estimation in state-of-the-art MCMC problems. However, it is hoped that a detailed understanding of basic MCMC algorithms will help guide us to the construction of robust and reliable methods, as well as identifying algorithms that are likely to exhibit unstable estimation properties.

Many questions remain unanswered in the theory of MCMC, even for simple problems. The results described in this article for the random walk Metropolis algorithm extend fairly easily to certain similar algorithms such as Langevin algorithms (Roberts and Tweedie 1996a). However, more sophisticated classes of algorithms involving the addition of auxiliary variables such as Hybrid Monte Carlo (Duane *et al.* 1987) or simulated tempering (Marinari and Parisi 1992) are considerably more difficult to handle by any theoretical methods, including the techniques of this article.

One important issue not addressed in any detail in this article is the behaviour of algorithms on high-dimensional distributions. Drift condition analysis can also be ineffective in high-dimensional examples. Here it is an over-simplification to approximate the algorithm's behaviour in terms of a one-dimensional function that spends most of its time shrinking towards a small set. The approximation affects computable bounds most severely. In fact, the qualitative results of Section 4 apply in arbitrary state spaces, although it might be more difficult to identify an appropriate drift function in more complex spaces.

In a similar way, drift conditions are not a very natural way to model very flat and diffuse distributions that spend very little time in any modal area of the target density. In this situation, computable bounds based on drift conditions can be poor, although all the qualitative theory can in principle still be applied. Therefore, these techniques are difficult to apply, for instance, to the simulation of many point processes used in stochastic geometry, such as Strauss and cluster processes (Mecke and Stoyan 2000).

In high dimensions, the performance of algorithms is closely connected to the scaling properties of proposal variances. For instance, if in a d-dimensional problem we use a d-dimensional proposal distribution with constant variance in each dimension, how should that variance, $\sigma^2(d)$, depend upon d? It turns out that for distributions that do not exhibit 'too much' dependence or discontinuity, it is optimal to take $\sigma^2(d) \propto d^{-1}$, and in this case it is possible to demonstrate that the convergence time of the algorithm is $O(d)$ (Roberts *et al.* 1997; Breyer and Roberts 1998). The techniques used to prove results of this kind are very different from the methods outlined here, using instead weak convergence ideas, followed by analysis of the properties of the infinite dimensional limiting process.

In summary, the theory described in this article strongly influences the behaviour of practically implemented MCMC algorithms. Even in examples where precise conditions such as those in (4.2) cannot be written down, it is useful to have an understanding of what could go wrong, and how problems manifest themselves in observed simulation output. Hopefully the article will make some contribution towards the practical understanding of the theory of MCMC methods.

Acknowledgements

I am indebted to Soren Jarner, Omiros Papaspiliopoulos, Jeffrey Rosenthal, and Richard Tweedie for numerous discussions on the subject-matter of this article. I would also like to thank Peter Green and a very helpful referee for many constructive suggestions.

References

Breyer, L. and Roberts, G. O. (1998). From Metropolis to diffusions: Gibbs states and optimal scaling. *Stochastic Processes and their Applications*, **90**, 181–206.

Brooks, S. P. and Roberts, G. O. (1998). Assessing convergence of Markov chain Monte Carlo algorithms. *Statistics and Computing*, **8**, 319–35.

Chan, K. S. (1993). Asymptotic behaviour of the Gibbs sampler. *Journal of the American Statistical Association*, **88**, 320–6.

Cowles, K. and Carlin, B. P. (1996). Markov chain Monte Carlo convergence diagnostics: a comparative review. *Journal of the American Statistical Association*, **91**, 883–904.

Diaconis, P. and Saloff-Coste, L. (1998). What do we know about the Metropolis algorithm? *Journal of Computer and System Sciences*, **57**, 20–36.

Duane, S., Kennedy, A. D., Pendleton, B. J., and Roweth, D. (1987). Hybrid Monte Carlo. *Physics Review Letters B*, **195**, 217–22.

Fort, G. and Moulines, E. (2000a). V-subgeometric ergodicity for a Hastings–Metropolis algorithm. *Statistics and Probability Letters*, **49**, 401–10.

Fort, G. and Moulines, E. (2000b). Computable bounds for subgeometrical and geometrical ergodicity. Unpublished, available on the MCMC preprint server.

Frieze, A., Kannan, R., and Polson, N. G. (1994). Sampling from log-concave distributions. *Annals of Applied Probability*, **4**, 812–37.

Frigessi, A., Martinelli, F., and Stander, J. (1996). Computational complexity of Markov chain Monte Carlo methods for finite Markov random fields. *Biometrika*, **84**, 1–19.

Gammerman, D. (1997). *Markov chain Monte Carlo*. Chapman & Hall, London.

Gilks, W. R., Richardson, S., and Spiegelhalter, D. J., eds (1996). *Markov chain Monte Carlo in Practice.* Chapman & Hall, London.

Green, P. J. (1995). Reversible jump Markov chain Monte Carlo computation and Bayesian model determination. *Biometrika,* **82**, 711–32.

Hastings, W. K. (1970). Monte Carlo sampling methods using Markov chains and their applications. *Biometrika,* **57**, 97–109.

Jarner, S. and Hansen, E. (2000). Geometric ergodicity of Metropolis algorithms. *Stochastic Processes and their Applications,* **85**, 341–61.

Jarner, S. and Roberts, G. O. (2000). Convergence of heavy tailed MCMC algorithms. Unpublished, available on the MCMC preprint server.

Jarner, S. and Roberts, G. O. (2002). Polynomial convergence rates of Markov chains. *Annals of Applied Probability,* **12**, 224–47.

Jerrum, M. and Sinclair, A. (1996). The Markov chain Monte Carlo method: an approach to approximate counting and integration. In *Approximation Algorithms for NP-hard Problems* (ed. D. Hochbaum), pp. 482–520.

Lund, R. B., Meyn, S. P., and Tweedie, R. L. (1996). Computable exponential convergence rates for stochastically ordered Markov processes. *Annals of Applied Probability,* **6**, 218–37.

Lund, R. B. and Tweedie, R. L. (1996). Geometric convergence rates for stochastically ordered Markov chains. *Mathematics of Operations Research,* **21**, 182–94.

Marinari, E. and Parisi, G. (1992). Simulated tempering. *Europhysics Letters,* **19**, 451–8.

Mecke, K. R. and Stoyan, D., eds (2000). *Statistical Physics and Spatial Statistics.* Springer-Verlag, Berlin.

Metropolis, N., Rosenbluth, A., Rosenbluth, M., Teller, A., and Teller, E. (1953). Equations of state calculations by fast computing machines. *Journal of Chemical Physics,* **21**, 1087–91.

Meyn, S. P. and Tweedie, R. L. (1993). *Markov Chains and Stochastic Stability.* Springer-Verlag, London.

Meyn, S. P. and Tweedie, R. L. (1994). Computable bounds for convergence rates of Markov chains. *Annals of Applied Probability,* **4**, 981–1011.

Møller, J., Mira, A., and Roberts, G. O. (2001). Perfect slice samplers. *Journal of the Royal Statistical Society,* B, **63**, 593–606.

Neal, R. (2002). Slice sampling. *Annals of Statistics.* (In press).

Polson, N. G. (1995). Convergence of Markov chain Monte Carlo algorithms. In *Bayesian Statistics 5* (eds J. Bernardo, J. Berger, A. P. Dawid, and A. F. M. Smith), pp. 297–321. Oxford University Press, UK.

Propp, J. G. and Wilson, D. B. (1996). Exact sampling with coupled Markov chains and applications to statistical mechanics. *Random Structures and Algorithms*, **9**, 223–52.

Robert, C. P. (1995). Convergence control techniques for Markov chain Monte Carlo algorithms. *Statistical Science*, **10**, 231–53.

Robert, C. P. and Casella, G. (1999). *Monte Carlo Statistical Methods*. Springer-Verlag, New York.

Roberts, G. O. and Rosenthal, J. S. (1997). Geometric ergodicity and hybrid Markov chains. *Electronic Communications in Probability*, **2**, 13–25.

Roberts, G. O. and Rosenthal, J. S. (1998). Optimal scaling of discrete approximations to Langevin diffusions. *Journal of the Royal Statistical Society*, B, **60**, 255–68.

Roberts, G. O. and Rosenthal, J. S. (2000). The polar slice sampler. Unpublished, available on the MCMC preprint server.

Roberts, G. O. and Tweedie, R. L. (1996a). Exponential convergence of Langevin diffusions and their discrete approximations. *Bernoulli*, **2**, 341–64.

Roberts, G. O. and Tweedie, R. L. (1996b). Geometric convergence and central limit theorems for multidimensional Hastings and Metropolis algorithms. *Biometrika*, **83**, 95–110.

Roberts, G. O. and Tweedie, R. L. (1999). Bounds on regeneration times and convergence rates of Markov chains. *Stochastic Processes and their Applications*, **80**, 211–29.

Roberts, G. O. and Tweedie, R. L. (2000). Rates of convergence of stochastically monotone and continuous time Markov models. *Journal of Applied Probability*, **37**, 359–73.

Roberts, G. O. and Tweedie, R. L. (2003). *Understanding MCMC*. In preparation.

Roberts, G. O., Gelman, A., and Gilks, W. R. (1997). Weak convergence and optimal scaling of random walk Metropolis algorithms. *Annals of Applied Probability*, **7**, 110–20.

Rosenthal, J. S. (1995). Minorization conditions and convergence rates for Markov chain Monte Carlo. *Journal of the American Statistical Association*, **90**, 558–66.

Saloff-Coste, L. (1997). Lectures on finite Markov chains. *Springer Lecture Notes in Mathematics*, **1665**, 301–413.

Stramer, O. and Tweedie, R. L. (1999). Langevin-type models II: self-targeting candidates for MCMC algorithms. *Methodology and Computing in Applied Probability*, **1**, 307–28.

5A
Advances in MCMC: a discussion

Christian P. Robert
Université Paris Dauphine, France

1   Introduction

While Gareth Roberts presents a very good coverage of the latest developments in convergence of MCMC algorithms, and provides the reader with an excellent reminder of the essential notions such as drift functions and minorization conditions, this discussion presents some thoughts on reparameterization and perfect sampling issues. A more general review of the current state of MCMC methods can be found in Cappé and Robert (2000).

2   Parameterization and reparameterization

The simulation of a distribution does not, from a probabilistic point of view, depend on the parameterization adopted for this distribution: if I know how to simulate $x \sim \pi$, I also know how to simulate $y \sim \tilde{\pi}(y) = \pi(\Psi(y))|\mathrm{d}\Psi(y)/\mathrm{d}y|$, when $x = \Psi(y)$! This invariance under reparameterization transfers in practice when considering simulation methods such as cumulative distribution function (CDF) inversion or accept–reject algorithms (since the Jacobian then cancels) or ratio-of-uniform techniques. This is not exactly the case with MCMC algorithms.

For a given Markov kernel, some quantities are invariant to the choice of the parameterization, such as the total variation distance to the stationary distribution, the rates of convergence, the minorization and drift conditions (with appropriate changes of the factors ϵ, ν, C, and V), or the central limit theorem (CLT). But some conditions in Roberts' article are not: for instance, the boundedness of π does not transfer to $\tilde{\pi}$, the use of $V(x) = \pi^u(x)$ is not equivalent to $\tilde{V}(x) = \tilde{\pi}^u(x)$, and the tail behaviours of Proposition 5.1 are not preserved under transformations.

Moreover, as already shown by Gelfand *et al.* (1995) in a hierarchical setting, the choice of the parameterization of the statistical model may drastically affect the convergence of a Gibbs sampler. The same applies to Metropolis–Hastings algorithms, in the sense that the choice of a standard move such as the random walk Metropolis–Hastings algorithm creates a dependency on the parameterization: for instance, Figs. 3(a) and 3(c) in this article illustrate this dependence. In both cases, the proposal is $x^{(t)} + \sigma\epsilon_t$, but in the first case it uses the regular parameterization of the Cauchy distribution, while in the second case it uses the log parameterization. (Similar features occur for the Langevin Metropolis–Hastings algorithm.)

This being stated, different convergence properties will result from different choices of parameterization, given that the Markov kernels differ not only because of a Jacobian term of the form, $|\mathrm{d}\Psi(y)/\mathrm{d}y|$, but also because of different proposal distributions. For instance, the Metropolis–Hastings ratios are

$$\frac{\pi(x')}{\pi(x)} \quad \text{and} \quad \frac{\pi(x')}{\pi(x)} \frac{|\mathrm{d}\Psi(y')/\mathrm{d}y|}{|\mathrm{d}\Psi(y)/\mathrm{d}y|}$$

and will thus lead to different outcomes for the same (x, x').

It is then of interest to consider whether special parameterizations lead to better convergence properties, although a general result of this nature is difficult to fathom! For instance, the improvement from Fig. 3(a) to Fig. 3(c) results from a move from $\mathbb{R}_+$ to $\mathbb{R}$ by the log transform, which is a more natural setting for a random walk since there is no reflecting barrier at 0.

We can however propose an opposite move as a general reparameterization device: bring the support of the distribution down to $(0, 1)^d$ by a componentwise logit transform

$$\theta_i(\omega_i) = \log\left(\frac{\omega_i}{1 - \omega_i}\right)$$

if the support of ω_i is $\mathbb{R}$, leading to the density

$$\tilde{\pi}(\omega) = \pi(\theta(\omega)) \prod_{i=1}^{d} \frac{1}{\omega_i(1 - \omega_i)},$$

with obvious substitutes if the support is $\mathbb{R}_+$ or an interval $[a, b]$.

Consider, for instance, the half-Cauchy distribution of Roberts, which now appears as a distribution on $(0, 1)$ with density

$$\tilde{\pi}(\omega) = \frac{4}{\pi} \frac{(1 - \omega)}{\omega(1 + 4(\omega - 0.5)^2)},$$

or the sticky normal-gamma example, which reduces to

$$\tilde{\pi}(\omega_1, \omega_2) = \left(\frac{\omega_2}{1 - \omega_2}\right)^{n/2}$$

$$\times \exp\left\{-0.5\frac{\omega_2}{1 - \omega_2}\left[S + n\left(\log\frac{\omega_1}{1 - \omega_1} - \bar{x}\right)^2\right]\right\} \frac{1}{(1 - \omega_1)^2} \frac{1}{\omega_2(1 - \omega_2)}.$$

One reason for proposing this reparameterization is that it is associated with a natural random walk, already used by Hastings (1970), based on a $U(-a, a)^d$ proposal, $\omega^{(t+1)} = \omega^{(t)} + a\,\epsilon$. The value of a can then be scaled, based on the acceptance rate, as in standard normal proposals. For instance, for the Cauchy distribution, $a = 2.85$ leads to an average acceptance rate of $\rho = 0.31$ while, for the sticky normal-gamma distribution, $\rho = 0.32$ for $a = 0.544$. (These values of *a* were found by trial-and-error, aiming at an average acceptance rate between 0.3 and 0.5.)

The second reason for this proposal is that the support becomes (relatively) compact and should therefore more likely lead to acceptable performances for the resulting chains. For one thing, there should be no need for a tail study on $(0,1)^d$, even though the set is not compact. For instance, for the Cauchy example, Fig. 1 shows there is no problem with a starting value $x_0 = 50\,000$, since the path does not linger around this extreme value and the sample CDF reproduces the theoretical CDF (for the 'thick pen' test!).

Similarly, in the sticky normal-gamma example, the algorithm is able to recover from a starting value as extreme as $\mu_0 = 15$ and $\tau_0 = 0.001$.

While these are only two simple examples and while higher-dimension problems may enjoy slower convergence or implementation difficulties (like the scaling of a), it does make sense to study in a systematic way the effects of the reparameterization to $(0,1)^d$ on the convergence and mixing (correlation) properties of the corresponding Markov chain. The relative compactness of the support makes excursions into the tails of the original distribution π more conceivable since, when $\omega < a$, the proposed ω' takes values in $(0, \omega + a)$. (In other words, compared with a standard Cauchy random walk on $\mathbb{R}$, ∞ is simply 'one bit away', since the uniform generator on $(0,1)$ takes values on $\{1/M, 2/M, \ldots, (M-1)/M\}$.) The sampler is thus more likely to explore different regions than regular sam-

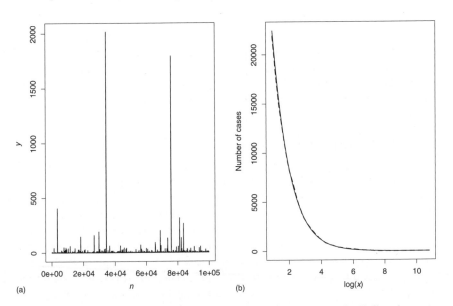

FIG. 1. (a) A path of the Metropolis–Hastings chain for the half-Cauchy target and a random walk on $(0,1)$ using the reparameterization on $(0,1)$. (b) Comparison of the empirical and theoretical tail probabilities.

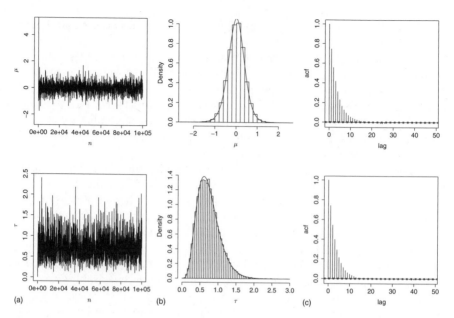

FIG. 2. (a) Path, (b) histogram, and (c) autocorrelation curve of the Metropolis–
Hastings chain for the sticky normal-gamma target for μ (top) and σ (bottom).

plers, even if it misses finer details in the central regions. Given the inherent
simplicity of its implementation, we thus recommend its implementation as a
checking device: while it may converge more slowly than the sampler of interest,
it may also exhibit features that show that the sampler of interest has not yet
converged.

3 Perfect sampling

Since perfect sampling is missing from this volume, and since this cannot be an
oversight from the editors(!), I wonder if this is to be taken as a signal that perfect
sampling is not yet achievable for highly structured statistical systems even if it
is for other highly complex stochastic systems. Given the space limitation, I will
not go into a detailed description (see, for example, Dimakos (2001)), but my
current feeling is that the increasing resemblance of perfect sampling algorithms
to their elder accept–reject algorithms implies that we cannot gain much in terms
of ease and speed of implementation, if at all. Take, for instance, the case when
π, or more generally π/h, is bounded, when h is the proposal distribution, as in
Mengersen and Tweedie (1996) or Corcoran and Tweedie (2002): in such cases,
accept–reject is also feasible and I suspect its efficiency is close (or equivalent)
to the efficiency of the best perfect sampling algorithm.

Another example of this connection with accept–reject is the representation of the stationary measure

$$\pi(A) = \sum_{t=1}^{\infty} \Pr(N_t \in A)\Pr(T^* = t)$$

exhibited by Hobert and Robert (2000), under a minorization condition. Here, T^* denotes the tail process associated with the renewal time T, $\Pr(T^* = t) \propto \Pr(T \geqslant t)$, and N_t is distributed as X_t under a no-renewal condition. When the chain is uniformly ergodic, this is the multigamma coupler of Green and Murdoch (1998). More importantly, this representation, while inspired by perfect sampling arguments, does produce a simulation from π with no call to coupling.

4 Conclusion

At this evolutionary stage of the MCMC methods it may be useful to reconsider the results contained in Roberts' article about speeds of convergence. On the one hand, they brilliantly analyse why the standard half-Cauchy sampler does not work, and also provide some amazing computable bounds. They clearly illustrate the difference between polynomial and geometric rates. On the other hand, they only cover what one could call the most regular problems. For instance, tail conditions are not helpful in identifying a lack of convergence (or proper mixing) due to limited identifiability features or to multimodality, as in mixture models, or yet in (relatively) compact spaces, as those imposed by computer representations. Although these conditions do depend on the dimension of the parameter space, it is not clear how useful they are in high-dimension problems, where the normal-like structure of the posterior distribution is more likely to disappear. More bluntly, total variation distance, and similar measures, grasp only one aspect of MCMC convergence and are not necessarily related to 'good' mixing properties.

This being said, I also believe that more should be done towards error assessment for MCMC methods, such as CLT variance evaluation, and that the current work of Roberts and co-authors is fairly promising in that regard. Moreover, an evaluation of the type of convergence may point out stronger dependence on initial conditions and call for a stricter convergence assessment, thus acting as a pre-diagnostic tool. There remains much to be done though towards automation of such processes.

Acknowledgements

Research partially supported by the EU TMR network ERB-FMRX-CT96-0095 on 'Computational and statistical methods for the analysis of spatial data' and by ESF through the HSSS network and programme.

5B
On some current research in MCMC

Arnoldo Frigessi
Norwegian Computing Centre, Oslo

1 Introduction

MCMC algorithms used for inference in multivariate statistical models with complex dependency structure are often far too slow and their convergence is difficult to predict and monitor. MCMCs are run many times to debug the code, which is in itself difficult, and to perform model choice and validation. For comparison, MCMC-based inference is much more difficult than maximum likelihood inference using Newton–Raphson, due to the un-normalized form of the involved densities, the ambition to approximate the whole distribution instead of only exploring a mode, and the type of convergence of the chain. Of course, in the end we need to trust our inferential results, *even those* obtained with MCMC!

I congratulate the author of this article for his presentation of important aspects of the MCMC theory. He shows that simple MCMC procedures can be completely untrustworthy. In order to make MCMC-based inference more reliable, Roberts describes current research in estimating the number of needed iterations, before starting the MCMC or during the run. As seen, this is in general extraordinarily difficult but can be achieved for specific MCMC procedures and for certain families of models. For these, we can probably trust our results. However, very often these MCMCs are much too slow and useless in practice, and I share some of the words of caution expressed by Roberts in his conclusions. Another direction of research aims to invent new sampling methods whose convergence is still difficult to control a priori, but which are provably faster than other standard algorithms. In the rest of this discussion I shall concentrate on this second research stream, and provide some links to a few approaches I believe to be promising.

2 Adaptive MCMC

In the Metropolis–Hastings framework, the proposal density $q(\mathbf{x}_n, \cdot)$ depends, in step n, at most on the current state $\mathbf{x}_n$. It would be attractive to learn more from the past, i.e. to sample a proposed state $\mathbf{y}$ from a density that depends on the whole sampled trajectory $H_n = \{\mathbf{x}_0, \mathbf{x}_1, \ldots, \mathbf{x}_n\}$. Intuitively, the goal is to extract from H_n information to speed up convergence. For example, if all variables are updated at every step, their proposal could be a current empirical estimate of the target density, in which case the proposal adapts itself while the chain is running. There are also other sampling strategies that perform some type of learning about the target before running the final chain used for inference. The most used one consists in restarting a new MCMC with the 'best' proposal density after many pilot runs with several test proposals. Other approaches use many chains in parallel with state switching mechanisms, as in Chauveau and Vandekerkhove (1999). I consider the term *adaptive* to characterize algorithms that perform learning on the basis of their own, single trajectory.

If the proposal step is allowed to depend on more than just $\mathbf{x}_n$, then the process is not a Markov chain, and its convergence to the target π is not guaranteed. In fact, the usual Markov theory for establishing convergence and speed is no longer valid.

There are several interesting papers on adaptive Hastings sampling, which follow three main ideas. The first is to allow for a limited type of adaptation, which maintains the Markov property, as with delayed rejection (Tierney and Mira 1999; Tjelmeland and Hegstad 2001). The second approach is to assume that the process undergoes some sort of regeneration, so that convergence can be established by guaranteeing that enough regeneration happens, as in Gilks *et al.* (1998), Holden (2000), Gåsemyr (2003), and Sahu and Zhigljavsky (2003). The third approach is to construct a proposal density that depends on the whole past, but converges in law, as in Haario *et al.* (2000) and Holden (2000).

Delayed rejection goes like this. At step n, a first state $\mathbf{y}_1$ is sampled from a proposal density $q(\mathbf{x}_n, \cdot)$. If $\mathbf{y}_1$ were rejected using the usual acceptance probability (3.1), then a state $\mathbf{y}_2$ is sampled from a second proposal transition kernel $q_2((\mathbf{x}_n, \mathbf{y}_1), \cdot)$ and $\mathbf{y}_2$ is accepted with a certain probability that ensures reversibility with respect to π. See Tierney and Mira (1999) for details. Limited adaptation only happens on the basis of the current state and the rejected ones. The reversible Markov chain setting is preserved. In practice, delayed rejection seems to work well, for rather simple and intuitively reasonable choices of q_2, even when taking into account effective CPU time. The integrated autocorrelation time of the delayed algorithm can be shown to be smaller than that of the not delayed one. The delay can be extended to $k \geqslant 2$ rejections, at a cost of complicating the acceptance probabilities. A more general delayed rejection scheme is proposed in Green and Mira (2001), which includes a reversible jump version.

If the state space has an atom, then adaptation can happen at regeneration points. Examples in Gilks *et al.* (1998) consider an adaptive random walk Metropolis algorithm with a Gaussian kernel. Gåsemyr (2003) and Sahu and Zhigljavsky (2003) suggest also a non-parametric variant. Holden (2000) uses, for example, a mixture with components centred in local modes plus a flat component, able to 'discover' new modes. All the proofs of convergence rest on a uniform boundedness condition of, for example, $\pi(\mathbf{y})/q_\theta(\mathbf{y})$ or a Doeblin condition of the transition density. These are strong assumptions, which essentially reduce the process to one with little memory, contradicting the motivating requirements. Boundedness does not seem to be necessary, intuitively.

Haario *et al.* (2000) suggest a process with a multivariate Gaussian scaled approximation of π as a proposal, centred in the current state and with a co-variance matrix that is estimated using H_n. The authors are able to prove that $\sum_{i=1}^{n} f(\mathbf{X}_i)/n$ converges to $E_\pi(f(\mathbf{X}))$ almost surely, for any bounded and measurable function f. The estimated covariance matrices converge to a certain perturbation of the covariances described by π. Here it is assumed that π is supported by a bounded set.

As can be seen, all these approaches have a limited amount of adaptation, and the theory they are based on is too independent of what is learned from H_n. The problem is that adaptation easily leads to processes that do not converge to the correct target π. If, however, the equilibrium measure were not too far from π, then this would often be satisfactory, given the underlying uncertainty about models. For example, Haario *et al.* (1998) first suggested estimating the covariance matrix using only the last k iterates. This gives rise to a kth-order Markov chain whose stationary distribution is slightly biased. The fact that it is easier to handle processes based on the whole H_n leads to another probabilistic tool that I think is worth exploring in this context: variable length Markov chains, proposed by Bühlmann and Wyner (1999). Here $k(n)$ is the memory index at step n, also called the context function in Rissanen (1983), and the proposal would depend on $\{\mathbf{x}_n, \mathbf{x}_{n-1}, \ldots, \mathbf{x}_{k(n)}\}$. This should allow a substantial reduction in computations. If there is a representation as a Markov chain, then this would have a state space of varying dimension. The challenge would be to suggest a way to automatically identify $k(n)$ while running. A final caveat: if π has, say, two separated modes between which transitions are difficult, then adapting very well to only one of them, which had been discovered so far, would delay overall convergence even further. Adaptation should also lead to non-explored areas.

3 Non-reversible Markov chains

Statistical physicists often have good reasons to postulate that their dynamical processes are time reversible. In statistics, reversibility is simply making the analysis of MCMCs easier. Non-reversible MCMCs are more difficult to design. In the literature non-reversible chains are suggested by Hwang *et al.* (1993), Diaconis *et al.* (2000), and Mira and Geyer (2000). In these papers several non-reversible samplers perform better than their reversible versions. Mira and Geyer (2000) study the possibility of modifying a reversible sampler to obtain a better non-reversible one. Reversibility is a rather unnatural requirement, since it is equivalent to $p(\mathbf{x}_1, \mathbf{x}_2)p(\mathbf{x}_2, \mathbf{x}_3)p(\mathbf{x}_3, \mathbf{x}_1) = p(\mathbf{x}_1, \mathbf{x}_3)p(\mathbf{x}_3, \mathbf{x}_2)p(\mathbf{x}_2, \mathbf{x}_1)$. Why should this assumption speed up weak convergence or reduce variances? Furthermore, if A and B are two π-reversible kernels, the product AB rarely is reversible, though π-invariant. More non-reversible samplers should become available in our toolbox.

4   Post-processing

By post-processing I mean every operation that is performed after $(\mathbf{X}_1, \mathbf{X}_2, \ldots,$ $\mathbf{X}_n)$ is sampled, to obtain better estimates of $\mathrm{E}_\pi(f(\mathbf{X}))$. There are several forms of Rao–Blackwellization, for instance $n^{-1} \sum_{i=1}^{n} \mathrm{E}_\pi(f(\mathbf{X}_i)|h(\mathbf{X}_i))$, where h is a sufficient statistic, and $n^{-1} \sum_{i=1}^{n} \mathrm{E}_\pi(f(\mathbf{X}_i)|\mathbf{X}_{i-1} = \mathbf{x}_{i-1})$, where the expectation often can be computed. See, among others, Geyer (1995), Robert and Casella (1999), McKeague and Wefelmeyer (2000), and Greenwood and Wefelmeyer (2000). These forms give a reduction in asymptotic variance for reversible Markov chains. A different idea is un-systematic subsampling, which is a form of random thinning, see MacEachern and Peruggia (2000). Post-processing should be seen also in the framework of model criticism and improvement (see Chapter 14 in this volume). From the sampled trajectories we should be able to find indications for reparameterization and model improvement.

5   More ideas

Here I briefly indicate some further recent literature. Important progress is happening in the area of blocking variables for updating: see Roberts and Sahu (1997), Carter and Kohn (1998), Hobert and Geyer (1998), Knorr-Held and Rue (2000), and Wilkingson and Yeung (2000). Looking to our neighbouring sciences is always inspiring: the influence of modern numerical analysis is visible in Rue (2001), Liu and Sabatti (2000), Neal (2002), and Liu et al. (2001). Operations research and combinatorics invent MCMCs to sample or to count objects, as for example in Jacobson and Matthews (1996), Luby and Vigoda (1999), and Rasmussen (1997). Statistical physics provides us with new algorithms (see, for example, Aldous (1998)) and convergence studies, as in Cooper and Frieze (1999) and Van den Berg and Brouwer (2000). Coupling is not only the basis of exact sampling, it leads also to new algorithms, as in Neal (2002) and Frigessi et al. (2000), and to convergence diagnostics, as in Johnson (1998).

Acknowledgements

Research partially supported by the European Science Foundation through the HSSS programme and by the Norwegian Research Council project no. 121144/ 420.

Additional references in discussion

Aldous, D. (1998). A Metropolis-type optimisation algorithm on the infinite tree. *Algorithmica*, **22**, 388–412.

Bühlmann, P. and Wyner, A. J. (1999). Variable length Markov chains. *Annals of Statistics*, **27**, 480–513.

Cappé, O. and Robert, C. P. (2000). MCMC: Ten years and still running! *Journal of the American Statistical Association*, **95**, 1282–6.

Carter, C. K. and Kohn, R. (1998). Block sampling for Markov random fields.

Unpublished, available on the MCMC preprint server.

Chauveau, D. and Vandekerkhove, P. (1999). Improving convergence of the Hastings–Metropolis algorithm with a learning proposal. Submitted.

Cooper, C. and Frieze, A. M. (1999). Mixing properties of the Swendsen–Wang process on a class of graphs. *Random Structures and Algorithms*, **15**, 242–61.

Corcoran, J. and Tweedie, R. L. (2002). Perfect sampling from independent Metropolis–Hastings chains. *Journal of Statistical Planning and Inference*, **104**, 297–314.

Diaconis, P., Holmes, S., and Neal, R. M. (2000). Analysis of a non-reversible Markov chain sampler. *Annals of Applied Probability*, **10**, 726–52.

Dimakos, X. K. (2001). A guide to exact simulation. *International Statistical Review*, **69**, 27–48.

Frigessi, A., Gåsemyr, J., and Rue, H. (2000). Antithetic coupling of two Gibbs sampler chains. *Annals of Statistics*, **28**, 1128–49.

Gåsemyr, J. (2003). On an adaptive Metropolis–Hastings algorithm with independent proposal. To appear in *Scandinavian Journal of Statistics*.

Gelfand, A. E., Sahu, S. K., and Carlin, B. P. (1995). Efficient parametrization for normal linear mixed models. *Biometrika*, **82**, 479–88.

Geyer, C. J. (1995). Conditioning in Markov Chain Monte Carlo. *Journal of Computational and Graphical Statistics*, **4**, 148–54.

Gilks, W. R., Roberts, G. O., and Sahu, S. K. (1998). Adaptive Markov Chain Monte Carlo through regeneration. *Journal of the American Statistical Association*, **93**, 1045–54.

Green, P. J. and Mira, A. (2001). Delayed rejection in reversible jump Metropolis–Hastings. *Biometrika*, **88**, 1035–53.

Green, P. J. and Murdoch, D. (1998). Exact sampling for Bayesian inference: towards general purpose algorithms. In *Bayesian Statistics 6* (eds J. M. Bernardo, J. O. Berger, A. P. Dawid, and A. F. M. Smith). Oxford University Press, UK.

Greenwood, P. and Wefelmeyer, W. (2000). *Empirical estimators based on MCMC data*. Technical Report. Submitted.

Haario, H., Saksman, E., and Tamminen, J. (1998). An adaptive Metropolis algorithm. Technical Report, University of Helsinki.

Haario, H., Saksman, E., and Tamminen, J. (2000). Adaptive proposal distribution for random walk Metropolis algorithm. *Computational Statistics*, **14**.

Hobert, P. and Geyer, C. J. (1998). Geometric ergodicity of Gibbs and block Gibbs samplers for a hierarchical random effects model. Technical Report, University of Florida.

Hobert, J. P. and Robert, C. P. (2000). Moralizing perfect sampling. Technical

Report, CREST, Paris.

Holden, L. (2000). *Adaptive Chains*. Submitted.

Hwang, C. R., Hwang Ma, S., and Sheu, S. (1993). Accelerating Gaussian diffusions. *Annals of Applied Probability*, **3**, 897–913.

Jacobson, M. T. and Matthews, P. (1996). Generating uniformly distributed random Latin squares. *Journal of Combinatorial Design*, **4**, 405–37.

Johnson, V. E. (1998). A coupling–regeneration scheme for diagnosing convergence in Markov Chain Monte Carlo algorithms. *Journal of the American Statistical Association*, **93**, 238–48.

Knorr-Held, L. and Rue, H. (2000). Block sampling in Markov random field models for disease mapping. Statistics reports, Department of Mathematical Sciences, Norwegian University of Science and Technology, Trondheim, Norway.

Liu, J. S. and Sabatti, C. (2000). Generalized Gibbs sampler and multigrid Monte Carlo for Bayesian computation. *Biometrika*, **87**, 353–69.

Liu, J. S., Liang, F., and Wong, W. H. (2001). The use of multiple-try method and local optimization in Metropolis sampling. *Journal of the American Statistical Association*, **96**, 561–73.

Luby, M. and Vigoda, E. (1999). Fast convergence of the Glauber dynamics for sampling independent sets. *Random Structures and Algorithms*, **15**, 229–41.

MacEachern, S. N. and Peruggia, M. (2000). Subsampling the Gibbs sampler: variance reduction. *Statistics and Probability Letters*, **47**, 91–8.

McKeague, I. W. and Wefelmeyer, W. (2000). Markov chain Monte Carlo and Rao–Blackwellization. *Journal of Statistics, Planning and Inference*, **85**, 171–82.

Mengersen, K. L. and Tweedie, R. L. (1996). Rates of convergence of the Hastings and Metropolis algorithms. *Annals of Statistics*, **24**, 101–21.

Mira, A. and Geyer, C. J. (2000). On non-reversible Markov chains. In *Fields Institute Communications: Monte Carlo Methods* (ed. N. Madras), **26**, pp. 93–108.

Rasmussen, L. E. (1997). Approximately counting cliques. *Random Structures and Algorithms*, **11**, 395–411.

Rissanen, J. (1983). A universal data compression system. *IEEE Transactions on Information Theory*, **29**, 656–64.

Roberts, G. O. and Sahu, S. K. (1997). Updating schemes, correlation structure, blocking and parameterization for the Gibbs sampler. *Journal of the Royal Statistical Society*, B, **59**, 291–317.

Rue, H. (2001). Fast sampling of Gaussian Markov random fields. *Journal of the Royal Statistical Society*, B, **63**, 325–38.

Sahu, S. K. and Zhigljavsky, A. A. (2003). Self regenerative Markov chain Monte Carlo with Adaptation. To appear in *Bernoulli*.

Tierney, L. and Mira, A. (1999). Some adaptive Monte Carlo methods for Bayesian inference. *Statistics in Medicine*, **18**, 2507–15.

Tjelmeland, H. and Hegstad, B. K. (2001). Mode jumping proposals in MCMC. *Scandinavian Journal of Statistics*, **28**, 205–24.

Van den Berg, J. and Brouwer, R. (2000). Random sampling for the monomer-dimer model on a lattice. *Journal of Mathematical Physics*, **41**, 1585–97.

Wilkinson, D. J. and Yeung, S. K. H. (2000). Conditional simulation from highly structured Gaussian systems, with application to blocking-MCMC for the Bayesian analysis of very large linear models. Technical Report, Department of Statistics, University of Newcastle.

6
Trans-dimensional Markov chain Monte Carlo

Peter J. Green
University of Bristol, UK

1 Introduction

Readers of this book will need no further convincing of the importance of Markov chain Monte Carlo (MCMC) in numerical calculations for highly structured stochastic systems, and in particular for posterior inference in Bayesian statistical models. Another chapter (Chapter 5 in this volume) is devoted to a discussion of some of the currently important research directions in MCMC generally. This article is more narrowly focused on MCMC methods for what can be called 'trans-dimensional' problems, to borrow a nicely apt phrase from Roeder and Wasserman (1997): those where the dynamic variable of the simulation, the 'unknowns' in the Bayesian set-up, does not have fixed dimension.

Statistical problems where 'the number of things you don't know is one of the things you don't know' are ubiquitous in statistical modelling, both in traditional modelling situations such as variable selection in regression, and in more novel methodologies such as object recognition, signal processing, and Bayesian non-parametrics. All such problems can be formulated generically as a matter of joint inference about a model indicator k and a parameter vector θ_k, where the model indicator determines the dimension n_k of the parameter, but this dimension varies from model to model. Almost invariably in a frequentist setting, inference about these two kinds of unknown is based on different logical principles, but, at least formally, the Bayes paradigm offers the opportunity of a single logical framework – it is the joint posterior $p(k, \theta_k|Y)$ of a model indicator and parameter given data Y that is the basis for inference. How can this be computed?

We set the joint inference problem naturally in the form of a simple Bayesian hierarchical model. We suppose, given a prior $p(k)$ over models k in a countable set $\mathcal{K}$, and for each k, a prior distribution $p(\theta_k|k)$ and a likelihood $p(Y|k, \theta_k)$ for the data Y. For definiteness and simplicity of exposition, we suppose that $p(\theta_k|k)$ is a density with respect to n_k-dimensional Lebesgue measure, and that there are no other parameters, so that where there are parameters common to all models these are subsumed into each $\theta_k \in \mathcal{R}^{n_k}$. Additional parameters, perhaps in additional layers of a hierarchy, are easily dealt with. Note that in this article, all probability distributions are proper.

The joint posterior

$$p(k, \theta_k | Y) = \frac{p(k)p(\theta_k | k)p(Y | k, \theta_k)}{\sum_{k' \in \mathcal{K}} \int p(k')p(\theta'_{k'} | k')p(Y | k', \theta'_{k'}) d\theta'_{k'}}$$

can always be factorized as

$$p(k, \theta_k | Y) = p(k | Y)p(\theta_k | k, Y),$$

that is as the product of posterior model probabilities and model-specific parameter posteriors. This identity is very often the basis for reporting the inference,
and in some of the methods mentioned below is also the basis for computation.

It is important to appreciate the generality of this basic formulation. In
particular, note that it embraces not only genuine model-choice situations, where
the variable k indexes the collection of discrete models under consideration,
but also settings where there is really a single model, but one with a variable
dimension parameter, for example a functional representation such as a series
whose number of terms is not fixed. In the latter case, arising sometimes in
Bayesian non-parametrics, for example, k is unlikely to be of direct inferential
interest.

It can be argued that responsible adoption of a Bayesian hierarchical model
of the kind introduced above presupposes that, for example, parameter priors
$p(\theta_k | k)$ should be compatible in the sense that inference about functions of parameters that are meaningful in several models should be approximately invariant to k. Such compatibility could in principle be exploited in the construction
of MCMC methods, although I am not aware of general methods for doing so.
However, it is philosophically tenable that no such compatibility is present, and
we shall not assume it.

Trans-dimensional MCMC has many applications other than to Bayesian
statistics. Much of what follows will apply equally to them all; however, for
simplicity, I shall use the Bayesian motivation and terminology throughout.

In Section 2, reversible jump MCMC is discussed, and this is related to other
model-jumping approaches in Section 3. The following section treats alternatives
to model-jumping, and Section 5 discusses and analyses some of the issues involved in choosing between the within- and across-model approaches. In Section
6, a simple fully-automated reversible jump sampler is introduced, and finally
Section 7 notes some recent methodological extensions.

2   Reversible jump MCMC

In the direct approach to computation of the joint posterior $p(k, \theta_k | Y)$ via
MCMC we construct a single Markov chain simulation, with states of the form
(k, θ_k); we might call this an *across-model* simulation. We address other approaches in later sections.

The state space for such an across-model simulation is $\bigcup_{k \in \mathcal{K}} (\{k\} \times \mathcal{R}^{n_k})$;
mathematically, this is not a particularly awkward object, and our construction

involves no especially challenging novelties. However, such a state space is at least a little non-standard! Formally, our task is to construct a Markov chain on a general state space with a specified limiting distribution, and as usual in Bayesian MCMC for complex models, we use the Metropolis–Hastings paradigm to build a suitable reversible chain. As we see in the next subsection, on the face of it, this requires measure-theoretic notation, which may be unwelcome to some readers. The point of the 'reversible jump' framework is to render the measure theory invisible, by means of a construction using only ordinary densities. In fact, in the formulation given below, different and I hope improved from that of Green (1995), even the fact that we are jumping dimensions becomes essentially invisible!

2.1 Metropolis–Hastings on a general state space

We wish to construct a Markov chain on a state space $\mathcal{X}$ with invariant distribution π. As usual in MCMC we will consider only reversible chains, so the transition kernel P satisfies the detailed balance condition

$$\int_{(x,x')\in A\times B} \pi(\mathrm{d}x)P(x,\mathrm{d}x') = \int_{(x,x')\in A\times B} \pi(\mathrm{d}x')P(x',\mathrm{d}x) \qquad (2.1)$$

for all Borel sets $A, B \subset \mathcal{X}$. In Metropolis–Hastings, we make a transition by first drawing a candidate new state x' from the proposal measure $q(x,\mathrm{d}x')$ and then accepting it with probability $\alpha(x,x')$, to be derived below. If we reject, we stay in the current state, so that $P(x,\mathrm{d}x')$ has an atom at x. This contributes the same quantity $\int_{A\cap B} P(x,\{x\})\pi(\mathrm{d}x)$ to each side of (2.1); subtracting this leaves

$$\int_{(x,x')\in A\times B} \pi(\mathrm{d}x)q(x,\mathrm{d}x')\alpha(x,x') = \int_{(x,x')\in A\times B} \pi(\mathrm{d}x')q(x',\mathrm{d}x)\alpha(x',x). \quad (2.2)$$

It can be shown (Green 1995; Tierney 1998) that $\pi(\mathrm{d}x)q(x,\mathrm{d}x')$ is dominated by a symmetric measure μ on $\mathcal{X}\times\mathcal{X}$; let its density (Radon–Nikodym derivative) with respect to this μ be f. Then (2.2) becomes

$$\int_{(x,x')\in A\times B} \alpha(x,x')f(x,x')\mu(\mathrm{d}x,\mathrm{d}x') = \int_{(x,x')\in A\times B} \alpha(x',x)f(x',x)\mu(\mathrm{d}x',\mathrm{d}x)$$

and, using the symmetry of μ, this is clearly satisfied for all Borel A, B if

$$\alpha(x,x') = \min\left\{1, \frac{f(x',x)}{f(x,x')}\right\}.$$

This might be written more informally in the apparently familiar form

$$\alpha(x,x') = \min\left\{1, \frac{\pi(\mathrm{d}x')q(x',\mathrm{d}x)}{\pi(\mathrm{d}x)q(x,\mathrm{d}x')}\right\}. \qquad (2.3)$$

2.2 A constructive representation in terms of random numbers

Fortunately, the apparent abstraction in this prescription can be circumvented in most cases. By considering how the transition will be implemented in a computer program, the dominating measure and Radon–Nikodym derivatives can be generated implicitly. Take the case where $\mathcal{X} \subset \mathcal{R}^d$, and suppose π has a density (also denoted π) with respect to the d-dimensional Lebesgue measure. At the current state x, we generate, say, r random numbers u from a known joint density g, and then form the proposed new state as some suitable deterministic function of the current state and the random numbers: $x' = h(x, u)$, say. The left-hand side of (2.2) can then be written as an integral with respect to (x, u):

$$\int_{(x,x') \in A \times B} \pi(x)g(u)\alpha(x, x') \, dx \, du.$$

The reverse transition from x' to x would be made with the aid of random numbers $u' \sim g'$ giving $x = h'(x', u')$. If the transformation from (x, u) to (x', u') is a diffeomorphism (the transformation and its inverse are differentiable), then we can first write the right-hand side of (2.2) as an integral with respect to (x', u'), and then apply the standard change-of-variable formula. We then see that the $(d + r)$-dimensional integral equality (2.2) holds if

$$\pi(x)g(u)\alpha(x, x') = \pi(x')g'(u')\alpha(x', x) \left| \frac{\partial(x', u')}{\partial(x, u)} \right|,$$

where the last factor is the Jacobian of the diffeomorphism from (x, u) to (x', u'). Thus, a valid choice for α is

$$\alpha(x, x') = \min \left\{ 1, \frac{\pi(x')g'(u')}{\pi(x)g(u)} \left| \frac{\partial(x', u')}{\partial(x, u)} \right| \right\}, \tag{2.4}$$

involving only ordinary joint densities.

While this reversible jump formalism perhaps is a little indirect, it proves a flexible framework for constructing quite complex moves using only elementary calculus. In particular, the possibility that $r < d$ covers the case, typical in practice, that given $x \in \mathcal{X}$, only a lower-dimensional subset of $\mathcal{X}$ is reachable in one step. (The Gibbs sampler is the best-known example of this, since in that case only some of the components of the state vector are changed at a time, although the formulation here is more general as it allows the subset not to be parallel to the coordinate axes.) Separating the generation of the random innovation u and the calculation of the proposal value through the deterministic function $x' = h(x, u)$ is deliberate; it allows the proposal distribution $q(x, B) = \int_{\{u : h(x, u) \in B\}} g(u) du$ to be expressed in many different ways, for the convenience of the user.

2.3 The trans-dimensional case

However, the main benefit of this formalism is that expression (2.4) applies, without change, in a variable dimension context, if we use the same symbol $\pi(x)$

for the target density whatever the dimension of x in different parts of $\mathcal{X}$. Provided that the transformation from (x, u) to (x', u') remains a diffeomorphism, the individual dimensions of x and x' can be different. The dimension-jumping is indeed 'invisible'.

In this setting, suppose the dimensions of x, x', u, and u' are d, d', r, and r' respectively, then we have functions $h : \mathcal{R}^d \times \mathcal{R}^r \to \mathcal{R}^{d'}$ and $h' : \mathcal{R}^{d'} \times \mathcal{R}^{r'} \to \mathcal{R}^d$, used respectively in $x' = h(x, u)$ and $x = h'(x', u')$. For the transformation from (x, u) to (x', u') to be a diffeomorphism requires that $d + r = d' + r'$, so-called 'dimension-matching'; if this equality failed, then the mapping and its inverse could not both be differentiable.

2.4 Details of application to the model-choice problem

Returning to our generic model-choice problem, we wish to use these reversible jump moves to sample the space $\mathcal{X} = \bigcup_{k \in \mathcal{K}} (\{k\} \times \mathcal{R}^{n_k})$ with an invariant distribution π, which here is $p(k, \theta_k | Y)$.

Just as in ordinary MCMC, we typically need multiple types of moves to traverse the whole space $\mathcal{X}$. Each move is a transition kernel reversible with respect to π, but only in combination do we obtain an ergodic chain. The moves will be indexed by m in a countable set $\mathcal{M}$, and a particular move m proposes to take $x = (k, \theta_k)$ to $x' = (k', \theta'_{k'})$ or vice versa for a specific pair (k, k'); we denote $\{k, k'\}$ by $\mathcal{K}_m$. The detailed balance equation (2.2) is replaced by

$$\int_{(x,x') \in A \times B} \pi(\mathrm{d}x) q_m(x, \mathrm{d}x') \alpha_m(x, x') = \int_{(x,x') \in A \times B} \pi(\mathrm{d}x') q_m(x', \mathrm{d}x) \alpha_m(x', x)$$

for each m, where now $q_m(x, \mathrm{d}x')$ is the joint distribution of move type m and destination x'. The complete transition kernel is obtained by summing over m, so that for $x \notin B$, $P(x, B) = \sum_M \int_B q_m(x, \mathrm{d}x') \alpha_m(x, x')$, and it is easy to see that (2.1) is then satisfied.

The analysis leading to (2.3) and (2.4) is modified correspondingly, and yields

$$\alpha_m(x, x') = \min \left\{ 1, \frac{\pi(x')}{\pi(x)} \frac{j_m(x')}{j_m(x)} \frac{g'_m(u')}{g_m(u)} \left| \frac{\partial(x', u')}{\partial(x, u)} \right| \right\}.$$

Here $j_m(x)$ is the probability of choosing move type m when at x, the variables x, x', u, and u' are of dimensions d_m, d'_m, r_m, and r'_m respectively, with $d_m + r_m = d'_m + r'_m$, we have $x' = h_m(x, u)$ and $x = h'_m(x', u')$, and the Jacobian has a form correspondingly depending on m.

Of course, when at $x = (k, \theta_k)$, only a limited number of moves m will typically be available, namely those for which $k \in \mathcal{K}_m$. With probability $1 - \sum_{m:k \in \mathcal{K}_m} j_m(x)$ no move is attempted.

2.5 Some remarks and ramifications

In understanding the reversible jump framework, it may be helpful to stress the key role played by the joint state-proposal equilibrium distributions. The fact that the degrees of freedom in these joint distributions are unchanged when x and

x' are interchanged allows the possibility of reversible jumps across dimensions, and these distributions directly determine the move acceptance probabilities.

Note that the framework gives insights into Metropolis–Hastings that apply quite generally. State-dependent mixing over a family of transition kernels in general infringes the detailed balance, but is permissible if, as here, the move probabilities $j_m(x)$ enter properly into the acceptance probability calculation. Note also the contrast between this randomized proposal mechanism, and the related idea of mixture proposals, where the acceptance probability does not depend on the move actually chosen; see the discussion in Besag *et al.* (1995, Appendix 1). Contrary to some accounts that connect it with the jump in dimension, the Jacobian comes into the acceptance probability simply through the fact that the proposal destination $x' = h(x, u)$ is specified indirectly.

Finally, note that in a large class of problems involving nested models, the only dimension change necessary is the addition or deletion of a component of the parameter vector (think of polynomial regression, or autoregression of variable order). In such cases, omission of a component is often equivalent to setting a parameter to zero. These problems can be handled in a seemingly more elementary way, through allowing proposal distributions with an atom at zero: the usual Metropolis–Hastings formula for the acceptance probability holds for densities with respect to arbitrary dominating measures, so the reversible jump formalism is not explicitly needed. Nevertheless, it leads to exactly the same algorithm.

Other authors have provided different pedagogical descriptions of reversible jump. Waagepetersen and Sorensen (2001) provide a tutorial following the lines of Green (1995) but in much more detail, and Besag (1997, 2000) gives a novel formulation in which variable dimension notation is circumvented by embedding all θ_k within one compound vector; this has something in common with the product-space formulations in the next subsection.

3   Relations to other across-model approaches

Several alternative formalisms for across-model simulation are more or less closely related to reversible jump.

3.1   Jump diffusion

In addressing challenging computer vision applications, Grenander and Miller (1994) proposed a sampling strategy they termed jump diffusion. This comprised two kinds of move – between-model jumps, and within-model diffusion according to a Langevin stochastic differential equation. Since in practice continuous-time diffusion has to be approximated by a discrete-time simulation, they were in fact using a trans-dimensional Markov chain. Had they corrected for the time discretization by a Metropolis–Hastings accept/reject decision (giving a so-called Metropolis-adjusted Langevin algorithm or MALA) (Besag 1994), this would have been an example of reversible jump.

Phillips and Smith (1996) applied jump diffusion creatively to a variety of

Bayesian statistical tasks, including mixture analysis, object recognition, and variable selection.

3.2 Point processes, with and without marks

Point processes form a natural example of a distribution with variable-dimension support, since the number of points in view is random; in the basic case, a point has only a location, but more generally may be accompanied by a *mark*, a random variable in a general space.

A continuous-time Markov chain approach to simulating certain spatial point processes, by regarding them as the invariant distributions of spatial birth-and-death processes, was suggested and investigated by Preston (1977) and Ripley (1977). More recently, Geyer and Møller (1994) proposed a Metropolis–Hastings sampler as an alternative to using birth-and-death processes; their construction is a special case of reversible jump.

Stephens (2000) notes that various trans-dimensional statistical problems can be viewed as abstract marked point processes: in these models, the items of which there are a variable number are regarded as marked points. For example, in a normal mixture model the points represent the mean–variance pairs of the components, marked with the component weights. Stephens borrows the birth-and-death simulation idea to develop a methodology for finite mixture analysis, and also suggests that the approach appears to have much wider application, citing change point analysis and regression variable selection as partially worked examples. The key feature of these three settings that allows the approach to work is the practicability of integrating out latent variables so that the likelihood is fully available. See also Hurn *et al.* (2003) for an application to mixtures of regressions. Cappé *et al.* (2001) have recently given a rather complete analysis of the relationship between reversible jump and continuous-time birth-and-death samplers.

3.3 Product-space formulations

Several relatives of reversible jump work in a product-space framework, that is, one in which the simulation keeps track of all θ_k, not only the 'current' one. The state space is therefore $\mathcal{K} \times \otimes_{k \in \mathcal{K}} \mathcal{R}^{n_k}$ instead of $\bigcup_{k \in \mathcal{K}} (\{k\} \times \mathcal{R}^{n_k})$. This has the advantage of circumventing the trans-dimensional character of the problem, at the price of requiring that the target distribution be augmented to model all θ_k simultaneously. For some variants of this approach, this is just a formal device, for others it leads to significantly extra work.

Let θ_{-k} denote the composite vector consisting of all $\theta_l, l \neq k$, catenated together. Then the joint distribution of $(k, (\theta_l : l \in \mathcal{K}), Y)$ can be expressed as

$$p(k)p(\theta_k|k)p(\theta_{-k}|k, \theta_k)p(Y|k, \theta_k), \qquad (3.1)$$

since we make the natural assumption that $p(Y|k, (\theta_l : l \in \mathcal{K})) = p(Y|k, \theta_k)$. It is easily seen that the third factor $p(\theta_{-k}|k, \theta_k)$ has no effect on the joint posterior $p(k, \theta_k|Y)$; the choice of these conditional distributions, which Carlin and Chib (1995) call 'pseudo-priors', is entirely a matter of convenience, but may influence the efficiency of the resulting sampler.

Carlin and Chib (1995) adopted pseudo-priors that were conditionally independent: $p(\theta_{-k}|k,\theta_k) = \prod_{l\neq k}p(\theta_l|k)$, and assumed $p(\theta_l|k)$ does not depend on k for $k \neq l$. They used a Gibbs sampler, updating k and all θ_l in turn. This evidently involves sampling from the pseudo-priors, and they therefore propose to design these pseudo-priors to ensure reasonable efficiency, which requires their approximate matching to the posteriors: $p(\theta_l|k) \approx p(\theta_l|l,Y)$.

Green and O'Hagan (1998) pointed out both that Metropolis–Hastings moves could be made in this setting, and that in any case there was no need to update $\{\theta_l, l \neq k\}$ to obtain an irreducible sampler. In this form the pseudo-priors are only used in computing the update of k. Dellaportas *et al.* (2002) proposed and investigated a 'Metropolised Carlin and Chib' approach, in which joint model indicator/parameter updates were made, and in which it is only necessary to resample the parameter vectors for the current and proposed models.

Godsill (2001) introduces a general 'composite model space' framework that embraces all of these methods, including reversible jump, facilitating comparisons between them. He devised the formulation (3.1), or rather, a more general version in which the parameter vectors θ_k are allowed to overlap arbitrarily, each θ_k being identified with a particular sub-vector of one compound parameter. This framework helps to reveal that a product-space sampler may or may not entail possibly cumbersome additional simulation, updating parameters that are not part of the 'current' model. It also gives useful insight into some of the important factors governing the performance of reversible jump, and Godsill offers some suggestions on proposal design.

Godsill's formulation deserves further attention, as it provides a useful language for comparing approaches, and in particular examining one of the central unanswered questions in trans-dimensional MCMC. Suppose the simulation leaves model k and later returns to it. With reversible jump, the values of θ_k are lost as soon as we leave k, while with some versions of the product-space approach the values are retained until k is next visited. Intuitively either strategy has advantages and disadvantages for sampler performance, so which is to be preferred?

4 Alternatives to joint model-parameter sampling

The direct approach of a single across-model simulation is in many ways the most appealing, but alternative indirect methods that treat the unknowns k and θ_k differently should not be neglected.

4.1 Integrating out the parameters

If in each model k the prior is conjugate for the likelihood, then $p(\theta_k|k,Y)$ may be explicitly available, and thence can be calculated the *marginal likelihoods*

$$p(Y|k) = \frac{p(\theta_k|k)p(Y|k,\theta_k)}{p(\theta_k|k,Y)}$$

and finally the posterior probabilities $p(k|Y) \propto p(k)p(Y|k)$. In the very limited cases where this is possible, Bayesian inference about k, and about θ_k given k, can be conducted separately, and trans-dimensional simulations are not needed.

The approach has been taken a little further by Godsill (2001), who considers cases of 'partial analytic structure', where some of the parameters in θ_k may be integrated out, and the others left unchanged in the move that updates the model, to give an across-model sampler with probable superior performance.

4.2 Within-model simulation

If samplers for the within-model posteriors $p(\theta_k|Y,k)$ are available for each k, then the joint posterior inference for (k, θ_k) can be constructed by combining separate simulations conducted within each model. See Carlin and Louis (1996, Section 6.3.1) for a more detailed discussion.

The posterior $p(\theta_k|Y,k)$ for the parameters θ_k is in any case a within-model notion, and is the target for an ordinary Bayesian MCMC calculation for model k. Since

$$\frac{p(k_1|Y)}{p(k_0|Y)} = \frac{p(k_1)}{p(k_0)} \frac{p(Y|k_1)}{p(Y|k_0)}$$

(the second factor being the *Bayes factor* for model k_1 vs. k_0), to find the posterior model probabilities $p(k|Y)$ for all k it is sufficient to estimate the marginal likelihoods

$$p(Y|k) = \int p(\theta_k, Y|k) \, d\theta_k$$

separately for each k, using individual MCMC runs. Several different methods have been devised for this task.

Noting that $p(Y|k)$ can be expressed as $\{\int [p(\theta_k|k,Y)/p(Y|k,\theta_k)] \, d\theta_k\}^{-1}$ or more directly as $\int p(Y|k,\theta_k)p(\theta_k|k) \, d\theta_k$, leads respectively to the estimates

$$\hat{p}_1(Y|k) = N \left/ \sum_{t=1}^{N} \left\{ p(Y|k, \theta_k^{(t)}) \right\}^{-1} \right. \quad \text{and} \quad \hat{p}_2(Y|k) = N^{-1} \sum_{t=1}^{N} p(Y|k, \theta_k^{(t)}),$$

based on MCMC samples $\theta_k^{(1)}, \theta_k^{(2)}, \ldots$ from the posterior $p(\theta_k|Y,k)$ and the prior $p(\theta_k|k)$, respectively. Both of these are simulation-consistent, but have high variance, with possibly few terms contributing substantially to the sums in each case. Composite estimates, based like $\hat{p}_1$ and $\hat{p}_2$ on the importance sampling identity $E_p(f) = E_q(fp/q)$, perform better, including those of Newton and Raftery (1994) and Gelfand and Dey (1994). For example, Newton and Raftery propose to simulate from a mixture $\tilde{p}(\theta_k; Y, k)$ of the prior and posterior, and use

$$\hat{p}_3(Y|k) = \frac{\sum_{t=1}^{N} p(Y|k, \theta_k^{(t)})w(\theta_k^{(t)})}{\sum_{t=1}^{N} w(\theta_k^{(t)})},$$

where $w(\theta_k) = p(\theta_k|k)/\tilde{p}(\theta_k; Y, k)$.

Chib (1995) has introduced new, indirect estimates of the marginal likelihood based on the identity $p(Y|k) = p(Y|k, \theta_k^\star)p(\theta_k^\star|k)/p(\theta_k^\star|k, Y)$ for any fixed parameter point $\theta_k^\star$. The factors in the numerator are available, and in contexts where the parameter can be decomposed into blocks with explicit full conditionals, the denominator can be estimated using simulation calculations that use the same Gibbs sampling steps as the posterior simulation. Note, however, that Neal (1999) has demonstrated that Chib's application of this idea to mixture models is incorrect. Chib and Jeliazkov (2001) extend the idea to cases where Metropolis–Hastings is needed.

5 Some issues in choosing a sampling strategy

Several studies have addressed the strengths and weaknesses of reversible jump MCMC and the other trans-dimensional setups above compared with within-model simulations that compute marginal likelihoods and thence Bayes factors. Particularly noteworthy are Dellaportas *et al.* (2002), Godsill (2001), and Han and Carlin (2001). Each of these discusses some of the issues involved and provides comparisons of implementations and performance on test problems, although, understandably in the present state of our knowledge with these methods, it is hard to see any of these as entirely definitive.

One of the key matters influencing the choice here is the number of models to be entertained, taking account of the degree of homogeneity between them. The ideal situation for the 'within-model' strategy would be a case where the models are all of a different character, and fully-tested samplers with acceptable performance are already available for each. In such a case, building an across-model sampler could be very laborious compared with adding marginal likelihood calculations to each model separately.

Some authors have recorded poor performance with reversible jump methods. Since reversible jump algorithms embrace all Metropolis–Hastings methods for the across-model state space, it is hard to believe that there are no methods in this huge class that would give acceptable performance. It would be fairer to say that existing examples of reversible jump implementations may be poor templates for constructing samplers in some new situations. A difficulty is that the across-model state space may be hard to visualize so that some of the intuition that guides the construction of samplers in simpler spaces is not available.

Others have deemed reversible jump methods cumbersome to construct and difficult to tune. There seems to be a need for further methodological work, developing broader classes of across-model samplers, with associated visualization techniques, to assist in construction and tuning. Very recent work by Brooks *et al.* (2003) may be a good step in this direction; see Section 7.2. Of course, as in other domains for MCMC, fully-automated sampler construction would be a tremendous advantage: a very limited step towards this is introduced in Section 6 below.

Finally, the across-model approach does have another potential benefit – the possibility that jumping models can improve mixing. This is discussed next.

5.1   Is it good to jump?

There are various not entirely substantiated claims in the literature to the effect that jumping between parameter subspaces is either inherently damaging to MCMC performance and should therefore be avoided where possible, or alternatively that it is helpful for performance, and might even be attempted when it is not strictly necessary.

For example, Richardson and Green (1997) describe a simple experiment, illustrated in their Fig. 9, demonstrating that in a particular example of a mixture problem with a strongly multimodal posterior, mixing is clearly improved by using a trans-dimensional sampler, while Han and Carlin (2001) claim to have 'intuition that some gain in precision should accrue to MCMC methods that avoid a model space search'.

In truth, the proper answer is 'it depends', but some simple analysis does reveal some of the issues. There are three main situations that might be considered: in the first, we require full posterior inference about (k, θ_k). A second possibility is that we wish to make within-model inference about θ_k separately, for each of a (perhaps small) set of values of k. The third case is where k is really fixed, and the other models are ruled out a priori. This third option is clearly the least favourable for trans-dimensional samplers: visits of the (k, θ_k) chain to the 'wrong' models are wasted from the point of view of extracting useful posterior information; let us try to analyse whether superior mixing in these other models can nevertheless make it worthwhile to use a trans-dimensional sampler.

5.2   The two-model case

For simplicity, we suppose there are just two models, $k = 1$ and 2, and let π_k denote the distribution of θ_k given k: only π_1 is of interest. We have transition kernels Q_{11} and Q_{22}, with $\pi_k Q_{kk} = \pi_k$ for each k (we use a notation apparently aimed at the finite state space case, but it is quite general: for example, πQ means the probability measure $(\pi Q)(B) = \int \pi(\mathrm{d}x) Q(x, B)$). We now consider the option of also allowing between-model transitions, with the aid of kernels Q_{12} and Q_{21}; for realism, these are improper distributions, integrating to less than 1, reflecting the fact that in practice across-model Metropolis–Hastings moves are frequently rejected. When a move is rejected, the chain does not move, contributing a term to the 'diagonal' of the transition kernel; thus we suppose there exist diagonal kernels D_1 and D_2, and we have the global balance conditions for the across-model moves: $\pi_1 D_1 + \pi_2 Q_{21} = \pi_1$ and $\pi_2 D_2 + \pi_1 Q_{12} = \pi_2$.

Assuming that we make a random choice between the two moves available from each state, α and β being the probabilities of choosing to attempt the between-model move in models 1 and 2 respectively, the overall transition kernel for the across-model sampler is

$$P = \begin{pmatrix} (1-\alpha)Q_{11} + \alpha D_1 & \alpha Q_{12} \\ \beta Q_{21} & (1-\beta)Q_{22} + \beta D_2 \end{pmatrix}$$

using an obvious matrix notation. The invariant distribution is easily seen to be

$\pi = (\gamma\pi_1, (1-\gamma)\pi_2)$, where $\gamma = \beta/(\alpha+\beta)$.

Now suppose we run the Markov chain given by P, but look at the state only when in model 1. By standard Markov chain theory, the resulting chain has kernel $\widetilde{Q}_{11} = (1-\alpha)Q_{11} + \alpha D_1 + \alpha Q_{12}\{I - (1-\beta)Q_{22} - \beta D_2\}^{-1}\beta Q_{21}$. The comparison we seek is that between using Q_{11} or the more complicated strategy that amounts to using $\widetilde{Q}_{11}$, but we must take into account differences in costs of computing. Suppose that executing Q_{11} or Q_{22} has unit cost per transition, while attempting and executing the across-model moves has cost c times greater. Then, per transition, the equilibrium cost of using P is $\gamma(1-\alpha) + \gamma\alpha c + (1-\gamma)\beta c + (1-\gamma)(1-\beta)$, and this gives on average γ visits to model 1. The relative cost in computing resources of using $\widetilde{Q}_{11}$ instead of Q_{11} therefore simplifies to $(1-\alpha) + 2\alpha c + \alpha(1-\beta)/\beta$ (using the relationship $\gamma\alpha = (1-\gamma)\beta$).

If we choose to measure performance by asymptotic variance of a specific ergodic average, then we have integrated autocorrelation times τ and $\widetilde{\tau}$ for Q_{11} and $\widetilde{Q}_{11}$ respectively, and jumping models is a good idea if

$$\tau < \widetilde{\tau}\{(1-\alpha) + 2\alpha c + \alpha(1-\beta)/\beta\}.$$

Of course, $\widetilde{\tau}$ depends on α and β.

5.3 Finite state space example

It is interesting to compute these terms for toy finite state space examples where the eigenvalue calculations can be made explicitly. For example, taking $D_1 = D_2 = 0.8I$, corresponding to an 80% rejection rate for between-model moves, and all the Q matrices to be symmetric reflecting random walks on $m = 10$ states, with differing probabilities of moving, to model differently 'sticky' samplers, specifically $(Q_{11})_{i,i\pm 1} = 0.03$, $(Q_{12})_{i,i\pm 1} = 0.2 \times 0.1$, $(Q_{21})_{i,i\pm 1} = 0.2 \times 0.1$, and $(Q_{22})_{i,i\pm 1} = 0.3$, we find that model jumping is worthwhile for all c up to about 15, with optimal $\alpha \approx 1$ and $\beta \approx 0.1$. This is a situation where the rapid mixing in model 2 compared with that in model 1 justifies the expense of jumping from 1 to 2 and back again.

5.4 Tempering-by-embedding

Such considerations raise the possibility of artificially embedding a given statistical model into a family indexed by k, and conducting an across-model simulation simply to improve performance – that is, as a kind of simulated tempering (Marinari and Parisi 1992). A particular example of the benefit of doing so was given by Hodgson (1999) in constructing a sampler for the restoration of ion channel signals. A straightforward approach to this task gave poor mixing, essentially because of a high posterior correlation between the model hyperparameters and the hidden binary signal. This correlation is higher when the data sequence is longer, so a tempering-by-embedding solution was to break the data into blocks, with the model hyperparameters allowed to change between adjacent blocks. The part of the prior controlling this artificial model elaboration was adjusted empirically to give moderately high rates of visiting the real model, while spending sufficient time in the artificial heterogeneous models for the harmful correlation to be substantially diluted.

Further evidence that model-jumping can provide effective tempering, admittedly in a somewhat contrived setting, was provided by Richardson and Green (1997). They compared fixed-k and variable-k samplers for a normal mixture problem with k components, applied to a symmetrized bimodal data set. In this case, there was substantial posterior support for $k = 2$ and 4; MCMC-based inference about parameters conditional on $k = 3$ was greatly superior using the variable-k sampler.

6 An automatic generic trans-dimensional sampler

The possibility of automating the construction of a MCMC sampler for any given target distribution is attractive but elusive. It would be a tremendous practical advantage if the user could just specify the target in algebraic form, perhaps together with a few numerical constants such as starting values, and leave the computer both to construct an algorithm and then run it to create a reliable sample.

The nearest we can come to this ideal at present, for sampling from a fixed-dimensional density, is the random-walk Metropolis (RWM) sampler (see Roberts, Chapter 5 in this volume), in the most simple form where all variables are simultaneously updated. Other possibilities, requiring a little more user input, are Langevin methods, or the hybrid samplers of Duane *et al.* (1987). RWM is not a panacea. From a theoretical perspective it is imperfect since even geometric ergodicity is not guaranteed, as it requires conditions on the relative size of the tails of the target and proposal densities. In fact, no kind of ergodicity is certain, since there may be holes in the support of the target and/or the proposal density which could prevent irreducibility, but such pathologies are easily avoided. There is also the important practical consideration that updating all variables at once prevents the exploitation of factorizations of the target that make the acceptance probabilities for lower-dimensional updates particularly cheap to compute.

In spite of these drawbacks, the RWM methods are useful and it would be valuable to have an analogous class of methods for trans-dimensional problems, particularly for exploratory use. In this section we propose a rather naive approach to this quest, but as experiments show, the results are quite promising.

Suppose that for each model k, we are given a fixed n_k-vector μ_k and a fixed $n_k \times n_k$-matrix B_k. Consider the situation where we are currently in state (k, θ_k) and have proposed a move to model k', drawn from some transition matrix $(r_{k,k'})$. The form of the proposed new parameter vector depends on whether $n_{k'}$ is less than, equal to, or more than n_k. We set:

$$
\theta'_{k'} =
\begin{cases}
\mu_{k'} + B_{k'}[RB_k^{-1}(\theta_k - \mu_k)]_1^{n_{k'}} & \text{if } n_{k'} < n_k, \\[2mm]
\mu_{k'} + B_{k'}RB_k^{-1}(\theta_k - \mu_k) & \text{if } n_{k'} = n_k, \\[2mm]
\mu_{k'} + B_{k'}R\begin{pmatrix} B_k^{-1}(\theta_k - \mu_k) \\ u \end{pmatrix} & \text{if } n_{k'} > n_k.
\end{cases}
$$

Here $[\cdots]_1^m$ denotes the first m components of a vector, R is a fixed orthogonal matrix of order $\max\{n_k, n_{k'}\}$, and u is an $(n_{k'} - n_k)$-vector of random numbers with density $g(u)$.

Note that if $n_{k'} \leqslant n_k$, then the proposal is deterministic (apart from the choice of k'). Since everything is linear, the Jacobian is trivially calculated: if $n_{k'} > n_k$, we have

$$
\left| \frac{\partial(\theta_{k'})}{\partial(\theta_k, u)} \right| = \frac{|B_{k'}|}{|B_k|}.
$$

Thus the acceptance probability is $\min\{1, A\}$, where

$$
A = \frac{p(k', \theta'_{k'}|y)}{p(k, \theta|y)} \frac{r_{k',k}}{r_{k,k'}} \frac{|B_{k'}|}{|B_k|} \times
\begin{cases}
g(u) & \text{if } n_{k'} < n_k, \\[1mm]
1 & \text{if } n_{k'} = n_k, \\[1mm]
g(u)^{-1} & \text{if } n_{k'} > n_k.
\end{cases}
$$

Since it is orthogonal, the matrix R plays no role in this calculation.

If the model-specific targets $p(\theta_k|k, y)$ were normal distributions, with means μ_k and variances $B_k B_k^T$, if the innovation variables u were standard normal, and if we could choose $r_{k,k'}/r_{k',k} = p(k'|Y)/p(k|Y)$, these proposals would already be in detailed balance, with no need to compute the Metropolis–Hastings accept/reject decision. This is the motivation for the idea.

This suggests that, provided the $p(\theta_k|k, y)$ are reasonably unimodal, with mean and variance approximately equal to μ_k and $B_k B_k^T$, this simple sampler may be effective. A simple modification, likely to give performance more robust to heavy tails in the targets, would be to use t-distributions in place of the normals for u. Another modification, plausibly likely on general grounds to improve mixing, is to randomize over the orthogonal matrix R, or, more simply, take R to be a random permutation matrix. By the usual argument about randomized proposals (Besag et al. 1995, Appendix 1), this randomization can be ignored when calculating the acceptance probability.

In applications, we are only likely to have approximations to the mean and variances of $p(\theta_k|k, y)$ when we can conduct pilot runs within each model separately – thus limiting the idea to cases where the set of models $\mathcal{K}$ is finite and small. In our implementation, we loop over these models and perform a short run of RWM on each to estimate the means μ_k and variances $B_k B_k^T$. We then find the lower triangular square root of the variance B_k (and its determinant) by Cholesky decomposition; the advantage of using a lower triangular B_k is that we can use forward substitution to multiply B_k^{-1} into a vector.

Finally, the idea might have broader applicability if the pilot runs were used also to detect and correct gross departures from normality – perhaps a transformation to reduce skewness could be estimated, for example. We have not explored such modifications.

6.1 Examples

This method has been implemented as a stand-alone Fortran program, available from the author by email (`P.J.Green@bris.ac.uk`), which calls a function written by the user to compute $\log p(k, \theta_k, y)$. The only other information required about the problem, also provided by this function, are the number of models, their dimensions, and rough settings for the centre and spread of each variable, used for initial values and spread parameters for the RWM moves. The code is set up to alternate between model-jumping moves as described above, and within-model moves by RWM.

We have tried the approach on two non-trivial examples; the codes for these two were identical apart from the information just described.

6.1.1 *Variable selection in a small logistic regression problem*

Dellaportas *et al.* (2002) illustrate their comparisons between model-jumping algorithms on a small data set. This is a 2×2 factorial experiment with a binomially distributed response variable. All five interpretable models are entertained, with numbers of parameters (n_k) equal to 1, 2, 2, 3, and 4 respectively. We follow Dellaportas *et al.* exactly in terms of prior settings, etc. One million sweeps of the automatic sampler, many more than is needed for reliable results, takes about 18 seconds on a 800 MHz PC. The acceptance rate for the model-jumping moves was 29.4%, and the integrated autocorrelation time for estimating $E(k|y)$ was estimated by Sokal's method (Green and Han 1992) to be 2.90. The posterior model probabilities were computed to be (0.005, 0.493, 0.011, 0.439, 0.052), consistent with the results of Dellaportas *et al.*

6.1.2 *Change point analysis for a point process*

We revisit the change point analysis of the coal mine disaster data used in Green (1995). In this illustration, we condition on the number of change points k lying in the set $\{1, 2, 3, 4, 5, 6\}$ (which covers most of the posterior probability). All the prior settings, etc. are as in Green (1995). There are $2k + 1$ parameters in model k. For this problem, 1 million sweeps takes about 28 seconds on a 800 MHz PC. The posterior for the number of change points was estimated to be (0.058, 0.251, 0.294, 0.236, 0.117, 0.044) for the values $k = 1$ to 6. Note that this differs somewhat from the results reported in Green (1995); in fact, if the sampler derived there is run for 200 000 sweeps instead of 40 000, the results become very similar. On this problem, the automatic sampler mixes much less well: the acceptance rate for model-jumping is 5.9%, while the integrated autocorrelation time estimate rises to 118. This decline in performance is presumably due to the extreme multimodal character of many of the parameter posteriors.

For comparison, the sampler described in Green (1995) takes 14 seconds for 1 000 000 sweeps on this computer, with an acceptance rate of 21% and estimated autocorrelation time of 67.8. On this basis, the relative efficiency of the automatic sampler is only $(14 \times 67.8)/(28 \times 118) \approx 29\%$, but of course the implementation time was far less.

6.2   Limitations of this approach

I have stressed that this automatic sampler cannot be expected to have very broad applicability. However, its successful use on the second example above shows that it can be surprisingly tolerant to multimodality in the model-specific targets. (As seen in Figs. 3 and 4 of Green (1995), multimodality is evident even for $k = 2$, and it rapidly becomes very severe for larger k.) In such cases it is necessary for the proposal spread factors provided to be sufficiently large that there is adequate jumping between modes in the pilot runs within each model.

This approach is unlikely to be useful for more than a small set of models, so that, for example, variable selection between many variables is probably out of reach. It may, however, be worth exploring whether quite crude approximations to the means and variances of each target give adequate performance, and whether such approximations can be generated for variable selection problems without conducting pilot runs on all models.

7   Methodological extensions

7.1   Delayed rejection

An interesting modification to Metropolis–Hastings is the splitting rejection idea of Tierney and Mira (1999), which has recently been extended to the reversible jump setting by Green and Mira (2001), who call it delayed rejection.

The idea is simple: if a proposal is rejected, instead of 'giving up', staying in the current state, and advancing time to the next transition, we can instead attempt a second proposal, usually from a different distribution, and possibly dependent on the value of the rejected proposal. It is possible to set the acceptance probability for this second-stage proposal so that a detailed balance is obtained, individually within each stage. The idea can be extended to further stages.

By the results of Peskun (1973), generalized in Tierney (1998), such a strategy is always advantageous in terms of reducing asymptotic variances of ergodic averages, on a sweep-by-sweep basis, since the probability of moving increases by stage. Whether it is actually worth doing will depend on whether the reduction in Monte Carlo variance compensates for the additional computing time for the extra stages; the experiments reported in Green and Mira (2001) suggest that this can be the case.

The second-stage acceptance probability is calculated by an argument along the same lines as that in Section 2 above. We use two vectors of random numbers u_1 and u_2, drawn from densities g_1 and g_2, respectively, and two deterministic functions mapping these and the current state into the proposed new states, $y = h_1(x, u_1)$ and $z = h_2(x, u_1, u_2)$, respectively. Both u_1 and u_2 appear in the expression for z to allow this second-stage proposal to be dependent on the rejected first-stage candidate y; for example, z may be a move in a different 'direction' in some sense.

The first-stage proposal is accepted with probability $\alpha_1(x, y)$ calculated as usual:

$$\alpha_1(x, y) = \min\left\{1, \frac{\pi(y)g_1'(u_1')}{\pi(x)g_1(u_1)}\left|\frac{\partial(y, u_1')}{\partial(x, u_1)}\right|\right\},$$

where u_1' is such that $x = h_1'(y, u_1')$.

Consider the case where the move to y is rejected. We need to find an acceptance probability $\alpha_2(x, z)$ giving a detailed balance for the second-stage proposal z. As in the single-stage case, we set up a diffeomorphism between (x, u_1, u_2) and $(z, \widetilde{u_1}, \widetilde{u_2})$, where $\widetilde{u_1}$ and $\widetilde{u_2}$ would be the random numbers used in the first- and second-stage attempts from z. Then $x = h_2'(z, \widetilde{u_1}, \widetilde{u_2})$ and the first-stage move, if accepted, would have taken us to $y^\star = h_1'(z, \widetilde{u_1})$.

Completing the argument as in Section 2.2, equating integrands after making the change of variable, we find that a valid choice for the required acceptance probability is

$$\alpha_2(x, z) = \min\left\{1, \frac{\pi(z)}{\pi(x)}\frac{\widetilde{g_1}(\widetilde{u_1})\widetilde{g_2}(\widetilde{u_2})}{g_1(u_1)g_2(u_2)}\frac{[1 - \alpha_1(z, y^\star)]}{[1 - \alpha_1(x, y)]}\left|\frac{\partial(z, \widetilde{u_1}, \widetilde{u_2})}{\partial(x, u_1, u_2)}\right|\right\}. \tag{7.1}$$

In a model-jumping problem, we would commonly take y and z to lie in the same model, and $y^\star$ to be in the same model as x, although as discussed by Green and Mira, other choices are possible. For example, where models are ordered by complexity, z might lie between x and y, so that the second-stage proposal is less 'bold'.

7.2 Efficient proposal choice for reversible jump MCMC

The most substantial recent methodological contribution to reversible jump MCMC generally is work by Brooks *et al.* (2003) on the efficient construction of proposal distributions.

This is focused mainly on the quantitative question of selecting the proposal density ($g(u)$ in Section 2.2) well, having already fixed the transformation ($x' = h(x, u)$) into the new space. The qualitative choice of such a transformation h is perhaps more elusive and challenging.

Brooks, Giudici, and Roberts propose several new methods, falling into two main classes. The first is concerned with analysis of the acceptance rate (2.3) as a function of u for small u (on an appropriate scale of measurement). The second class of methods works in a product-space formulation somewhat like that in Section 3, including some novel formulations with autoregressively constructed auxiliary variables.

Their methods are implemented and compared on examples including choice of autoregressive models, graphical Gaussian models, and mixture models.

7.3 Diagnostics for reversible jump MCMC

Monitoring of MCMC convergence on the basis of empirical statistics of the sample path is important, while not of course a substitute for a good theoretical understanding of the chain. There has been some concern that across-model chains are intrinsically more difficult to monitor, perhaps implying that their use should be avoided.

In truth, the degree of confidence that convergence has been achieved provided by 'passing' a diagnostic convergence test declines very rapidly as the dimension of the state space increases. In more than, say, a dozen dimensions, it is difficult to believe that a few, even well-chosen, scalar statistics give an adequate picture of convergence of the multivariate distribution. It is high, rather than variable, dimensions that are the problem.

In most trans-dimensional problems in Bayesian MCMC it is easy to find scalar statistics that retain their definition and interpretation across models, typically those based on fitted and predicted values of observations, and these are natural candidates for diagnostics, requiring no special attention to the variable dimension.

However, recognizing that there is often empirical evidence that a trans-dimensional simulation stabilizes more quickly within models than it does across models, there has been recent work on diagnostic methods that addresses the trans-dimensional problem more specifically. A promising approach by Brooks and Giudici (2000) is based on analysis of sums of squared variation in sample paths from multiple runs of a sampler. This is decomposed into terms attributable to between- and within-run, and between- and within-model variation.

Acknowledgements

Much of the work discussed here benefited strongly through HSSS funding, both of workshops and conferences where I had a chance to discuss it, and of individual visits to pursue research collaborations. I am grateful to the ESF, and to all my collaborators on reversible jump MCMC research problems: Carmen Fernández, Paolo Giudici, Miles Harkness, Matthew Hodgson, Antonietta Mira, Agostino Nobile, Marco Pievatolo, Sylvia Richardson, Luisa Scaccia, and Claudia Tarantola.

References

Besag, J. (1994). Contribution to the discussion of paper by Grenander and Miller. *Journal of the Royal Statistical Society*, B, **56**, 591–2.

Besag, J. (1997). Contribution to the discussion of paper by Richardson and Green. *Journal of the Royal Statistical Society*, B, **59**, 774.

Besag, J. (2000). Markov chain Monte Carlo for statistical inference. Working paper, No. 9. Center for Statistics and the Social Sciences, University of Washington.

Besag, J., Green, P. J., Higdon, D., and Mengersen, K. (1995). Bayesian computation and stochastic systems (with discussion). *Statistical Science*, **10**, 3–66.

Brooks, S. P. and Giudici, P. (2000). MCMC convergence assessment via two-way ANOVA. *Journal of Computational and Graphical Statistics*, **9**, 266–85.

Brooks, S. P., Giudici, P., and Roberts, G. O. (2003). Efficient construction of reversible jump MCMC proposal distributions. *Journal of the Royal Statistical Society*, B. To appear.

Cappé, O., Robert, C. P., and Rydén, T. (2001). Reversible jump MCMC converging to birth-and-death MCMC and more general continuous time samplers. Preprint no. 2001:27, Centre for Mathematical Sciences, Lund University.

Carlin, B. P. and Chib, S. (1995). Bayesian model choice via Markov chain Monte Carlo. *Journal of the Royal Statistical Society*, B, **57**, 473–84.

Carlin, B. P. and Louis, T. A. (1996). *Bayes and empirical Bayes methods for data analysis*. Chapman & Hall, London.

Chib, S. (1995). Marginal likelihood from the Gibbs output. *Journal of the American Statistical Association*, **90**, 1313–21.

Chib, S. and Jeliazkov, I. (2001). Marginal likelihood from the Metropolis–Hastings output. *Journal of the American Statistical Association*, **96**, 270–81.

Dellaportas, P., Forster, J. J., and Ntzoufras, I. (2002). On Bayesian model and variable selection using MCMC. *Statistics and Computing*, **12**, 27–36.

Duane, S., Kennedy, A. D., Pendleton, B, J., and Roweth, D. (1987). Hybrid Monte Carlo. *Physical Letters*, B, **195**, 216–22.

Gelfand, A. E. and Dey, D. K. (1994). Bayesian model choice: asymptotics and exact calculations. *Journal of the Royal Statistical Society*, B, **56**, 501–14.

Geyer, C. J. and Møller, J. (1994). Simulation procedures and likelihood inference for spatial point processes. *Scandinavian Journal of Statistics*, **21**, 359–73.

Godsill, S. J. (2001). On the relationship between MCMC model uncertainty methods. *Journal of Computational and Graphical Statistics*, **10**, 230–48.

Green, P. J. (1995). Reversible jump Markov chain Monte Carlo computation and Bayesian model determination. *Biometrika*, **82**, 711–32.

Green, P. J. and Han, X.-L. (1992). Metropolis methods, Gaussian proposals and antithetic variables. In *Stochastic Models, Statistical Methods and Algorithms in Image Analysis*, Lecture Notes in Statistics, No. 74 (eds A. Frigessi, P. Barone, and M. Piccioni), pp. 142–64. Springer-Verlag, Berlin.

Green, P. J. and O'Hagan, A. (1998). *Model Choice with MCMC on Product Spaces Without Using Pseudo-priors*. Department of Mathematics, University of Nottingham.

Green, P. J. and Mira, A. (2001). Delayed rejection in reversible jump Metropolis–Hastings. *Biometrika*, **88**, 1035–53.

Grenander, U. and Miller, M. I. (1994). Representations of knowledge in complex systems. *Journal of the Royal Statistical Society*, B, **56**, 549–603.

Han, C. and Carlin, B. P. (2001). MCMC methods for computing Bayes factors: a comparative review. *Journal of the American Statistical Association*, **96**, 1122–32.

Hodgson, M. E. A. (1999). A Bayesian restoration of an ion channel signal. *Journal of the Royal Statistical Society*, B, **61**, 95–114.

Hurn, M. A., Justel, A., and Robert, C. P. (2003). Estimating mixtures of regressions. *Journal of Computational and Graphical Statistics*. To appear.

Marinari, E. and Parisi, G. (1992). Simulated tempering: a new Monte Carlo scheme. *Europhysics Letters*, **19**, 451–8.

Neal, R. M. (1999). Erroneous results in 'Marginal likelihood from the Gibbs output'. http://www.cs.utoronto.ca/~radford.

Newton, M. A. and Raftery, A. E. (1994). Approximate Bayesian inference with the weighted likelihood bootstrap (with discussion). *Journal of the Royal Statistical Society*, B, **56**, 3–48.

Peskun, P. H. (1973). Optimum Monte Carlo sampling using Markov chains. *Biometrika*, **60**, 607–12.

Phillips, D. B. and Smith, A. F. M. (1996). Bayesian model comparison via jump diffusions. In *Practical Markov Chain Monte Carlo* (eds W. R. Gilks, S. Richardson, and D. J. Spiegelhalter), pp. 215–39. Chapman & Hall, London.

Preston, C. J. (1977). Spatial birth-and-death processes. *Bulletin of the International Statistical Institute*, **46**, 371–91.

Richardson, S. and Green, P. J. (1997). On Bayesian analysis of mixtures with an unknown number of components (with discussion). *Journal of the Royal Statistical Society*, B, **59**, 731–92.

Ripley, B. D. (1977). Modelling spatial patterns (with discussion). *Journal of the Royal Statistical Society*, B, **39**, 172–212.

Roeder, K. and Wasserman, L. (1997). Contribution to the discussion of paper by Richardson and Green. *Journal of the Royal Statistical Society*, B, **59**, 782.

Stephens, M. (2000). Bayesian analysis of mixture models with an unknown number of components – an alternative to reversible jump methods. *Annals of Statistics*, **28**, 40–74.

Tierney, L. (1998). A note on Metropolis–Hastings kernels for general state spaces. *Annals of Applied Probability*, **8**, 1–9.

Tierney, L. and Mira, A. (1999). Some adaptive Monte Carlo methods for Bayesian inference. *Statistics in Medicine*, **18**, 2507–15.

Waagepetersen, R. and Sorensen, D. (2001). A tutorial on reversible jump MCMC with a view toward QTL-mapping. *International Statistical Review*, **69**, 49–61.

6A
Proposal densities and product-space methods

Simon J. Godsill
University of Cambridge, UK

1 Introduction

In this article, Peter Green has provided the most informative and complete survey currently available of the issues surrounding Bayesian model uncertainty using MCMC methods. Naturally he has focused on the reversible jump methods which have dominated the field over recent years, although he has pointed out the close relationships with the product-space formulations of Besag (1997), Carlin and Chib (1995), Godsill (2001), and Dellaportas *et al.* (2002).

Practitioners have readily adopted reversible jump methods for use in complex Bayesian problems, and yet even after several years in the literature the methods have a reputation for being somehow 'difficult' to understand and still more difficult to implement successfully. Green's article helps further to demystify the reversible jump methodology by providing some useful new discussion material and a very transparent derivation of the basic results. The article also discusses recent developments in proposal design and introduces a novel proposal mechanism for general models.

So, is there any methodological work still to be done in the field? Green's article is very clear on this issue: the basic frameworks, whether pure reversible jump or combined with product-space ideas, are well established; however, the specifics of a generic implementation are not, and it is clear that it is these areas that can most benefit from renewed research effort. In fact, given the general interest from a wide variety of disciplines in this topic, there have been surprisingly few methodological developments in the area up to now. In the following sections I will focus on just two developing topics: automatic proposal generation and product-space methods.

2 Construction of proposal densities

Key to the effective operation of reversible jump methods is the choice of proposal distributions. Most applications to date have constructed proposals on an *ad hoc* basis, attempting to place the proposed parameters in regions of high probability mass in the new model's parameter space. This can be successful in some cases, but it is tempting to seek an automatic procedure that does not require the tuning and pilot runs often required in these *ad hoc* settings. There have been some recent advances in this direction, as discussed in Green's article. I will attempt to interpret Green's new proposal mechanism in the light of more standard Gaussian approximation methods for reversible jump.

In Godsill (2001) it was suggested that an optimal choice of proposal would be the full conditional posterior probability for the parameters in the new model, i.e. set $q(\theta_{k'}) = p(\theta_{k'}|k', y)$, in which case the acceptance ratio simplifies to

$$\frac{p(k'|y)q(k|k')}{p(k|y)q(k'|k)},$$

where $q(k'|k)$ is the probability that model k' is proposed from model k. We note that this highly idealized setting leads to an acceptance probability which is constant for all values of $\theta_{k'}$. This is in agreement with the objectives of the 'higher order methods' proposed in Brooks *et al.* (2000), in which proposals are specifically designed so that one or more derivatives of the acceptance ratio are set to zero locally at a chosen representative 'centring' point. Brooks *et al.* (2000) lend some theoretical weight to the suggestion that $p(\theta_{k'}|k', y)$ is a good proposal density by proving that the *capacitance* of the Markov chain is optimized by this choice of proposal in a simple two-model setting, and I would conjecture that the result is also valid in much more general model selection settings. This suggestion leads to the much-used idea that the proposal distribution, while in practice never *equal* to $p(\theta_{k'}|k', y)$, should be designed to approximate the full conditional if possible. A natural starting point here is a Gaussian proposal matching the first- and second-order moments of the target conditional distribution. Using the same notation as Green, we propose an $n_{k'}$-dimensional vector v from the standard normal density, and generate the proposed parameter as $\theta_{k'} = \mu_{k'} + B_{k'}v$, giving an acceptance ratio

$$A = \frac{p(k', \theta_{k'}|y)q(k|k')q(v')|B_k'|}{p(k, \theta_k|y)q(k'|k)q(v)|B_k|}. \tag{2.1}$$

In the case that the target parameter conditional is indeed Gaussian with moments $\mu_{k'}$ and $B_{k'}B_{k'}^T$ this simplifies to

$$A = \frac{p(k'|y)q(k|k')}{p(k|y)q(k'|k)},$$

and we have perfectly adapted Metropolis–Hastings on the marginal model index space. Thus, in the case of a Gaussian target with correctly specified Gaussian proposals, the acceptance probabilities of this and Green's proposed method are identical and hence the two samplers explore the model indexing space equally rapidly. The interesting possibilities with Green's proposal arise when the targets are non-Gaussian, since the acceptance ratio of Green's method then appears to eliminate some of the variability in the acceptance ratio by replacing $q(v')/q(v)$ in (2.1) with a single term $q(u)$, which is the density of a generally much lower-dimensional Gaussian than either $q(v)$ or $q(v')$. The question then arises as to how the target ratios $p(k', \theta_{k'}|y)/p(k, \theta_k|y)$ compare between the two approaches, and it is clear that when the target is strongly non-Gaussian, either method may well lead to high acceptance probabilities. However, this is qualitative thinking and it would be very interesting to discover how these two related approaches fared relative to one another in the examples of Section 6.1 of Green's article. Clearly, the approximation of each candidate model, even with a Gaussian, will require a great deal of work for large model spaces. However, one can envisage hybrid approaches in which a substantial proportion of the parameters remain fixed in model-jumping proposals, as in many standard reversible jump implementations to date, while a Gaussian approximation is applied to a more manageable subset of parameters conditional on those fixed parameters.

3   Product-space methods

Product-space methods provide another interesting viewpoint on model uncertainty, since they allow simulation to be performed, at least conceptually, on a fixed dimension space. Various authors have shown that reversible jump algorithms can be obtained as special cases of product space methods (and vice versa); see Besag (1997), Godsill (2001), and Dellaportas *et al.* (2002).

Very general classes of model space sampling can be written in the composite model space framework of Godsill (2001), which is a product-space representation, allowing for any overlap between parameters of different models that is computationally convenient (for example, nested models and variable selection models are easily encoded within the framework). Consider a 'pool' of N parameters $\theta = (\theta_1, \ldots, \theta_N)$. A candidate model k can be described in terms of this pool of parameters by means of an indexing set $\mathcal{I}(k) = \{i_1(k), i_2(k), \ldots, i_{l(k)}(k)\}$ which contains $l(k)$ distinct integer values between 1 and N. The parameters $\theta_{\mathcal{I}(k)}$ of model k are then defined as $\theta_{\mathcal{I}(k)} = (\theta_i;\ i \in \mathcal{I}(k))$. In the simplest case we have $\mathcal{I}(k) = k$, which leads to a straightforward model selection scenario with no overlap between model parameters. In other cases, such as variable selection or nested models, it may be convenient to 'share' parameters between more than one model. The posterior distribution for the composite model space can now be expressed as

$$p(k, \theta|y) = \frac{p(y|k, \theta_{\mathcal{I}(k)})\, p(\theta_{\mathcal{I}(k)}|k)\, p(\theta_{-\mathcal{I}(k)}|\theta_{\mathcal{I}(k)}, k)\, p(k)}{p(y)}, \qquad (3.1)$$

where $\theta_{-\mathcal{I}(k)} = (\theta_i;\ i \in \{1, \ldots, N\} - \mathcal{I}(k))$ denotes the parameters *not* used by model k. All of the terms in this expression are defined explicitly by the chosen likelihood and prior structures except for $p(\theta_{-\mathcal{I}(k)}|\theta_{\mathcal{I}(k)}, k)$, the 'prior' for the parameters in the composite model which are not used by model k. It is easily seen that any proper distribution can be assigned arbitrarily to these parameters without affecting the required marginals for the remaining parameters. This fixed dimensionality distribution can now be used as the target for an MCMC algorithm. One of the possible benefits of such a scheme, as suggested in Godsill (2001), is that parameters from models other than the current model can in principle be stored and used to construct effective proposals when those other models are proposed again. There are, however, some basic pitfalls which can beset this type of approach. The first is storage: one would not wish to store all parameters of all models in memory if the pool of parameters θ is large. The second is tractability. Consider the pure model selection scenario in which there is no overlap between parameters, i.e. $\mathcal{I}(k) = k$. Now, it might seem sensible to set the target density for some or all of the unused parameters equal to the data conditional posterior, in which case they can be updated at each iteration according to any suitable MCMC scheme and they will always be generating useful values for future model-jumping proposals. This can be achieved by choosing the arbitrary prior distribution for these parameters as follows:

$$p(\theta_{-k}|\theta_k, k) = \prod_{j \neq k} p(\theta_j|j, y).$$

However, it is easily verified that model-jumping proposals under such a scheme require the marginal model probabilities in the acceptance ratio, and hence the method is self-destroying since it requires us to know exactly one of the quantities we wish to estimate! Clearly, the arbitrary prior probability should not be chosen in this intuitively reasonable way.

Another approach that might have similar benefits would be to assign some reasonable distributions for the arbitrary priors, such as a tractable approximation to the data conditional posterior distribution for those parameters, but to apply a very slowly mixing Markov chain when updating these parameters. This would allow the parameters of each model to retain some memory of their earlier configuration when that model was last selected by the MCMC. A promising approach related to this concept has been devised by Brooks *et al.* (2000). In it they assume a nested structure to the models, and augment the parameter space with sufficient auxiliary variables to make the total parameter space equal in dimensionality to the most complex candidate model. These auxiliary variables are then slowly updated at each iteration according to an autoregressive Markov chain with a standard Gaussian stationary distribution. The auxiliary variables are then used directly to generate deterministic model-jumping proposals to higher-order models. The extra memory and persistence introduced into the chain in this way is shown to induce a better exploration of the tails of the model order distribution for a graphical model example.

More general schemes with this flavour can easily be devised based on the general product-space framework. It may be reasonable, for example, to use one or more of the auxiliary parameters to help construct a random proposal rather than a deterministic one. Another extension would address the memory storage problems: rather than update all the auxiliary parameters using a slowly mixing Markov chain, update only those parameters within some suitably chosen 'neighbourhood' of the currently selected model. The remaining auxiliary parameters are sampled independently directly from their target distribution, which would be carefully chosen for tractability, and hence do not need to be sampled until their corresponding model number is proposed.

It seems reasonable that ideas of this sort can lead to improved performance of reversible jump algorithms. There will usually however be an increased burden of computational load and memory storage requirements, so it must remain to be seen whether performance improvements are sufficient to merit the extra work.

6B
Trans-dimensional Bayesian non-parametrics with spatial point processes

Juha Heikkinen
Finnish Forest Research Institute, Helsinki, Finland

1 Introduction

Point processes are a class of models where the notion of a variable dimension is inherent. The main part of this discussion is concerned with the application of marked point processes as prior models in non-parametric Bayesian function estimation, reformulating and revising earlier joint work with Elja Arjas and listing some other related work (Section 2). Accordingly, the discussion is centred on trans-dimensional *modelling* rather than on the simulation techniques themselves, and connects to some of the material in Chapters 8 and 10. I shall end, however, with an example illustrating the role of the dimension-matching requirement (Section 3). The point made there is rather marginal to Green's main message, but hopefully interesting and/or instructive to modellers working with constraints.

2 Non-parametric Bayesian function estimation

Heikkinen and Arjas (1998) introduced a (trans-dimensional) non-parametric Bayesian approach to the estimation of the intensity function of a spatial Poisson process. The approach is similar to that in the change point and image analysis examples of Green (1995), and can be directly generalized to a wide variety of function estimation problems (Heikkinen 1998). It has been applied to a problem involving simultaneous interpolation, regression, and intensity estimation (Heikkinen and Arjas 1999), and closely related methods have been developed for image analysis (Nicholls 1998; Møller and Skare 2001), multivariate regression and classification (Denison *et al.* 2002b, Chapter 7), and disease mapping (Knorr-Held and Rasser 2000; Denison and Holmes 2001). The following paragraphs show how I would now prefer to introduce the method.

Consider the estimation of real-valued surfaces $f : S \to \mathcal{R}$ defined on a bounded support $S \subset \mathcal{R}^2$. A trans-dimensional approximation of f is obtained through its parameterization by marked point pattern $\theta = \{(x_1, y_1), \ldots, (x_k, y_k)\}$, in which the locations are a simple point pattern $x = \{x_1, \ldots, x_k\}$ on S with a variable number k of randomly located points, and the marks represent values $y_i = f_\theta(x_i)$ of the approximating function. To complete the approximation, we apply some rough and simple inter/extrapolation rule to determine $f_\theta(s)$, $s \in S \setminus x$. By pointwise averaging over a large number of such rough approximations, varying the number and locations of the points in x, we can then obtain a smooth estimate of f. With unbounded k the parameter space is effectively infinite-dimensional and hence the inference honestly non-parametric, yet the computations can be handled by trans-dimensional MCMC.

The inter/extrapolations of Heikkinen and Arjas (1998) were step functions on the Voronoi tessellations of S generated by the location patterns x (see Hurn *et al.*, Chapter 8 in this volume, Section 2.1.2; S. Richardson, Chapter 10 in this volume, Section 3.2.1). However, Voronoi tessellations could be replaced by the Delaunay or other more general triangulations (cf. Nicholls 1998). In addition to the more flexible geometry, triangular partitions offer the opportunity of making the function approximations piecewise linear instead of piecewise constant (see below). In the estimation of smooth functions, I would prefer the computationally simpler Delaunay triangulations, the greater flexibility of other triangulations being more valuable in problems like segmentation (Nicholls 1998).

Our prior of x was the homogeneous Poisson process, and large differences between nearby function values were penalized by a Markov random field prior for $y|x$. This led to unnecessary complications with the normalizing constants, which could have been avoided by modelling the marked point pattern θ directly as a nearest-neighbour Markov point process with correlated marks, as did Møller and Skare (2001). Then the marginal prior of x is no longer a Poisson process, but that seems like a small price to pay for an otherwise more tractable model. Although Denison and Holmes (2001) deem this smoothing unnecessary in the first place, I think that it should lead to qualitatively more reasonable individual approximations of f, and thereby to more realistic inferences on its shape, for example. For extrapolations beyond the convex hull of the data, dependence priors seem essential.

Motivated by such considerations, let me then sketch an approach I would currently suggest. Assuming S to be a polygon, let θ be a marked point process including locations on the edges and vertices of S as in the model of Nicholls (1998). Define the prior density of θ with respect to the distribution of the appropriate marked Poisson process by something like

$$p(\theta) \propto \lambda^k \exp \left\{ -\tau \sum_{x_i \sim_x x_j} (y_i - y_j)^2 \right\},$$

where $x_i \sim_x x_j$, if the tiles $S(x; x_i) \ni x_i$ and $S(x; x_j) \ni x_j$ of the Voronoi tessellation generated by x are adjacent. Finally, define f_θ as that unique surface that passes through all points (x_i, y_i) of θ and is linear within each triangle in the Delaunay tessellation generated by x. If f can only take a finite (and small) number of distinct values, as in image classification, for example, then I would follow Møller and Skare (2001) in using the Voronoi step functions and an extension like

$$p(\theta) \propto \lambda^k \exp\left\{ -\tau \sum_{x_i \sim_x x_j} \mathbf{1}(y_i \neq y_j) \right\}$$

of the Potts model, where $\mathbf{1}$ denotes the indicator function.

Theoretically, this approach works regardless of the dimension of S. However, the effort needed both for the implementation and for the computations increase rapidly with the dimension; for an example, see the three-dimensional problem in reservoir modelling tackled by Møller and Skare (2001). Independence priors allow for a computationally feasible approach for moderate dimensional S (Denison *et al.* 2002b), but Denison *et al.* (2002a) have found that they do not work well in very high dimensions, either.

3   On constraints and dimension matching

In most applications of trans-dimensional MCMC, the major problem seems to be finding *efficient* proposal distributions. When there are constraints in the parameter space, however, even the choice of *valid* proposals may not be trivial. A typical case in the function estimation context are problems involving interpolation (Heikkinen and Arjas 1999), of which a toy example is given below.

Consider the function estimation problem of Section 2 with the constraint $f(s_0) = f_0$ for some $s_0 \in S$, and step function approximations f_θ taking constant value y_i on each Voronoi tile:

$$f_\theta(s) = \sum_{i=1}^{k} y_i \mathbf{1}\{s \in S(x; x_i)\}.$$

Suppose we wish to implement the simplest possible sampler with two kinds of move proposal: death of one random point in the current θ and birth of a new point (ξ, η) with a uniform random location $\xi \in S$ and mark η sampled from some distribution on $\mathcal{R}$. In Green's formalism, we would then have $r = 2k$, $r' = 2k + 2$, $u = (\xi, \eta)$, and $d' = 0$ for the birth move from θ with k points to $\theta' = \theta \cup \{u\}$. But if $s_0 \in S(x \cup \{\xi\}; \xi)$, then the only proposal yielding a positive acceptance probability would be the (one-dimensional) $\theta' = \theta \cup (\xi, f_0)$. This would leave the mark η of u unused and hence violate the dimension-matching requirement.

The concrete consequences of the failure in dimension-matching are revealed only when trying to work out the acceptance probability for the death of (x_i, y_i) with $s_0 \in S(x; x_i)$. For positive chances of acceptance, we are forced to propose the function value f_0 on that tile $S(x \setminus x_i; x_j)$ which contains s_0 in the proposed tessellation. In other words, the death proposal is

$$\theta' = \theta \setminus \{(x_i, y_i), (x_j, y_j)\} \cup (x_j, y_j'),$$

where $y'_j = f_0$. But if $y_j \neq f_0$, then our simple sampler cannot reverse this move, because it proposes $y''_j = y'_j = f_0$ in the birth move from θ'.

Returning to the birth move $x' = x \cup \xi$ with $s_0 \in S(x'; \xi)$, the considerations above lead to the conclusion that the dimension we cannot use in proposing the function value on tile $S(x'; \xi)$, that is, the random mark η of u, must be used to perturb the current function value $y_j(= f_0)$ on the tile $S(x; x_j)$ containing s_0. See Heikkinen and Arjas (1999) for the details of one such sampler.

One might ask: Why not just keep a fixed generating point (s_0, f_0) and avoid the whole difficulty? But this would result in different smoothing around s_0 than elsewhere.

Acknowledgements

I wish to express my thanks to Elja Arjas for fruitful collaboration, to Peter Green for constructive comments, and to Jesper Møller for inspiring discussions. Support from the Academy of Finland (projects 51471 and 52303) is gratefully acknowledged.

Additional references in discussion

Denison, D. G. T. and Holmes, C. C. (2001). Bayesian partitioning for estimating disease risk. *Biometrics*, **57**, 143–9.

Denison, D. G. T., Adams, N. M., Holmes, C. C., and Hand, D. J. (2002a). Bayesian partition modelling. *Computational Statistics and Data Analysis*, **38**, 475–85.

Denison, D. G. T., Holmes, C. C., Mallick, B. K., and Smith, A. F. M. (2002b). *Bayesian Methods for Nonlinear Classification and Regression*. Wiley, Chichester.

Heikkinen, J. (1998). Curve and surface estimation using dynamic step functions. In *Practical Nonparametric and Semiparametric Bayesian Statistics* (eds D. Dey, P. Müller, and D. Sinha), pp. 255–72. Springer-Verlag, New York.

Heikkinen, J. and Arjas, E. (1998). Non-parametric Bayesian estimation of a spatial Poisson intensity. *Scandinavian Journal of Statistics*, **25**, 435–50.

Heikkinen, J. and Arjas, E. (1999). Modeling a Poisson forest in variable elevations: a nonparametric Bayesian approach. *Biometrics*, **55**, 738–45.

Knorr-Held, L. and Rasser, G. (2000). Bayesian detection of clusters and discontinuities in disease maps. *Biometrics*, **56**, 13–21.

Møller, J. and Skare, Ø. (2001). Coloured Voronoi tessellations for Bayesian image analysis and reservoir modelling. *Statistical Modelling*, **1**, 213–32.

Nicholls, G. (1998). Bayesian image analysis with Markov chain Monte Carlo and colored continuum triangulation models. *Journal of the Royal Statistical Society*, B, **60**, 643–59.

7

Particle filtering methods for dynamic and static Bayesian problems

Carlo Berzuini

Dipartimento di Informatica e Sistemistica, University of Pavia, Italy

Walter R. Gilks

MRC Biostatistics Unit, Cambridge, UK

1   Introduction

Markov chain Monte Carlo (MCMC) methods are often used in *static* problems, where we have a single, fixed distribution of interest, for example a Bayesian posterior distribution. This article considers the use of Monte Carlo methods in problems where the distribution of interest, or *target* distribution, evolves as we analyse it, typically as a consequence of new incoming data. These we call *dynamic* problems. When the target is an evolving posterior measure, as occurs within a Bayesian approach to data analysis, the task is to update posterior estimates of unknown parameters every time the posterior distribution changes. In missile tracking, for example, an estimate of the current position of the missile must be repeatedly updated in the light of the accumulating missile observations. Similar problems arise in on-line time series forecasting, on-line signal processing, real-time recognition, and automatic control.

MCMC methods of a standard type may not be appropriate in dynamic problems, partly because there may not be enough time between two successive updatings for a fully converged Markov chain to generate the required samples. This shifts our interest towards a more general class of Monte Carlo approaches, called *particle filters* (PFs), which are designed to react rapidly to changes in the target distribution. PFs have a long history in the engineering literature, notably in automatic control, dating back to the work of Handschin and Mayne (1969), Handschin (1970), Zaritskii *et al.* (1975), and Akashi and Kumamoto (1977). A recent survey of this area can be found in Doucet *et al.* (2001).

This article traces the development of the main ideas in particle filtering, but also tries to highlight current research directions in the area. We take as our starting point the Sampling Importance Resampling (SIR) filter by Gordon *et al.* (1993), and then move to a more sophisticated methodology involving a combination of particle filtering and MCMC ideas. We illustrate particle filters with an example in target tracking. Finally, we discuss the potential role of PFs in those static Bayesian problems which are usually tackled via MCMC.

2   Dynamic Bayesian problems

Particle filters are especially useful in problems where the observed data become available sequentially in time and one is interested in performing inference in an on-line fashion. For concreteness, we focus on the class of *dynamic Bayesian problems*, where the aim is to perform on-line inference based on an evolving posterior distribution. PFs may also be useful in non-Bayesian problems, and even in problems where the target is not a probability distribution. But this is left to the reader's intuition.

In a *dynamic* problem we observe a stochastic process, $X_1, X_2, \ldots, X_t, \ldots$, where the symbol $t \in \mathbb{N}$ represents time, and x_t represents the observed value for the random variable X_t. Let $x_{1:t} := (x_1, \ldots, x_t)$ represent the data available at time t, and assume that $x_{1:t}$ is generated by a model expressed in terms of a vector θ_t of unknown parameters, with

$$\theta_t = (\theta_{t-1}, \eta_t). \tag{2.1}$$

Here the (possibly empty) vector $\eta_t \in \mathbb{R}^p$ represents unknown parameters incorporated in the model at time t, as in hidden Markov modelling, for example, where η_t is the state parameter associated with each new data observation x_t. Within a Bayesian approach to data analysis, the posterior (target) distribution just after t, denoted by $\pi_t(\theta_t)$, or simply π_t, is the probability distribution of θ_t given $x_{1:t}$. We have the following relationship between target distributions at consecutive times, π_{t-1} and π_t:

$$\pi_t(d\theta_t) = c_t \cdot \pi_{t-1}(d\theta_{t-1}) \cdot p(d\eta_t|\theta_{t-1}, x_{1:t-1}) \cdot p(x_t|\theta_t), \tag{2.2}$$

where c_t denotes a generally unknown normalizing constant. As time progresses, the following sequence of distributions is generated by the process:

$$\pi_0(d\theta_0), \quad \pi_1(d\theta_1), \ldots, \pi_t(d\theta_t), \ldots. \tag{2.3}$$

Note that the above sequence evolves in an expanding parameter space as a consequence of (2.1). Also note that π_t is generally known up to a normalizing constant, $\{c_t\}$, except for π_0, which will generally represent our uncertainty about θ_0 prior to considering the observations $\{x_{1:t}\}$.

Just after time t, inference will focus on a specified *target function* $g_t(\theta_t)$, defined on the support of $\pi_t(d\theta_t)$. One is generally interested in the filtered expectation

$$\mathrm{E}_{\pi_t} g_t(\theta_t) := \int g_t(\theta_t)\, \pi_t(d\theta_t), \tag{2.4}$$

where E_π represents expectation with respect to a distribution π. If the data are modelled by a linear Gaussian state space model (where η_t represents the unknown state at time t), then the posterior $\pi_t(\theta_t)$ can be analytically evaluated in a recursive way by means of the Kalman filter, so that (2.4) can be calculated straightforwardly. This is possible also when the data are assumed to follow a hidden Markov model in which the unknown state at time t, represented by η_t, has a finite discrete number of possible values. But when the state space is not finite, and the data depart from linearity and/or Gaussianity, analytical tractability is lost. A possible remedy may be to apply an (approximate) Kalman filter to a (recursively constructed) linear–Gaussian approximation of the given model, but this extended Kalman filter approach often turns out to be unreliable (see Andrieu *et al.* 1999). Alternative approaches include neural networks and MCMC, although these latter will often be inappropriate in a dynamic context, as noted in the introduction. Our interest in what follows focuses on the very general class of Monte Carlo methods called *particle filters* (PFs), which allow the updating to be performed rapidly. A PF maintains in θ_t space a set of points, called *particles*, and evolves them in such a way that at every time t their empirical distribution closely follows π_t, and can therefore be used for rapid calculation of Monte Carlo approximations of integrals of the form (2.4).

3 Motivating example

As a concrete motivating example, consider a version of the classical *bearings-only tracking* problem (Gordon *et al.* 1993; Carpenter *et al.* 1998, 1999; Bergman 1999). The problem, which we further develop in Section 11, is described by Fig. 1. In this figure, the abscissa and the ordinate represent the east and the north coordinates of a ship, respectively. The ship is moving along an unobserved trajectory represented by small circles, starting at $(0,1)$. Noisy angular measurements of the ship's position, represented in the figure by dotted lines, are taken at regular intervals by a stationary observer. We wish to use prior information about the ship's initial coordinates, plus the gradually incoming measurements, to track the ship. At time t, this means we want to use the available information to update our estimate of the current ship's position, or our forecast of the future path of the ship.

Let ϵ_t and ν_t denote the (unobserved) ship's east and north coordinates at time t, and let the corresponding (unobserved) velocities be denoted as $\dot{\epsilon}_t$ and $\dot{\nu}_t$, respectively. Then consider the following model of the ship's motion:

$$
\begin{aligned}
\dot{\epsilon}_t &\sim \mathrm{N}(\dot{\epsilon}_{t-1}, \tau^{-1}), & \epsilon_t &= \epsilon_{t-1} + \dot{\epsilon}_{t-1}, \\
\dot{\nu}_t &\sim \mathrm{N}(\dot{\nu}_{t-1}, \tau^{-1}), & \nu_t &= \nu_{t-1} + \dot{\nu}_{t-1},
\end{aligned}
\tag{3.1}
$$

where $\mathrm{N}(c,d)$ denotes a normal distribution with mean c and variance d. If $\tan^{-1}(\nu_t/\epsilon_t)$ is the ship's bearing at time t, then we assume the measurement data x_t to be generated according to the following equation:

$$
x_t = \tan^{-1}(\nu_t/\epsilon_t) + \mathrm{N}(0, \kappa^2),
\tag{3.2}
$$

which is non-linear, and therefore prevents recursive updating via analytic methods. The parameter κ will henceforth be treated as a known quantity. The con-

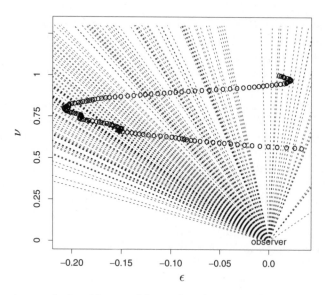

FIG. 1. Bearings-only tracking problem: the observer takes repeated imprecise angular measurements of the position of a ship moving along an unobserved trajectory.

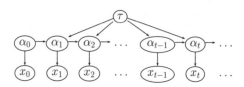

FIG. 2. Conditional independence structure of the model used in the bearings-only example.

ditional independence structure of the model at time t is depicted by the directed acyclic graph reported in Fig. 2, where $\alpha_t := (\dot{\epsilon}_t, \dot{\nu}_t, \epsilon_t, \nu_t)$. The parameter κ is not represented in the graph. A Bayesian approach to the problem requires specification of a prior distribution on the parameters representing the ship's initial conditions $(\epsilon_0, \nu_0, \dot{\epsilon}_0, \dot{\nu}_0)$ and the average smoothness τ of the ship's trajectory. It may be reasonable to assume that this prior has the form $p(\epsilon_0, \nu_0, \dot{\epsilon}_0, \dot{\nu}_0)\,p(\tau)$.

The set of parameters contained in the model at time t is $\theta_t = (\tau, \alpha_0, \ldots, \alpha_t)$. Then (2.1) specializes to $\theta_t = (\theta_{t-1}, \alpha_t)$, where α_t in this example plays the role of the symbol η_t used elsewhere in this article. The target measure π_t in this problem is defined on the σ-algebra generated in $\mathbb{R}^{2t+3}$ by θ_t. The Bayes recursion (2.2) takes the specific form

$$\pi_t(\mathrm{d}\theta_t) = c_t \cdot \pi_{t-1}(\mathrm{d}\theta_{t-1}) \cdot p(\mathrm{d}\alpha_t|\alpha_{t-1}, \tau) \cdot p(x_t|\alpha_t), \qquad (3.3)$$

where the normalizing constants $\{c_t\}$ have no closed form analytic expression. At every new time $t \in \mathbb{N}$, the model and the graph expand by the addition of a new node pair (x_t, α_t). At time t we might, for example, be interested in the filtered expectation for the ship position, (ϵ_t, ν_t). We might also be interested in some form of prediction about the future path of the ship. In both cases the computational task focuses on an integral of the form (2.4). Berzuini and Gilks (2001) examine the case where the ship has an unknown *type*, the different types corresponding to different laws of motion of the ship. In this case, the information contained in the accumulating data can be used to infer both about the unknown position and type of the ship.

Dynamic problems are encountered in many areas of science and technology, for example in medical monitoring, wherever rapid, on-line decisions need to be taken on the basis of rapidly accumulating patient data. Further areas include blind deconvolution of digital communications channels (Liu and Chen 1995), digital enhancement of speech and audio signals (Godsill and Rayner 1998), financial forecasting (Pitt and Shephard 1999), gene prediction in DNA sequences, prediction of the spatial conformation of biopolymers (Liu and Lawrence 1999), analysis of pedigrees in genetics (Kong *et al.* 1994; Irving *et al.* 1994), on-line analysis of image sequences (Sutherland and Titterington 1994; Isard and Blake 1996), and expert systems (Kanazawa *et al.* 1995). Also, prequential model validation (Dawid 1984) can be approached as a dynamic problem, in which the data are incorporated sequentially, in some (real or artificial) order, so as to generate the sequence of predictive and posterior distributions that provide the basis for criticizing the model.

4   General particle filtering strategy

Particle filters rest upon the idea that the target distribution π_t, where t indexes any specific stage of the evolution process, can be approximated through a set of N points in θ_t space, each ith point in this set being denoted as $\theta_{t,i}$ and being associated with a corresponding weight $w_{t,i}$. These weighted points are often called *particles*. At stage t of the evolution process, the particle filter yields a set of particles and weights which we collectively denote as

$$S_t = \{\theta_{t,i}, w_{t,i}\}_{i=1...N}. \tag{4.1}$$

This set is said to *properly approximate* π_t if, for any bounded continuous function $r(\theta_t)$, we have

$$\lim_{N \to \infty} \frac{\sum_{i=1}^{N} w_{t,i} r(\theta_{t,i})}{\sum_{i=1}^{N} w_{t,i}} = E_{\pi_t} r(\theta_t), \tag{4.2}$$

which means that were we able at time t to obtain from the particle filter a huge number N of particles, then, as $N \to \infty$, the left-hand side of (4.2) would tend to the expectation of $r(\theta_t)$ with respect to π_t. This implies that at time t we can approximate the quantity of interest (2.4), $E_{\pi_t} g_t(\theta_t)$, by a weighted average of the target function $g_t(\theta_t)$ over the current set of particles.

GENERAL TEMPLATE

Step 1: Particles in S_{t-1} are taken from their current θ_{t-1} space to the new, higher-dimensional, θ_t space. This involves adding new components to each coordinate vector $\theta_{t-1,i}$ to obtain an extended vector $\tilde{\theta}_{t,i}$. The particle weights in this step are usually left unchanged. The resulting set of particles is then

$$S'_{t-1} = \{\tilde{\theta}_{t,i}, w_{t-1,i}\}_{i=1...N}.$$

Depending on the specific algorithm, the set of particles S'_{t-1} will usually be regarded as a proper approximation of a distribution $\tilde{\pi}_t$ in θ_t space, called the *importance distribution*.

Step 2: The set of particles S'_{t-1} is turned into a new set of particles

$$S_t = \{\theta_{t,i}, w_{t,i}\}_{i=1...N},$$

which properly approximates π_t. This may involve various sorts of manipulation, such as updating the weights, or resampling, or changing the particle positions in the parameter space.

A particle filter may be defined as an algorithm which iteratively updates the currently available set of particles, S_{t-1}, into a new set S_t, each time the target distribution π_{t-1} evolves into π_t, in such a way that S_t properly approximates π_t. Although a diversity of updating schemes have been devised for particle filters, it is possible to describe the generic tth iteration of the algorithm using a general 'template' which fits many of the well-known particle filters. Such a template is reported in the box above. It is assumed that at the beginning of Step 1 (see box), we have available data $x_{1:t}$ and a set S_{t-1} of weighted particles that properly approximates π_{t-1}.

At the end of Step 2, the expected value of the target function $g_t(\theta_t)$ with respect to π_t can be estimated on the basis of the particles as

$$\hat{g}_t = \frac{\sum_{j=1}^{N} w_{t,j} g_t(\theta_{t,j})}{\sum_{j=1}^{N} w_{t,j}}. \tag{4.3}$$

Here, and in what follows, we assume that a constant number N of particles are maintained at each stage of the evolution process. This is not really essential, and could be easily relaxed, for example by allowing the user to fix a different number of particles for each new stage. One might also consider schemes in which the number of particles is allowed to vary across stages in a random fashion, but this would raise the problem how to prevent this number from becoming too big or too small.

We have kept the template description of a generic particle filter deliberately vague. In the following sections we shall consider in more detail a number of special cases of the template, which correspond to well-known particle filter algorithms.

5 The Sampling Importance Resampling filter

The *Sampling Importance Resampling* (SIR) filter (Smith and Gelfand 1992; Gordon *et al.* 1993; Isard and Blake 1996) assumes at time $t-1$ that we have a set of particles S_{t-1} with unit weights that properly approximates π_{t-1}. See the description of this algorithm in the box below (following the general structure set out in the template of the previous section).

The correspondence between the *Extend* step and Step 1 of the template in the previous section should be clear to the reader. Equation (5.1) specifies the way new components $\eta_{t,i}$ are added to the coordinate vector of each current ith particle. These added components are sampled from the conditional prior specified by (5.2), which is part of the model specification. In hidden Markov structures, such as the one represented by the directed graph of Fig. 2, the conditional prior simplifies to $p(\mathrm{d}\eta_t|\theta_{t-1,i})$. Each generated $\tilde{\theta}_{t,i}$ will follow the distribution

$$\tilde{\pi}_t(\mathrm{d}\eta_t, \mathrm{d}\theta_{t-1}) = p(\mathrm{d}\eta_t|\theta_{t-1}, x_{1:t-1})\pi_{t-1}(\mathrm{d}\theta_{t-1}), \qquad (5.4)$$

which is then, by definition, the importance distribution for this algorithm. Think of this distribution as a reasonable approximation for π_t based on data

---

SIR

1. **(After $t-1$) Extend.** For $i=1\ldots N$ set

$$\tilde{\theta}_{t,i} = (\theta_{t-1,i}, \eta_{t,i}), \qquad (5.1)$$

where $\eta_{t,i}$ is sampled from

$$p(\mathrm{d}\eta_t|\theta_{t-1,i}, x_{1:t-1}). \qquad (5.2)$$

2. **(Just after t) Weight and resample.** For $i=1\ldots N$, set

$$\tilde{w}_{t,i} = \frac{\pi_t(\mathrm{d}\tilde{\theta}_{t,i})}{\pi_{t-1}(\mathrm{d}\theta_{t-1,i}) \cdot p(\eta_t|\mathrm{d}\theta_{t-1,i}, x_{1:t-1})}, \qquad (5.3)$$

then create a new set $\{\theta_{t,i}\}_{i=1,\ldots,N}$ of particles by *resampling* from the set $\{\tilde{\theta}_{t,j}\}_{j=1,\ldots,N}$ with replacement, with resampling probabilities proportional to the respective weights $\{\tilde{w}_{t,j}\}_{j=1,\ldots,N}$. Finally, set

$$S_t = \{\theta_{t,i}, w_{t,i} = 1\}_{i=1\ldots N}.$$

information available at time $t - 1$.

The *Weight and Resample* step corresponds to Step 2 of the template in the previous section. Each particle $\tilde{\theta}_{t,i}$ in this step is assigned a weight $\tilde{w}_{t,i}$ indicating the degree to which that particle is representative in the current posterior π_t, according to (5.3). In the motivating example of Section 3, analogously to what we find in many other applications, the analytical expression of the weight (5.3) will simplify to the conditional likelihood of x_t. That is, in the specific example, to $\tilde{w}_{t,i} = p(x_t|\alpha_t)$. In the remaining part of the *Weight and Resample* step, a new set of particle coordinates is created by resampling from the set $\{\tilde{\theta}_{t,j}\}_{j=1,\dots,N}$ with replacement, with resampling probabilities proportional to the weights $\{\tilde{w}_{t,j}\}$, causing the elimination of unrepresentative particles. The resampling is discussed further in Section 7. The weight of each resampled particle $\theta_{t,i}$ is then reset to 1.

The SIR algorithm is triggered at time 0 by drawing independent samples directly from π_0. In most Bayesian applications involving an expanding directed acyclic graph, as in the above bearings-only example, π_0 would be the prior distribution on the *founder* nodes of the graph, θ_0, which will typically admit straightforward independent sampling. If the latter condition is not satisfied, suppose without loss of generality that the data start arriving at time 1. Then set π_0 to a convenient approximation of the prior, and π_1 equal to the prior.

Berzuini *et al.* (1997) establish a central limit theorem for the estimator (4.3) obtained when the SIR algorithm is applied at every stage of the updating. A more general and comprehensive theory of convergence of the SIR filter is provided by Del Moral and Guionnet (1999). The SIR algorithm can be efficiently implemented on a parallel computer, by assigning each particle an independent processor, and by allowing the N processors to operate in parallel. Both the particle extension and the particle weighting can be tackled in this way. A significant degree of parallelism can also be achieved in the resampling.

6   Why resample?

One might question the need to resample at Step 2 of the SIR algorithm. After all, we know that (i) the resampling step is not necessary for the consistency of the estimator, and (ii) while the resampling step reduces the variance of the weights, it also increases the variance of the estimator. The reason we introduce resampling is that without it we would eventually be left with a set of particles where only a few particles (in the limit, one) have an important weight. Resampling offers this 'important' subset of particles an opportunity to amplify itself. Ultimately, this may reduce the variance of the estimator. An interesting possibility is occasionally (at some iterations of the algorithm) to modify Step 2 of the SIR algorithm in such a way to skip the resampling. This would entail setting $S_t = \{\theta_{t,i} = \tilde{\theta}_{t,i}, w_{t,i} = w_{t-1,i}\,\tilde{w}_{t,i}\}_{i=1,\dots,N}$. Some heuristics proposed by Liu and Chen (1998) and Carpenter *et al.* (1998) may be taken as a basis for deciding, in an on-line fashion, how often in a given application the skipping should be done.

7   Efficient resampling

We now examine computationally efficient ways of doing the resampling involved in Step 2 of the SIR algorithm. We continue to restrict attention to schemes in which a fixed constant number N of particles is kept across the updating steps. The resampling at stage t determines the (possibly null) number $m_{t,j}$ of times particle $\tilde{\theta}_{t,j}$ is selected to be part of the updated set S_t. If we let

$$\tilde{w}_{t,j}^* = \frac{\tilde{w}_{t,j}}{\sum_k \tilde{w}_{t,k}},$$

then the resampling algorithm is generally designed to be unbiased, that is, to ensure $\mathrm{E} m_{t,j} = N \, \tilde{w}_{t,j}^*$. One possible approach to the resampling at Step 2 of the SIR algorithm is to draw the N numbers $\{m_{t,j}\}_{j=1,\ldots,N}$ from a multinomial distribution with parameter $(\tilde{w}_{t,1}^*, \ldots, \tilde{w}_{t,N}^*)$ (Gordon *et al.* 1993). This can be performed in $\mathcal{O}(N)$ operations using an algorithm reported in Ripley (1987). With this procedure we have $\mathrm{var}(m_{t,j}) = N \cdot \tilde{w}_{t,j}^* \cdot (1 - \tilde{w}_{t,j}^*)$. However, in most practical situations it will be faster to use a systematic approach of the kind proposed by Carpenter *et al.* (1999) or Kitagawa (1996), for example. One possibility is the *binning* algorithm. This unbiased resampling algorithm generates a set U of N points in the interval $[0, 1]$, each of the points a distance N^{-1} apart, and then takes $m_{t,j}$ to be the number of points in U that lie between $\sum_{i=1}^{j-1} \tilde{w}_{t,i}^*$ and $\sum_{i=1}^{j} \tilde{w}_{t,i}^*$.

Crisan *et al.* (1999) propose the following *fractional updating* resampling algorithm. For each particle $\tilde{\theta}_{t-1,j}$ compute $s_{t,j} = N \cdot \tilde{w}_{t,j}^*$, and then set $m_{t,j} = \mathrm{int}(s_{t,j}) + \mathrm{Bernoulli}(s_{t,j} - \mathrm{int}(s_{t,j}))$, where the notation $\mathrm{int}(r)$ represents the integer part of a real number r, and the notation $\mathrm{Bernoulli}(r)$ represents a random variable following a Bernoulli distribution with mean r. With this algorithm, as with the previous binning algorithm, we have variance $\mathrm{var}(m_{t,j}) = N \cdot (s_{t,j} - \mathrm{int}(s_{t,j})) \cdot (1 - s_{t,j} + \mathrm{int}(s_{t,j}))$.

8   Impoverishment

Repeated resampling stages tend to progressively *impoverish* the set of particles, by decreasing the number of distinct values of θ_t represented in that set at a given stage t of the evolution process. This is illustrated by Fig. 3, in the simple situation where $\theta_t = \theta$ is scalar. The shaded profiles represent the distributions π_{t-1} and π_t, respectively, at abscissa $(t-1)$ and t. Circles represent particle coordinates for S_{t-1} and S_t, respectively, at abscissa $(t-1)$ and t. The figure shows the actual result of one iteration of the SIR filter. Note that only those (few) particles of S_{t-1} that lie nearer the mode of π_t will be resampled and be passed to the subsequent generation. The consequence is a severe impoverishment of the set of particles, with a negative impact upon the variance of the estimator, especially when π_t differs considerably from π_{t-1}. In those situations where the target evolves in an expanding space, the impoverishment will tend to be more

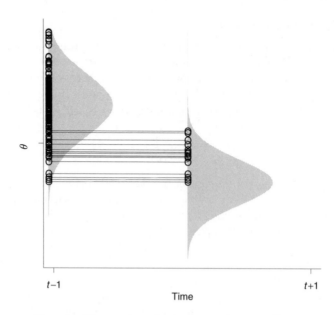

FIG. 3. Illustrating the impoverishment effect.

severe along the directions corresponding to static unknown parameters, such as the τ parameter in the ship tracking example of Section 3.

The tendency of the SIR filter to produce impoverishment is partly attributable to the choice of the importance distribution. The predictive distribution (5.4) is usually a poor approximation to π_t. Because of this, many of the particles generated at the *Extend* step will be highly unlikely under the new posterior, and consequently eliminated by the resampling. In principle, this can be remedied by a better choice of the importance distribution within the SIR scheme, but unfortunately this will generally introduce an intractable integral in the expression for the weights $\tilde{w}_{t,i}$.

9 Introducing Markov iterations

Let us return to the SIR algorithm of Section 5. We have seen that Step 1 consists of drawing samples from the importance distribution (5.4). This distribution is based on data $x_{1:t-1}$ available at $t-1$: it does not avail itself of the information contained in the more recent data x_t. As a consequence, this importance distribution may sometimes turn out to be a rather poor approximation of π_t. We have seen that, in this case, severe impoverishment may follow.

In the light of this consideration, a better strategy might be to use the particles available just after time $t-1$ *and* the information contained in x_t to construct an improved approximation to π_t, then let samples drawn from such an approximation form the new updated set of particles. This is the idea underlying the

---

MHIR

At time t, just after receiving x_t data: use the available particles $\{\theta_{t-1,i}\}_{i=1,...,N}$ to approximate the target distribution π_t, as given by (2.2), with the following mixture distribution:

$$c_t \sum_{i=1}^{N} w_{t-1,i} p(\eta_t | \theta_{t-1,i}, x_{1:t-1}) \, p(x_t | \theta_{t-1,i}, \eta_t) \, I_{\theta_{t-1} = \theta_{t-1,i}}. \qquad (9.1)$$

Use MCMC to draw samples $\{\theta_{t,i}\}_{i=1,...,N}$ from the above distribution. Then set

$$S_t = \{\theta_{t,i}, w_{t,i} = 1\}_{i=1...N}$$

to be the updated set of particles.

Metropolis–Hastings Importance Resampling (MHIR) filter of Berzuini *et al.* (1997). Iteration t of this algorithm is described in the box above. A characteristic feature of this algorithm is that Step 1 and Step 2 of the general template of Section 4 are fused into a single step. The normalizing constants $\{c_t\}$ will generally be unknown, but this is not a problem with MCMC. Because (9.1) has the form of a mixture, it can be sampled via MCMC by exploiting a suggestion of Besag and Green (1993), using the mixture component index as an auxiliary variable. Pitt and Shephard (1999) refine this idea and incorporate it into their *auxiliary particle filter* (APF). Berzuini *et al.* (1997), in a medical monitoring application, note that MHIR performs better than SIR. Also Pitt and Shephard demonstrate a performance of their APF superior to SIR, and to MHIR as well. However, as noted by Carpenter *et al.* (1998), MHIR and APF lose the ability of SIR to exploit the methods of systematic resampling discussed in Section 7. For this reason, and because MCMC sampling from (9.1) involves in any case a burn-in of the Markov chain, MHIR and APF may turn out to be slow in certain applications. Finally, in common with SIR, MHIR and APF can do very little to prevent particle depletion along the directions of the θ_t space corresponding to static parameters whose values affect inference over the entire process, and whose marginal posterior distribution significantly evolves throughout the process. One example of such a parameter is the τ parameter in the bearings-only example. A more radical approach to this problem is discussed in the next section.

10   Enhancing particle diversity (resampling and moving)

Although the approach of the previous section may reduce the rate of impoverishment, it does not eliminate the problem. A more radical approach to the problem consists of systematically introducing *diversity* in the set of particles. One way of achieving this is to move each resampled particle from its current position to a new random position of the parameter space, at some or all stages of updating.

Sutherland and Titterington (1994), for example, replace the resampling step with resampling from a kernel density estimate (KDE) constructed on the current set of particles. Similarly, West (1993) proposes resampling through *adaptive importance sampling*, where the bandwidth of the KDE is reduced on successive iterations of the adaptation to reduce the bias in the variance of the target distribution. Liu and Chen (1998) propose sampling from a Rao–Blackwellized reconstruction of the target distribution. Within an application of particle filters in the bearings-only problem of Section 3, Carpenter *et al.* (1999) propose an approach in which the resampled particles are moved by using Markov chain iterations.

A general methodology that combines resampling and moving within a unified particle filtering framework is the Resample–Move algorithm (Gilks and Berzuini 2001). In this algorithm the swarm of particles is adapted to an evolving target distribution by periodic resampling steps and through occasional, diversity-enhancing, Markov chain moves which lead each individual particle from its

---

RESAMPLE–MOVE

1. **(After $t-1$) Extend.** For $i = 1 \ldots N$ set

$$\tilde{\theta}_{t,i} = (\theta_{t-1,i}, \eta_{t,i}),$$

where $\eta_{t,i}$ is sampled from

$$p(\mathrm{d}\eta_t | \theta_{t-1,i}, x_{1:t-1}).$$

2. **(Just after t) Weight, resample and move.** For $i = 1 \ldots N$, set:

$$\tilde{w}_{t,i} = \frac{\pi_t(\mathrm{d}\tilde{\theta}_{t,i})}{\pi_{t-1}(\mathrm{d}\theta_{t-1,i}) \cdot p(\eta_t | \mathrm{d}\theta_{t-1,i}, x_{1:t-1})},$$

then create a new set $\{\theta_{t,i}^*\}_{i=1,\ldots,N}$ of particles by resampling from the set $\{\tilde{\theta}_{t,j}\}_{j=1,\ldots,N}$ with replacement, with resampling probabilities proportional to the weights $\{\tilde{w}_{t,j}\}_{j=1,\ldots,N}$. Next, each ith particle is moved from its current position $\theta_{t,i}^*$ to a new random position $\theta_{t,i}$ in θ_t space via one or more iterations of a Markov chain with a transition kernel $q_{t,i}$:

$$\theta_{t,i} \sim q_{t,i}(\mathrm{d}\theta_t | \theta_{t,i}^*), \qquad (10.1)$$

with the constraint that $q_{t,i}$ preserves stationarity with respect to π_t. Set

$$S_t = \{\theta_{t,i}, w_{t,i} = 1\}_{i=1\ldots N}$$

to be the final updated set of particles.

current position to a new random position of the parameter space. Resample–Move is described in the box above. Note that the only (although important) difference from the above SIR algorithm lies in the particle moving at the end of each iteration.

Any MCMC method, such as the Gibbs sampler or Metropolis–Hastings (Metropolis *et al.* 1953; Hastings 1970), can be used to design the kernel $q_{t,i}$. The fact that $q_{t,i}$ may depend on t and/or i opens a variety of possible strategies, in which the particle-moving mechanism varies across updating stages and across particles. Contrary to standard MCMC applications, the number of iterations of the chain performed on a single particle in a single Move step is arbitrary: no burn-in is required. The reason for this is that, before the moving, at time t, each particle has (approximate) marginal distribution π_t, which is preserved regardless of how many times we apply the transition kernel. For the same reason, the transition kernel need not be ergodic. Thus, for example, at any tth updating stage, it is possible to save computer time by leaving some of the coordinates of the particle unchanged. In practice, in most applications much insight is needed to exploit this possibility without incurring severe particle depletion along certain directions of the parameter space. Berzuini and Gilks (2001) discuss such issues within the context of the bearings-only application. The combined effect of the resampling and of the moving is illustrated by Fig. 4 in a completely analogous situation as that described by Fig. 3, so as to highlight the advantages of Resample–Move with respect to SIR.

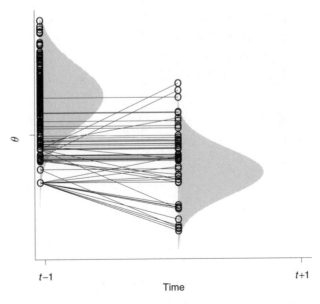

FIG. 4. Resampling *and* moving.

Gilks and Berzuini (2001) have established a central limit theorem for the estimator (4.3) obtained when the Resample–Move algorithm is applied at every stage of the updating. Crisan and Doucet (2000) give more extensive and general convergence proofs.

The resampling step could in principle be eliminated, provided the weight of each particle is multiplicatively updated from one iteration to the next, rather than reset to 1 after each resampling. However, Fig. 5 illustrates the possible negative consequences of this. The figure repeats the experiment of Fig. 4 *without* performing the resampling. Comparison of the two figures shows that, without resampling, adaption to the target change is very poor. Those particles (the great majority) which at time $t-1$ lie far outside the region of strong support under π_t cannot make big enough jumps to reach such a region. This will eventually result in heavy weights accumulating on just a few particles, exerting a progressively negative impact on the variance of the estimator. The resampling avoids all this by selecting and then moving only those particles that are really promising, in that they lie near the mode of π_t. In summary, good performance depends on a *cooperation* of resampling and moving.

The Move step can be efficiently implemented on a parallel computer, by allowing each particle to be taken care of by a separate processor, and by allowing the N processors to operate in parallel.

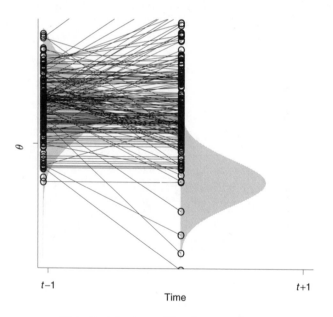

FIG. 5. Moving *without* resampling.

Crisan and Doucet (2000) generalize Resample–Move by replacing (10.1) with

$$\theta_{t,i} \sim Q_{t,i}(\mathrm{d}\theta_t | \{\theta_{t,k}^*\}_{k=1,\ldots,N}), \tag{10.2}$$

where $Q_{t,i}$ is a conditional distribution we call a *transition distribution*, which preserves stationarity with respect to π_t. This allows all the information about π_t which is contained in $\{\theta_{t,k}^*\}_{k=1,\ldots,N}$ to be exploited in the construction of the updated set of particles S_t.

11   An illustration

We use the bearings-only problem of Section 3 to compare the performance of a SIR filter with that of the Resample–Move algorithm. The SIR filter and Resample–Move were applied to the angular time series depicted in Fig. 1, assuming the model described in Section 3, with a prior $p(\epsilon_0, \nu_0, \dot{\epsilon}_0, \dot{\nu}_0) = \mathrm{N}(0.01, 0.04) \times \mathrm{N}(1.03, 0.1) \times \mathrm{N}(0.002, 0.003) \times \mathrm{N}(-0.013, 0.003)$, representing prior belief that at time 0 the ship was moving slowly and at close distance from the point $(0, 1)$. Because the data were simulated from an initial position $\epsilon_0 = 0, \nu_0 = 1$, the above $p(\epsilon_0, \nu_0)$ prior margin is slightly displaced with respect to the simulated 'truth'. By contrast, the above $p(\dot{\epsilon}_0, \dot{\nu}_0)$ prior margin is centred around the 'true' values of the initial velocities. The parameter τ was assigned a gamma prior, with an expected value slightly higher than the value used in simulating the data. The parameter κ was taken to be equal to 0.05 throughout. The number N of particles was kept equal to 100 over the entire filtering process. The implementation details, for the two algorithms, and the types of move employed are described in Berzuini and Gilks (2001).

The smoothness parameter τ turns out not to be estimable on the basis of the data alone, due to a lack of identifiability. However, under the above described proper prior, this parameter has a proper posterior distribution.

Figure 6 compares the performance of the two filters. The ship's true trajectory follows the solid line, starting from near the upper border of the plot, and ending near the lower border. The black triangles, representing the filtered estimates of the ship's position generated by the SIR filter, are largely discrepant from the true trajectory, revealing a very poor tracking. By contrast the circles, representing the corresponding Resample–Move estimates, reveal a satisfactory tracking. The problem with SIR is that the set of particles, as it is propagated over the filtering process, is eventually left without low values of τ. Such a loss prevents the method from adapting to the gradual decline of the posterior mean of τ. A side-effect of this is severe underestimation of the uncertainty of the forecasts of the future path of the ship. The superiority of Resample–Move in this example is largely attributable to the fact that the resampled values of τ are occasionally moved, so that particle depletion along the τ subspace is avoided.

In the context of the present application, Berzuini and Gilks (2001) discuss the use of particle filtering to simultaneously perform on-line forecasting and model selection, in the presence of uncertainty about the data-generating model.

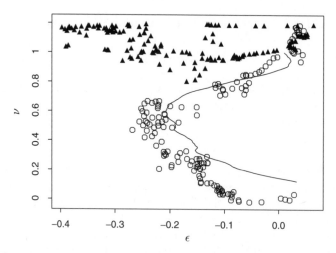

FIG. 6. The solid line represents the ship's true trajectory towards the lower right corner of the plot. The circles represent the Resample–Move filter's output. The black triangles represent the output of a SIR filter.

12   Particle filtering for static problems

To this point, we have focused on dynamic problems. As we have discussed, maintaining an ensemble of particles allows rapid assimilation of new data, through resampling and through adjusting the particle weights. In a static context, where the target distribution is fixed, and real-time inference or prediction is not required, it is less obvious that PFs have a useful role. Other methods, in particular MCMC, are capable of delivering samples from a posterior distribution without the overheads of PFs. However, we argue here that the PF framework offers great opportunities for solving static problems. This is an extremely important selling point of particle filters, as the majority of applications of Bayesian theory are static.

MCMC has now been established as the dominant mode of Bayesian analysis of complex static problems. However, for many static problems, severe mixing problems are encountered in the application of MCMC. That is, the MCMC algorithm is very slow to converge. Within the MCMC framework, the only solution is to devote considerable, expert, human time in devising better-mixing proposal distributions, often of the order of months. The alternative of simply waiting for a poorly mixing algorithm to converge is often not feasible, since each run may take several days, and in most applications the algorithm will need to be run many times under slightly different assumptions or data. Even worse, running a poorly mixing algorithm is dangerous, as it can lead to premature assertions of convergence.

By contrast, suppose we want to explore a static target $\pi(\theta)$ by means of a Resample–Move filter. The following two-stage procedure may be employed.

In the first stage a large sample of particles, $\{\theta_{0,i}\}$, is drawn independently from an initial distribution, π_0. This could be either a prior distribution for the parameters, θ, or a rough guess at π. These particles, equipped with weights $w_{0,i} = \pi(\theta_{0,i})/\pi_0(\theta_{0,i})$, can be used immediately to make inferences or predictions based on π. Thus, unlike MCMC, we do not have to wait an indefinite period (for convergence of the Markov chain) before obtaining results. In the second stage, the particles $\{\theta_{0,i}\}$ are resampled and moved to new positions $\{\theta_{1,i}\}$ by one or more Markov iterations, to locate any mode that is not reflected in π_0. Although a good mixing of the Markov chain in the second stage is advantageous, we argue that the above PF approach to a static target is more robust than MCMC with respect to a poor design of the proposal.

In Fig. 7 we apply the above approach to a static bimodal target π. This example uses a Gaussian π_0 which does not reflect the two modes of the target, and hence relies on the move step to locate the modes. Also, in a PF approach to a static target it may be advantageous to devise well-mixing move steps, and one bonus of the PF approach over MCMC is the possibility of designing effective moves through the imaginative use of concepts borrowed from genetic algorithms. Thus, for static problems, particle filters can combine the power of importance sampling, MCMC, and genetic algorithms.

The simple Resample–Move technique shown in Fig. 7 would probably fail to locate the two modes of π if these modes were too far apart. In such a case, one possibility is to consider a sequence of distributions in θ space, denoted as

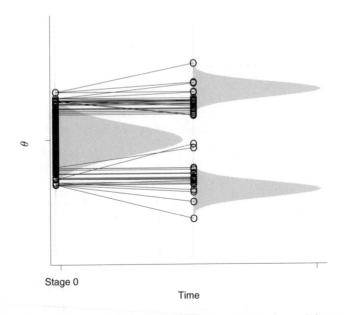

FIG. 7. Simple Resample–Move approach to a static bimodal target π.

$\{\pi_k(\theta); k = 0 \ldots K\}$, that gradually shifts from π_0 to π. This resulting *pseudo-target sequence* is then fed into the particle filter as if it were the target sequence arising in a dynamic problem, gradually easing the particles into the modes of the target. A way of generating an appropriate pseudo-target sequence, in a specific application, is to allow data information x, which by definition in a static context is made available as a batch, to be incorporated incrementally. One possibility is to specify an initial distribution, $\pi_0(\theta)$, and then gradually evolve it into π, for example as in

$$\pi_0(\theta) \left(\frac{\pi(\theta)}{\pi_0(\theta)} \right)^{\gamma_0}, \pi_0(\theta) \left(\frac{\pi(\theta)}{\pi_0(\theta)} \right)^{\gamma_1}, \ldots, \pi_0(\theta) \left(\frac{\pi(\theta)}{\pi_0(\theta)} \right)^{\gamma_K},$$

where γ is a vector of real numbers such that $\gamma_0 = 0$, $\gamma_k > \gamma_{k-1}$, and $\gamma_K = 1$. This approach we call *annealing-like* particle filtering.

Alternatively, consider the *pseudo-sequential* approach, in which the data, $x_{1:K} = (x_1 \ldots x_K)$, are incorporated in a gradual fashion, as if they were obtained sequentially, so as to generate the following target sequence:

$$p(\theta), \qquad p(\theta|x_1), \qquad p(\theta|x_{1:2}), \ldots, \qquad p(\theta|x_{1:K}), \qquad (12.1)$$

which gradually evolves from the prior, $p(\theta)$, into the target of interest, $p(\theta|x_{1:K})$. The sequence (12.1) is then gradually fed into the particle filter. In other words, we work through the entire data set in a sequential fashion, in some way pretending that the sets $x_1, x_2, \ldots$ are acquired sequentially. A similar approach has been used by Chopin (2000).

Acknowledgements

We would like to thank Peter Clifford, Peter Green, and Geir Storvik for valuable comments on earlier drafts of this article.

References

Akashi, H. and Kumamoto, H. (1977). Random sampling approach to state estimation in switching environments. *Automatica*, **13**, 429–34.

Andrieu, C., de Freitas, N., and Doucet, A. (1999). Sequential MCMC for Bayesian model selection. In *IEEE Signal Processing Workshop on Higher Order Statistics, June 14–16, 1999*. Caesarea, Israel.

Bergman, N. (1999). *Recursive Bayesian Estimation, Navigation and Tracking Applications*. Dissertation No. 579, Department of Electrical Engineering, Linkoeping University.

Berzuini, C. and Gilks, W. R. (2001). RESAMPLE–MOVE filtering with cross-model jumps. In *Sequential Monte Carlo Methods in Practice* (eds A. Doucet, N. de Freitas, and N. Gordan), pp. 117–38. Springer-Verlag, New York.

Berzuini, C., Best, N., Gilks, W. R., and Larizza, C. (1997). Dynamic conditional independence models and Markov chain Monte Carlo methods. *Journal of the American Statistical Association, Theory and Methods*, **92**, 1403–12.

Besag, J. and Green, P. J. (1993). Spatial statistics and Bayesian computation. *Journal of the Royal Statistical Society*, B, **55**, 25–102.

Carpenter, J. R., Clifford, P., and Fearnhead, P. (1998). Building robust simulation-based filters for evolving data sets. Technical Report, Department of Statistics, University of Oxford, UK.

Carpenter, J. R., Clifford, P., and Fearnhead, P. (1999). An improved particle filter for nonlinear problems. *IEEE Proceedings in Radar, Sonar and Navigation*, **146**, 2–7.

Chopin, N. (2000). A sequential particle filter method for static models. Unpublished manuscript.

Crisan, D. and Doucet, A. (2000). Convergence of sequential Monte Carlo methods. Technical Report CUED–F–INFENG/TR 381, Department of Engineering, University of Cambridge, UK.

Crisan, D., Del Moral, P., and Lyons, T. (1999). Discrete filtering using branching and interacting particle systems. *Markov Processes and Related Fields*, **5**, 293–318.

Dawid, P. (1984). Statistical theory: the prequential approach, with discussion. *Journal of the Royal Statistical Society*, A, **147**, 278–92.

Del Moral, P. and Guionnet, A. (1999). A central limit theorem for non-linear filtering using interacting particle systems. *Annals of Applied Probabability*, **9**, 275–97.

Doucet, A. (1998). On sequential simulation-based methods for Bayesian filtering. Technical Report CUED/F–INFENG/TR 310, Department of Engineering, University of Cambridge, UK.

Doucet, A., de Freitas, N., and Gordon, N. (eds) (2001). *Sequential Monte Carlo Methods in Practice*. Springer-Verlag, New York.

Gilks, W. R. and Berzuini, C. (2001). Following a moving target – Monte Carlo inference for dynamic Bayesian models. *Journal of the Royal Statistical Society*, B, **63**, 1–20.

Godsill, S. J. and Rayner, P. J. W. (1998). *Digital Audio Restoration – A Statistical Model-Based Approach*. Springer-Verlag, New York.

Gordon, N. J., Salmond, D. J., and Smith, A. F. M. (1993). Novel approach to nonlinear non-Gaussian Bayesian state estimation. *IEEE Proceedings in Radar, Sonar and Navigation*, **140**, 107–13.

Handschin, J. E. (1970). Monte Carlo techniques for prediction and filtering of non-linear stochastic processes. *Automatica*, **6**, 555–63.

Handschin, J. E. and Mayne, D. Q. (1969). Monte Carlo techniques to estimate the conditional expectation in multi-stage non-linear filtering. *International Journal of Control*, **9**, 547–59.

Hastings, W. K. (1970). Monte Carlo sampling methods using Markov chains and their applications. *Biometrika*, **57**, 97–109.

Irving, M., Cox, N., and Kong, A. (1994). Sequential imputation for multilocus linkage analysis. *Proceedings of the National Academy of Sciences of the U.S.A.*, **91**, 11684–8.

Isard, M. and Blake, A. (1996). Contour tracking by stochastic propagation of conditional density. In *Proceedings of the European Conference on Computer Vision, Cambridge, 1996*, 343–56.

Kanazawa, K., Koller, D., and Russell, S. J. (1995). Stochastic simulation algorithms for dynamic probabilistic networks. *Proceedings of the Eleventh Annual Conference on Uncertainty in Artificial Intelligence (UAI–95)*, pp. 346–51. Morgan Kaufmann, San Francisco.

Kitagawa, G. (1996). Monte Carlo filter and smoother for non-Gaussian nonlinear state space models. *Journal of Computational and Graphical Statistics*, **5**, 1–25.

Kong, A., Liu, J. S., and Wong, W. H. (1994). Sequential imputations and Bayesian missing data problems. *Journal of the American Statistical Association*, **9**, 278–88.

Liu, J. S. and Chen, R. (1995). Blind deconvolution via sequential imputation. *Journal of the American Statistical Association*, **90**, 567–76.

Liu, J. S. and Chen, R. (1998). Sequential Monte Carlo Methods for dynamic systems. *Journal of the American Statistical Association*, **93**, 1032–44.

Liu, J. S. and Lawrence, C. E. (1999). Bayesian inference on biopolymer models. *Bioinformatics*, **15**, 38–52.

Metropolis, N., Rosenbluth, A., Rosenbluth, M., Teller, A., and Teller, E. (1953). Equations of state calculations by fast computing machines. *Journal of Chemical Physics*, **21**, 1087–91.

Pitt, M. K. and Shephard, N. (1999). Filtering via simulation: auxiliary particle filters. *Journal of the American Statistical Association*, **94**, 590–9.

Ripley, B. D. (1987). *Stochastic Simulation*. Wiley, New York.

Rubin, D. B. (1988). Using the SIR algorithm to simulate posterior distributions. In *Bayesian Statistics 3* (eds J. M. Bernardo, M. H. DeGroot, D. V. Lindley, and A. F. M. Smith), pp. 395–402. Oxford University Press, UK.

Smith, A. F. M. and Gelfand, A. E. (1992). Bayesian statistics without tears: a sampling–resampling perspective. *The American Statistician*, **46**, 84–8.

Sutherland, A. I. and Titterington, D. M. (1994). Bayesian analysis of image sequences. Technical Report No. 94-3, Department of Statistics, University of Glasgow, UK.

West, M. (1993). Mixture models, Monte Carlo, Bayesian updating and dynamic models. In *Computing Science and Statistics*, **24**, pp. 325–33. Interface Foundation of North America, Fairfax Station, VA.

Zaritskii, V. S., Shimelevich, L. I., and Svetnik, V. B. (1975). Monte Carlo technique in problems of optimal data processing. *Automation and Remote Control*, **12**, 95–103.

7A
Some further topics on Monte Carlo methods for dynamic Bayesian problems

Geir Storvik
University of Oslo and Norwegian Computing Centre

1　Introduction

Algorithms for Monte Carlo filtering seem to have been rediscovered many times in different areas; see Doucet (1998) and the references therein. Although the basic ideas are similar, aspects of the detailed level can be of great importance concerning the behaviour of the filter. An important contribution of Berzuini and Gilks through this and other articles is a coherent framework for such algorithms.

In this discussion I will focus on the following topics: (1) when to resample, (2) the theoretical properties of dynamic Monte Carlo algorithms, and (3) what to do with static parameters.

2　When to resample

Two different approaches have usually been applied when constructing particle filters. The first approach (Doucet 1998; Liu and Chen 1998) is motivated from importance sampling. Here, at each time point, η_t is generated from some importance distribution h_t (which can depend on θ_{t-1} as well as $x_{1:t}$) while the importance weight is updated according to

$$w_t = w_t \frac{\pi_t(\mathrm{d}\theta_t)}{\pi_{t-1}(\mathrm{d}\theta_{t-1})h_t(\mathrm{d}\eta_t)}.$$

In order to avoid impoverishment, resampling can be included, resulting in the SIR algorithm defined by Berzuini and Gilks.

The second approach (Gordon *et al.* 1993; Pitt and Shephard 1999) is based on density approximation. Assume $\{(\theta_{t-1,i}, w_{t-1,i}), i = 1, \ldots, N\}$ is an approximate weighted sample from $\pi_{t-1}(\mathrm{d}\theta_{t-1})$. Using the transition distribution $p(\mathrm{d}\eta_t|\theta_{t-1})$, the sample can be used to obtain an estimate of the predictive distribution

$$\pi_{t-1}(\mathrm{d}\theta_t) \approx \sum_{j=1}^{N} w_{t-1,i} p(\mathrm{d}\eta_t|\theta_{t-1,j}) I(\theta_{t-1} = \theta_{t-1,j}). \tag{2.1}$$

Ignoring the approximation error above, the updated distribution when a new observation x_t arrives is given by

$$\pi_t(d\theta_t) \propto \sum_{j=1}^{N} w_{t-1,j} p(x_t|\theta_t) p(d\eta_t|\theta_{t-1,j}) I(\theta_{t-1} = \theta_{t-1,j}). \tag{2.2}$$

A new sample at time t can be obtained by sampling from this distribution. In most situations, exact sampling from (2.2) is not possible, but approximate methods such as sampling/importance resampling or Markov chain Monte Carlo methods can be applied. The MHIR algorithm of Berzuini and Gilks is a special case of this approach.

Merging the two approaches into a single class of algorithms can be of interest to see the similarities (and the differences). Consider the following algorithm (where for simplicity the prior $p(d\eta_t|\theta_{t-1})$ is used as the proposal distribution).

(1) **Sampling**: For $j = 1, \ldots, N$:
    (a) Sample I_j from $\{1, \ldots, N\}$ with probability s_{j,I_j}.
    (b) Draw $\tilde{\eta}_{t,j} \sim p(d\eta_t|\theta_{t-1,I_j})$.
    (c) Put $\tilde{\theta}_{t,j} = (\theta_{t-1,I_j}, \tilde{\eta}_{t,j})$ and $q_j = p(x_t|\tilde{\theta}_{t,j})/s_{j,I_j}$.
(2) **Importance Resampling**: For $j = 1, \ldots, N$:
    (a) Sample $J_j = i$ with probability r_{j,J_j}.
    (b) Put $\theta_{t,j} = \tilde{\theta}_{t,J_j}$ and $w_{t,j} = q_{J_i}/r_{j,i}$.

The extension from typically used algorithms is the possibility of resampling both before and after the generation of $\tilde{\eta}_{t,j}$. The weights $s_{j,i}$ may depend on $(\theta_{t-1,j}, w_{t-1,j})$ and x_t, but not on $\eta_{t,j}$. $r_{j,i}$ will typically depend on q_j as well. Note that in this framework we have allowed the weights to vary with j. $s_{j,i} = I(j = i)$ and $r_{j,i} = q_i$ correspond to the SIR algorithm while $s_{j,i} = s_i$ and $r_{j,i} = I(j = i)$ correspond to the density approximation approach. The approaches therefore differ with respect to when resampling is performed.

The two resampling strategies are illustrated in Fig. 1 for the model $\eta_t \sim N(\eta_{t-1}, 0.5^2)$, $x_t \sim N(\eta_t^2, 0.2^2)$. The first column of particles corresponds to $\eta_{t-1,i}$ The second column of particles is the ones resampled in step 1(a). The third column is the $\tilde{\eta}_{t,i}$ while the fourth column contains $\eta_{t,i}$. In the left panel, no resampling is performed in step 1(a), resulting in columns 1 and 2 being identical. In the right panel, no resampling is performed in step 2(a), resulting in the two last columns being identical. For both approaches, resampling reduced the samples to the same number (7) of distinct values. The second approach does however have many more distinct values in the end because of its possibility of changes after resampling.

When both the resampling steps 1(a) and 2(a) are replaced by systematic selection ($s_{i,j} = r_{i,j} = I(i = j)$), an advantage is that independent samples are obtained. Resampling at either step introduces dependence between particles and bias in Monte Carlo estimates, so should be avoided as much as possible.

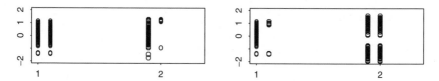

FIG. 1. Evolution of particles when resampling is performed in step 2(a) (left
   panel) and in step 1(a) (right panel).

On the other hand, as stated by Berzuini and Gilks and many others, resampling
is, in practice, necessary in order to avoid impoverishment. On the basis of the
above comparison of resampling strategies, a question then is: *When should one
resample?* My opinion is that if resampling is to performed, it should be done
in step 1(a) and not in step 2(a). There are three reasons for this:

(1) Considering the samples as a basis for the approximation (2.1), this estimate
   can be interpreted as a Rao–Blackwellized estimate when no resampling is
   performed at step 2a.

(2) When resampling is performed at step 1(a), the new observation x_t can be
   utilized to choose a component from the mixture (Pitt and Shephard 1999).

(3) When resampling is done in step 1(a) it is actually done *later* than if it is
   done in step 2(a). This is because sampled values of η_t will be resampled at
   time $t + 1$ when resampling is performed in step 1(a).

3   Theoretical properties of Monte Carlo filters

So far, very few theoretical results about the Monte Carlo filters are available.
Doucet *et al.* (2001) contains the disappointing proposition saying that the variance of the weights in the SIS algorithm increase over time, implying that impoverishment is unavoidable. This result is a strong motivation for including
resampling in the algorithm. Berzuini *et al.* (1997) are able to prove a central
limit theorem for estimates based on particles obtained from the SIR algorithm
using resampling at each stage. More general results on convergence are given
in Crisan and Doucet (2000), which shows that most algorithms proposed will
converge properly as $N \to \infty$. The limit is however on increasing the number
of simulations N for fixed time points t. In applications with long time series,
which frequently occur in economics and engineering, for example, of more practical interest would be the behaviour of the filter for *fixed* N as time increases.
In Künsch (2001) it is proven that if N grows like t^2 as t increases, then the
error in the approximate distribution remains fixed over time. The order t^2 can
probably be improved by introducing additional conditions or by constructing
more efficient filters. The conditions presented in Künsch (2001) are already
quite restrictive, for instance not allowing linear Gaussian processes for $\{\eta_t\}$.

What kind of results might we expect? In order to get some insight, consider
the simple linear Gaussian model $\eta_t = \eta_{t-1} + w_t$, $x_t = \eta_t + v_t$, where both $\{w_t\}$

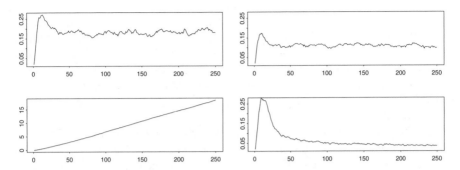

FIG. 2. Monte Carlo estimates of $\mathrm{E}[(\tilde{\eta}_t - \hat{\eta}_t)^2]/\mathrm{var}[\hat{\eta}_t]$ for the model described in the text. Both upper panels have the same scale on the y-axis.

and $\{v_t\}$ are zero-mean white Gaussian noise processes. Assume we are interested in predicting η_t on-line by $\hat{\eta}_t = \mathrm{E}[\eta_t|x_{1:t}]$. In this example, this quantity can of course be calculated by the Kalman filter, which makes it possible to compare the true estimate with the approximate estimate obtained by a Monte Carlo filter: $\tilde{\eta}_t = N^{-1}\sum_j \eta_{t,j}$.

In Fig. 2, Monte Carlo estimates of $\mathrm{E}[(\tilde{\eta}_t - \hat{\eta}_t)^2]/\mathrm{var}[\hat{\eta}_t]$ are plotted as a function of time. The upper panels are for $\mathrm{var}[w_t] = 0.1^2$ while for the lower panels $\mathrm{var}[w_t] = 0$. In both cases $\mathrm{var}[v_t] = 1$. The left panels correspond to the SIR filter (as defined by Berzuini and Gilks) without a **Move** step, while for the right panels a **Move** step (one Metropolis iteration) is included in the algorithm.

Concentrating first on the left panels, with the presence of noise in the underlying process the error seems to stabilize when time increases. This is in contrast to the other case, where the error seems to continue to grow. The properties of such filters therefore seem to depend on the dynamic nature of the underlying process.

Similar plots are given when the SIR filter including a **Move** (one Metropolis iteration) is included (right panels of Fig. 2). The improvement is striking. Not only do we get smaller errors, but the second case now seems to have stabilized! When **Move** steps are included, it is therefore a possibility that the relative error might stabilize even in the presence of static variables in the state vector.

To the author's knowledge, no theoretical results are yet available to support these results. The approach given by Künsch (2001) seems a promising direction in this respect.

4   Static parameters

A serious problem with the Monte Carlo filtering methods is their inability to handle static variables. The usual trick when an unknown static parameter, ψ say, is present, is to include this parameter in the state vector, $\tilde{\theta}_t = (\theta_t, \psi)$, and perform Monte Carlo filtering on this extended state space. This is also the approach applied by Berzuini and Gilks. Because of the non-dynamic structure of ψ, this part of the state vector will rapidly degenerate. Although the inclusion of the **Move** step in the SIR algorithm seems to resolve this problem, a Markov transition with respect to π_t will in general require storage of the whole vector $\theta_{t,i}$ in addition to all observed data $x_{1:t}$. For long time series, such storage can be a serious problem. Furthermore, the problem of choosing an appropriate Markov transition kernel will include all the difficulties involved in the more standard use of MCMC.

A different approach is considered by Storvik (2002). For many models, the time t posterior distribution of the static parameter ψ, given θ_t, only depends on some low-dimensional statistic $T_t = T_t(\theta_t)$. In such cases, impoverishment of the state vector can be avoided by including T_t in the state vector instead of ψ. The **Extend** in the Monte Carlo filtering procedures can then be replaced by a three-step procedure: (1) sample ψ from $f_{t,\psi}(\psi|T_{t-1,j})$, (2) sample η_t from $f_{t,\theta}(\eta_t|\theta_{t-1}, \psi)$, and (3) update $T_t = T(T_{t-1}, \eta_t)$. Appropriate modifications on the calculations of weights are necessary, but trivial. The impoverishment is avoided, in that samples of ψ at previous time points are not used; only the sufficient statistics are preserved.

7B
General principles in sequential Monte Carlo methods

Peter Clifford
Oxford University, UK

1 Particle filters

Simulation-based filters, such as the condensation algorithm and the Bayesian bootstrap or sampling importance resampling filter, aim to represent the joint posterior distribution of the unknown parameters and state variables in a dynamical statistical model by a discrete system of particles that evolves and adapts recursively as new information becomes available. For a review, see Doucet *et al.* (2001). In the simplest case, the particles are collectively constructed so that they approximate a random sample from the joint posterior distribution. In practice, large numbers of particles may be required to provide adequate approximations and, for certain applications, after a series of recursive updates, the particle system will often collapse to a single point at the resampling stage.

The idea of sampling importance resampling (SIR) dates back to Rubin (1988). For a standard Bayesian problem with a set of observations x, a vector of parameters θ, a likelihood $L(\theta, x)$, and a prior density $\pi(\theta)$, Rubin devised a simple way of obtaining an approximate sample from the posterior distribution.

- Simulate a sample $\tilde{\theta}^1, \tilde{\theta}^2, \ldots, \tilde{\theta}^n$ from $\pi(\theta)$.

- Calculate weights $q_i \propto L(\tilde{\theta}^i, x)$; $\sum q_i = 1$.
- Resample n times (with replacement) from the discrete θ-distribution with

$$\pi(\theta = \tilde{\theta}^i) = q_i.$$

The resulting sample $\theta^1, \theta^2, \ldots, \theta^n$ is an 'approximate' sample from $\pi(\theta|x)$.

The method is clearly related to the Monte Carlo method of importance sampling and at first glance there would be seem to be no advantage in performing the resampling stage, since posterior expectations could instead be estimated by the usual weighted calculations.

Suppose now that the observational framework expands dynamically, giving additional data and an expanded parameter set.

- Data: x, x^+
- Parameters: θ, θ^+, functions of interest $g(\theta, \theta^+)$
- Likelihood: $L(\theta, \theta^+; x, x^+)$
- Joint prior density: $\pi(\theta)\pi(\theta^+|\theta)$

Now the problem is to simulate from $\pi(\theta, \theta^+|x, x^+)$.

The expansion of the parameter space poses questions for the Markov chain Monte Carlo (MCMC) method since it is not obvious how to efficiently reuse earlier simulations from $\pi(\theta|x)$ to facilitate simulation from $\pi(\theta, \theta^+|x, x^+)$. This remains an open problem. For the SIR filter, the way forward is much clearer, as was identified by Gordon *et al.* (1993). The SIR values obtained for the $\pi(\theta|x)$ distribution at the first stage are simply treated as authentic samples from that distribution and then individually propagated by the $\pi(\theta^+|\theta)$ density. They are subsequently weighted by the relevant likelihood and resampled.

A basic weakness of the SIR method is immediately apparent, specifically the risk that the support of the sample becomes concentrated at a few points with relatively high likelihood. It is important to be able to assess the effect of such *sample impoverishment* on inferences about the posterior distribution. The remedy for this impoverishment is to simulate the initial sample from a distribution that is very close to the posterior density, adjusting the weights $\{q_i\}$ appropriately. In practice this may be difficult to achieve.

Of course, similar problems occur when using more traditional MCMC methods to explore the posterior distribution. Suppose that a Markov chain, using (say) the Metropolis algorithm, has been used to obtain successive values $\theta^1, \theta^2, \ldots, \theta^n$ simulated from the equilibrium density $\pi(\theta|x)$. The quantity $\mathrm{E}\{g(\theta)|x\}$ can be estimated by the sample mean $\bar{g}$. If $\{\theta^i\}$ were independent then $\mathrm{var}(\bar{g}) = \sigma_g^2/n$. However, typically, owing to positive autocorrelation in the sequence $\{\theta^i\}$, the variance $\mathrm{var}(\bar{g}) = T\sigma_g^2/n$, where $T > 1$. The sample has been 'impoverished' by the autocorrelation. The quantity T, termed the *correlation time*, is such that the sample size is effectively n/T.

In earlier papers, Carpenter *et al.* (1998, 1999) have introduced a method for assessing the efficiency of filtering algorithms, by replication of the filter and subsequent comparison of the replicates by simple ANOVA techniques. The effective sample size can be used to monitor the performance of the filter. Decisions about when to resample or whether to augment the population of particles are then based on maintaining the sample size at an acceptable level. Various proposals have been made about how to monitor the performance of the SIR filter and its generalizations. These are based upon estimates of the effectiveness of importance sampling. Liu (1996) suggested an approximation to the effective sample size that is independent of the function $g(\theta)$. However, just as with MCMC sampling, the effective sample size depends on the function $g(\theta)$ of interest and examples can be found where Liu's approximation is misleading (Carpenter *et al.* 1998).

2 Generalizations and enhancements

2.1 Monte Carlo integration

It is important to remember that Bayesian updating is essentially an integration problem and that the purpose is to provide a compact and efficient representation of certain posterior distributions. Obtaining random samples from these posterior distributions can be viewed as a secondary goal.

It follows that the nodes θ_t^i and weights q_i need not be a sample of independent values. More generally, the set of values can be viewed as a random measure which has the property that, as $n \to \infty$,

$$\sum_{i=1}^{n} q_i g(\theta^i) \xrightarrow{P} \mathrm{E}\{g(\theta)|x\}.$$

The objective is to choose the measure so as to reduce the variance of the estimator. For example, at the resampling stage, we can do better than multinomial sampling by using a systematic scheme (Carpenter *et al.* 1999).

2.2 Regularized sampling and effective dimension

Many of the difficulties that are encountered in the Bayesian analysis of evolving data sets are difficulties which are commonly encountered in practical high-dimensional Monte Carlo integration.

Quasi-random or super-regular sampling is known theoretically to produce increased efficiency in low dimensions and there are empirical studies that support its use in high dimensions. However, it appears that these empirical findings apply when the *effective dimension* of the problem is low, that is when most of the variation in the integrand is expressed in a few dimensions. For specific problems such as the bearings-only tracking problem, considered by Berzuini and Gilks in their article, it is important to identify these directions and make sure that they are sampled efficiently.

2.3 Parametric particle filters

In certain applications, the statistical model fluctuates between a number of different regimes. In many cases, once the regime is known, Bayesian updating can be carried out explicitly (using Kalman filtering). For this type of problem, it is convenient to tag the particles with parametric values (means and covariances) rather than specific values of the state variable. They must also be tagged by the regime indicator.

Regime switching is assumed to be governed by a hidden Markov model. The method proceeds by generating new particles for each possible regime switch. However, the number of particles grows geometrically with time and it is necessary to introduce a culling mechanism.

The problem can be formulated as that of selecting a subset S of n particles with modified weights $\tilde{q}$ from a larger set of particles with weights q so that

$$E \sum_{i \in S} (q^i - \tilde{q}^i)^2$$

is a minimum. The solution is given by Fearnhead (1998). An $O(N)$ algorithm for the selection problem is given in Fearnhead and Clifford (2001).

2.4   Self-avoiding particles

For a particle filter the objective is to choose a measure so that

$$\sum_{i=1}^{n} q_i g(\theta^i) \xrightarrow{P} E\{g(\theta)|x\},$$

as $n \to \infty$, for typical functions g of the state space. In terms of variance reduction, more effective random measures can be obtained by using more sophisticated Monte Carlo integration techniques, involving *sample regularization*. A few weighted particles which are spread effectively throughout the state space will represent a probability density much better than a large simple random sample.

In general, weighted particle systems carry greater information than unweighted ones. Weighted particle systems, and in particular systems in which weights are carried forward across the update stage, can be made to work effectively, even in highly non-linear problems. A generic weakness of the simplest particle filters is that they 'commit too soon'. Lagged filters that work on subsequences of observations with a correspondingly expanded state space will have better performance and increased computational costs. Experience with Monte Carlo smoothing filters, in the bearings-only problem, indicates that short transient data sequences can be effectively analysed with Metropolis samplers, provided that block moves are incorporated. For this example, the crucial move involves rescaling the trajectory. Extending the particle description, or *signature*, to incorporate this information enables block moves to be added to the repertoire of particle updates. Similar strategies can be employed in the analysis of models with a hidden Markov structure.

3   Bearings-only tracking

Bearings-only tracking is a classic non-linear filtering problem. An object moves in two dimensions. Its position and velocity are unknown. The state variable is assumed to be updated by a known linear Gaussian process. At each time point, a noisy bearing, or angle, is observed. The problem is to reconstruct the trajectory given the observed bearings and a prior distribution on the initial position and velocity. An initial investigation of the problem by MCMC methods shows which directions must be sampled thoroughly (Carpenter *et al.* 1996). Scaling the track by multiplying each range variable by a constant does not affect the likelihood since none of the angles changes. The changing factors can be incorporated into the filter by extending the *signature* of each particle. This is accomplished by storing, for each particle, a five-dimensional vector of sufficient statistics. The MCMC scale move, when it is made, is a Gibbs move, sampling from a modified gamma distribution. It produces an efficiency gain of up to 40-fold. For a detailed exposition of the method, see Fearnhead (1998, 2001).

Additional references in discussion

Carpenter, J. R., Clifford, P., and Fearnhead, P. (1996). Sampling strategies for Monte Carlo filters of non-linear systems. In *IEEE Colloquium Digest, Target Tracking and Control, DERA, Malvern* (ed. D. P. Atherton), Vol. 6, 253, pp. 1–3.

Fearnhead, P. (1998). *Sequential Monte Carlo Methods in Filter Theory.* PhD thesis, Oxford University, UK. Available from http://www.stats.ox.ac.uk/~fhead.

Fearnhead, P. (2001). *MCMC, Sufficient Statistics and Particle Filters.* Available from http://www.stats.ox.ac.uk/~fhead.

Fearnhead, P. and Clifford, P. (2001). The detection of change points in online well logging. Technical Report, Statistics Department, Oxford University, UK.

Künsch, H. R. (2001). State space and hidden Markov models. In *Complex Stochastic Systems*, Monographs on Statistics and Applied Probability, No. 87 (eds O. E. Barndorff-Nielsen, D. R. Cox, and C. Klüppelberg), pp. 109–73. Chapman & Hall, London.

Liu, J. S. (1996). Metropolized independent sampling with comparisons to rejection sampling and importance sampling. *Statistics and Computing*, 6, 113–9.

Storvik, G. (2002). Particle filters in state space models with the presence of unknown static parameters. *IEEE Transactions on Signal Processing*, 50, 281–9.

8

Spatial models in epidemiological applications

Sylvia Richardson
Imperial College, London, UK

1   Introduction

Spatial models have been at the heart of the early development of Bayesian hierarchical models and associated computation techniques. In varied fields such as image analysis, agriculture, ecology, and disease mapping, standard analysis techniques proved inadequate. Stimulated by these applications, hierarchical models involving a random field, on which was superimposed a noise process, have been built. Such schemes were used successfully to uncover interesting features of the underlying spatial structure as well as addressing a variety of regression questions while accounting for spatial dependence. The size, layout of the data, and the aims are very different in image analysis, ecology, and geographical epidemiology. Despite this, the building blocks making up the hierarchical structure have many points in common. In this article we focus principally on the class of spatial models that have been commonly used in epidemiological applications; models in image analysis and ecology are treated in Chapters 10 and 9.

Spatial analyses abound in the epidemiological literature. Geographical variations of chronic diseases have long been recognized as being able to suggest important aetiological clues. The monograph by Doll (1980) describes some of the early hypotheses concerning the influence of environment and life-style on cancer mortality. Recent studies have investigated many other health events, such as coronary and ischaemic heart diseases, respiratory illnesses, and perinatal outcomes, and have ranged in geographical scales from international comparisons to within-region analyses. More recently, the availability of geographically-indexed health and population data at a small scale, with advances in computing and geographical information systems, have opened the way for serious exploration of small area health statistics based on routinely collected data. The recent book by Elliott *et al.* (2000) gives a series of examples. Such analyses have been used to raise hypotheses, for example about disease clusters, and to address specific questions concerning, for example, perinatal health in relation to environmental sources of pollution. They are particularly suited to situations where the time lag between exposure and impact on health is short.

The statistical models developed are closely linked to the declared aims of small area analyses. There are broadly three types of analyses: disease mapping studies, geographic correlation studies, and point source studies.

Disease mapping is carried out to summarize spatial and spatio-temporal variation in risk (Clayton and Kaldor 1987; Clayton and Bernardinelli 1992; Mollié 1996; Knorr-Held and Besag 1998). The information derived may be used for simple descriptive purposes, for example by comparing patterns between different diseases and/or different time periods, as well as to provide a reference for further studies. In particular, it may be useful to characterize the clustering pattern of the disease to provide background information for local public health enquiries that may arise. Most commonly, the heterogeneity highlighted by disease mapping exercises suggests covariates that might be added to the model.

Geographic correlation studies are aimed at exploiting geographical variations in exposure to environmental variables (such as air pollution, background radiation, water quality) again in order to gain clues as to disease aetiology. The statistical models that are used for disease mapping and geographic correlation studies may be similar, but the aims of the analyses are different (Clayton *et al.* 1993): disease mapping studies are essentially descriptive, whilst geographic correlation studies focus on the role of explanatory variables.

Point source type studies are carried out when an increased risk close to a 'source' (e.g. a nuclear installation or a waste-disposal site) is suspected. These studies are local by nature because any increased exposure due to the potential source is unlikely to extend far. We will not dwell on the spatial models used in these studies, since they arise from case-specific physical and biological hypotheses on the exposure and the risk. Morris and Wakefield (2000) provide a review of the assessment of disease risk in relation to point/line sources.

When dealing with observational studies, it is important to consider the interplay between the regression modelling of explanatory variables and the structure adopted for the residual variability. Failure to model overdispersion and correlation in the residuals leads to invalid inference for the regression parameters.

In geographical correlation studies, residual heterogeneity and overdispersion are typically present, even when all the potentially relevant covariates have been included in the model. In this non-experimental framework, it cannot be hoped that all sources contributing to it will have been identified, measured, and suitably incorporated into a regression-like equation. Moreover, the residual heterogeneity often has a spatial structure, inherited from a number of geographically varying factors, such as genetic characteristics of populations, socio-cultural habits, or climatic conditions. Early references to spatial regression in a non-hierarchical framework are those of Cliff and Ord (1981) and Cook and Pocock (1983), while Best *et al.* (1999) adopt a Bayesian perspective. While all the authors agree that residual variability has to be accounted for, the choice of model for the spatial component of the residuals and its possible influence on the estimation of covariate effects remains a delicate issue.

In the next section, the different types of spatial data used in epidemiology are reviewed since they lead to different styles of modelling. Models for point data are treated in Section 3, followed in Section 4 by a discussion of how these models aggregate. Models directly specified at an area level are reviewed in Section 5. Finally, Section 6 considers the way covariates are introduced in spatial models and illustrates the issues of confounding between the spatial structure of covariates and residuals on simulated data. Much of the recent work in spatial modelling has adopted a Bayesian perspective. This will be reflected in the content of the article, but not to the exclusion of other points of view.

2 Level of spatial modelling for disease outcomes

2.1 Types of data

Disease mapping exercises require an exhaustive recording of cases and an assessment of the population at risk. We distinguish between point data where the 'exact' location of the case is known and area-referenced count data where cases are aggregated over geographically defined areas. The form in which disease data are made available in a country usually follows from the public health procedures and confidentiality rules adopted there. Data on exposure characteristics may also be available at the individual level, on an arbitrarily fine grid of locations, or as summaries aggregated over predefined areas. Point data give the closest link to the conceptual framework of a population at locations $\{s_i, i = 1 \ldots n\}$, on which risk factors are acting (at possibly different geographical levels), leading to disease cases.

Such data are not available routinely in many countries, but the increasing use of Geographical Information Systems (GIS) is encouraging the development of databases at the point level. In the UK, for example, places of residence of cases can be indexed at the postcode level. Demographic data are usually available at a coarser geographic resolution, for example by Enumeration Districts (EDs) in each census year. Thus, there could be considerable uncertainties in the assessment of the number of people at risk in small areas, uncertainties that warrant specific modelling (Best and Wakefield 1999). We are in a situation of measurement error, which is more severe at a small geographical scale. Part of this error is due to migration, which is more frequent when small areas are considered.

Point data are also available through case–control studies, which record detailed information (including location) on cases and a set of controls. These studies require epidemiological expertise to address the selection and other sources of biases and a substantial investment of time and expense. Thus, they do not constitute a major data source for spatial modelling. Nevertheless, they have proved valuable in investigations around point sources (Diggle *et al.* 2000).

The most common type of data in spatial epidemiology is count data corresponding to administrative areas, an aggregation level that is often artificial for epidemiological investigations. The size of the population at risk can vary greatly between the areas analysed. The consequent differential variability is artificial and must not be confused with real underlying heterogeneity. In aggregating, sampling variability and, in some cases, measurement error is reduced, but the essential link between exposure and health for each individual is lost.

2.2   Hierarchical models

Whatever the type of data, realistic spatial analyses have to deal with many sources of complexity. There are difficulties in defining the background population rate and allowing for confounding factors. Mismeasurements of important environmental determinants are more the rule than the exception. Residual spatial dependence is difficult to characterize. This (non-exhaustive) enumeration shows clearly that it would be inadvisable to directly specify a model of the observations that accounts for all sources of variability. Instead, the Bayesian hierarchical framework offers a flexible and modular way to decompose these and to incorporate, at any level of the model, relevant factors contributing to the heterogeneity. For example, individual attributes and area-level covariates can be included in the same model. Moreover, this framework leads to a coherent propagation of the different components of uncertainty onto the posterior distribution of the parameters of interest. Much of the work that is reviewed in this article is thus formulated in this context.

3   Models for point data

3.1   Framework

The natural statistical paradigm for modelling data consisting of records of events at point locations is the point process framework. The interest is in the modelling of the intensity measure associated with the point process. Attribute vectors associated with the point locations, usually referred to a 'marks', can also be introduced. One of the most commonly used point process model is the *inhomogeneous Poisson process*. Informally, if an 'event' process follows an inhomogeneous Poisson process over a region R with intensity measure $\Lambda(ds), s \in R$, then the number of events observed in a subregion A of R follows a Poisson distribution with parameter $\Lambda(A) = \int_A \Lambda(ds)$. In some cases, the measure will have a density with respect to the Lebesgue measure, called the intensity function of the Poisson process. By abuse of notation, we will denote this density $\Lambda(s)$. In this case, we have $\Lambda(A) = \int_A \Lambda(s)\,ds$. It will always be clear from the context whether we are referring to the intensity measure $\Lambda(ds)$ or the intensity function $\Lambda(s)$. This latter quantity represents the expected number of events per unit area in a small neighbourhood around s. In disjoint sub-areas, the number of events are independent.

In spatial epidemiology, a number of authors (Diggle and Elliott 1995; Lawson 2000) have proposed inhomogeneous Poisson process as empirical schemes for describing the locations of cases of specific diseases or those of a random set of controls selected from the population at risk. This assumes fundamentally that an event occurrence is not influenced by that of other events nearby. This model is thus inappropriate for describing the locations of cases of contagious diseases. Because spatial phenomena for contagious and chronic diseases are radically different, in this article we restrict our discussion of spatial models to the context of chronic disease epidemiology. However, inhomogeneous Poisson processes might also be deemed inappropriate for describing the spatial distributions of population at risk or that of chronic disease cases, since these distributions commonly exhibit some spatial clustering. As noted by Diggle (2000) in his reference to Bartlett (1964), it is hard to distinguish statistically a point process with constant intensity and dependent points from one consisting of independent points but with varying intensity. We shall favour the latter formulation and incorporate explanatory variables and/or latent random effects into the modelling of the varying intensity, to account for the clustering of cases.

With this in mind, we thus adopt the framework of an inhomogeneous Poisson process for the population at risk, from which cases are drawn according to a disease process. To be precise, we consider the location of the individuals at risk as a Poisson process $N(\mathrm{d}s)$ with intensity measure $w(\mathrm{d}s)$, and a thinning function that 'picks' out the cases independently with probability

$$p(s) = \mathrm{Pr}(\text{individual at } s \text{ is a case}).$$

The distribution of cases of disease thus corresponds to a Poisson process with intensity measure $p(s)w(\mathrm{d}s)$ (Cressie 1993). For simplicity and because we want to focus our discussion on $p(s)$, we shall assume that $w(\mathrm{d}s)$ is known. The expected number of cases in an area A is given by $\int_A p(s)w(\mathrm{d}s)$ and the expected disease incidence rate over A is $\int_A p(s)w(\mathrm{d}s)/\int_A w(\mathrm{d}s)$, thus giving $p(s)$ the interpretation of a disease rate. It is the spatial variability of the rate $p(s)$ that is of primary interest.

3.2 Inhomogeneous Poisson processes

In this section we review different approaches for estimating the intensity function $\Lambda(s)$ of an inhomogeneous Poisson process. We will return to the specific epidemiological formulation involving the disease rate $p(s)$ and the population intensity measure $w(\mathrm{d}s)$ in Section 3.4.

In general, $\Lambda(s)$ can be treated as a deterministic function of unknown form or as a realization of a stochastic process. In the former case, non-parametric methods using kernel type smoothing methods have been extensively employed, with applications in forestry or ecology in particular (Diggle 1985). In the latter case, the framework becomes that of a doubly stochastic (Cox) process.

3.2.1 *Non-parametric Bayesian estimation*

In the spirit of much recent work on flexible Bayesian modelling, a hierarchical framework for non-parametric Bayesian estimation of Poisson intensities has been developed by Heikkinen and Arjas (1998), extending some of the seminal work of Green (1995). There, it is assumed that the intensities are piecewise constant, with an unknown number of pieces, a formulation that has strong connections with the mixture or partition models for aggregated data described in Section 5.3. To be precise, $\Lambda(s)$ is taken a priori to be a step function on a random Voronoi tessellation of the domain R. The tiles $\{A_k\}$ are defined via a random number of generating points $\{g_k\}$, corresponding to realizations of a homogeneous Poisson process with a fixed intensity, that influences the degree of smoothing. Thus, conditionally on the partition $\{A_k\}$ of R, $\Lambda = \sum_k \Lambda_k \mathbf{1}_{A_k}$. Different choices can be made for the process defining the 'heights' or levels Λ_k of the steps on each Voronoi tile. Heikkinen and Arjas impose the reasonable assumption of favouring small steps for neighbouring tiles. They use a conditional autoregressive Gaussian structure for the $\log(\Lambda_k)$ (Besag 1974), the parameters of the conditional variance influencing the amount of smoothing. This local definition is exploited by the Markov chain Monte Carlo (MCMC) algorithm. The number of tiles is updated by birth–death type moves.

3.3 Cox processes

Among the vast class of point processes, we restrict our discussion here to two types that have been used in epidemiological applications. These are both (doubly stochastic) Cox processes, that is the intensity of the Poisson process is treated as a random measure. They differ in their parametric specifications for the intensity measure.

3.3.1 *Log Gaussian Cox processes*

In log Gaussian Cox (LGC) processes, the intensity surface is related to a Gaussian process. It is assumed that $\log \Lambda(s) = Z(s)$, where $Z(s)$ is a real-valued stationary Gaussian process with mean $\mu = EZ(s)$, variance $\sigma^2 = \text{var}(Z(s))$, and correlation function $r(s_1 - s_2) = \text{cov}(Z(s_1), Z(s_2))/\sigma^2$. This correlation function is usually chosen to be isotropic and is parameterized in terms of a scale parameter. These processes enjoy nice mathematical properties which have been derived by Møller *et al.* (1998). By using different forms for the spatial covariance function, LGC processes can account for a wide variety of point patterns.

Møller *et al.* recommend carrying out estimation of the mean and covariance parameters by using moment-based estimators rather than the likelihood, which is usually intractable. What is particularly relevant to epidemiological applications is how this process can be approximated on a small grid. Basically, once a discretization grid s_{ij} is chosen, a Gaussian field is generated on the grid and, conditional on the values $Z(s_{ij})$, a Poisson process with intensity $\exp(Z(s_{ij}))$ is generated for each cell of the grid. There is an interplay between the smoothness of the Gaussian field and the suitable level of discretization. At a coarse level of discretization, such approximation of the Gaussian field might become unsatisfactory, with the need to consider aggregated versions of LGC processes.

3.3.2 *Gamma random fields*

Rather than going to the log link, gamma random field models construct the intensity measure as a mixture of gamma process distributions. Their properties were developed in Wolpert and Ickstadt (1998) and Ickstadt and Wolpert (1999). The intensity Λ is defined as $\Lambda(s) = \int_R k(s, u)\Gamma(du)$, where $\Gamma(du)$ is a gamma random field, $\Gamma(du) \sim Ga(\alpha(du), \tau(u))$ with shape measure $\alpha(du)$ and inverse scale $\tau(u)$, and $k(s, u)$ is a chosen non-negative two-dimensional smoothing kernel centred on the point s. Details of the mathematical properties of gamma random fields are given in the references above. Their realizations are almost surely discrete with jumps, or impulses, of size γ_j at countably infinite many places u_j. Therefore, the intensity $\Lambda(s)$ can be written as $\sum_j k(s, u_j)\gamma_j$. This spatial model can thus be loosely described as a smooth version of these impulses, with smoothness dependent on the 'decay' parameter of the kernel and the parameters that control the shape and scale of the gamma random field. These can be treated as unknown in a Bayesian implementation if necessary.

Algorithms for sampling inhomogeneous gamma random fields are not straightforward and are discussed in Wolpert and Ickstadt (1998) and Walker and Damien (2000). Wolpert and Ickstadt give details of posterior inference via MCMC, and illustrate the implementation on a data set concerning point locations of Hickory trees. An epidemiological application relating atmospheric pollution to health events is described in Best *et al.* (2000b), and a discrete version of the process is employed in Best *et al.* (2000a).

3.4 Spatial variability of the disease rate

In epidemiology, we consider the cases of disease as arising from a Poisson process with intensity measure $p(s)w(ds)$. One difficulty is that the demographic point process $N(ds)$, with intensity measure $w(ds)$, is not usually observed at the same spatial resolution as are the cases. Typically, the population counts are smoothed on a fine grid to produce an estimate of $w(ds)$, which is assumed to be known. The estimation of $p(s)$ using the location of the cases will thus be dependent on the way $w(ds)$ was estimated. In an analysis performed at the resolution of a grid of $750m \times 750m$ quadrats, Best *et al.* (2000a) took area-weighted averages of ED-level population values to estimate $w(ds)$. It would be natural to extend this framework to bivariate point process models for population and disease events.

To model the spatial variability of the disease rate $p(s)$ using any of the formulations described in Sections 3.2.1, and 3.3.1–3.3.2 for the intensity function $\Lambda(s)$, it is natural to consider a logit transform of $p(s)$, since $p(s) \leqslant 1$. This method was chosen by Diggle *et al.* (1998) when modelling, at a point level, the probability of a type of gastro-intestinal infection in North-West England. In some problems, however, interest is focused on integrating the point process over an area and, in this case, the logistic transformation causes analytical difficulties. Representations such as $p(s) = \sum_j k(s, u_j)\gamma_j$ (gamma random fields) or $p(s) = \exp(Z(s))$ (LGC), are also used. In practice, one should ensure that the constraint $p(s) \leqslant 1$ is enforced when carrying out the estimation using such formulations. In the case of a rare disease, log or logit transformation would be similar.

Instead of relying on imperfect measures of population intensity, Bithell (1990) and subsequent authors use the locations of a set of controls when such a sample is available. To interpret the results and obtain a spatial odds function, it is necessary to hypothesize that the selection of controls among the population of non-cases has not been influenced by their spatial location. An interesting avenue that has not been explored would be to perform non-parametric Bayesian estimation in the manner of Heikkinen and Arjas (1998) *separately* for the intensity of the cases and the controls, and obtain as a by-product the posterior distribution of the spatial odds function at any point of the study region.

4 Aggregation of point level models

There have been two general approaches to modelling counts of events aggregated over a set of predefined areas. In the first and most common, statistical models are directly formulated at the aggregated level. We will return to this class of models in the next section. The second approach starts by positing a model at the point level and then derives the aggregated level model by integrating the point level model over the spatial partition, at least approximately. This has the advantage that the parameters have meanings independently of the partition, leading to inference that is robust to the level of aggregation. On the other hand, except in particular cases, the resulting processes are not easy to handle computationally. We talk of 'aggregation coherence' when the modelling framework allows the possibility of relating the parameters of the process at different scales.

Assuming that the distribution of the cases of a disease follow a Poisson process with intensity measure $p(s)w(\mathrm{d}s)$ implies immediately that, given $p(\cdot)$, the number of cases $Y(A_i)$ in area A_i is a Poisson random variable with mean

$$\mu(A_i) = \mathrm{E}(Y(A_i) \mid p) = \int_{A_i} p(s)w(\mathrm{d}s). \qquad (4.1)$$

Rather than considering the disease rate $p(s)$, some authors introduce the relative risk function $\theta(s) = p(s)/\bar{p}$, where $\bar{p} = \int p(s)w(\mathrm{d}s)/\int w(\mathrm{d}s)$ is the overall probability of disease in the region R. A quantity of epidemiological interest at the aggregated level, the area-level relative risk $\theta(A_i)$ for the area A_i, is then defined as $\theta(A_i) = \int_{A_i} \theta(s)w(\mathrm{d}s)/N_i$, where $N_i = \int_{A_i} w(\mathrm{d}s)$. Thus, we have an equivalent expression to (4.1): $\mu(A_i) = \bar{p}N_i\theta(A_i)$.

4.1 Stratification

Epidemiologically, it is often important to separate the cases into more homogeneous categories or strata. Thus, $w(\mathrm{d}s)$ and $p(s)$ become respectively $w_j(\mathrm{d}s)$ and $p_j(s)$, where j indexes a specific risk group strata. If we make the strong assumption that $p_j(s)$ is piecewise constant in the spatial domain, i.e. that $p_j(s) = p_{ij}$ for all s in A_i, then we find that marginally, $Y_{ij} \sim \mathrm{Poisson}(p_{ij}N_{ij})$. Here, Y_{ij} is the number of cases of disease in area A_i, stratum j, and N_{ij} corresponds to the population at risk in area A_i, stratum j: $N_{ij} = \int_{A_i} w_j(\mathrm{d}s)$.

Aggregation of cases over strata is possible if we assume that the Poisson processes for the different age classes are independent and that a multiplicative decomposition $p_{ij} = \theta_i\bar{p}_j$ holds, where θ_i is the relative risk associated with area A_i and $\bar{p}_j$ denote reference probabilities related to stratum j. If $Y_i = Y(A_i)$ is the total number of cases in area A_i, we then derive: $Y_i \sim \mathrm{Poisson}(E_i\theta_i)$, where $E_i = \sum_j N_{ij}\bar{p}_j$ denotes the expected number of cases.

The assumption that the disease rate $p_j(s)$ has a constant value in each area A_i is too restrictive. Hence, in the following sections we relax it and discuss some of the characteristics of the aggregated level models that could be derived by integrating point process models for $p(s)$. For the sake of simplicity, we do not explicitly stratify the disease rate in the remainder of this section. Including a stratum index and aggregating over strata would be straightforward.

4.2 Models linked to Gaussian random fields (GRFs)

When the link between the point process and the underlying Gaussian field is non-linear, the distribution of the resulting integrated process is not tractable mathematically. Indeed, if $Z(s)$ is a Gaussian field, then the linear functional $\int_A q(s)Z(s)\,\mathrm{d}s$ is a Gaussian random variable for any weight function $q(\cdot)$, but this is no longer true for non-linear functionals.

Starting from a LGC process for $p(s)$, the aggregated process, $\mu(A) = \int_A \exp(Z(s))w(\mathrm{d}s)$, is a natural candidate model for the expected number of cases in disease mapping applications (Hjort 1999; Cressie and Richardson 2001). The expectation and further moments of $\mu(A)$ can be computed. Determining the distribution of $\log \mu(A)$ remains intractable, but as indicated by Møller in the following discussion (Chapter 8B), numerical techniques might be used to study this process. It would thus be interesting to explore further the use of integrated LGC processes in epidemiological applications.

Some progress can be made by resorting to linearization by first-order approximations. This line is taken by Kelsall and Wakefield (2002) in deriving a model for the distribution of $\log(\theta(A_i))$ based on LGC representation of the relative risk function, $\theta(s) = \exp(Z(s))$, where $Z(s)$ is a stationary isotropic Gaussian field. The expectation and the covariances of $\theta(A_i)$ are easily expressed in terms of the parameters of the GRF, $Z(s)$. Kelsall and Wakefield approximate the mean and covariance of $\log(\theta(A_i))$ using first-order Taylor expansions and assume a multivariate normal distribution for $\{\log \theta(A_i)\}$, relying on an argument of 'permanence of lognormality' used in geostatistics. A full Bayesian implementation of this approximate model, using MCMC algorithms, is described by Kelsall and Wakefield (2002) for a particular class of Gaussian fields.

4.3 Models linked to gamma random fields

An alternative to LGC processes for the disease rate $p(s)$ is to consider gamma random fields, so that $p(s) = \sum_j k(s, u_j)\gamma_j$. In this case, the aggregated process

$$\mu(A) = \mathrm{E}(Y(A) \,|\, p) = \sum_j \gamma_j \int_A k(s, u_j)w(\mathrm{d}s)$$

is again a linear function of the gamma impulses, with kernels scaled to the aggregated level. The distribution of the aggregated process $\mu(A)$ is intractable in general unless restrictive conditions are imposed (see discussion 8B by Møller).

A discrete version of this process has been used by Best *et al.* (2000a) in an application concerning childhood respiratory problems. A predefined partition

$\{B_k\}$ of the region R is chosen (that can be much finer than that of $\{A_i\}$) and the gamma impulses γ_k become latent quantities associated with this partition.

5 Models directly specified at the aggregated level

In view of the computational difficulties outlined in the previous section, much of the work on spatial modelling of count data has used hierarchical schemes directly defined at the chosen level of aggregation. The detailed specification of these schemes has been guided empirically by important features of the data to be captured, e.g. local variability, overdispersion, and spatial dependence. Such models, built at a particular aggregation level, are not easily transformed onto a different scale. This is not necessarily a drawback as health questions are often formulated at a specific geographical scale, but it is important to be aware that inference for such schemes is only relevant for that specific level of aggregation.

Here the basic setup is that outlined in Section 4.1, with the data consisting of numbers of events $Y_i = Y(A_i)$ occurring in area A_i, the set of areas $\{A_i, i = 1, 2, \ldots, n\}$, representing a partition of the region R of interest, and E_i denoting the expected number of cases after adjustment for strata related to age, sex, and possibly socio-economic characteristics (e.g. deprivation index). The classes of models that we describe share the common purpose of estimating the underlying relative risks θ_i in a hierarchical framework, originally proposed by Clayton and Kaldor (1987). At the lowest level of the hierarchy, a Poisson model is commonly adopted, $Y_i \sim \text{Poisson}(E_i \theta_i)$, independently for $i = 1, 2, \ldots, n$.

The models differ in the choice of prior structure for $\{\theta_i, i = 1, 2, \ldots, n\}$ at the next level of the hierarchy. The θ_is are attempting to capture different phenomena creating *overdispersion* of the Poisson counts; see Wakefield *et al.* (2000). In particular, their structure has to accommodate the effect of known risk factors measured at the area level as well as the possibility of residual spatial variability. In this section we discuss ways to model the spatial variability of the θ_is. Covariates will be introduced in these models in Section 6.

As with models of point processes, both parametric specification via loglinear Gaussian models and semiparametric formulations involving partition or mixture models have been proposed. Most formulations of the spatial structure use a neighbourhood graph. This is a prescribed undirected graph in which two areas i and j are said to be neighbours, written either as $i \sim j$, as $i \in \partial j$, or as $j \in \partial i$, if they are adjacent with respect to this graph.

5.1 Markov random fields models

The pioneering work of Besag *et al.* (1991) and Clayton and co-workers (1992, 1993), introduces loglinear Gaussian models for $\{\theta_i, i = 1, 2, \ldots, n\}$ using a conditional formulation. A commonly used model, referred to as the intrinsic conditional autoregressive (CAR) model, specifies the distribution of $\nu_i = \log(\theta_i)$ by

$$p(\nu_i | \nu_j, j \neq i) \sim \text{N}(\bar{\nu}_i, (n_i \kappa)^{-1}),$$

where κ is an unknown precision parameter, $\bar{\nu}_i = \sum_{j \in \partial i} \nu_j / n_i$, and n_i denotes the number of neighbours of area i. An equivalent expression for the joint distribution of $\{\nu_i, i = 1, 2, \ldots, n\}$ can be derived.

This is an example of the general class of Markov random field (MRF) models that were discussed by Besag (1974). A general formulation and properties of CAR models are detailed in Besag and Kooperberg (1995). To account for a non-spatial component of overdispersion, Besag *et al.* (1991) propose to model the distribution of $\log(\theta_i)$ as the sum of a CAR process and an independent identically distributed (i.i.d.) $N(0, \tau^{-1})$ process. These models have been commonly used in much of the recent work in disease mapping. The resulting estimates of the relative risks, which borrow strength from the neighbouring areas and are smoothed towards a local mean, have been described in a number of publications; see, for example, Bernardinelli *et al.* (1995) and Mollié (1996). When the data is sparse, it is not always possible to identify the two components, and sensitivity to prior specifications for the hyperparameters κ and τ has been noted (Richardson and Monfort 2000). Extensions and variants of the CAR model have been investigated. They include using more general weights than the simple 0–1 contiguities on a graph (e.g. Stern and Cressie 1999), and formulating alternative distributions to the Gaussian (as discussed in Besag *et al.* (1995) and Best *et al.* (1999)).

Recent work by Knorr-Held and Rue (2002) gives efficient algorithms for joint updating of $\{\nu_i\}$ in a CAR model, and shows that single site algorithms for the $\{\nu_i\}$ can exhibit poor mixing.

5.2 Direct modelling of covariance structure

In MRF models, the model specification concerns the conditional distributions, the joint distribution being derived from these. An alternative route followed by some authors is to start with a multivariate Gaussian model for $\{\log \theta_i, i = 1, 2, \ldots, n\}$ and to introduce a spatial structure directly in the covariance matrix Σ. This is referred to as *simultaneous modelling*. Various spatial structures ensuring that Σ is positive definite have been proposed (Ripley 1981), the most common ones involving exponential or Bessel functions (the Matérn class). Under the assumption of isotropy, the (i, j)th element of Σ is modelled in terms of the distance $d(i, j)$ between representative points of areas A_i and A_j.

Choosing the form of $\Sigma = \Sigma(\phi)$ among allowable parameterizations is a delicate question. The choice depends on empirical considerations, sometimes guided by variogram plots. Wakefield and Morris (1999) and Diggle *et al.* (1998) use the form of exponential decrease in $d(i, j)$. The influence of the choice of spatial models for the residuals in spatial regression problems has been investigated by Richardson *et al.* (1992). Rather than choosing a single model, it could be appealing to employ Bayesian model averaging over a set of models.

One important drawback of this approach is that implementing joint models can be computationally unwieldy when the dimension of Σ is large. Estimation of such models relies on inverting $\Sigma(\phi)$, which is usually not sparse. In contrast, the sparseness of the precision matrices associated with MRFs can be exploited algorithmically (Rue 2001).

5.3 Partition, allocation, and mixture models

In the parametric models described above, the parameters characterizing the spatial dependence are constant for the whole set of areas and have a global effect on the amount of smoothing performed. Consequently, there is a risk of over-smoothing. Alternative formulations that allow a degree of *adaptive local smoothing* have been developed. We refer specifically to partition and mixture models where the continuous random field model for $\{\theta_i\}$ is replaced by a spatially structured allocation model, with each cluster or component having a constant unknown relative risk. The aim is to allow for the possibility of discontinuities, while effecting nevertheless a necessary amount of smoothing and borrowing of strength by allowing areas to be allocated to the same cluster. Estimates of $\{\theta_i\}$ are obtained by averaging over a large number of configurations weighted by the corresponding posterior probabilities.

5.3.1 *Cluster and partition models*

The basic idea is to assume that the region $R = \bigcup_i A_i$ can be partitioned into a set of k non-overlapping clusters, $C_j, j = 1, \ldots, k$, each composed of areas A_i which are contiguous. A constant relative risk θ_j is associated with each cluster. The number and the shape of the clusters are unknown.

Different ways of creating spatially structured partitions have been proposed. The first model used in epidemiology was proposed by Knorr-Held and Rasser (2000). Among the areas A_i, k of them are chosen at random as so-called 'cluster centres'. Conditional on these, the remaining areas are associated with the cluster centre that is closest in terms of the minimal number of area boundaries that have to be crossed to reach it. Conditional on k, each configuration of centres is assumed equally likely a priori. The log relative risks, $\log \theta_i$, are given Gaussian priors, with unknown hyperparameters.

Denison and Holmes (2001) make use of the idea of Voronoi tessellations as a clustering device for allocating areas A_i into clusters. This requires representing each area A_i using, say, a centre point, in effect treating the area as a point location. The tessellation-generating points T are assumed to be located at any point of the region R of interest. Conditional on a set of k generating points, the partition element C_j is then composed of areas A_i with centre points closer to the jth generating point than to any other generating point. Denison and Holmes choose independent gamma priors for the θ_j conditional on T, in order to exploit the Poisson–gamma conjugacy. Thus, the marginal likelihood $p(Y|T)$ has a closed form expression that is used in the Metropolis algorithm for updating the tessellation. The Poisson–gamma conjugacy results in a fast algorithm, but this benefit would be lost if covariates were included.

A different class of priors for clustering areas has been suggested by Gangnon and Clayton (2000), using a class of Markov connected component field priors that define the probability of a configuration of clusters using associated 'scores'.

5.3.2 *Hidden Markov random fields with Potts allocations*

A class of hidden discrete state MRF models has been proposed by Green and Richardson (2002) as flexible schemes for disease mapping. The underlying process is that of a finite mixture model that allows spatial dependence. The model adopted replaces θ_i by θ_{z_i}, where $\{z_i, i = 1, 2, \ldots, n\}$ are *allocation variables* taking values in $\{1, 2, \ldots, k\}$ and $\{\theta_j, j = 1, 2, \ldots, k\}$ characterize k different components. As in Richardson and Green (1997), the number of components k is treated as unknown. Given k, the allocations $\{z_i\}$ follow a spatially-correlated process, the Potts model, that has been used in image processing and other spatial applications (Besag 1986; Högmander and Møller 1995). The Potts model involves an interaction parameter ψ that controls the degree of spatial dependence by favouring probabilistically those allocation patterns where like-labelled locations are neighbours. In this way the prior knowledge that areas close-by tend to have similar risks is modelled through the allocation structure. The parameter ψ is treated as unknown. Independent gamma priors are placed on the $\{\theta_j, j = 1, 2, \ldots, k\}$. Note that the model does not require that mixture components are spatially connected.

Detailed specifications can be found in Green and Richardson (2002), where it is also shown how to implement this model, using split/merge reversible jump moves. Computation of a needed normalizing constant that depends on the neighbourhood graph and ψ is done off-line. Results reported in Green and Richardson (2002) indicate that this model is successful in detecting discontinuities, while giving also good estimates in cases of smooth variations of the underlying $\{\theta_i\}$. Alternative mixture model specifications for similarly aggregated spatial data were proposed by Fernández and Green (2002).

The approximate coherence of partition and mixture models as the data are considered at different levels of aggregation has not been investigated. Apart from Denison and Holmes, all models treat the space discretely, with a fixed set of areas. Denison and Holmes have transposed the Voronoi point process partitioning of Heikkinen and Arjas at an aggregated level, but this is an approximation that will depend on the size of the areas.

6 Covariate effects in spatial models

We revisit some of the models discussed previously in order to review how covariate effects are introduced. We emphasize that in many instances the central question concerns covariate effects, and as pointed out in the introduction, the residual spatial structure of the disease rate can be thought of as a proxy for many unidentified confounders.

Looking at the literature, the overwhelming majority of geographical epidemiology studies investigating the effect of environmental exposure (in the wide sense) are carried out at the level of groups, defined as sets of people living in different areas, and fall under the category of 'ecological design'. In interpreting ecological studies, a typical problem is how to relate inference on exposure effect from the group level down to the individual or the point level, even though the individual link between exposure and effect is lost. The difference between risk estimates at the individual or point level and at the group level is referred to as *cross-level* or *ecological bias*. We will not here discuss all the possible components contributing to this bias; see Richardson (1992), Greenland and Robins (1994), and Richardson and Monfort (2000) and references therein. We will restrict our discussion to some of the modelling issues.

6.1 Point process models

Let us suppose that $\{X_l(s)\}$ denote generically a set of covariates that are related to exposure at location s. Two formulations, based on multiplicative or additive effects, have been proposed to introduce covariate effects into the modelling of the Poisson process.

In the first one, covariate effects are incorporated in the disease rate $p(s)$ through a loglinear model of the form

$$p(s) = \exp\left(\sum_l \alpha_l X_l(s)\right)\Lambda(s), \tag{6.1}$$

where the intensity function $\Lambda(s)$ models the residual spatial variation of the disease rate. This expression corresponds to the traditional multiplicative risk model which, on the basis of strong empirical observations, has been adopted in epidemiology as a useful parsimonious representation of joint effects for many risk factors (Breslow and Day 1980). As can be anticipated, the log link creates difficulties when aggregating the effect of covariates over an area.

Alternatively, an additive formulation was suggested by Ickstadt and Wolpert (1999), with covariate effects introduced by

$$p(s) = \sum_l \beta_l X_l(s) + \Lambda(s). \tag{6.2}$$

Data-dependent constraints on the β_l have to be imposed to ensure the positivity of $p(s)$. The effects of covariates is now assumed incremental and also detrimental in most cases by imposing $\beta_l > 0$. This formulation assumes a priori that different sources of damage are contributing independently to the risk. In terms of public health, it has been argued that the excess of risk is a useful quantity.

Different models for $\Lambda(s)$ can be used. Cressie and Richardson (2001) suggest LGC processes coupled with the multiplicative formulation (6.1), while Ickstadt and Wolpert (1999) use gamma processes and the additive formulation (6.2). Rather than assuming a specific form for the effects of covariates on the point pattern, Heikkinen and Arjas (1999) employ a non-parametric form (piecewise constant). The example treated concerns the effect of altitude on tree density. This framework could be adapted to epidemiological problems.

6.2 Aggregation and ecological bias

For simplicity of exposition, we formulate the problem with one covariate, $l = 1$ in (6.1) or (6.2), and suppress the index l. The average exposure over the area A will be denoted by $\bar{X}(A) = \int_A X(s)\,\mathrm{d}s/|A|$, where $|A|$ is the size of area A. We assume, as usual, that the population intensity measure $w(\mathrm{d}s)$ is known. We distinguish the multiplicative and additive cases. Taking the expectation in (4.1), we find for the multiplicative case (6.1):

$$E(Y(A)) = \int_A \exp(\alpha X(s))E(\Lambda(s))w(\mathrm{d}s). \tag{6.3}$$

From expression (6.3), one can see that, even if the population intensity were constant over A, there would be no simple relation between $E(Y(A))$ and $\bar{X}(A)$. In particular, there is no constant c such that $E(Y(A)/|A|) = c\exp(\alpha\bar{X}(A))$ for all A. Nevertheless, in most geographical correlation studies carried out at the aggregated level, Poisson means are related to average exposure through an expression of the form $\exp(\alpha^*\bar{X}(A))$. There is no interpretation of the coefficient α^* in terms of the point level model. This is one aspect of *ecological bias*. This form of bias is related to the misspecification of the relation at the aggregated level due to the difficulty of integration of the multiplicative model.

To carry out the integration in (6.3) and obtain an integrated-level expression that involves the point level coefficient α, it is necessary to know the distribution of $X(s)$ within the area A. Usually, this distribution is not known and only bounds or area-level summaries like $\bar{X}(A)$ or the variance are available. Lasserre *et al.* (2000) suggest a workable approximation for the case of dichotomous risk factors. Cressie and Richardson (2001) discuss how to incorporate partial knowledge on $X(s)$ within the framework of maximum entropy to derive an approximation of the within-area distributions of risk factors. Another way of acquiring such knowledge is to use results of within-area surveys on subgroups of individuals when these are available. Different estimation methods using survey results have been recently investigated by Wakefield and Salway (2001). This is an interesting area for further research.

If instead of (6.1), one uses (6.2), then the additivity is crucially useful in integrating. We find

$$E(Y(A)) = \beta \int_A X(s)w(\mathrm{d}s) + \int_A E(\Lambda(s))w(\mathrm{d}s). \tag{6.4}$$

With the additive risk model (6.2), the same coefficient β is estimated with data given at the aggregated or at the point level, so long as one can measure the population weighted average exposure. In recent work, Best *et al.* (2000b) have formulated risk functions where covariates are introduced at the point level both in an additive and multiplicative fashion. The residual spatial structure is modelled additively using gamma processes. This extension is useful, but integration again becomes difficult if the covariates with multiplicative effect are spatially varying.

In summary, outside the framework of additive risk factors (recorded or latent) at the point level, the shape of the exposure relationship is different at the point and aggregated levels, and integrating up from the point level to obtain *aggregation coherent models* requires additional knowledge about within-area distribution of risk factors besides measuring their mean level over the area. Since this distribution is elusive in many cases, most aggregated level models resort to empirical considerations for specifying the effect of average exposure to several environmental risk factors on the relative risk θ_i. The regression coefficients estimated are thus dependent on the aggregation level at which the ecological study is performed.

6.3 Covariate effects in models specified at the aggregated level

Let us suppose that measures $\{\bar{X}_l(A_i)\}$ of average exposure to several risk factors are available for each area A_i. At the aggregated level, covariates have been traditionally introduced in a loglinear fashion, starting with Clayton *et al.* (1993). For example, in MRF models, the risk in area A_i is often represented as

$$\theta_i = \alpha \exp\left(\sum_l \beta_l \bar{X}_l(A_i) + \nu_i + \delta_i \right), \qquad (6.5)$$

where ν_i follows a CAR model and δ_i is an exchangeable Gaussian process, both constrained to have zero mean, as discussed in Section 5.1. Similarly, in the mixture model with Potts allocations, the risk is represented as

$$\theta_{z_i} \exp\left(\sum_l \beta_l \bar{X}_l(A_i) \right). \qquad (6.6)$$

Departing from the loglinear tradition, Guidici *et al.* (2000) include effects of categorical covariates in a non-parametric way. The risk θ_i is decomposed into the product of a spatial component and a covariate component. Partition models with an unknown number of 'clusters' or 'levels' are used both for the spatial component and the covariate effect. The example considered by Guidici *et al.* has only a small number of categories. Further work would be needed to assess the performance of the method in a more challenging situation.

6.4 Confounding between covariates and random effects

Spatial processes are introduced into the modelling of the relative risks alongside the covariates to reflect the effect of latent confounders that have a spatial structure. It is therefore natural to suspect that the inclusion of these spatial random effects will influence the estimation of the coefficients $\{\beta_l\}$. Usually, ecological analyses are reported in two stages. First, a simple Poisson regression is carried out, with no allowance for spatially structured overdispersion. Secondly, spatial random effects are included (possibly with other unstructured components). Changes, and in particular decreases, in the regression parameter estimates between the two stages have been reported (Clayton *et al.* 1993; Best *et al.* 1999).

In a discussion of this phenomenon, Clayton *et al.* talk of possible 'dilution' of the covariate effect, when spatially structured random effects are included, while their omission leaves the possibility of confounding by location.

The change is affected by the similarity between the spatial distribution of the covariate(s) and that of the random effects. To investigate this further, we have carried out a simple simulation experiment.

6.4.1 *Simulation set-up*

We have considered two basic cases.

Case I In Ia, a smoothly varying covariate is generated from a CAR model, then fixed. In Ib, a covariate with a simple North–South contrast is generated, then fixed. The Poisson counts are simulated with mean $\theta_i E_i$, where $\theta_i = \exp(\beta X_i)$ and X denotes generically the covariate. This is a simple Poisson regression setup.

Case II The Poisson counts are simulated with mean $\theta_i E_i$, where $\theta_i = \exp(\beta X_i + \nu_i)$. The covariate X is generated from a CAR model as previously. In IIa and IIb, the term ν_i is generated independently of the covariate X. In IIa, ν_i has a CAR structure, while in IIb, $\exp(\nu_i)$ fluctuates randomly around a simple North–South contrast. In IIc, the term ν_i is now generated as positively spatially correlated with X, by adding a perturbation to X.

In all cases, the spatial layout of the $n = 94$ mainland French départements was used. The simulation setup is similar to the one used in Green and Richardson (2002). The E_is correspond to the expected numbers of deaths from larynx cancer in females for the period 1986–1993 and range from 2 to 58. The true β used throughout is 1.2. These data sets were analysed following either the model displayed in (6.5) and referred to as MRF in Table 1, or that given by (6.6), referred to as MIX. The MCMC implementation described in Green and Richardson (2002), modified to include estimation of the parameter β, was used with the same prior settings and a uniform prior on β.

6.4.2 *Results*

Each row of Table 1 corresponds to the average over the 10 independent replications of the data pattern, everything held fixed except Y. The posterior mean

TABLE 1. Estimation of covariate effect in different cases of residual spatial structure. True value of $\beta = 1.2$.

Data sets	Case Ia		Case Ib		Case IIa		Case IIb		Case IIc	
	MIX	MRF	MIX	MRF	MIX	MRF	MIX	MRF	MIX	MRF
$E(\beta\|Y)$	1.18	1.16	1.19	1.17	1.14	1.19	1.25	1.18	2.60	2.65
$SD(\beta\|Y)$	0.11	0.14	0.07	0.11	0.18	0.19	0.14	0.23	0.29	0.28
RAMSE	0.07	0.20	0.08	0.21	0.28	0.29	0.21	0.30	0.50	0.48

and standard deviation of β are given in the first two rows. For simulated data, we know the 'true' underlying risks, and these will be denoted by $\{\theta_i^t\}$. We then calculate for each area A_i, $\mathrm{MSE}_i = \mathrm{E}((\theta_i^t - \theta_i)^2 | Y)$, the (posterior) mean square error, where θ_i corresponds to the estimate given either by the mixture model or by the MRF model. To summarize the performance over the whole map, we compute $\mathrm{RAMSE} = \sqrt{(\sum_i \mathrm{mse}_i / n)}$, the root averaged mean square error.

Case I We observe immediately that in both cases Ia and Ib, there is no notable dilution of the covariate effect, with the mean estimate very close to the 'true' value of 1.2. For either spatial structure of the covariates, having unnecessarily included in the model spatially structured random effects has not biased the estimation of the ecological regression. There is a difference between the MRF and the mixture model concerning the posterior standard deviation, which is larger for the MRF on average. This is related to the difference in the structure of the two models. In the cases of no overdispersion, the mixture model finds essentially only one component. Over the 10 simulations, the posterior probability for $k = 1$ was 0.76 in (a) and 0.74 in (b). Thus, the performance of the mixture model is in keeping with that of a simple Poisson regression with no overdispersion (adjusting a simple Poisson regression for either Ia or Ib gives identical posterior mean and standard deviation to the MIX column, with an RAMSE of 0.06). On the other hand, the MRF model is over-parameterized in this case. This leads to an increase of the posterior standard deviation of β as well as a deterioration of the RAMSE which is larger than for the mixture model.

Case II In Cases IIa and IIb, the covariate and the additional random effects were simulated independently. The results for the mixture model show a little more fluctuation from the value of 1.2 than in Case I, but this is not substantial. The MRF model performs well, with the posterior mean close to the true value. In terms of RAMSE, the two models are comparable, with a small improvement for the mixture model in the case where the random effects $\exp(\nu_i)$ have a spatial discontinuity. We see no 'dilution' of the covariate effect in this case. Further work is needed to understand potential dilution effects in cases when the mean function is misspecified. In case IIc on the other hand, there is strong confounding between the spatial structure of the covariate and the random effects, with a coefficient equal to 1.42 when regressing $\exp(\nu_i)$ on $\exp(X)$. This confounding is directly reflected in the substantial bias of the ecological regression coefficient, which is inflated additively by around 1.4. The two models have a similar performance. In terms of bias, there is no improvement gained by the inclusion of a spatially structured component. Here, we would not talk of dilution, but of bias due to confounding, bias which can go either way.

7   Concluding remarks

There are many interesting avenues to be explored. Some of the challenges lie in modelling population processes together with the disease process, investigating the flexibility and computational feasibility of deriving integrated processes from different point process formulations, implementing flexible non-parametric models for covariate effects, and characterizing group-level (contextual) effects operating on top of individual-level covariate effects using data sets that include both group and individual level covariates and health data.

Acknowledgements

I wish to acknowledge the benefits to my research from my participation in all the ESF–HSSS activities. I wish to express my thanks to my recent and long-standing colleagues with whom I've had many enlightening discussions on spatial epidemiology and spatial statistics: Julian Besag, Nicky Best, Noel Cressie, Chantal Guihenneuc-Jouyaux, Peter Green, Denis Hémon, Katja Ickstadt, Alex Lewin, Leo Knorr-Held, Virginie Lasserre, Christine Monfort, Annie Mollié, and Jon Wakefield. Support from INSERM, ITM contract 4TM05F, and the Small Area Health Statistics Unit is gratefully acknowledged.

References

Bartlett, M. S. (1964). The spectral analysis of two-dimensional point processes. *Biometrika*, **51**, 299–311.

Bernardinelli, L., Clayton, D., and Montomoli, C. (1995). Bayesian estimates of disease maps: how important are the priors? *Statistics in Medicine*, **14**, 2411–31.

Besag, J. (1974). Spatial interaction and the statistical analysis of lattice systems (with discussion). *Journal of the Royal Statistical Society*, B, **36**, 192–225.

Besag, J. (1986). On the statistical analysis of dirty pictures (with discussion). *Journal of the Royal Statistical Society*, B, **48**, 259–302.

Besag, J. and Kooperberg, C. (1995). On conditional and intrinsic autoregressions. *Biometrika*, **82**, 733–46.

Besag, J., York, J., and Mollié, A. (1991). Bayesian image restoration with applications in spatial statistics (with discussion). *Annals of the Institute of Mathematical Statistics*, **43**, 1–59.

Besag, J., Green, P. J., Higdon, D., and Mengersen, K. (1995). Bayesian computation and stochastic systems (with discussion). *Statistical Science*, **10**, 3–66.

Best, N. G. and Wakefield, J. C. (1999). Accounting for inaccuracies in population counts and case registration in cancer mapping studies. *Journal of the Royal Statistical Society*, A, **162**, 363–82.

Best, N. G., Arnold, R. A., Thomas, A., Waller, L. A., and Conlon, E. M. (1999). Bayesian models for spatially correlated disease and exposure data (with discussion). In *Bayesian Statistics 6* (eds J. M. Bernardo, J. O. Berger, A. P. Dawid, and A. F. M. Smith), pp. 131–56. Oxford University Press, UK.

Best, N. G., Ickstadt, K., Wolpert, R. L., and Briggs, D J. (2000a). Combining models of health and exposure data: the SAVIAH study. In *Spatial Epidemiology: Methods and Applications* (eds P. Elliott *et al.*), pp. 393–414. Oxford University Press, UK.

Best, N. G., Ickstadt, K., and Wolpert, R. L. (2000b). Spatial Poisson regression for health and exposure data measured at disparate level. *Journal of the American Statistical Association*, **95**, 1076–88.

Bithell, J. F. (1990). An application of density estimation to geographical epidemiology. *Statistics in Medicine*, **9**, 691–701.

Breslow, N. E. and Day, N. E. (1980). *Statistical Methods in Cancer Research*, Vol. 1. IARC Scientific Publication No.32, International Agency for Research on Cancer, Lyon, France.

Clayton, D. and Bernardinelli, L. (1992). Bayesian methods for mapping disease risk. In *Geographical and Environment Epidemiology: Methods for Small Area Studies* (eds P. Elliott, J. Cuzick, D. English, and R. Stern), pp. 205–20. Oxford University Press, UK.

Clayton, D. and Kaldor, J. (1987). Empirical Bayes estimates of age-standardised relative risks for use in disease mapping. *Biometrics*, **43**, 671–81.

Clayton, D., Bernardinelli, L., and Montomoli, C. (1993). Spatial correlation in ecological analysis. *International Journal of Epidemiology*, **22**, 1193–202.

Cliff, A. D. and Ord, J. K. (1981). *Spatial processes: Models and Applications*. Pion, London.

Cook, D. G. and Pocock, S. J. (1983). Multiple regression in geographical mortality studies, with allowance for spatially correlated errors. *Biometrics*, **39**, 361–71.

Cressie, N. (1993). *Statistics for Spatial Data* (rev. edn). Wiley, New York.

Cressie, N. and Richardson, S. (2001). Approximating ecological bias via maximum entropy distributions. Available at `http://ifr69.vjf.inserm.fr/~u170/sylvia`

Denison, D. G. T. and Holmes, C. C. (2001). Bayesian partitioning for estimating disease risk. *Biometrics*, **57**, 143–9.

Diggle, P. J. (1985). A kernel method for smoothing point process data. *Applied Statistics*, **34**, 138–47.

Diggle, P. J. (2000). Overview of statistical methods for disease mapping and its relationship to cluster detection. In *Spatial Epidemiology: Methods and Applications* (eds P. Elliott *et al.*), pp. 87–103. Oxford University Press, UK.

Diggle, P. J. and Elliott, P. (1995). Disease risk near point sources: statistical issues for analyses using individual or spatially aggregated data. *Journal of Epidemiology and Community Health*, **49**(Suppl. 2), S20–S27.

Diggle, P. J., Tawn, J. A., and Moyeed, R. A. (1998). Model-based geostatistics (with discussion). *Applied Statistics*, **47**, 299–350.

Diggle, P. J., Morris, S. E., and Wakefield, J. C. (2000). Point-source modelling using matched case–control data. *Biostatistics*, **1**, 89–105.

Doll, R. (1980). The epidemiology of cancer. *Cancer*, **45**, 2475–85.

Elliott, P., Wakefield, J. C., Best, N. G., and Briggs, D. J., eds (2000). *Spatial Epidemiology: Methods and Applications*. Oxford University Press, UK.

Fernández, C. and Green, P. J. (2002). Modelling spatially correlated data via mixtures: a Bayesian approach. *Journal of the Royal Statistical Society*, B, **64**, 805–26.

Gangnon, R. E. and Clayton, M. K. (2000). Bayesian detection and modeling of spatial disease clusters. *Biometrics*, **56**, 922–35.

Giudici, P., Knorr-Held L., and Raßer, G. (2000). Modelling categorical co-variates in Bayesian disease mapping by partition structures. *Statistics in Medicine*, **19**, 2579–93.

Green, P. J. (1995). Reversible jump Markov chain Monte Carlo computation and Bayesian model determination. *Biometrika*, **82**, 711–32.

Green, P. J. and Richardson, S. (2002). Hidden Markov models and disease mapping. *Journal of the American Statistical Association*, **97**, 1055–70.

Greenland, S. and Robins, J. (1994). Ecological studies: biases, misconceptions, and counterexamples. *American Journal of Epidemiology*, **139**, 747–60.

Heikkinen, J. and Arjas, E. (1998). Non-parametric Bayesian estimation of a spatial Poisson intensity. *Scandinavian Journal of Statistics*, **25**, 435–50.

Heikkinen, J. and Arjas, E. (1999). Modeling a Poisson forest in variable elevations: a nonparametric approach. *Biometrics*, **55**, 738–45.

Hjort, N. L. (1999). Discussion on Bayesian models for spatially correlated disease and exposure data. In *Bayesian Statistics 6* (eds J. M. Bernardo, J. O. Berger, A. P. Dawid, and A. F. M. Smith), pp. 150–1. Oxford University Press, UK.

Högmander, H. and Møller, J. (1995). Estimating distribution maps from atlas data using methods of statistical image analysis. *Biometrics*, **51**, 393–404.

Ickstadt, K. and Wolpert, R. L. (1999). Spatial regression for marked point processes (with discussion). In *Bayesian Statistics 6* (eds J. M. Bernardo, J. O. Berger, A. P. Dawid, and A. F. M. Smith), pp. 323–41. Oxford University Press, UK.

Kelsall, J. and Wakefield, J. C. (2002). Modeling spatial variation in disease risk: A geostatistical approach. *Journal of the American Statistical Association*, **97**, 692–701.

Knorr-Held, L. and Besag, J. (1998). Modelling risk from a disease in time and space. *Statistics in Medicine*, **17**, 2045–60.

Knorr-Held, L. and Rasser, G. (2000). Bayesian detection of clusters and discontinuities in disease maps. *Biometrics*, **56**, 13–21.

Knorr-Held, L. and Rue, H. (2002). On block updating in Markov random field models for disease mapping. *Scandinavian Journal of Statistics*, **29**, 597–614.

Lasserre, V., Guihenneuc-Jouyaux, C., and Richardson, S. (2000). Biases in ecological studies: utility of including within-area distribution of confounders. *Statistics in Medicine*, **19**, 45–59.

Lawson, B. (2000). Cluster modelling of disease incidence via RJMCMC methods: a comparative evaluation. *Statistics in Medicine*, **19**, 2361–75.

Mollié, A. (1996). Bayesian mapping of disease. In *Markov Chain Monte Carlo in Practise* (eds W. R. Gilks, S. Richardson, and D. J. Spiegelhalter), pp. 359–79. Chapman & Hall, London.

Møller, J., Syversveen, A. R., and Waagepetersen, R. P. (1998). Log Gaussian Cox processes. *Scandinavian Journal of Statistics*, **25**, 451–82.

Morris, S. E. and Wakefield, J. C. (2000). Assessment of disease risk in relation to a pre-specified source. In *Spatial Epidemiology: Methods and Applications* (eds P. Elliott *et al.*), pp. 153–84. Oxford University Press, UK.

Richardson, S. (1992). Statistical methods for geographical correlation studies. In *Geographical and Environment Epidemiology: Methods for Small Area Studies* (eds P. Elliott, J. Cuzick, D. English, and R. Stern), pp. 181–204. Oxford University Press, UK.

Richardson, S. and Green, P. J. (1997). On Bayesian analysis of mixtures with an unknown number of components (with discussion). *Journal of the Royal Statistical Society*, B, **59**, 731–92.

Richardson, S. and Monfort, C. (2000). Ecological correlation studies. In *Spatial Epidemiology: Methods and Applications* (eds P. Elliott *et al.*), pp. 205–20. Oxford University Press, UK.

Richardson, S., Guihenneuc, C., and Lasserre, V. (1992). Spatial linear models with autocorrelated error structure. *The Statistician*, **41**, 539–57.

Ripley, B. D. (1981). *Spatial Statistics*. Wiley, Chichester.

Rue, H. (2001). Fast sampling of Gaussian Markov random fields. *Journal of the Royal Statistical Society*, B, **63**, 325–38.

Stern, H. and Cressie, N. (1999). Inference for extremes in disease mapping. In *Disease Mapping and Risk Assessment for Public Health* (eds A. Lawson, A. Biggeri, D. Bohning, E. Lesaffre, J.-F. Viel, and R. Bertolini). Wiley, Chichester.

Wakefield, J. C. and Morris, S. E. (1999). Spatial dependence and errors-in-variables. In *Bayesian Statistics 6* (eds J. M. Bernardo, J. O. Berger, A. P. Dawid, and A. F. M. Smith). Oxford University Press, UK.

Wakefield, J. C. and Salway, R. (2001). A statistical framework for ecological and aggregate studies. *Journal of the Royal Statistical Society*, A, **164**, 119–37.

Wakefield, J. C., Best, N. G., and Waller, L. (2000). Bayesian approaches to disease mapping. In *Spatial Epidemiology: Methods and Applications* (eds P. Elliott *et al.*), pp. 104–27. Oxford University Press, UK.

Walker, S. and Damien, P. (2000). Representation of Lévy processes without Gaussian components. *Biometrika*, **87**, 477–83.

Wolpert, R. L. and Ickstadt, K. (1998). Poisson/gamma random field models for spatial statistics. *Biometrika*, **85**, 251–67.

8A
Some remarks on Gaussian Markov random field models for disease mapping

Leonhard Knorr-Held
Lancaster University, UK

In this contribution, I will comment briefly on some issues arising in disease mapping, in particular on Gaussian Markov random field (GMRF) models. The overall success of the GMRF approach in disease mapping is remarkable. It is a useful and flexible tool that can easily be combined with other formulations such as measurement error (Bernardinelli *et al.* 1997) or dynamic models (Knorr-Held and Besag 1998; Knorr-Held 2000) in a hierarchical Bayesian framework. Software developed within the last decade already allows for the routine use of such extended models. However, there are certain limitations and problems associated with this class of models and I will describe some of them now, indicating, wherever possible, potential remedies.

GMRF models as proposed in Besag *et al.* (1991) are defined through the improper prior density

$$p(\nu|\kappa) \propto \exp\left(-\frac{\kappa}{2}\sum_{i\sim j}(\nu_i - \nu_j)^2\right) = \exp\left(-\frac{1}{2}\nu'K\nu\right) \qquad (1)$$

for the log-relative risk parameters ν. Note that, for simplicity, we use here the simple form of an intrinsic GMRF without weights, i.e. the precision matrix K has non-diagonal entries $k_{ij} = -\kappa$ if $i \sim j$ and zero elsewhere, while the diagonal elements are chosen such that all row (and column) sums are equal to zero. Model (1) is non-adaptive as it contains one global spatial smoothing parameter κ. However, if the number of regions n is fairly large, then it seems desirable to use *adaptive* smoothing methods. I will now discuss possible adaptive extensions of GMRFs.

Inspired by the literature on stochastic volatility models for time series (e.g. Shephard and Pitt 1997, and references therein) and on statistical image analysis (e.g. Clifford 1986; Aykroyd 1998), a possible adaptive modification of the intrinsic GMRF model is to assume that the (log) precision κ varies spatially as well. More specifically, one could replace $\kappa \sum_{i\sim j}(\nu_i - \nu_j)^2$ with $\sum_{i\sim j}\kappa_{ij}(\nu_i - \nu_j)^2$ in (1) and let $\xi_{ij} = \log \kappa_{ij}$ follow another GMRF model in a second hierarchy, but now on the *graph of links between regions*, not on the original graph of regions. For example, one could consider links as adjacent if they share a common region, i.e. link $i \sim j$ is adjacent to link $i \sim k$ if $j \neq k$.

There are potential problems to be expected with such a formulation, due to the increased complexity of the model. First, the new graph of links may become rather large, with as many nodes as there are links in the original graph; however, information in the data about the spatially varying precision will typically be limited. Therefore care needs to be taken when specifying the prior for the precision of the second GMRF and prior sensitivity can be expected. Furthermore, for reliable MCMC simulation, it will be necessary to block update the ξ vector. Note that due to the log-link, there are non-linearities in the full conditional for ξ, so a Metropolis–Hastings proposal based on some approximation has to be used, along similar lines as in Rue (2001) and Knorr-Held and Rue (2002). Also, the 'normalizing constant' of (1), the square root of the product of the non-zero eigenvalues of K (Besag and Higdon 1999), needs to be taken into account. Note that the computation of the non-zero eigenvalues can be done analytically only on a regular grid (Künsch 1986). Finally, it might be advisable to use a proper (stationary) GMRF model for ξ to avoid problems with improper posteriors as such problems already arise in the non-adaptive case with an improper prior for κ.

Simpler formulations that avoid some of the above problems could be based on the factorization $\kappa_{ij} = \kappa \cdot w_{ij}$ with a gamma prior for κ and independent priors for each w_{ij}, for example beta priors or discrete priors on $\{0,1\}$. Both approaches still allow for a varying precision, but without any spatial structure. The latter approach has the slight disadvantage that the study region may be split into two or even more completely separate pieces. This would cause further rank deficiencies of the precision matrix K. Incidentally, there are relationships here to formulations with independent gamma-distributed parameters $w_{ij} \sim G(\phi/2, \phi/2)$. Marginally (integrating with respect to w_{ij}) this leads to t-distributed errors with ϕ degrees of freedom (e.g. Besag and Higdon 1999) and might be considered as a further alternative.

To summarize, it seems that adaptive versions of GMRF models are possible, but rather hard to implement successfully. An overall comparison with mixture and partition models, which are by default adaptive, seems difficult. From a modelling point of view, both partition and mixture models have the additional advantage that the 'null hypothesis' of constant disease risk in the whole study region is an explicit part of the formulation, whereas in the GMRF case it is only a limiting case ($\kappa \rightarrow \infty$). However, there is only limited experience with these models in more extended settings (e.g. with covariates), partly because the implementation is typically more challenging and computation times will be considerably longer.

I will now turn to the question of how to specify the 'normalizing constant' in the intrinsic GMRF model (1). The factor is important if κ needs to be estimated as well, which is always the case in practice. There seems to be no exact recommendation in the literature about this issue; the paper by Besag *et al.* (1991) uses the additional multiplicative factor $\kappa^{n/2}$, while other arguments suggest $\kappa^{(n-1)/2}$ instead. A motivation for the former is the derivation of the intrinsic GMRF as a limiting case of a stationary GMRF with a regular precision matrix, hence with n degrees of freedom. On the other hand, the precision matrix K in (1) has only rank $n-1$ and hence $n-1$ non-zero eigenvalues, the square root of the product of these non-zero eigenvalues (see above) is therefore proportional to $\kappa^{(n-1)/2}$.

In practice one would not expect any major differences between n or $n-1$ degrees of freedom, at least if n is rather large. However, if the data are sparse (i.e. there are not many cases), the relative risk estimates can be rather different, even for large n. For example, in a re-analysis of the Sardinia data ($n = 366$, total of 619 cases) (Bernardinelli *et al.* 1997), estimates of relative risk with $n-1$ degrees of freedom show considerably less shrinkage to unity than those obtained with n degrees of freedom. This can be seen from Fig. 1, which compares posterior mean relative risk estimates and corresponding posterior probabilities of a relative risk above 1.0. Both were obtained using the block-updating algorithm described in Knorr-Held and Rue (2002) with a 'vague' gamma G(0.25, 0.0005) prior for κ as recommended by Bernardinelli *et al.* (1995). Incidentally, very similar results can be obtained using the original Besag *et al.* model with additional unstructured parameters.

A possible explanation for the discrepancies is the following: suppose $n-1$ is 'correct' and we place a gamma G(0.25, 0.0005) prior on κ. A G(0.25, 0.0005) prior with n degrees of freedom is now equivalent to a G(0.75, 0.0005) prior and $n-1$ degrees of freedom. Figure 2 displays the two density functions of the corresponding *inverted gamma* (IG) distributions of the *variance* κ^{-1}. The IG(0.75, 0.0005) prior gives considerably more weight on very small values of the variance κ^{-1}. Note that both densities are normalized to unity – the apparent

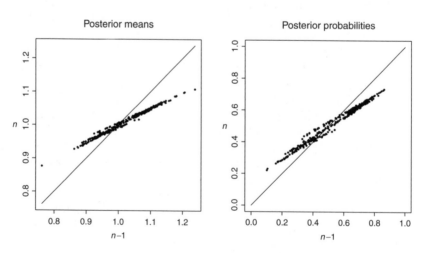

FIG. 1. A comparison of posterior mean relative risk estimates and corresponding posterior probabilities with n and $n-1$ degrees of freedom, respectively.

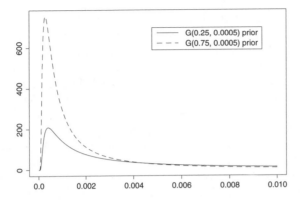

FIG. 2. A comparison of the prior on κ^{-1} implied by $n - 1$ (solid line) and n (broken line) degrees of freedom.

dominance of the IG(0.75, 0.0005) prior is balanced outside of the range shown for $\kappa^{-1} > 0.01$. Since there is not much information in the data about spatial variation, the posterior reproduces these large discrepancies, with considerably more shrinkage of the relative risk estimates for n degrees of freedom. Note that these considerations do not depend on the actual value of n.

These thoughts, however, leave us with some slightly disturbing consequences about the propriety of the posterior distribution implied by model (1) with unknown κ: the common approach to avoid an improper posterior is to use a proper prior for κ; see, for example, the discussion in Besag *et al.* (1991). However, if the choice of the degrees of freedom for κ is completely arbitrary then, for example, a proper gamma prior G(0.25, 0.0005) for κ with $n - 1$ degrees of freedom is equivalent to an improper 'G(−0.25, 0.0005)' prior with n degrees of freedom. Does this prior now imply a proper posterior? Further research into the definition of the degrees of freedom in (1) and the possible impropriety of posterior distributions in such models seems necessary.

Incidentally, the above analysis of the Sardinia data is based on an adjacency-based neighbourhood structure, which is rather irregular and one might think that this structure is at least partly responsible for the discrepancies. However, we have found similar results for sparse data on a regular grid. Nevertheless, it is worth pointing out that adjacency-based neighbourhoods (for irregularly distributed areas) are problematic in general, as there is typically a strong dependence of the prior amount of smoothing on the number of neighbours, especially if that number is small (Bernardinelli *et al.* 1995). If more appropriate weights are not available, a rough but still more satisfactory approach could be based on choosing all adjacent regions but *at least three* of the nearest (not necessarily adjacent) regions as neighbours as in Ranta and Penttinen (2000).

As a final note, I agree with S. Richardson that in epidemiological applications the main goal is not only to describe the spatial pattern present in the data, but also to include covariates in order to remove the spatial pattern and explain the spatial variation, at least to some extent. Indeed, as already mentioned, one of the nice features of the GMRF approach is that additional covariate effects can easily be added. However, most of the current scientific discussion seems to be about modelling spatial variation, not about modelling covariate effects. From a methodological perspective, more work on flexible models for covariate effects seems necessary. In particular, non-parametric Bayesian models for covariate effects (for example Fahrmeir and Lang 2001) could replace restrictive linear assumptions. Also, more work on including individual-level (in contrast to area-level) covariate information seems necessary (Best *et al.* 2000b). In fact, omitting covariates or misspecifying covariate effects may cause serious artifacts in the estimated (residual) risk surface, regardless of how flexible the chosen spatial model is.

Acknowledgements

The author expresses thanks to Julian Besag, Nicky Best, Sylvia Richardson, Håvard Rue and the editors for helpful comments.

8B
A comparison of spatial point process models in epidemiological applications

Jesper Møller
Aalborg University, Denmark

As pointed out in Sylvia Richardson's article (hereafter referred to as [SR]), point process models provide a natural setting in epidemiological applications. The aims of this contribution are to review the advantages and problems when using such models for aggregated data, compare the properties of different model classes, clarify some of the properties of these model classes in connection to epidemiological applications (which sometimes seem to have been misunderstood), propose some alternative models which are not covered in [SR], and point out some open problems. The focus is on three particular model classes: the Heikkinen and Arjas (1998) models, log Gaussian Cox (LGC) processes (Møller *et al.* 1998), and shot-noise Gaussian Cox (SNGC) processes (Brix 1999). Poisson/gamma models (Daley and Vere-Jones 1988; Wolpert and Ickstadt 1998) are special cases of SNGC processes.

1   Brief description of models

As in [SR], the point process Y of exact locations of cases observed in a bounded planar region R, is assumed to be a Cox process driven by a random intensity function $\Lambda(s)$, $s \in R$, i.e. $Y|\Lambda$ is a Poisson process with intensity measure $\Lambda(A) = \int_A \Lambda(s)\,\mathrm{d}s$, $A \subseteq R$. When dealing with aggregated data $Y(A_1), \ldots, Y(A_n)$, where $Y(A_i)$ is the number of cases in $A_i \subset R$, we assume $A_1, \ldots, A_n$ to be a subdivision of R.

In Heikkinen and Arjas (1998), $\Lambda = \sum_k \Lambda_k \mathbf{1}_{A_k}$, where $\{A_k\}$ is the Voronoi tessellation generated by a point process of nuclei $\{g_k\} \subset R$, i.e. A_k is the set of points in R closer to g_k than to any other nuclei. The nuclei follow a homogeneous Poisson process on R and, conditionally on $\{g_k\}$, $\{\log \Lambda_k\}$ is modelled by a conditional autoregression (Besag 1974).

For an LGC model, $\Lambda(s) = \exp(Z(s))$, where Z is a Gaussian process with mean function $\mu(s) = \mathrm{E}Z(s)$ and covariance function $c(s,t) = \mathrm{cov}(Z(s), Z(t))$. To obtain a well-defined Cox process, smoothness properties of c are required (Møller *et al.* 1998). Note that stationarity of Z is not required, but it may be convenient to assume that $c(s,t)$ depends only on the distance $||s-t||$.

The construction of an SNGC process is a bit more complicated: $\Lambda(s) = \sum_j k(s, u_j)\gamma_j$, where k is a kernel (for simplicity we assume that $k(\cdot, u)$ is a density function for a continuous random variable), and $\{(u_j, \gamma_j)\} \subset E \times [0, \infty)$, where E is a given planar region. Typically in applications, $E = R$, or in order to reduce edge effects, E is much larger than R. Furthermore, $\{(u_j, \gamma_j)\}$ is a Poisson process with intensity measure

$$\nu(A \times B) = (\alpha(A)/\Gamma(1-\kappa)) \times \int_B \gamma^{-\kappa-1} \exp(-\tau\gamma)\mathrm{d}\gamma, \quad A \subseteq E, \ B \subseteq [0, \infty),$$

where $\kappa < 1$ and $\tau \geqslant 0$ are parameters with $\tau > 0$ if $\kappa \leqslant 0$, and α is a non-negative and non-zero Radon measure. If $\kappa < 0$, we obtain a kind of modified Neyman–Scott process as $\{u_j\}$ is a Poisson process with intensity measure $\tau\alpha/|\kappa|$, and $\{u_j\}$ is independent of the 'marks' $\{\gamma_j\}$, which in turn are mutually independent and follow a common gamma distribution $\Gamma(-\kappa, \tau)$ (in a usual Neyman–Scott process, all marks are equal and deterministic). The situation is less simple for $\kappa \geqslant 0$ as there are infinite many points $\{u_j\}$. For $\kappa = 0$, we have a Poisson/gamma model. As noted in Wolpert and Ickstadt (1998), we may extend the model by replacing the parameter τ with a positive function $\tau(u), u \in E$, and redefining

$$\nu(A \times B) = (1/\Gamma(1-\kappa)) \int_A \int_B \gamma^{-\kappa-1} \exp(-\tau(u)\gamma)\alpha(\mathrm{d}u)\mathrm{d}\gamma, \quad A \subseteq E, \ B \subseteq [0, \infty).$$

Extensions to time–space models are studied in Brix and Møller (2001) and Brix and Diggle (2001) for LGC processes, and in Brix and Chadoeuf (1998) for SNGC processes.

2 General properties of the models

The Heikkinen–Arjas and SNGC models can be used for non-parametric Bayesian modelling. LGC processes provide easily interpretable Cox process models with varied degrees of smoothing in the intensity surface $\exp(Z)$, as determined by the smoothness properties of μ and c. In what follows we discuss the general properties of the varied models, particularly how flexible they are, and how they may be checked.

Two fundamental characteristics of the Cox process Y are the intensity function $\rho(s) = \mathrm{E}\Lambda(s)$, $s \in R$, and the pair correlation function $g(s,t) = \mathrm{E}(\Lambda(s)\Lambda(t))/(\rho(s)\rho(t))$, $s,t \in R$ (see, for example, Stoyan *et al.* 1995). Non-parametric estimation of ρ is discussed in Diggle (1985). If $g(s,t)$ depends only on $s-t$, then there is a simple relationship between the pair correlation function and the extension to inhomogeneous point processes of Ripley's K-function introduced in Baddeley *et al.* (2000), who also describe non-parametric estimation of K and g.

To the best of my knowledge, there has been little exploration of the question of model assessment for the Heikkinen–Arjas model. Heikkinen and Arjas (1998) stress that although Λ may be expected to be a smooth function but is modelled as a step function, smoothness may be obtained in the posterior mean $\mathrm{E}(\Lambda(s)|Y)$. The specification of hyperparameters in the Heikkinen–Arjas model is a delicate matter. It is not clear to me to what extent the much more involved Heikkinen–Arjas construction is beneficial as compared with ordinary non-parametric kernel estimation (Diggle 1985). However, the former may better identify rapid or sharp changes in the intensity surface, and posterior mean estimates, uncertainties, and other things of interest may be determined by Markov chain Monte Carlo (MCMC) methods. The method may become even more useful if the Voronoi tessellation $\{A_k\}$ is replaced by one of Nicholls' (1998) triangulation models.

LGC processes provide some flexibility in modelling aggregated point patterns when the aggregation is due to spatial heterogeneity. As discussed in Møller *et al.* (1998), certain Neyman–Scott processes have many properties in common with LGC processes. Note that the distribution of an LGC process restricted to a subregion is easily obtained, so problems with edge effects can be avoided, in particular when Y and $\exp(Z)|Y$ are to be simulated. Another very useful property is that the distribution of Z and hence Y is uniquely determined by the intensity and pair correlation functions, since $\rho(s) = \exp(\mu(s) + c(s,s)/2)$ and $g(s,t) = \exp(c(s,t))$. Owing to these simple relationships, non-parametric estimates of ρ and g may suggest an appropriate choice of a parametric model class for μ and c, and model validation may conveniently be based on these and other summary statistics, cf. Møller *et al.* (1998) and Brix and Møller (2001). Since likelihood-based methods are intractable for LGC processes, Møller *et al.* (1998) advocate the use of minimum contrast methods for estimating the unknown parameters of μ and c. Fully Bayesian analysis of LGC models with covariate information is considered in Benes *et al.* (2002) for a point pattern of cases showing where humans in Central Bohemia have been infected by ticks.

Even more flexibility may be obtain by using SNGC processes. Closed form expressions of summary statistics are rare; an expression for the pair correlation function is known when the kernel k is Gaussian; see Brix (1999) and Wolpert and

Ickstadt (1998). The distribution of a SNGC process restricted to a subregion is in general intractable, and some care may be needed when dealing with edge effects. It remains to compare the applicability of Poisson/gamma and other SNGC processes with LGC processes for Bayesian inference problems based on MCMC methods (see below). Possibly, this is easiest for modified Neyman–Scott processes (i.e. those with $\kappa < 0$) because of their simple construction.

3   Aggregation coherence and natural scaling

In epidemiological applications, $\Lambda(\mathrm{d}s) = p(s)w(\mathrm{d}s)$, where w is an intensity measure for the individuals at risk, while p is a random risk function of primary interest, which in principle should satisfy the condition $0 \leqslant p \leqslant 1$, cf. [SR]. In general, when p is modelled in some way by using any of the models considered above, aggregation coherence is lost in the sense that there is no natural scaling at different levels of aggregation, and care is needed when results at different levels are to be compared, not least when incorporating covariates. As in [SR], assume that w is known and (Y, p) is unknown except that aggregated data $Y(A_1), \ldots, Y(A_n)$ are available. Then all we know is that given p, the $Y(A_i)$ are conditionally independent and Poisson distributed with means $\Lambda(A_i) = \int_{A_i} p(s)w(\mathrm{d}s)$, and the generation and evaluation of such integrals will normally require both simulation methods and numerical techniques. In fact, the joint and marginal distributions of $Y(A_1), \ldots, Y(A_n)$ are usually intractable. In order to overcome this and the above-mentioned problems it may be preferable not to 'integrate out' over the missing data but treat it more as a missing data problem.

The article by Richardson discusses cases of a log Gaussian process, $p(s) = \exp(Z(s))$, and a Poisson/gamma model, $p(s) = \sum_j \gamma_j k(s, u_j)$, noting that restrictions are needed to ensure that the risk function takes values in $[0, 1]$ (recall that $p(s)$ is interpreted as a thinning probability). However, numerical evaluation of the integral $\int_{A_i} p(s)w(\mathrm{d}s)$ appears not to be more complicated for many other forms of p, including the possibly more natural choice $p(s) = \exp(Z(s))/(1 + \exp(Z(s)))$, where it is ensured that $0 < p < 1$. Similarly, when models with covariates are constructed, from a computational view point it would not make much difference which kind of model is used.

Considering for instance the Poisson/gamma model $p(s) = \sum_j \gamma_j k(s, u_j)$, rather restrictive assumptions are needed to obtain a known distribution: assuming $\{B_l\}$ is a countable subdivision of E so that $\tau(u) = \tau_l$ on each B_l, and $k(s, u) = k_{il}$ on each $A_i \times B_l$, then $\{\Lambda(A_i)\}$ is a kind of moving average gamma field as $\Lambda(A_i) = w(A_i) \sum_l k_{il} \Gamma_l$, where $\Gamma_l \equiv \sum_{u_j \in B_l} \gamma_j \sim \Gamma(\alpha(B_l), \tau_l)$ and the Γ_l are independent.

4   Computational aspects

For all the above-mentioned models we need to resort to simulations no matter whether the distribution of (Λ, Y), or the conditional distribution of Λ (and possible hyperparameters) given Y, or the conditional distribution of Y, Λ (and possible hyperparameters) given aggregated data $Y(A_1), \ldots, Y(A_n)$ are considered. Since the MCMC algorithms for the different models are rather complicated, it would not make much sense to give the details here. Note that some kind of approximation is used in many of the papers: Heikkinen and Arjas (1998) use a simple local approximation when they calculate Hastings ratios involving the normalizing constants in the conditional autoregressions corresponding to two models of different dimensions (the techniques in Rue (2000) may possibly overcome this problem); Møller *et al.* (1998) discretize the Gaussian field, using a fine grid; Wolpert and Ickstadt (1998) consider only those (u_j, γ_j) with $\gamma_j > \epsilon$, where $\epsilon > 0$ is a given threshold (e.g. the machine precision). Apart from the LGC case, the theoretical properties of the algorithms have so far not been studied in detail. In particular, Møller *et al.* (1998) establish geometric ergodicity for a truncated Langevin–Hastings algorithm for simulating $\Lambda | Y$.

5 Concluding remarks

All the models considered have their advantages and disadvantages. None makes life easy for spatial epidemiologists, and for that reason direct modelling of aggregated data as described in [SR] may be hard to beat. Treating the analysis of aggregated data as a missing data problem, MCMC methods may be feasible so that 'aggregation coherence' is ensured. Moreover, computational problems with integrals and varied kinds of approximations/discretizations/truncations are typical in the existing MCMC algorithms. Alternative models such as modified Neyman–Scott processes seem not yet used in epidemiological applications, and they leave the hope for developing simpler and yet flexible models.

Acknowledgements

Research supported by the European Union's research network 'Statistical and Computational Methods for the Analysis of Spatial Data, ERB-FMRX-CT96-0096' and by the Centre for Mathematical Physics and Stochastics (MaPhySto). Funding by a grant from the Danish National Research Foundation and by the Danish Natural Science Research Council is gratefully acknowledged.

Additional references in discussion

Aykroyd, R. G. (1998). Bayesian estimation for homogeneous and inhomogeneous Gaussian random fields. *IEEE Transactions on Pattern Analysis and Machine Intelligence*, **20**, 533–9.

Baddeley, A. J., Møller, J., and Waagepetersen, R. (2000). Non- and semi-parametric estimation of interaction in inhomogeneous point patterns. *Statistica Neerlandica*, **54**, 329–50.

Benes, V., Bodlak, K., Møller, J., and Waagepetersen, R. P. (2002). Log Gaussian Cox process models for disease mapping. Technical Report R-02-2001, Department of Mathematical Sciences, Aalborg University.

Bernardinelli, L., Pascutto, C., Best, N. G., and Gilks, W. R. (1997). Disease mapping with errors in covariates. *Statistics in Medicine*, **16**, 741–52.

Besag, J. and Higdon, D. (1999). Bayesian analysis of agricultural field experiments (with discussion). *Journal of the Royal Statistical Society*, B, **61**, 691–746.

Brix, A. (1999). Generalized gamma measures and shot-noise Cox processes. *Advances in Applied Probability*, **31**, 929–53.

Brix, A. and Chadoeuf, F. (1998). Spatio-temporal modeling of weeds by shot-noise G Cox processes. Report 98-7, Department of Mathematics and Physics, Royal Veterinary and Agricultural University, Copenhagen.

Brix, A. and Diggle, P.J. (2001). Spatio-temporal prediction for log-Gaussian Cox processes. *Journal of the Royal Statistical Society*, B, **63**, 823–41.

Brix, A. and Møller, J. (2001). Space-time multitype log Gaussian Cox processes with a view to modelling weed data. *Scandinavian Journal of Statistics*, **28**, 471–88.

Clifford, P. (1986). Contribution to the discussion of the paper by Besag (1986). *Journal of the Royal Statistical Society*, B, **48**, 284.

Daley, D. J. and Vere-Jones, D. (1988). *An Introduction to the Theory of Point Processes*. Springer-Verlag, New York.

Fahrmeir, L. and Lang, S. (2001). Bayesian inference for generalized additive mixed models based on Markov random field priors. *Applied Statistics*, **50**, 201–20.

Knorr-Held, L. (2000). Bayesian modelling of inseparable space-time variation in disease risk. *Statistics in Medicine*, **19**, 2555–67.

Künsch, H.-R. (1986). Intrinsic autoregressions and related models on the two-dimensional lattice. *Biometrika*, **74**, 517–24.

Nicholls, G. K. (1998) Bayesian image analysis with Markov chain Monte Carlo and coloured continuum triangulation models. *Journal of the Royal Statistical Society*, B, **60**, 643–59.

Ranta, J. and Penttinen, A. (2000). Probabilistic small area risk assessment using GIS-based data: a case study on Finnish childhood diabetes. *Statistics in Medicine*, **19**, 2345–59.

Shephard, N. and Pitt, M. K. (1997). Likelihood analysis of non-Gaussian measurement time series. *Biometrika*, **84**, 653–67.

Stoyan, D., Kendall, W. S., and Mecke, J. (1995). *Stochastic geometry and its applications* (2nd edn). Wiley, New York.

9

Spatial hierarchical Bayesian models in ecological applications

Antti Penttinen
University of Jyväskylä, Finland

Fabio Divino
Università di Roma 'La Sapienza', Italy

Anne Riiali
Pohjola, Finland

1   Introduction

The spatial distribution of species, the suitability of sites, and association between species are matters of importance in biological and ecological research. These issues have entered also into modern forest research where the objectives of planning strategies are not only in timber production but increasingly in the multi-purpose use of forest areas, including the ability to host plant and animal species.

Biogeography is the science of observing, recording, and explaining the geographical distribution of all living things (Pielou 1979; Heikkinen and Högmander 1997). In biogeography, one objective is the estimation of the range of a species, that is the area where the species can be discovered. Other objectives exist, such as interspecific competition and island biogeography, the latter being important in the study of gene flows between populations.

Large-scale variation in the spatial distribution of a population is studied in *atlas surveys*. In this approach, the study area is divided into sub-areas and the presence or absence of a species of interest are recorded. Typically, the estimated range is presented as a map; hence the name 'biogeographical mapping'. Interesting Bayesian applications can be found dealing with the difficulty in modelling data that do not necessarily result from probability sampling in Heikkinen and Högmander (1994), Weir and Pettitt (2000), and in Högmander and Møller (1996). Though atlas surveys are concerned with the occurrence of a species at the sites, the abundances may be included. Specific methods exist such as line transect surveys (Buckland *et al.* 1993).

Within the range, local conditions affect the small-scale variation in the distribution of species. The smallest observational units are called here 'sites', which are typically area units, or something more specific, e.g. fixed points. The sites may turn out to be heterogeneous in the sense that they have variation in the suitability for the occupation of a species. A current tendency in data collection is to find covariate information on sites in order to explain heterogeneity. Among other methods, remote sensing and airborne images are commonly applied to derive such covariates. Statistical modelling is developing from logistic regression (Buckland and Elston 1993) and autologistic regression (Augustin *et al.* 1996) to hierarchical models (Riiali *et al.* 2001).

Estimating association between species is an important task in ecology. Ordination methods such as correspondence analysis or canonical correspondence analysis have been applied widely in plant ecology for finding communities of species and for classifying the sites according to plant communities. Ordination is descriptive, however. If the spatial aspect of association is emphasized, then statistical inference will be complex because the effect of the range of a species and the suitability of the sites affect the species' distribution simultaneously.

Quantitative ecological research has adopted Geographical Information Systems (GIS) for storing, manipulating, analysing, and visualizing spatial or spatio-temporal data. In addition, development of the observational techniques allows data which are extensive and complex, possibly resulting from a multi-source observational system. The consequences are of two kinds. First, improved possibilities exist for extensive data, potentially rich in covariate information. On the other hand, extensive data tend to be spatially heterogeneous, controlled only partially by the observed covariates. This is why advanced non-stationary modelling tools have been called for. It should be noted that 'spatial analysis' offered by standard GIS frameworks does not contain enough possibilities for rigorous statistical inference. However, hierarchical Bayesian modelling extends the inferential possibilities for GIS, see Mugglin *et al.* (1999).

The difficulty in modelling spatial or spatio-temporal biogeographical data is often due to the data themselves: the design is poor or the data are observational when the argument of probability sampling is not fulfilled. In addition, the quality of data can be poor in large-scale studies. Another difficulty is in the interpretation of the results due to unobserved confounding factors. Furthermore, observations may be made on several scales. For example, the responses and covariates may be observed using different resolution or covariates entered through point sampling, as in Diggle *et al.* (1998).

In this article, we deal with modelling of suitability of sites for a species and association between species, where the data are recorded at all sites of the study area. The emphasis is on the analysis of multi-species data, with smoothing of the marginal maps for each species, explaining suitability of the sites by covariates, and estimating the association between the species as specific objectives. The joint modelling is also expected to be more informative for statistical inference on single species than the corresponding marginal models.

2 Modelling the suitability of sites

Let us consider a fixed finite set of sites $S = \{s_1, \ldots, s_n\}$, identified with their spatial coordinates. These can be area units (labelled according to their centres) or points, for example. Our motivation has been the modelling of lichens, epiphytes growing on tree-trunks. In this case S is assumed to be a fixed population of host trees with known locations $s_1, \ldots, s_n$. Let us further define a neighbourhood relation $\sim$ in S. For modelling purposes, it may be sufficient to know the neighbourhood for each tree, but spatial coordinates are needed in mapping and also in interpretation.

We associate a binary random variable y_s to the site s indicating the presence ($y_s = 1$) or absence ($y_s = 0$) of the species under consideration. It is natural to characterize the suitability of s for the species by means of the occupation probability $\theta_s = \Pr(y_s = 1 | \theta_s)$. The collection $\theta = (\theta_1, \ldots, \theta_n)$ is called the *suitability map*.

We make the following assumption: the species' distribution is affected mainly by external factors such as the properties of the sites, the environment, and the landscape history, but the propagation is non-contagious. Then, it is biologically relevant to assume that the y_s are conditionally independent given θ. This assumption is realistic for the lichen species under consideration. The likelihood for θ based on the observation $y = (y_1, \ldots, y_n)$ is given by

$$L(y|\theta) = \prod_{s=1}^{n} \theta_s^{y_s} (1 - \theta_s)^{1-y_s}. \tag{2.1}$$

Assume that a covariate vector $z_s = (z_{s1}, \ldots, z_{sm})$ is known at each site s and that it describes the measured properties of the site. This information should be included in the model. Heterogeneity (clustering) due to unobserved external variation is modelled by a hidden variable, which is assumed to be 'smooth', or in other words, positively spatially correlated. Then we obtain a hierarchical model

$$\xi_s = \log\left(\frac{\theta_s}{1 - \theta_s}\right) = \alpha + \sum_{j=1}^{m} \beta_j z_{sj} + v_s, \tag{2.2}$$

where $v = (v_1, \ldots, v_n)$ is a spatially structured random field. A prior distribution for v can be chosen to be an intrinsic Gaussian random field in S defined by

$$p(v_1, \ldots, v_n | \kappa) \propto \kappa^{n/2} \exp\left\{ -\tfrac{1}{2}\kappa \sum_{s \sim s'} w_{ss'} (v_s - v_{s'})^2 \right\}, \tag{2.3}$$

where κ is an unknown positive parameter affecting the smoothing level, $w_{ss'}$ is a non-negative weight, and the sum is over all neighbouring pairs of sites. In cases where repetitions at each site exist, a spatially unstructured component u_s, modelling unexpected variation (heterogeneity) of the site s, can be added to the right side of (2.2). This model specification is the one suggested by Besag *et al.* (1991), and also mentioned in S. Richardson (Chapter 8 in this volume). The model defined by (2.2) and (2.3) exploits two features of the data. First, the covariates introduced explain the observed heterogeneity and their effect will be estimated from the whole data. Secondly, the random field prior smooths the local variation using the contextual information of the data.

The distinction between the ordinary logistic model and the hierarchical model specified by (2.2) and (2.3) is obvious. The first mentioned model does not contain the spatial random component v_s and the responses y_s are treated as independent. The autologistic model, suggested by Besag (1974) and applied in ecological studies, for example by Preisler (1993) and Augustin *et al.* (1996), is an extension of the ordinary logistic model, specified conditionally by

$$\log\left\{\frac{\theta_s(y_s)}{1-\theta_s(y_s)}\right\} = \alpha + \sum_{j=1}^{m}\beta_j z_{sj} + \sum_{s\sim s'}\delta_{ss'}y_{s'}, \tag{2.4}$$

where $\theta_s(y_s) = \Pr(y_s = 1 | y_{-s})$ and y_{-s} is defined as $y_{-s} = y \setminus \{y_s\}$. This auto-logistic model also has two components: the covariate effects and the interaction part that models the interaction between the sites. While the hierarchical model defined by (2.2) and (2.3), and the autologistic model (2.4) both allow for correlated responses, their interpretations are different. In the hierarchical model, the clustering of attributes results from environmental heterogeneity, whereas (2.4) is a Markov random field modelling interaction, e.g. contagious propagation. Both can be used as statistical models for spatial dependence. One should note, however, that the autologistic model (2.4) is not valid for highly clustered patterns.

Riiali *et al.* (2001) apply the hierarchical model in the estimation of suitability maps for four lichen species in an area with 1253 host trees. One covariate, diameter at breast height (DBH), which is related to the age of the tree, has been used to explore the heterogeneity of the sites. The approach is fully Bayesian and the calculation of the posterior distribution of the unknowns is done by means of Markov chain Monte Carlo (MCMC).

If counts of rare events at the sites are observed, then the Poisson model $y_s | \theta_s \sim \text{Poisson}(\theta_s)$ is a convenient choice together with the log-linear model

$$\log\theta_s = \alpha + \sum_{j=1}^{m}\beta_j z_{sj} + v_s + u_s \tag{2.5}$$

for the mean θ_s, where the v_s are spatially structured and the u_s are exchangeable variables. This model was suggested by Besag *et al.* (1991). The map $(\theta_1, \ldots, \theta_n)$ now explains spatial variation of the mean and is subject to estimation. This is a link to mapping of disease incidence or risk in epidemiology; see Elliott *et al.* (2000) and S. Richardson (Chapter 8 in this volume).

3   Extension of the suitability model to two species

We next turn to the problem of analysing simultaneously two species observed in S. The objective is to construct a model for vector-valued binary responses, where the sites are heterogeneous and the species are allowed to be dependent. This gives information to an ecologist on the spatial structure of association between the species. In addition, it is another advanced means for constructing marginal suitability maps for the species. This approach was suggested by Penttinen *et al.* (2000).

To be more specific, let the species be labelled 1 and 2 and let y_{ls} now stand for the indicator of the occurrence of species l ($l = 1, 2$) at s ($y_{ls} = 1$, if species l is present at s, 0 otherwise). Then the parameter of interest is the joint occurrence probability

$$\theta_{ijs} = \Pr(y_{1s} = i, y_{2s} = j \mid \theta_{ijs}), \quad i, j = 0, 1. \tag{3.1}$$

These probabilities define the association table at s. We apply the log odds ratio $\phi_{12s} = \log((\theta_{11s}\theta_{00s})/(\theta_{10s}\theta_{01s}))$ as a measure of association. The map $(\phi_{12s}, s \in S)$ is called the *association map* of the two species.

It is convenient to introduce the loglinear representation for the joint occurrence probabilities:

$$\begin{aligned}
\log \theta_{11s} &= \alpha_s + \phi_{1s} + \phi_{2s} + \phi_{12s}, \\
\log \theta_{10s} &= \alpha_s + \phi_{1s}, \\
\log \theta_{01s} &= \alpha_s + \phi_{2s}, \\
\log \theta_{00s} &= \alpha_s,
\end{aligned} \tag{3.2}$$

where ϕ_{1s} and ϕ_{2s} are the marginal effects of species 1 and 2, respectively, and ϕ_{12s} is the effect of the species' association. The terms ϕ_{1s}, ϕ_{2s}, and ϕ_{12s} are further modelled as in (2.2), each with different fixed-effect coefficients and different mutually independent random fields. In particular,

$$\phi_{12s} = \alpha_{12} + \sum_{j=1}^{m} \beta_{12j} z_{sj} + v_{12s}$$

defines the spatial model for the association parameters.

Penttinen *et al.* (2000) apply this model in the fully Bayesian framework and estimate the association map for two lichen species as well as their marginal suitability maps. The marginal maps for species 1 and 2 can be obtained from $\theta_{10s} + \theta_{11s}$ and $\theta_{01s} + \theta_{11s}$, respectively.

The joint modelling for aggregated data, suggested by Knorr-Held and Best (2001), gives an extension of the Poisson model with shared latent components. Their model is different from our individual-level specification. In spatial epidemiology, the interest is not in the simultaneous occurrence of two diseases in the same individual but in the joint geographical variation of the disease incidence or mortality in the population.

4   A conditional model for association of species

If the interest is only in the association between two species and one has counts in multiple sub-areas (or sites), then simplified modelling based on conditioning by the marginal counts is possible.

Assume that the study area is divided into K sub-areas A_k, $k = 1, \ldots, K$, and the 2×2 association tables of counts for each subarea are of the form:

		Species 2		
		Present	Absent	Total
Species 1	Present	n_{11k}	n_{10k}	$n_{1 \cdot k}$
	Absent	n_{01k}	n_{00k}	$n_{0 \cdot k}$
	Total	$n_{\cdot 1k}$	$n_{\cdot 0k}$	n_k

The species' abundances and mutual association are now assumed to be constant within each sub-area A_k but can vary from one sub-area to another. Then the marginal counts $(n_{1 \cdot k}, n_{0 \cdot k})$ and $(n_{\cdot 1k}, n_{\cdot 0k})$ carry sample information on the abundances of species 1 and 2, respectively. The key idea is to consider the marginal occupation probabilities as nuisance parameters for the association and they will be avoided through conditioning by the marginal counts.

We present the model by means of a case study on the association between two lichen species *Lobaria pulmonaria* and *Nephroma parile*, in short Lpulm and Npari, respectively. These species can be found on the trunks of old-growth deciduous trees, which are mainly aspens (*Populus tremula*). The forest area is of size 25 ha and situated in Southern Finland. Occurrences of the lichen species are recorded on tree-trunks as binary outcomes. The total number of potential host trees in the data is 1253. The trees are aggregated into 7×7 contiguous quadrats (sub-areas) of size 90×90 m^2 and a 2×2 association table is calculated for each of the quadrats.

We include into our analysis those 26 quadrats that have non-zero marginal counts and have at least two neighbours with a common edge. These quadrats are shown as bold numbers in Table 1. In addition to the lichen occurrences, the mean DBH of the trees in each sub-area is calculated. This is an important factor in lichen occupation because it is related to the age of the trees. Further biological reasoning supports the inclusion of the DBH into the model (Kuusinen 1994, 1995; Kuusinen and Penttinen 1999; Riiali *et al.* 2001). This covariate information is shown in Table 1 (third row in each cell).

One can calculate the 'raw estimate' of association for each sub-area as

$$\hat{\phi}_k = \log \left\{ \frac{n_{11k} n_{00k}}{n_{10k} n_{01k}} \right\},$$

the log odds ratio of the contingency table in sub-area A_k. Here for simplicity we suppress the indexing by each species, so $\phi_{12k} = \phi_k$. This is an empirical local measure of association. Its asymptotic variance can be estimated by $n_{11k}^{-1} + n_{10k}^{-1} + n_{01k}^{-1} + n_{00k}^{-1}$ (when lichen occupations are assumed to be independent). This variance is large for relatively small counts and therefore the raw estimates

TABLE 1. Observed counts of Lpulm and Npari in trees in 90×90 m^2 quadrats. In each cell, the counts are n_{11k}, n_{10k} (first row) and n_{01k}, n_{00k} (second row). The mean DBH of the cell in centimetres is in the third row. The cells with bold numbers are used in the analysis.

0	0	**1**	**5**	**4**	**12**	**7**	**2**	0	0	0	0	0	0
0	1	**4**	**16**	**0**	**8**	**0**	**3**	0	0	0	0	0	0
50.0		**39.3**		**35.2**		**42.6**		–		–		–	
0	3	0	4	**5**	**11**	**16**	**18**	**1**	**0**	0	0	0	0
0	5	4	13	**2**	**9**	**16**	**51**	**0**	**3**	0	0	0	0
39.0		41.1		**45.1**		**39.0**		**49.5**		–		–	
0	0	0	0	**2**	**5**	**3**	**11**	**24**	**11**	**4**	**1**	0	0
0	9	1	10	**1**	**10**	**4**	**62**	**9**	**30**	**2**	**4**	0	0
34.4		39.9		**37.8**		**26.2**		**34.3**		**33.4**		–	
0	0	**1**	**0**	**5**	**8**	**5**	**8**	**14**	**8**	**6**	**4**	**0**	**1**
0	2	**0**	**5**	**7**	**14**	**0**	**22**	**7**	**16**	**2**	**11**	**0**	**20**
45.0		**39.5**		**38.1**		**28.4**		**30.1**		**28.6**		**26.0**	
0	0	0	0	**1**	**0**	**3**	**4**	**2**	**0**	**2**	**3**	**0**	**2**
0	0	0	0	**0**	**27**	**7**	**62**	**7**	**29**	**6**	**68**	**1**	**39**
–		–		**32.7**		**28.0**		**28.7**		**33.8**		**24.7**	
0	0	0	0	**1**	**1**	**2**	**6**	**2**	**2**	**4**	**8**	0	0
0	0	0	0	**4**	**23**	**3**	**28**	**8**	**41**	**4**	**37**	0	2
–		–		**39.1**		**33.7**		**27.0**		**28.7**		31.5	
0	0	0	0	**1**	**0**	**4**	**7**	**9**	**10**	**0**	**1**	0	0
0	0	0	0	**1**	**7**	**3**	**31**	**14**	**87**	**0**	**41**	0	5
–		–		**31.4**		**38.4**		**32.0**		**24.8**		27.4	

are usually not valid for association mapping. In what follows we give Bayesian estimates for the association map $(\phi_k, \ k = 1, \ldots, K)$.

Assume that the presence and absence of the two species are recorded at sites $s_{k1}, \ldots, s_{kn_k}$ in each sub-area A_k but the exact coordinates of the sites are not known. Assume further that the colonizations take place independently. The covariates available are 'ecological', observed at the sub-area level. If the presence and the absence of a species are denoted by 1 and 0, respectively, then let θ_{ijk} stand for the probability of the event i ($i = 0, 1$) for species 1 and j ($j = 0, 1$) for species 2 at a randomly chosen site in A_k. As before, we apply the log odds ratio $\phi_k = \log((\theta_{11k} \, \theta_{00k})/(\theta_{10k} \, \theta_{01k}))$ as the measure of association: $\phi_k = 0$ corresponds to the lack of association in sub-area A_k, whereas $\phi_k > 0$ means positive and $\phi_k < 0$ negative association. The marginal distribution for the occurrence of species 1, defined by $\theta_{1 \cdot k} = \sum_{j=0}^{1} \theta_{1jk}$ as a function of k, $k = 1, \ldots, K$, characterizes the marginal behaviour of this species; similarly $\theta_{\cdot 1k}$, $k = 1, \ldots, K$, is the map of the distribution of species 2. In biogeographical mapping these marginals are allowed to be heterogeneous. Instead of determining the spatial range and map of the abundance of species, our objective is to estimate the association map $(\phi_k, \ k = 1, \ldots, K)$. Hence the marginals $(\theta_{1 \cdot k}, \ k = 1, \ldots, K)$ for species 1 and $(\theta_{\cdot 1k}, \ k = 1, \ldots, K)$ for species 2 are considered to be spatial nuisance parameters.

A way of eliminating the marginals is by conditioning on the observed marginal counts $n_{c,k} = (n_{1\cdot k}, n_{0\cdot k}, n_{\cdot 1k}, n_{\cdot 0k})$ for each table. Then, assuming that the epiphytes occupy trees in area A_k independently of each other given the parameters of the corresponding association table, the inference on ϕ_k can be based on n_{11k}, the number of simultaneous occurrences of the two species. Indeed, conditional on the marginal counts, n_{11k} follows a non-central hypergeometric distribution (see, for example, Breslow and Day (1980, p. 269)). Let $y_k = n_{11k}$ stand for the response of the association k-table (in A_k). Then the k-table conditional likelihood is

$$
L(y_k \mid \phi_k, n_{c,k}) = \frac{\begin{pmatrix} n_{1\cdot k} \\ y_k \end{pmatrix} \begin{pmatrix} n_{0\cdot k} \\ n_{\cdot 1k} - y_k \end{pmatrix} \exp(y_k\,\phi_k)}{\displaystyle\sum_{x=L_k}^{U_k} \begin{pmatrix} n_{1\cdot k} \\ x \end{pmatrix} \begin{pmatrix} n_{0\cdot k} \\ n_{\cdot 1k} - x \end{pmatrix} \exp(x\,\phi_k)}\,,
\tag{4.1}
$$

with $L_k \leqslant y_k \leqslant U_k$, where $L_k = \max\{0, n_{\cdot 1k} - n_{0\cdot k}\}$ and $U_k = \min\{n_{\cdot 1k}, n_{1\cdot k}\}$. Assuming conditional independence of $y = (y_1, \ldots, y_K)$ given the association parameters $\phi = (\phi_1, \ldots, \phi_K)$, the marginal-conditioned likelihood is of the product form

$$
L(y \mid \phi_1, \ldots, \phi_K, n_{c,1}, \ldots, n_{c,K}) = \prod_{k=1}^{K} L(y_k \mid \phi_k, n_{c,k})\,.
\tag{4.2}
$$

We assume that $(\phi_k, k = 1, \ldots, K)$ is smooth in the sense that the association parameters of nearby sub-areas do not differ drastically from each other. Furthermore, the observed covariates $z_{k1}, \ldots, z_{km}$ could be used in the modelling and unstructured local variation may take place. Consequently, we suggest the model

$$
\phi_k = \alpha + \sum_{j=1}^{m} \beta_j z_{kj} + v_k + u_k, \quad k = 1, \ldots, K\,.
\tag{4.3}
$$

As in (2.2), $v = (v_k, k = 1, \ldots, K)$ is assumed to be an intrinsic Gaussian random field defined by the probability density (2.3) where the 'sites' correspond to sub-areas and have been provided with a neighbourhood relation. Finally, the u_ks are the unstructured random effects which are assumed to be exchangeable and $N(0, \sigma_u^2)$ distributed. Their role is to model unexpected non-spatial variation in the sub-tables.

In addition to the structural priors for the random components in (4.3), further prior assumptions on the regression coefficients and smoothing levels are needed. In our example, we have one covariate, DBH. Hence the fixed effect part contains two parameters, α and β, which are assumed to be independent and normally distributed with zero mean and variances σ_α^2 and σ_β^2, respectively. For the hyperpriors, we assume that the inverse variances $\kappa_\alpha = \sigma_\alpha^{-2}$, $\kappa_\beta = \sigma_\beta^{-2}$, $\kappa_u = \sigma_u^{-2}$, and $\kappa_v = \sigma_v^{-2}$ are gamma distributed with known parameters (C_α, D_α), (C_β, D_β), (C_u, D_u), and (C_v, D_v), respectively. (Here the mean and variance of the $\Gamma(C, D)$ distribution are C/D and C/D^2, respectively.) We assume further that they are independent of the previous hyperparameters. The choice of these parameters is specific to the problem. Commonly, the variances of the fixed effect parameters are chosen in a way that leads to weak priors. Note that the smoothing may be sensitive to this choice, because typically the data are not informative concerning the level of smoothing. To summarize, our hierarchical model consists of four levels: (i) the conditional model for the observation (conditional likelihood), (ii) the prior distribution for ϕ_k, $k = 1, \ldots, K$, (iii) priors for the fixed effect and random components, and (iv) hyperpriors for the parameters in the priors. The likelihood is considered to be conditional on the marginal counts of each sub-table. The posterior is then of the form

$$
\begin{aligned}
&p(\alpha, \beta, u, v, \kappa_\alpha, \kappa_\beta, \kappa_u, \kappa_v \mid y, n_c) \\
&= p(\kappa_\alpha)p(\kappa_\beta)p(\kappa_u)p(\kappa_v)p(\alpha \mid \kappa_\alpha)p(\beta \mid \kappa_\beta)p(u \mid \kappa_u)p(v \mid \kappa_v)L(y \mid \phi, n_c),
\end{aligned}
\tag{4.4}
$$

and the values of ϕ_k are determined by (4.3).

An important note should be added concerning the relevance of the conditional model (4.4). Let ϕ stand for the association parameters of the conditional likelihood $L(y \mid n_c, \phi)$ with prior $p(\phi)$. Then our conditional model is actually

$$
p(y, \phi \mid n_c) = p(\phi)L(y \mid n_c, \phi). \tag{4.5}
$$

If we denote the distribution of the marginal counts by $L(n_c \mid \psi)$, not specified above, then the corresponding joint model concerning the whole data (y, n_c) can be written as

$$
p(y, n_c, \phi, \psi) = p(\phi)p(\psi)L(n_c \mid \psi)L(y \mid n_c, \phi). \tag{4.6}
$$

Inference is based on the posterior $p(\phi \mid y, n_c)$, which is the same in both of the specifications (4.5) and (4.6). Hence using the conditional model (4.5) is justified in terms of the joint model (4.6). This requires in particular that the association parameters ϕ and the marginal count (or abundance) parameters ψ are a priori independent. Because there is one degree of freedom for association in each sub-table, conditional independence does not cause technical restrictions. Instead, the model does not allow the association to be a priori dependent on the abundances of the species. Note that a priori independence of the marginal and association parameters has been assumed also in the association model of Section 2.

When comparing (4.4) with the spatial model in Besag *et al.* (1991), two differences can be found. First, the model (2.2) is now set for the log odds ratios ϕ_k. Secondly, the likelihood (4.2) is conditioned by the sub-table marginals. Furthermore, our conditional model is a simplification of the full three-equation association model characterized by (3.1) and (3.2); the conditional model (4.3) is now a one-equation system. Liao (1999) suggests a similar model in the context of meta-analysis, similar in the sense that conditioning has been used in controlling heterogeneity between the association tables. The conditional likelihood part is the same as ours but the association tables in Liao (1999) do not have a spatial structure.

5   Empirical results

Owing to the intractable nature of the posterior distribution (4.4), inference on the log odds ratios ϕ_k, $k = 1, \ldots, K$, must be carried out using MCMC simulations. Let $\gamma = (\alpha, \beta, u_1, \ldots, u_K, v_1, \ldots, v_K, \kappa_\alpha, \kappa_\beta, \kappa_u, \kappa_v)$ represent the vector of parameters involved in the model. The MCMC algorithm simulates a Markov chain whose stationary distribution is the posterior $p(\gamma \mid y, n_c)$. The sequence of values drawn can be considered as a dependent sample from $p(\gamma \mid y, n_c)$, after stationarity is reached. In the construction of the Markov chain we use a Metropolis–Gibbs sampler approach (e.g. Besag *et al.* 1995). Let $\gamma_{-j} = (\gamma_1, \ldots, \gamma_{j-1}, \gamma_{j+1}, \ldots, \gamma_m)$ stand for the parameter vector omitting the component γ_j. The general idea of the Metropolis–Gibbs sampler algorithm is to sample successively from the conditional distribution $p(\gamma_j \mid \gamma_{-j}; y, n_c)$; when this does not have a simple closed form we use a Metropolis–Hastings step. Each of the parameters α, β, u_1, $\ldots$, u_K, v_1, $\ldots$, v_K is updated in turn within a Metropolis step by proposing a new value from a symmetric Gaussian distribution with the mean given by the current parameter value and with fixed variance. For these parameters the conditional distributions, needed in the computation of the acceptance probability can easily be derived from the posterior (4.4). Under our assumptions on the prior distributions, it can be shown that the full conditional distributions of the parameters κ_α, κ_β, κ_u, and κ_v are standard gamma distributions. Hence we can update them easily within a Gibbs sampler step by sampling directly from the full conditionals. The log odds ratios ϕ_k, $k = 1, \ldots, K$, are computed at each iteration using the current values of the parameters α, β, $u_1, \ldots, u_K$, and $v_1, \ldots, v_K$.

We now turn to computational results concerning association between the two lichen species *Lobaria pulmonaria* (Lpulm) and *Nephroma parile* (Npari) of Table 1. We present the results obtained by the MCMC algorithm described above. For the prior gamma distributions of the inverse variances κ_v and κ_u, the hyperparameters were fixed to $C = 1$ and $D = 0.005$. These choices lead to weak hyperpriors. In modelling the spatial term we use the conventional weights $w_{kl} = 1$ if A_k and A_l are adjacent sub-areas, 0 otherwise. Here the eight-neighbour system is applied except on the boundaries. The priors for α and β are Gaussian with mean 0 and variance σ^2, and the hyperprior for σ^{-2} is $\Gamma(1, 0.005)$.

In order to reduce the posterior correlation between α and β_{DBH} we worked with the centred DBH covariate. Convergence of the MCMC algorithm was first examined visually and the equilibrium state was obtained approximately after 3000 iterations. A safety margin of 2000 iterations was added and a fixed burn-in time of 5000 iterations was applied. The convergence of the chain was also monitored using the device of Gelman and Rubin (1992) based on parallel runs. Statistical inference was carried out using the next 10 000 iterations. All the parameters involved in the model as well as the log odds ratios were estimated as the corresponding marginal posterior sample means from the simulated chain. The simulated marginal posterior histograms of the parameters α, β_{DBH}, κ_α,

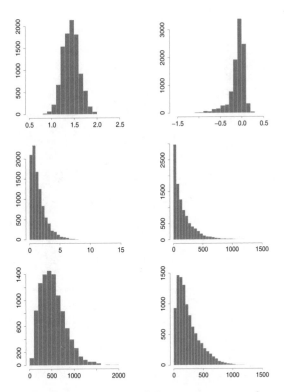

FIG. 1. Marginal posterior histograms of the parameters $\alpha, \beta_{\text{DBH}}, \kappa_\alpha, \kappa_{\beta_{\text{DBH}}}, \kappa_v$, and κ_u (from left to right and from top to bottom).

$\kappa_{\beta_{\text{DBH}}}$, κ_v, and κ_u are shown in Fig. 1. Histograms of the cellwise log odds ratios are reported in Fig. 2 and the corresponding posterior means and standard deviations can be found in Table 2 (upper row of each cell). A graphical presentation of the association map (cellwise posterior means) is shown in Fig. 3.

Table 2, Fig. 2, and Fig. 3 show that the Bayes estimator of the log odds ratios (the association map) is flat, typically around 1.4. This means that positive association between the two species (Lpulm, Npari) exists in the area analysed. The association is somewhat higher in the eastern part of the area. This observation has a biological explanation: the landscape forms a north-east slope which preserves humidity for a longer time giving better conditions for the lichen species to survive when colonized. It should be noted that the association may depend on the aggregation level used. In our case, however, the treewise association map is similar to the one obtained in Penttinen *et al.* (2000), explained in Section 2.

For comparisons, we calculated the crude estimates of the log odds ratios

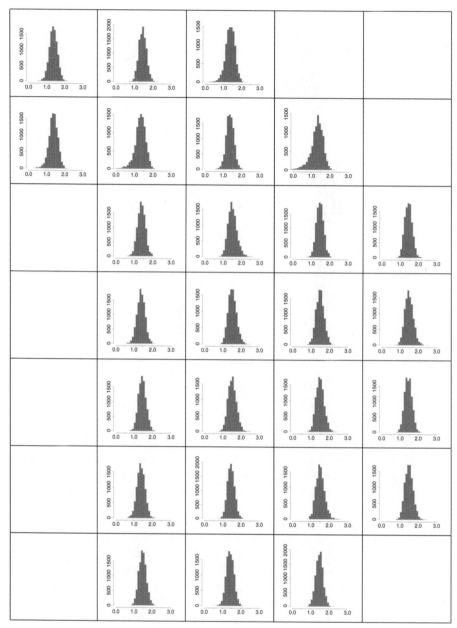

FIG. 2. Cellwise marginal posterior histograms of the log odds ratios. The ordering of the cells is the same as in Table 2. The range of the x-axis in each histogram is $[0, 3]$ with division of 0.5.

TABLE 2. Posterior means of the log odds ratios with posterior standard deviations for the sub-areas (first row of each cell); corresponding crude estimates of log odds ratios with approximate standard errors (second row of each cell).

1.37 (0.23)	1.43 (0.21)	1.37 (0.26)		
−0.17 (1.19)	3.31 (3.23)	4.65 (3.31)		
1.37 (0.25)	1.32 (0.29)	1.37 (0.21)	1.29 (0.36)	
−2.55 (3.25)	0.69 (0.93)	1.04 (0.45)	5.83 (4.61)	
	1.39 (0.21)	1.52 (0.25)	1.47 (0.20)	1.45 (0.21)
	1.33 (1.30)	1.44 (0.82)	1.97 (0.52)	1.98 (1.37)
	1.38 (0.22)	1.52 (0.23)	1.49 (0.22)	1.49 (0.24)
	0.22 (0.73)	4.94 (3.22)	1.37 (0.63)	2.06 (0.99)
	1.46 (0.21)	1.52 (0.23)	1.50 (0.23)	1.45 (0.21)
	8.00 (4.58)	1.89 (0.85)	4.46 (3.26)	2.02 (0.99)
	1.39 (0.23)	1.45 (0.20)	1.52 (0.24)	1.50 (0.23)
	1.73 (1.45)	1.14 (1.00)	1.62 (1.05)	1.52 (0.80)
	1.48 (0.21)	1.40 (0.21)	1.47 (0.20)	
	4.26 (3.46)	1.76 (0.86)	1.72 (0.54)	

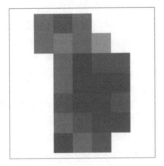

FIG. 3. Association map produced by Bayesian modelling. The grey levels are scaled according to the minimum and maximum values of ϕ in the map.

and the result is presented in Table 2 (second row in each cell). The constant $\varepsilon = 0.1$ is added to the observed counts to avoid the problem of zero frequencies. The following conclusions can be draw from Table 2: the crude estimates, which are based on small numbers of counts, are unstable, having large variation. The Bayesian estimates display the usual shrinkage, and the map is more stable. It is obvious that the crude estimates are generally not suitable for the construction of an association map.

The modelling above has used the model (4.3) with centred DBH as a covariate in each of the equations. We investigated also the influence of the two random components on the fixed effect, especially the need to include an unstructured component. Therefore we fitted three additional models. Table 3 gives the marginal posterior sample means of α and β_{DBH} and the corresponding standard errors calculated as sample standard deviations of the marginal pos-

TABLE 3. Posterior marginal means and standard errors for α and β_{DBH} corresponding to different fitted models.

Model	α	sd (α)	β_{DBH}	sd (β_{DBH})
$\alpha + \beta_{\text{DBH}} \text{ DBH}$	1.4393	0.0311	-0.1000	0.0332
$\alpha + \beta_{\text{DBH}} \text{ DBH} + v$	1.4427	0.0321	-0.0881	0.0250
$\alpha + \beta_{\text{DBH}} \text{ DBH} + u$	1.4459	0.0334	-0.0933	0.0306
$\alpha + \beta_{\text{DBH}} \text{ DBH} + v + u$	1.4305	0.0332	-0.1041	0.0370

terior samples. These results show that the DBH has only a negligible effect on the association between the two lichen species; however, there is biological evidence that the DBH explains the marginal occurrence probabilities for these two species. The influence of the random components part on the fixed effect is small for these particular data.

The choices of the prior distributions of the inverse variance parameters κ_u and κ_v are important because they modulate the strength of the different elements of the model. Following Pascutto *et al.* (2000), we studied the effect of prior choice: the alternatives of the $\Gamma(C, D)$ priors were (1) $C = 1$, $D = 0.001$; (2) $C = 1$, $D = 0.005$; and (3) $C = 1$, $D = 0.01$. Here alternative (1) gives a strong initial belief on small values of the variance parameter, alternative (2) is coherent with the one we have applied in data analysis, and alternative (3) is a more liberal choice. As a result, the (inverse) variance parameters are sensitive to the choice of the prior. Interestingly, the marginal posterior means of the association parameters are only weakly affected by the choice of the smoothing prior (typically less than 2%) but their posterior standard deviations are more sensitive (around 15%–20%). Hence the association maps based on posterior means are not strongly affected by the smoothing prior choice.

6   Discussion

The hierarchical model with a spatially structured random component, suggested by Besag *et al.* (1991), has been widely applied in the modelling of spatial data for scalar responses. The advantage of the model is that it allows different sources of heterogeneity arising from covariates, spatially structured variation, and unexpected local variation. The most common area of application has been in mapping the spatial distribution of disease incidences and risks.

The approach has several generalizations. The basic approach has been presented for Gaussian, binomial, and Poisson responses; see also S. Richardson (Chapter 8 in this volume). The family of distributions of the responses can be extended, e.g. in the frame of generalized linear modelling. Osnes and Aalen (1999) is an example where the frailty model of survival analysis is extended to a spatial context. Another direction is to extend the model to vector-valued responses. This type of question is of importance, for example, in joint mapping of diseases and in the study of suitability of sites for species in biogeography. The model helps in deducing the magnitude of covariate effects, but in most cases the final goal is to visualize the spatial distributions of parameters and the relationships between them. The model can be interpreted as a spatial smoother based on detailed structural and distributional information.

The empirical results derived from the lichen data are promising even when some of the sub-area counts are small. The results were similar to those obtained using the simultaneous site level modelling recalled in Section 2 (Penttinen *et al.* 2000) with respect to both the covariate effect and the derived association map. The posterior turned out to be sensitive to the choice of the smoothing parameter priors but this has only a small effect on the association maps based on the marginal posterior means. This sensitivity is also shared by other applications of latent Markov smoothers.

From the point of view of environmental studies, an important question is the prediction of the species' distribution over time, typically in a heterogeneous environment. This problem comprises all the aspects mentioned before: estimation of habitat suitability, evaluation of the effect of habitat changes on species, estimation of local extinction probabilities leading to changes in the range, and estimation of global extinction probabilities. Cross-sectional data give information on the structures of populations and on important relationships between species and between a species and the environment. However, if the purpose is to give predictions on population distributions from cross-sectional models, the results are sensitive to the assumptions. Furthermore, model validation may turn out to be impossible. Instead, dynamic modelling based on temporal data such as sequential atlas surveys (see Buckland *et al.* 1996) gives stronger evidence.

Our models assume the sites (trees) to consist of a fixed stand, leading to a regular or irregular lattice of sites. One could think also that the trees are random, a sample from a point process. The whole system is then described by a marked point process with tree characteristics and lichen occupation as marks. Biological reasons can be stated why lichens neither affect tree locations nor the properties of the trees. Our model is consistent with this more general model in the sense that the lichen occupation is examined as conditional given the forest stand. Conditioning has at least one drawback: the model cannot be used in experimenting with changes in occupation probabilities when some of the trees are removed. This would be a question of ecological relevance because, for example in this particular case, the old aspen trees will fall in the course of time. For this problem the full generality of the marked point process model is needed. A similar question concerns the relevance of many other conventional models. For example, a point process is considered to be a natural description of a real-world phenomenon and a model for the aggregated data is used to describe it. Consistency requirements mean that the model for aggregated data should result from the aggregation of the point process. This type of problem has been addressed in spatial epidemiology by Hjort (1999) and is discussed by S. Richardson (Chapter 8 in this volume).

In addition to the estimated map, hierarchical Bayesian modelling gives the ecologist an inferential tool for the evaluation of hypotheses of importance. For example, it is possible to estimate the probability that the suitability of species or their association is the largest in a particular sub-area selected on the grounds of biological knowledge.

Acknowledgements

This work was supported by the Academy of Finland through grant no. 37206 and a senior researcher's grant no. 385. Furthermore, the ESF/HSSS is acknowledged for a long-term visitor's grant. The availability of the data is due to Dr Mikko Kuusinen. Our thanks are also due to an anonymous referee and the editors for pointing out the correspondence between the conditional association model and a full model.

References

Augustin, N. H., Mugglestone, M. A., and Buckland, S. T. (1996). An autologistic model for the spatial distribution of wildlife. *Journal of Applied Ecology*, **33**, 339–47.

Besag, J. (1974). Spatial interaction and the statistical analysis of lattice systems (with discussion). *Journal of the Royal Statistical Society*, B, **36**, 192–236.

Besag, J., York, J., and Mollié, A. (1991). Bayesian image restoration with two applications in spatial statistics (with discussion). *Annals of the Institute of Statistical Mathematics*, **43**, 1–59.

Besag, J., Green, P., Higdon, D., and Mengersen, K. (1995). Bayesian computation and stochastic systems. *Statistical Science*, **10**, 3–66.

Breslow, N. E. and Day, N. E. (1980). *Statistical Methods in Cancer Research*, Vol. 1. IARC Scientific Publication No. 32, International Agency for Research on Cancer, Lyon, France.

Buckland, S. T. and Elston, D. A. (1993). Empirical models for the spatial distribution of wildlife. *Journal of Applied Ecology*, **30**, 81–102.

Buckland, S. T., Elston, D. A., and Beaney, S. J. (1996). Predicting distributional change of birds in north-east Scotland. *Global Ecology and Biogeography Letters*, **5**, 66–84.

Buckland, S. T., Anderson, D. R., Burnham, K. P., and Laake, J. L. (1993). *Distance Sampling. Estimating Abundance of Biological Populations*. Chapman & Hall, London.

Diggle, P. J., Tawn, J. A., and Moyeed, R. A. (1998). Model-based geostatistics (with discussion). *Applied Statistics*, **47**, 299–350.

Elliott, P., Wakefield, J. C., Best, N. G., and Briggs, D. J. (eds) (2000). *Spatial Epidemiology. Methods and Applications*. Oxford University Press, UK.

Gelman, A. and Rubin, D. (1992). Inference from iterative simulation using multiple sequences. *Statistical Science*, **7**, 457–511.

Heikkinen, J. and Högmander, H. (1994). Fully Bayesian approach to image restoration with an application in biogeography. *Applied Statistics*, **43**, 569–82.

Heikkinen, J. and Högmander, H. (1997). Statistics in biogeography. In *Encyclopedia of Statistical Sciences*, Update Vol. 1 (eds S. Kotz, C. B. Read, and D. L. Banks), pp. 56–61. Wiley, New York.

Hjort, N. L. (1999). Discussion of 'Bayesian models for spatially correlated disease and exposure data', by Best, N. G., Arnold, A., Thomas, A., Waller, L. A., and Conlon, E. M. In *Bayesian Statistics VI* (eds J. M. Bernardo, J. O. Berger, A. P. Dawid, and A. F. M. Smith), pp. 131–56. Oxford University Press, UK.

Högmander, H. and Møller, J. (1996). Estimating distribution maps from atlas data using methods of statistical image analysis. *Biometrics*, **51**, 393–404.

Knorr-Held, L. and Best, N. (2001). Shared component models for detecting joint and selective clustering of two diseases. *Journal of the Royal Statistical Society*, A, **164**, 73–85.

Kuusinen, M. (1994). Epiphytic lichen flora and diversity on *Populus tremula* in old-growth and managed forests of southern and middle boreal Finland. *Annales Botanici Fennici*, **31**, 245–60.

Kuusinen, M. (1995). Cyanobacterial macrolichens on *Populus tremula* as indicators of forest continuity in Finland. *Biological Conservation*, **75**, 43–9.

Kuusinen, M. and Penttinen, A. (1999). Spatial pattern of the threatened epiphytic bryophyte *Neckera pennata* at two scales in a fragmented boreal forest. *Ecography*, **22**, 729–35.

Liao, J. G. (1999). A hierarchical Bayesian model for combining multiple 2 × 2 tables using conditional likelihoods. *Biometrics*, **56**, 268–72.

Mugglin, A. S., Carlin, B. P., Zhu, L., and Conlon, E. (1999). Bayesian areal interpolation, estimation, and smoothing: an inferential approach for geographic information systems. *Environment and Planning A*, **31**, 1337–52.

Osnes, K. and Aalen, O. (1999). Spatial smoothing of cancer survival: a Bayesian approach. *Statistics in Medicine*, **18**, 2087–99.

Pascutto, C., Wakefield, J. C., Best, N. G., Richardson, S., Bernardinelli, L., Staines, A., and Elliott, P. (2000). Statistical issues in the analysis of disease mapping data. *Statistics in Medicine*, **19**, 2493–519.

Penttinen, A., Riiali, A., and Kuusinen, M. (2000). Bayesian modelling of biogeographical association in inhomogeneous environment. Manuscript.

Pielou, E. C. (1979). *Biogeography*. Wiley, New York.

Preisler, H. K. (1993). Modelling spatial patterns of trees attacked by bark-beetles. *Applied Statistics*, **42**, 501–14.

Riiali, A., Penttinen, A., and Kuusinen, M. (2001). Bayesian mapping of lichens growing on trees. *Biometrical Journal*, **43**, 1–20.

Weir, I. S. and Pettitt, A. N. (2000). Binary probability maps using a hidden conditional autoregressive Gaussian process with an application to Finnish common toad data. *Applied Statistics*, **49**, 473–84.

9A
Likelihood analysis of binary data in space and time

Julian Besag and Jeremy Tantrum
University of Washington, Seattle, USA

1 Introduction

Suppose that, located at each site i of a finite rectangular lattice $\mathcal{L}$, there is a plant, whose status, $x_{i,t} = 0$ if the plant is healthy or $x_{i,t} = 1$ if the plant is diseased, is recorded at equally spaced time points $t = 0, 1, \ldots, T$. Suppose also that diseased plants do not recover, so that $x_{i,t} \geqslant x_{i,t-1}$ for $t = 1, \ldots, T$. If we view the progress of the disease as a stochastic process, then $x_{i,t}$ is an observation on a binary random variable $X_{i,t}$. Correspondingly, we write X_t for the random array of the $X_{i,t}$s for all $i \in \mathcal{L}$. Then the simplest plausible family of probability models is Markov in time, in which case

$$\Pr(x_1, \ldots, x_T \,|\, x_0) = \prod_{t=1}^{T} \Pr(x_t \,|\, x_{t-1}). \tag{1.1}$$

It remains to model the terms in the product or, if we consider each site separately, to model

$$p_{i,t}(x_{i,t} \,|\, x_{-i,t}, x_{-i,t-1}) := \Pr(x_{i,t} \,|\, x_{i,t-1} = 0, \text{ all } x_{j,t}, x_{j,t-1} \text{ for } j \neq i), \tag{1.2}$$

$x_{i,t} = 0, 1$, for $i \in \mathcal{H}_{t-1}$, the set of sites containing healthy plants at time $t - 1$; here, $-i := \mathcal{L} \setminus i$.

The simplest formulation incorporating spatial dependence assumes that the probability (1.2) depends on $x_{-i,t-1}$ but not on $x_{-i,t}$. This approximation dates back at least to Bartlett (1975, p. 78) and can be motivated by a corresponding continuous-time model. Bartlett's analysis is concerned with the incidence of nettlehead virus in hop plants on an annual basis, for which the assumption is perhaps difficult to justify. The assumption is also made by Besag (1977a, Section 2.2) in analysing the progressive spread of a disease over a 179×14 approximately square-spaced lattice of endive (chicory) plants. The status of each plant was observed on four occasions at four-week intervals, relative to which the disease spread quite slowly, so that the approximation seemed plausible as a starting point, particularly since little was known about the underlying causal mechanism. The considerable simplification that occurs in this type of formulation is that $\Pr(x_t \,|\, x_{t-1})$, for $x_t \geqslant x_{t-1}$ elementwise, is a simple product of terms, each corresponding to a plant that is healthy at time $t - 1$ and which may or may not become diseased at t. It is then easy to implement maximum likelihood estimation for any corresponding parametric model. Otherwise, the distribution of X_t, given x_{t-1}, is generally a Markov random field with an intractable normalizing constant, so that maximum likelihood estimation is problematic.

In this discussion, we relax the assumptions to allow for contemporaneous effects in space–time modelling of binary data. We present a particular parametric formulation and describe the estimation of its parameters, both by maximum pseudo-likelihood and by Monte Carlo maximum likelihood. Implementation of the former is trivial, whilst, for the latter, we use coupling from the past (Propp and Wilson 1996) to generate perfect random samples from the target distribution. It is easy to extend the methods to more general formulations. As regards the endives data, we conclude that the omission of contemporaneous effects is inappropriate, despite the apparently slow spread of the disease.

2 Markov random field formulation

If we accept (1.1) but now allow contemporaneous dependence in (1.2), we might also postulate that $x_{-i,t}$ and $x_{-i,t-1}$ influence (1.2) only through the status of plants in the immediate vicinity of i. Specifically, we assume that the dependence is only on $d_{i,t}$ and $d_{i,t-1}$, the observed numbers of diseased plants directly adjacent to i at times t and $t-1$, respectively. Any assumption of this type implies stringent self-consistency conditions on (1.2). These are identified by the Hammersley–Clifford theorem; see, for example, Besag (1974) and, for historical commentary and the original 1971 proof, Clifford (1990). Here, we assume that

$$p_{i,t}(x_{i,t} \mid x_{-i,t}, x_{-i,t-1}) = \frac{\exp\{(\alpha_t + \beta_t d_{i,t} + \gamma_t d_{i,t-1})x_{i,t}\}}{1 + \exp(\alpha_t + \beta_t d_{i,t} + \gamma_t d_{i,t-1})}, \qquad (2.1)$$

where α_t, β_t, and γ_t are unknown parameters. Although valid mathematically, we relax this formulation in the statistical analysis, applying it only to the 177×12 interior of $\mathcal{L}$. We ignore this point for the moment and note instead that the model can be extended to include diagonally adjacent diseased plants or to add parameters that take account of the orientation of the adjacencies. We also note that the eventual interpretability of the parameters in (2.1) is helped if we rewrite $\alpha_t + \beta_t d_{i,t} + \gamma_t d_{i,t-1}$ as $\alpha_t + \beta_t a_{i,t} + \delta_t d_{i,t-1}$, where $\delta_t = \beta_t + \gamma_t$ and $a_{i,t} = d_{i,t} - d_{i,t-1}$ is the number of plants adjacent to i that become diseased between times $t-1$ and t.

Although (2.1) resembles an ordinary logistic regression and seems a natural starting point in its own right, such an interpretation is incorrect unless $\beta_t = 0$, because of the contemporaneous dependence on $d_{i,t}$. Rather, the formulation is autologistic (Besag 1974). Thus, let ∂i denote the set of sites adjacent to i and $i \sim j$ the set of all distinct pairs of adjacent sites. Then the distribution of X_t, conditional on x_{t-1}, is given by

$$\Pr(x_t \mid x_{t-1}) \propto \mathbf{1}[x_t \geqslant x_{t-1}] \exp(\alpha_t u_t + \beta_t v_t + \gamma_t w_t), \qquad (2.2)$$

where $1[\cdot]$ is an indicator function, operating elementwise, and

$$u_t = \sum_i x_{i,t}\,(1 - x_{i,t-1})\,, \tag{2.3}$$

$$v_t = \sum_{i \sim j} x_{i,t}\,x_{j,t}\,(1 - x_{i,t-1}x_{j,t-1})\,, \tag{2.4}$$

$$w_t = \sum_i \sum_{j \in \partial i} x_{i,t}\,x_{j,t-1}\,(1 - x_{i,t-1})\,. \tag{2.5}$$

However, (2.2) is very difficult to deal with directly, because the constant of proportionality involves summation over an enormous sample space, unless the lattice is very small. Hence, we resort to Monte Carlo methods to obtain maximum likelihood estimates of the parameters, as described below. Note that the problems are merely exacerbated if the parameters are assumed to be time invariant.

3   Parameter estimation

In this section we discuss the calculation of maximum likelihood estimates (MLEs) and maximum pseudo-likelihood estimates (MPLEs) for the parameters $\theta_t = (\alpha_t, \beta_t, \gamma_t)$. We also briefly consider the time homogeneous case, with the θ_ts constrained to be equal.

Let x_t^o denote the observed value of the array at time t. Then the MLE $\hat{\theta}_t$ of θ_t maximizes $\Pr(x_t^o \,|\, x_{t-1}^o)$ in (2.2). However, $\hat{\theta}_t$ cannot be calculated directly because the normalizing constant is intractable. In such circumstances, the MPLE $\tilde{\theta}_t$ provides a simple alternative, as we discuss below. Before doing so, we return to the problem of edge effects and the adoption of free boundary conditions, implicit in the above formulation. This constitutes an additional assumption, which we prefer to avoid by conditioning the analysis on the boundary at time t. We do this both for the MLE and the MPLE. The computations are almost the same, whether one conditions or not, and indeed, for these data, there is close numerical agreement between the two cases.

3.1   Maximum pseudo-likelihood estimates for the θ_ts

Quite generally, let $y^o = (y_1^o, \ldots, y_n^o)$ denote a single observation from an n-dimensional distribution $p(y; \theta)$ that is fully specified up to scale, apart from the value of the parameter vector θ. Then the corresponding *pseudo-likelihood* (Besag 1975) is simply the product of the n observed full conditionals (also known as local characteristics) $p(y_i^o \,|\, y_{-i}^o; \theta)$; and the MPLE $\tilde{\theta}$ maximizes this as a function of θ. Thus, the MPLE has the advantage that it requires only a one-dimensional normalization for each full conditional, analogous to the Gibbs sampler. Of course, there is no free lunch here and the MPLE may perform poorly in comparison with the MLE. Nevertheless, MPLEs and their asymptotic properties have been studied widely, especially for Markov random fields and Markov point processes; see, for example, Besag (1977a,b), Jensen and Møller

TABLE 1. Parameter estimates and standard errors.

		Time period		
		1 to 2	2 to 3	3 to 4
	MPLE	−3.78	−4.12	−4.47
α	MLE	−3.80	−4.12	−4.43
	SE	0.18	0.21	0.24
	MPLE	0.82	1.29	2.14
β	MLE	0.87	1.36	2.03
	SE	0.37	0.36	0.30
	MPLE	0.70	0.77	0.60
δ	MLE	0.72	0.74	0.60
	SE	0.17	0.16	0.18

(1991), Cométs (1992), Grenander (1993), Jensen and Künsch (1994), Mase (1995), Baddeley (2000), and Baddeley and Turner (2000).

Here, the MPLE of θ_t, based on x_t^o and x_{t-1}^o, maximizes the product over the interior sites $i \in \mathcal{H}_{t-1}$ of the $p_{i,t}(x_{i,t}^o \,|\, x_{-i,t}^o, x_{-i,t-1}^o)$ defined in (2.1). The maximization is trivial and can be carried out via a standard logistic regression program. The resulting MPLEs, with $\tilde{\gamma}_t$ replaced by $\tilde{\delta}_t$, are shown in Table 1. Of course, the standard errors produced by a logistic regression program do not apply in the present context, despite occasional claims to the contrary.

3.2   Maximum likelihood estimates for the θ_ts

Again quite generally, suppose we have a single observation y^o from a (discrete) multivariate distribution,

$$p(y; \theta) = h(y; \theta)/c(\theta), \tag{3.1}$$

where h is known, apart from the value of θ, but c, determined in principle by

$$c(\theta) = \sum_y h(y; \theta),$$

is intractable. Let $\breve{\theta}$ denote a current approximation to $\hat{\theta}$. Then, trivially, we can write

$$\hat{\theta} = \arg\max_\theta p(y^o; \theta) = \arg\max_\theta \ln \frac{p(y^o; \theta)}{p(y^o; \breve{\theta})} = \arg\max_\theta \left\{ \ln \frac{h(y^o; \theta)}{h(y^o; \breve{\theta})} - \ln \frac{c(\theta)}{c(\breve{\theta})} \right\}.$$

The first quotient on the right-hand side of this equation is known, and we can rewrite the second as

$$\frac{c(\theta)}{c(\breve{\theta})} = \sum_y \frac{h(y; \theta)}{c(\breve{\theta})} = \sum_y \frac{h(y; \theta)}{h(y; \breve{\theta})} p(y; \breve{\theta}),$$

which can be approximated by the empirical average,

$$\frac{1}{m} \sum_{k=1}^{m} \frac{h(y^k; \theta)}{h(y^k; \breve{\theta})},$$

for any θ in the neighbourhood of $\breve{\theta}$, where $y^1, \ldots, y^m$ is a random sample from $p(y; \breve{\theta})$. Derivatives can be handled similarly and hence a Newton–Raphson or other algorithm can be employed to find a better approximation to $\hat{\theta}$. Several stages of sampling may be needed to achieve convergence to $\hat{\theta}$.

Unfortunately, in most cases where one would like to use Monte Carlo MLE, it is not feasible to generate a random sample from $p(y; \hat{\theta})$ and one must turn instead to the Markov chain Monte Carlo (MCMC) version introduced by Penttinen (1984) in spatial statistics and by Geyer (1991) and Geyer and Thompson (1992) more generally. However, in our particular case, the target distribution for each t is closely related to the Ising model in statistical physics and so we can implement a perfect MCMC Gibbs sampler via monotone coupling from the past (Propp and Wilson, 1996) to generate random samples from (2.2), provided $\breve{\beta}_t$ is positive.

Some specific points are as follows. First, the Propp and Wilson theory embraces the need to condition on x_{t-1}^o (and on the boundary of x_t^o) because it applies also to non-homogeneous external magnetic fields. Secondly, Propp and Wilson's use of Sweeny's algorithm, rather than the Gibbs sampler (or heat bath algorithm), would be counterproductive here because the conditioning destroys any severe multimodality present in the symmetric Ising model. Thirdly, in running the algorithm, we can use x_{t-1}^o in place of the usual lower initialization of all zeros, because $X_t \geqslant x_{t-1}^o$ with probability one. Fourthly, note that, because (2.2) is an exponential family, the observed values u_t^o, v_t^o, and w_t^o match their conditional expectations at $\theta_t = \hat{\theta}_t$ and ultimately the corresponding Monte Carlo averages. Finally, Propp and Wilson's algorithm applies to more general autologistic formulations than (2.1), provided the contemporaneous parameters are positive.

As regards the endives data, we repeatedly apply coupling from the past, for each $t = 1, 2, 3$ and each successive $\hat{\theta}_t$, to generate random samples of size $m = 20\,000$ from the target distribution. Such a large sample size exceeds the needs of the practical application but also note that it is computationally inefficient to employ random samples. Instead, for any particular setting, one could use coupling from the past to produce a random draw from the target distribution and then use this to seed a single MCMC run from which the current working MLE would be updated. In the present context, coalescence by time zero is usually achieved in seven or fewer steps for $t = 1$ and $t = 2$ and in 31 or fewer for $t = 3$ and the loss of efficiency is unimportant. However, this would not always be the case.

Eventual MLEs of α_t, β_t, and δ_t, initiated by MPLE, are shown in Table 1, together with approximate standard errors. As one might anticipate, the transformation from γ_t to δ_t removes the strong negative correlation between $\hat{\beta}_t$ and $\hat{\gamma}_t$. Clearly, the β_ts play a significant role; their inclusion is also supported by formal MCMC goodness-of-fit tests, adapting Besag and Clifford (1989, Section 4). Indeed, because the $\hat{\gamma}_t$s are all negative, it appears that, for any fixed number of diseased neighbours at time t, a plant that is healthy at $t-1$ is more likely to be diseased at t the fewer the neighbours that are diseased at $t-1$. This is borne out by an appropriate tabulation of the raw data. Possible interpretations are given in Section 4.

3.3 Time homogeneity

Although not supported by the present data set, it is of some independent interest to consider the time-homogeneous version of the above formulation, in which $\theta_t = \theta$ for $t = 1, \ldots, T$. We do this specifically in the case of free boundary conditions. Thus, as regards the MPLE $\tilde{\theta}$, it is natural to redefine the pseudo-likelihood as

$$\prod_{t=1}^{T} \prod_{i \in \mathcal{H}_{t-1}} p_{i,t}(x_{i,t}^o \mid x_{-i,t}^o, x_{-i,t-1}^o),$$

given by (2.1) but without the parameter subscripts. Then again it is trivial to calculate $\tilde{\theta}$.

For the corresponding MLE $\hat{\theta}$, the likelihood function is given by (1.1) and hence is proportional to the product of (2.2) over $t = 1, \ldots, T$, again omitting the parameter subscripts. However, this is no longer an exponential family, because it matters now that the normalizing functions depend on $x_1, \ldots, x_{T-1}$, as well as on θ. It follows that, when $\theta = \hat{\theta}$, the observed $u_1^o + \ldots + u_T^o$ and so on should not match the sums of their expectations given x_0^o but instead the sums of their individual expectations given $x_0^o, \ldots, x_{T-1}^o$, respectively. Thus, the Monte Carlo needs to be run as before but with a single value of $\breve{\theta}$.

4 Discussion

When the spread of a disease over an array of plants seems slow, relative to the intervals between successive observations, it is tempting to assume that contemporaneous dependence can be ignored. Then MLE for a corresponding parametric model is straightforward. Unfortunately, such an assumption is invalid if, for example, diseased plants are infective but only for a short period of time, or disease results from sporadic attacks by an external agent, possibly a pest or short periods of harsh weather. Thus, there is no compelling reason for a disease to spread smoothly in continuous time, even when its progress in discrete time appears to be slow.

The inclusion of contemporaneous dependence in space–time formulations is straightforward but must respect the Hammersley–Clifford theorem. Parameter estimation for the resulting Markov random field can then be carried out by maximizing the corresponding pseudo-likelihood or, more satisfactorily, the likelihood itself. Implementation of the latter usually requires MCMC but it may be possible to produce perfect samples by coupling from the past, as exemplified in this discussion. Both methods of estimation can be applied to time-homogeneous formulations, when supported by the data.

Acknowledgements

I am grateful to the National Research Center for Statistics and the Environment for financial support.

9B
Some further aspects of spatio-temporal modelling

Alexandra Mello Schmidt
University of Sheffield, UK

1   Introduction

Spatio-temporal modelling has experienced substantial development in the last decade. Here I will concentrate on two aspects. The simplest kind of spatio-temporal model combines a standard homogeneous spatial covariance function with some standard time-series correlation structure, using the assumption of separability. These terms are defined below, where I will consider first approaches to modelling heterogeneous spatial processes, and then non-separable models.

2   Modelling of heterogeneous spatial processes

In some spatio-temporal problems we have time series data at each spatial location that give substantial information about spatial correlation structure, see Sampson and Guttorp (1992), S&G hereafter. It is then important to model the spatial covariance carefully and to accommodate possible inhomogeneity.

A spatial process is *homogeneous* when we have stationarity – the spatial distribution is unchanged when the origin of the index set is translated – and isotropy – the process is stationary under rotations about the origin. These are common assumptions that have been made in most of the applications in spatio-temporal modelling. However, these assumptions are not very realistic in environmental problems since local influences in the correlation structure of the spatial process may be clearly found in the data.

S&G pioneered an approach to modelling non-stationarity and anisotropy (heterogeneity). The main idea is based on a non-linear transformation of the original geographical locations (G space) into a latent space called D space, wherein stationarity and isotropy hold. Measuring the spatial correlations in a latent space is appealing, but in implementing this idea they make use of multidimensional scaling and thin-plate splines, which seem quite arbitrary.

Here I will describe three other models that tackle the problem of heterogeneous processes. The first approach follows the ideas introduced by S&G but in a Bayesian framework and was proposed by Schmidt and O'Hagan (2000). A similar model, developed independently, was proposed by Damian *et al.* (2000). The difference between the two lies in the specification of the prior distributions of the parameters and of the correlation model used to fit the observed correlations.

Suppose that time series data are available from each of n monitoring stations. Sample covariances between stations (possibly calculated after pre-processing to remove the temporal structure) provided information about the underlying spatial correlation structure. Letting the true correlation between observations at spatial positions $\mathbf{x}$ and $\mathbf{x}^*$ be $c(\mathbf{x}, \mathbf{x}^*)$, we follow S&G by modelling this as a function of $\|\mathbf{d}(\mathbf{x}) - \mathbf{d}(\mathbf{x}^*)\|$, which is the distance between $\mathbf{x}$ and $\mathbf{x}^*$ in D space, with $\mathbf{d}(\cdot)$ denoting the mapping of points in G space into their D space locations. Thus, homogeneity is assumed in D space, whereas the conventional geostatistical approach assumes homogeneity in G space by modelling $c(\mathbf{x}, \mathbf{x}^*)$ as a function of $\|\mathbf{x} - \mathbf{x}^*\|$.

Schmidt and O'Hagan (2000) assume that, a priori, $\mathbf{d}(\cdot)$ follows a Gaussian process with mean $\mathbf{m}(\cdot)$, equal to the coordinates in G space, and covariance function $\sigma_d^2 R_d(\cdot, \cdot)$. Here, σ_d^2 is a 2×2 matrix giving prior information about the coordinate system in D space and is modelled as a diagonal matrix where each component of the main diagonal has an inverse gamma prior distribution and R_d is a spatial correlation function describing prior beliefs about the smoothness of the mapping. This is represented by $R_d(\mathbf{x}, \mathbf{x}') = \exp(-b_d \|\mathbf{x} - \mathbf{x}'\|^2)$, where b_d is fixed according to some sensible value in the $\mathbf{x}$ scale. See Schmidt and O'Hagan (2000) for further discussion about how to assign the priors and the values for these parameters.

This is for the simplest case in which D space, like the G space, is two-dimensional. However, in some problems one might want to make use of covariates which are suspected to cause some of the anisotropy present in the data. A possible way to take this into account is by considering mappings to higher dimensions. We might consider, for instance, mapping from $R^2 \to R^3$, where the third axis of D space is the altitude of the location in D space. In the approach of Schmidt and O'Hagan (2000) mappings to higher dimensions are easily incorporated. S&G also allow a higher-dimensional D space, but do not consider this as a mechanism to make use of covariates.

Higdon *et al.* (1999) propose a different model for accommodating heterogeneity in spatial processes. It is based on a moving average specification of a Gaussian process. Suppose that $z(\mathbf{x})$ is a stationary Gaussian process with correlogram $\rho(\mathbf{s}) = \int_{R^2} k(\mathbf{x}) k(\mathbf{x} - \mathbf{s}) \, d\mathbf{x}$. The spatial process $z(\mathbf{x})$ can be described as the convolution of a Gaussian white noise process $\epsilon(\mathbf{x})$ with a convolution kernel $k(\mathbf{x})$,

$$z(\mathbf{x}) = \int_{R^2} k(\mathbf{x} - \mathbf{u})\epsilon(\mathbf{u}) \, d\mathbf{u}.$$

This idea of the convolution is extended by allowing the kernel to vary spatially, resulting in a non-stationary process $z(\cdot)$. They represent $k_{\mathbf{s}}(\mathbf{u})$ as a bivariate normal kernel centred at location $\mathbf{s}$ with spatially varying covariance matrix $\boldsymbol{\Sigma_{\mathbf{s}}}$. Therefore the shape of the kernel is a function of the location. Its parameterization aims to describe the scaling, orientation, and elongation at the different locations and this is expected to vary smoothly across the space. The covariance matrix of the kernel $k_{\mathbf{s}}(\cdot)$ at location $\mathbf{s}$, $\boldsymbol{\Sigma_{\mathbf{s}}}$, is defined geometrically using the one-to-one transformation from a bivariate normal distribution to its one standard deviation ellipse. Gaussian processes priors are assigned to the foci of these ellipses. See Higdon *et al.* (1999) for the specification of the prior distributions of the different levels of hierarchy of the model.

Fuentes and Smith (2000) also propose an alternative approach for modelling heterogeneous spatial processes. The basic idea of their model is that the process is represented by locally stationary and isotropic random fields. They model a non-stationary spatial process, $Y(\mathbf{x})$, as a convolution of local stationary processes, that is

$$Y(\mathbf{x}) = \int_G K(\mathbf{x} - \mathbf{s}) \, Z_{\boldsymbol{\theta}(\mathbf{s})}(\mathbf{x}) \, d\mathbf{s},$$

where $K(\cdot)$ is a kernel function with bandwidth h. The spatial process $Z_{\boldsymbol{\theta}}(\mathbf{x})$ comprises a family of independent stationary Gaussian processes indexed by $\boldsymbol{\theta}$. In this model the lack of stationarity is being modelled by the parameters of the stationary Gaussian processes $\boldsymbol{\theta}(\mathbf{s})$, as they are allowed to change slowly with the location $\mathbf{s}$. They claim that if $K(\cdot)$ is a sharply peaked function and $\boldsymbol{\theta}(\mathbf{s})$ varies slowly with the location $\mathbf{s}$, for $\mathbf{x}$ near $\mathbf{s}$, $Y(\mathbf{x})$ looks like a stationary process with parameter $\boldsymbol{\theta}(\mathbf{s})$. On the other hand, since $\boldsymbol{\theta}(\mathbf{s})$ may vary substantially over the whole space, it also allows significant non-stationarity. A Bayesian hierarchical approach is suggested in which $\boldsymbol{\theta}(\mathbf{s})$ is modelled as a spatial trend plus zero mean, spatially correlated noise. For more details about the prior distributions and implementation of this model, see Fuentes and Smith (2000).

3 Non-separable covariance matrices

A homogeneous spatio-temporal process is defined by a covariogram $C(d, t)$ specifying the covariance between two points differing by a distance d in space and t in time. If $C(d, t)$ is given by the product of the pure spatial covariance multiplied by the pure temporal covariance, then $C(d, t)$ is said to be separable. Separability is a widely used and convenient assumption (e.g. Le and Zidek (1992)) but it is quite restrictive because it does not model space–time interaction. It is possible to construct non-separable spatio-temporal models indirectly by hierarchical modelling. For instance, if a common (e.g. autoregressive) time series structure is assumed at all spatial locations, then its parameters will be allowed to vary spatially. Until very recently, there have been few attempts to model non-separability directly. Cressie and Huang (1999) give a methodology for developing classes of non-separable spatio-temporal stationary covariance functions directly in closed form. Their results are based on the relationship between covariance functions and their spectral representation. Based on these results

Storvik *et al.* (2002) compare a directly modelled covariance with one based on an autoregressive temporal model with spatial, correlated noise.

Brown *et al.* (2000) propose an alternative construction in which for each time t the spatial field is obtained by blurring the field at time $t - 1$ and adding a spatial random field.

Acknowledgement

The author is grateful to CNPq-Brazil for financial support during her PhD studies.

Additional references in discussion

Baddeley, A. J. (2000). Time-invariance estimating equations. *Bernoulli*, **6**, 1–26.

Baddeley, A. J. and Turner, R. (2000). Practical maximum pseudo-likelihood for spatial point patterns (with discussion). *Australian and New Zealand Journal of Statistics*, **42**, 283–322.

Bartlett, M. S. (1975). *The Statistical Analysis of Spatial Pattern*. Chapman & Hall, London.

Besag, J. E. (1975). Statistical analysis of non-lattice data. *The Statistician*, **24**, 179–95.

Besag, J. E. (1977a). Some methods of statistical analysis for spatial data. *Bulletin of the International Statistical Institute*, **47**, 77–92.

Besag, J. E. (1977b). Efficiency of pseudo-likelihood estimation for simple Gaussian fields. *Biometrika*, **64**, 616–18.

Besag, J. E. and Clifford, P. (1989). Generalised Monte Carlo significance tests. *Biometrika*, **76**, 633–42.

Brown, P. E., Kåresen, K., Roberts, G., and Tonellato, S. (2000). Blur-generated non-separable space-time models. *Journal of the Royal Statistical Society*, B, **62**, 847–60.

Clifford, P. (1990). Markov random fields in statistics. In *Disorder in Physical Systems* (eds G. Grimmett and D. J. Welsh). Clarendon Press, Oxford, UK.

Cométs, F. (1992). On consistency for a class of estimators for exponential families of Markov random fields on a lattice. *Annals of Statistics*, **20**, 455–68.

Cressie, N. and Huang, H.-C. (1999). Classes of non-separable, spatio-temporal stationary covariance functions. *Journal of the American Statistical Association*, **94**, 1330–40.

Damian, D., Sampson, P., and Guttorp, P. (2000). Bayesian estimation of semi-parametric non-stationary spatial covariance structures. *Environmetrics*, **12**, 161–78.

Fuentes, M. and Smith, R. (2000). Modelling nonstationary spatial processes as a convolution of local stationary processes. Technical Report, North Carolina State University (NCSU).

Geyer, C. J. (1991). Markov chain Monte Carlo maximum likelihood. In *Computing Science and Statistics: Proceedings of the Twenty-third Symposium on the Interface* (ed. E. M. Keramidas), pp. 156–63. Interface Foundation of North America, Fairfax Station, VA.

Geyer, C. J. and Thompson, E. A. (1992). Constrained Monte Carlo maximum likelihood for dependent data. *Journal of the Royal Statistical Society*, B, **54**, 657–99.

Grenander, U. (1993). *General Pattern Theory: A Mathematical Study of Regular Structures*. Clarendon Press, Oxford, UK.

Higdon, D., Swall, J., and Kern, J. (1999) Non-stationary spatial modelling. In *Bayesian Statistics 6* (eds J. M. Bernardo, J. O. Berger, A. P. Dawid, and A. F. M. Smith), pp. 761–8. Clarendon Press, Oxford, UK.

Jensen, J. L. and Künsch, H. R. (1994). On asymptotic normality of pseudo-likelihood estimates for pairwise interaction point processes. *Annals of the Institute of Statistical Mathematics*, **46**, 475–86.

Jensen, J. L. and Møller, J. (1991). Pseudolikelihood for exponential family models of spatial point processes. *Annals of Applied Probability*, **1**, 445–61.

Le, N. D. and Zidek, J. V. (1992). Interpolation with uncertain spatial covariance: a Bayesian alternative to kriging. *Journal of Multivariate Analysis*, **43**, 351–74.

Mase, S. (1995). Consistency of the maximum pseudo-likelihood estimator of continuous state space Gibbsian processes. *Annals of Applied Probability*, **5**, 603–12.

Pentinnen, A. (1984). Modelling interactions in spatial point patterns: parameter estimation by the maximum likelihood method. *Jyväskylä Studies in Computer Science, Economics and Statistics*, **7**, 1–107.

Propp, J. G. and Wilson, D. M. (1996). Exact sampling with coupled Markov chains and applications to statistical mechanics. *Random Structures and Algorithms*, **9**, 223–52.

Sampson, P. and Guttorp, P. (1992). Nonparametric estimation of nonstationary spatial covariance structure. *Journal of the American Statistical Association*, **87**, 108–19.

Schmidt, A. M. and O'Hagan, A. (2000). Bayesian inference for non-stationary spatial covariance structure via spatial deformations. Technical Report, University of Sheffield, UK.

Storvik, G., Frigessi, A., and Hirst, D. (2002). Space time Gaussian fields and their time autoregressive representation. *Statistical Modelling*, **2**, 139–61.

10

Advances in Bayesian image analysis

Merrilee Hurn
University of Bath, UK

Oddvar Husby and Håvard Rue
Norwegian University of Science and Technology, Trondheim, Norway

1 Introduction

Statistical interest in data in the form of images really took off in the 1980s following the papers by Geman and Geman (1984) and Besag (1986), although in fact Grenander had been working in the area of pattern recognition for some considerable time by then (Grenander 1976, 1978, 1981). At that stage, the work leaned towards pixel-based models and the restoration of images to remove degradation and provide a sharper picture. The past few years have seen changes in the emphasis and application of statistical image analysis techniques; the range of problems tackled has diversified and now encompasses more interpretation tasks such as object recognition and measurement (the so-called high-level tasks). New possibilities for Bayesian image analysis, and particularly object-based approaches, have opened up with the advances made in object modelling (Grenander *et al.* 1991; Grenander and Miller 1994) and the corresponding Markov chain Monte Carlo (MCMC) sampling of variable dimension distributions (Green 1995). It is now possible, in theory, to sample from a model that jumps around between various image interpretations (in terms of image content), although in practice models and algorithms must often be carefully tuned in order to attain reasonable levels of mixing (Rue and Hurn 1999).

We will attempt to provide an overview of recent work, highlighting the main directions, particularly in modelling in Section 2, and inference in Sections 3 and 4. Many of the ideas will be illustrated in Section 5 by an ultrasound application requiring careful physical modelling and recently devised sampling techniques.

2 Modelling

Since we are taking a Bayesian approach, our modelling takes two parts: defining a prior for a suitable representation of the scene being imaged, plus a likelihood model to capture the data acquisition process. We mention that there has been interesting statistical work in the computer vision literature that we are unable to include for reasons of space, for instance Jain *et al.* (1996) and Zhu and

Mumford (1997). Blake and Isard (1998) and Zhu *et al.* (1998) provide some further references and applications.

2.1   Prior modelling

In this subsection we describe three different approaches to modelling images. The distinctions arise in terms of the level of global or local information which one wishes to put into the model via the representation of the scene. There is here an issue of what can be expected of a prior in a high-dimensional image problem; often people use local and generic (low-level) priors merely as regularization terms. It is unrealistic to expect realizations from such priors to resemble typical images, but hopefully the local features are adequate for the kind of inference to be drawn from such models. On the other hand, high-level priors try to capture important features of the images, and carry significantly more structure; they can answer more global questions about the scene. Intermediate-level models fall between the two, in terms of input and output.

2.1.1   *Pixel-level modelling*

Pixel-level models are defined on the pixels in the image, most often using Markov random fields (MRFs). Traditionally, their use has been in restoration from imperfect and indirect observations, but they have also found applications as a part of a hierarchical model; there are connections here with work on spatial epidemiology (S. Richardson, Chapter 8 in this volume).

Categorical MRF models   Categorical MRF models such as the binary Ising model, and its multicolour extension, the Potts model, are frequently used as priors in tasks such as segmentation and classification. Although such models can contribute to visually acceptable posterior restorations, they do not provide good prior models for real scenes, and estimates of attributes such as the number of connected components may be poor. The difficulty is to design an MRF with a small neighbourhood in such a way that the global behaviour of samples, such as the size, shape, and number of compact objects against a background, is controlled. Recent progress has been made by Tjelmeland and Besag (1998), who also move away from the constraints of a square pixel lattice structure, using instead a hexagonal grid (which has benefits in terms of having fewer directional artefacts). The main idea in their approach is the careful definition of clique types, and their 26 possible configurations (up to rotation) are shown in the paper. Figure 1 shows some MCMC realizations from the model with three different sets of parameters. By blurring realizations like those in Fig. 1 and adding Gaussian noise, the posterior using the correct prior was compared with that using an Ising model. Considerations such as the posterior number of misclassified pixels (given the known true image), and visual appearance were slightly superior using the new model, but the real benefits came when studying more complex image functionals such as the number of distinct foreground objects, the number of objects larger than four pixels, or the edge length, with the Ising model performing much worse. It seems possible to model binary fields

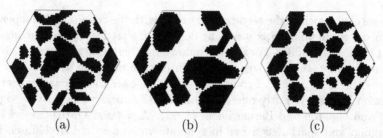

(a)   (b)   (c)

FIG. 1. Realizations from the fourth-order model of Tjelmeland and Besag (1998) with the three different sets of parameter values as in their Table 4.

using local cliques and to control to some extent the global properties; modelling is more delicate but it remains reasonable for design, coding, and computational efforts.

Grey-level models Grey-level models are often used in the context of visual restoration. To gain some insight into their behaviour, we consider the prior model described by Geman and Reynolds (1992), Geman and Yang (1995), and others:

$$\pi(\mathbf{x}) \propto \exp\left(-\beta \sum_m \omega_m \sum_c \phi(\mathcal{D}_c^{(m)}\mathbf{x})\right), \qquad (2.1)$$

where β is a smoothing parameter, $\{\omega_m\}$ is a set of positive weights, and the $\mathcal{D}_c^{(m)}\mathbf{x}$ are discrete approximations to the mth-order derivatives of the grey level for clique c (for example for $m = 1$ this is a simple pixel difference, for $m = 2$ this is the difference between first-order differences, and so on). The potential function $\phi(\cdot)$ is usually a symmetric function increasing on $[0, \infty)$ and such that $\phi(0) = 0$ and $\phi(\infty) < \infty$. A common choice is $\phi(u) = |u|/(1 + |u|)$. The motivation behind these conditions is to recapture discontinuities (in grey-level or higher derivatives of grey-level) without oversmoothing.

Further insight into the model's behaviour comes from considering (2.1) as the marginal of a model augmented with continuous 'edge variables' (Geman and Reynolds 1992). These edge variables play a similar, although implicit rather than explicit, role to the edge variables used in the seminal 1984 paper by Geman and Geman. In addition, this new augmented model also possesses some useful properties that allow for increased sampling speed. We will discuss this model further in Section 5, where we also explore the augmented model.

Gaussian MRF models Another important type of MRF are Gaussian MRFs which are particularly convenient both theoretically and computationally. Let $\mathbf{x}$ be Gaussian with zero mean and precision matrix $\mathbf{Q}$, then

$$\mathrm{E}(x_i \mid \mathbf{x}_{-i}) = -\sum_j \frac{Q_{ij}}{Q_{ii}} x_j \quad \text{and} \quad \mathrm{var}(x_i \mid \mathbf{x}_{-i}) = 1/Q_{ii}.$$

Hence we can associate the zero-patterns in $\mathbf{Q}$ with the conditional independence structure of $\mathbf{x}$. In imaging, $\mathbf{x}$ is usually defined on a lattice with neighbours those other pixels within a 3×3 or 5×5 window and this defines the graph. If i and j are neighbours, then we write $i \sim j$. In disease mapping applications say, $\mathbf{x}$ may be defined on a map of counties with neighbours those counties that share a common boundary. Early references on GMRFs are, for example, Besag *et al.* (1991) and Clayton and Bernardinelli (1992). A popular GMRF model for use in imaging, and spatial statistics in general, was considered by Künsch (1987) and Besag and Kooperberg (1995):

$$\pi(\mathbf{x} \mid \kappa) \propto \kappa^{(n-1)/2} \exp\left(-\frac{\kappa}{2}\sum_{i \sim j}(x_i - x_j)^2\right). \tag{2.2}$$

Note that (2.2) is improper, but that this is a strength (Besag and Kooperberg 1995), as we do not need to fix the overall level, for example. The precision matrix is $Q_{ij} = -\kappa$ if $i \sim j$ and $Q_{ii} = n_i\kappa$, where n_i is the number of neighbours of i. The conditional mean is simply the average value of the neighbours, while the conditional variance is $1/(n_i\kappa)$. Hence (2.2) will favour locally constant surfaces, with extensions to higher order surfaces also possible (Besag and Kooperberg 1995; Besag and Higdon 1999).

GMRF models are computationally convenient as their local characteristics are easy to compute and fast general simulation algorithms exist (Rue 2001), but again the question arises: Is it possible to construct a local GMRF in such a way that the global properties of the model are controlled? Here we have in mind smooth realizations with a given correlation function, such as the commonly used exponential, Gaussian, or members of the Matern family (Cressie 1993). If we can approximate such models well enough on a lattice using a GMRF with a small neighbourhood structure, then we have a model for which sampling is computationally fast, but which at the same time has nice global properties. Rue and Tjelmeland (2002) have recently demonstrated that it is indeed possible to derive GMRF substitutes for Gaussian fields on a lattice.

2.1.2 *Intermediate-level modelling*

Intermediate-level models fall between pixel-based descriptions and object-based methods; these models have the ability to capture some global features of the image without the need to specify the scene too exactly. Their role is to give a compact representation of the image, for example a polygonal segmentation of a satellite image where the regions are interpreted as crops and houses. We will now discuss models based on some (continuous) random partition of the plane into regions, with each of these regions given an intensity (or colour) to produce binary, categorical, or continuous models. The idea can be traced back to Arak *et al.* (1993). Other references are Clifford and Nicholls (1994), Green (1995), Møller and Waagepetersen (1996), Nicholls (1997, 1998), Heikkinen and Arjas (1998), Møller and Skare (2001); see also Richardson (Chapter 8 in this volume).

One simple way to construct a random partition of the plane is the Voronoi tessellation. Let Λ be the bounded region of interest, and let $\boldsymbol{\xi} = (\xi_1, \ldots, \xi_n)$ be a set of points from a homogeneous Poisson process on Λ. The Voronoi tessellation $\mathcal{V}(\boldsymbol{\xi}) = \{V_k(\boldsymbol{\xi})\}$ is constructed by letting each tile $V_k(\boldsymbol{\xi})$ consist of those parts of Λ that are closer to ξ_k than to any other point of $\boldsymbol{\xi}$ (see Fig. 2(a)). This defines a neighbourhood structure $\sim$ between contiguous tiles. As an alternative tessellation, Nicholls (1998) used a triangulation of the $\boldsymbol{\xi}$ and also points $\boldsymbol{\xi}_\partial$ on the boundary of Λ. This triangulation is formed by connecting points by straight edges until no more can be added without intersecting an edge already in place.

The Voronoi tessellation is relatively easy to manipulate as it is Markov (with respect to $\sim$ and $\boldsymbol{\xi}$), in the sense that adding a new point only requires the tiles around this new point to be recomputed. Similarly, removing a point ξ_k involves only the contiguous tiles. Similar comments apply for the triangulation approach. This property is obviously convenient for constructing MCMC algorithms.

Once a partitioning scheme is set up, a model is defined for the colourings of the tiles. For example, we can colour the tiles at random, black or white, to produce realizations like Fig. 2(b). The base measure for the Voronoi case is Poisson for $\boldsymbol{\xi}$ and uniform independent colouring for the colours. Relative to this, we define densities that tend to give neighbouring tiles similar colourings.

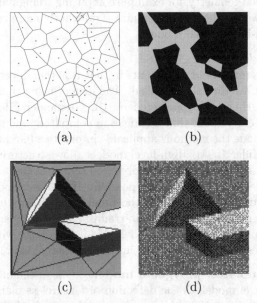

(a)   (b)

(c)   (d)

FIG. 2. (a) Example of a Voronoi tessellation; (b) a random colouring of (a); (c) shows an example of a triangulation of the image in (d).

Two natural choices are

$$f_1(\mathbf{x};\ \boldsymbol{\xi}) \propto \exp\left(-\beta \sum_{i \sim j} I[x_i \neq x_j]\right) \quad \text{and} \quad f_2(\mathbf{x};\ \boldsymbol{\xi}) \propto \exp(-\beta L(\mathbf{x}, \boldsymbol{\xi})),$$

where $L(\mathbf{x}, \boldsymbol{\xi})$ is the length of the edge between differently coloured neighbouring tiles. Both densities are similar in spirit to the Potts model, with f_2 allowing larger tiles more influence. A continuous colouring scheme is given by Heikkinen and Arjas (1998) using a GMRF with precision matrix $\beta\mathbf{Q}$, where $-Q_{ij}$ is the length of the edge between tiles i and j and Q_{ii} is the area of tile i; this structure encourages rapid changes when the tiles are small, and strong correlation between tiles sharing a long edge. If the normalization constant of the colouring with respect to $\boldsymbol{\xi}$ is known, then we can include it in the density to obtain a joint model with known marginals. This is possible with continuous GMRF colourings using the approach in Rue (2001), but not when using f_1 or f_2 above.

In combination with the likelihood, intermediate-level models can produce impressive restorations and reasonable interpretations of the data, even for quite high noise levels, see for example Fig. 2(c) and (d), and the experiments in Nicholls (1998) and Møller and Skare (2001). The implementation of this type of model can be eased by the existence of a huge variety of clever algorithms from computational geometry, for example computing triangulations and Voronoi tessellations. However, the coding remains significantly more complex than that required for pixel-based MRF models.

2.1.3 *High-level modelling*

High-level models make bold statements in that it is the objects under study, or their shapes, that are directly modelled. The idea of using compact, parametric models of object shape was pioneered by Ulf Grenander (1967, 1981), but it was not until the early 1990s that advances in computing and stochastic simulation algorithms made the methods applicable in practice (Grenander *et al.* 1991; Grenander and Miller 1994). High-level models allow structured understanding. The transformation from template to observed object makes tasks such as classification and detection of pathologies possible. High-level models also have the capacity to include uncertainty over the number of objects in view, requiring more complex sampling algorithms than traditional fixed dimension approaches. We will discuss just two classes of model; this field is huge with a wide variety of applications, see http://www.cis.jhu.edu/research.html for examples.

Deformable templates: polygonal models One of the most studied applications of high-level models is the detection of featureless planar objects (Baddeley and Van Lieshout 1993; Grenander and Manbeck 1993; Grenander and Miller 1994; Qian *et al.* 1996; Rue and Syversveen 1998; Rue and Hurn 1999; Stoica *et al.* 2000). Consider an object in the plane and suppose that its shape can be approximated either by a simple geometric shape such as an ellipse, or by an n-sided simple polygon with vertices $\mathbf{v}_0, \ldots, \mathbf{v}_{n-1}$ placed counterclockwise around the outline. This representation is the template for the object, and is

allowed to deform stochastically in order to describe the prior population variation of shape. There are two aspects to this type of model: first, how to allow
for stochastic deformation, and secondly how to construct a model that allows
for an unknown number of objects.

We concentrate first on the deformations of a single object. Considering a
geometric object such as an ellipse which is defined by a small number of shape
parameters, it is possible to specify prior distributions for these parameters. The
situation with a polygonal outline is more complicated, and generally work has
concentrated on modelling the edges rather than the vertices directly: $\mathbf{e}_j =
\mathbf{v}_{j+1} - \mathbf{v}_j$ for $j = 0, \ldots, n - 1$, where the indices are interpreted modulo n. The
usual approach is to consider an edge transformation matrix $\mathbf{s}$ acting on the
template edge vector $\mathbf{e}^0$:

$$ e_j = s_j e_j^0 = e_j^0 + r_j \begin{pmatrix} \cos\theta_j & \sin\theta_j \\ -\sin\theta_j & \cos\theta_j \end{pmatrix} e_j^0 = \begin{pmatrix} 1 + t_{0,j} & t_{1,j} \\ -t_{1,j} & 1 + t_{0,j} \end{pmatrix} e_j^0, $$

where the variables r_j and θ_j control the scale and angle in the change of edge
e_j. Imposing the condition that the deformed template is closed, $\sum_j s_j e_j^0 = 0$,
will introduce a dependency structure for the $t_{0,j}$ and $t_{1,j}$. Assuming a Gaussian
structure, the modelling issue becomes that of specifying the precision matrix $\mathbf{Q}$.
Grenander *et al.* (1991) let the unconstrained $\mathbf{t}_0$ and $\mathbf{t}_1$ be first-order cyclic Gaussian Markov random fields; then $\mathbf{t} = (\mathbf{t}_0, \mathbf{t}_1)$ is Gaussian with inverse covariance
matrix $\mathbf{Q_s} = \mathbf{I}_2 \otimes \mathbf{Q}$, where $\mathbf{Q}$ is a circulant matrix with entries $Q_{j,j} = \beta$ and
$Q_{j,j+1} = Q_{j,j-1} = \delta$. The circulant representation means that the model is in-
variant under cyclic permutations of the edges, a natural choice when the objects
have no apparent landmarks. Hobolth and Jensen (2000) consider second-order
Gaussian Markov random fields. Kent *et al.* (2000) consider the statistical properties of this model under more general covariance structures. They also discuss
alternative modelling strategies, treating scale and rotation conditionally rather
than jointly.

Note that representing the object through the edge vector $\mathbf{e}$ does not convey
any location information, and so in order to locate the template we must place
the first vertex $\mathbf{v}_0$ (or some other designated vertex) at a point $\mathbf{c} \in \mathbb{R}^2$. Similarly,
when using a geometrical template such as an ellipse, the centre must be specified
to locate the object. Hence we could think of the placement of the object as the
'point' and the parameters describing the object as a 'mark', motivating the
way in which a model can be constructed for an unknown variable number of
objects, by using a marked point process (Baddeley and Van Lieshout 1993).
Note that Grenander and Miller (1998) adopt an alternative formulation; see
some related comments in Green (1995). Here the points, representing locations,
have a density defined with respect to a Poisson point process in the plane. The
marks carried by each point are the values of the parameters describing that
object's deformation (plus any other information required). For example, Rue
and Hurn (1999) also use a label variable in order to distinguish two different
types of object. Models of this type can use the reversible jump MCMC algorithm

of Green (1995) to deal with the dimensionality change associated with different numbers of objects. Often sophisticated moves that merge and split objects are required (Grenander and Miller 1994; Rue and Syversveen 1998; Descombes *et al.* 1999b; Rue and Hurn 1999), but such moves are hard to construct and the convergence rate can be painfully slow.

Deformable templates: continuous models   In many situations the above-mentioned models are too simplistic to capture the detail and internal structure of the modelled objects. For instance, this is the case in many biomedical applications such as automatic computation of areas and volumes, and detection of pathologies, where one needs models that capture the detailed structure of anatomy. Objects are therefore more easily represented as images (Grenander and Miller 1994; Christensen *et al.* 1996), or as non-parametric curves and surfaces (Younes 1998; Bakircioglu *et al.* 1998); see, for example, Grenander and Miller (1998) for further references. Again, one does not model the objects themselves, but rather the transformations acting on a template object. For instance, if the object is an image, then transformations could be taken from the collection of smooth coordinate transforms of the image domain. Important modelling issues would then be the choice of boundary conditions, and the choice of prior on the set of transformations. Note that image matching is another application of the same methodology; see, for example, Glasbey and Mardia (1998) and, more recently, Glasby and Mardia (2001) for a review, and Sampson and Guttorp (1992) and Perrin and Senoussi (1999) for matching in different contexts.

One of the first works in a statistical setting was Amit *et al.* (1991), who consider the reconstruction of X-ray images of hands. Images are defined as mappings I from some fixed background space Ω to a range space $\mathcal{T}$. A particular image $I_0 \in \mathcal{I}$ is chosen as the template, and the variability in image appearance is modelled thorough the group $\mathcal{H}$ of diffeomorphic transformations $h : \Omega \ni \mathbf{x} \mapsto h(\mathbf{x}) = \mathbf{x} - u(\mathbf{x}) \in \Omega$, where $u(\mathbf{x})$ is the displacement field. The displacement field is assumed to be Gaussian distributed with a covariance kernel equal to the Green's function of a squared differential operator L. To be able to do inference the infinite-dimensional random variable u is approximated by a truncated orthonormal expansion of the eigenvectors of L. In this framework Amit *et al* (1991) implicitly assume that the deformations are small so that u can be approximately assumed to be in a Hilbert space $\mathcal{U}$ with norm $\|u\|_{\mathcal{U}}^2 = \|Lu\|^2$. In reality the group $\mathcal{H}$ of diffeomorphisms is a curved manifold and assuming a vector space structure leads to inconsistencies. In particular, as pointed out by Dupuis *et al.* (1998), the quadratic penalties $\|Lu\|^2$ imply restoring forces that are proportional to the displaced distance. This is called elastic deformation, since large deformations are severely penalized, being drawn back towards the template image. Whether this is undesirable or not depends on the application, but certainly the lack of symmetry is unnatural in applications where there is no obvious choice of template, and where images need to be interchangeable.

To construct large-deformation maps Christensen *et al.* (1996) and Dupuis *et al.* (1998) let $h(\mathbf{x})$ be the output $h(\mathbf{x}, 1)$ of a flow $h(\mathbf{x}, t)$ generated by a velocity field $v(\mathbf{x}, t)$:

$$\frac{\partial h(\mathbf{x}, t)}{\partial t} = -\left(\left(\nabla_{\mathbf{x}}^T h \right)(\mathbf{x}, t) \right) v(\mathbf{x}, t), \qquad h(\mathbf{x}, 0) = \boldsymbol{x}.$$

In this case the deformation stress is not accumulated, hence the term viscous deformations (Christensen *et al.* 1996). A similar approach can be found in Trouvé (1998). Since v is tangential to $\mathcal{H}$ we can assume it to be in a Hilbert space $\mathcal{V}$ with norm $\| \cdot \|_{\mathcal{V}}$. As in Amit *et al.* (1991) the norm is induced by a linear differential operator L. In most image matching applications the operator is of the form $L = a\nabla^2 + b\nabla\nabla + c$ chosen so that the solution obeys certain laws from continuum mechanics for deformable bodies, see Christensen *et al.* (1996) and Grenander and Miller (1998). The posterior energy for v becomes

$$\frac{1}{2} \int_0^1 \int_\Omega |Lv(\mathbf{x}, t)|^2 \, \mathrm{d}\mathbf{x} \, \mathrm{d}t + \frac{1}{2\sigma^2} \int_\Omega |I_0\left(h(\mathbf{x}, 1)\right) - I(\mathbf{x})|^2 \, \mathrm{d}\mathbf{x}, \qquad (2.3)$$

from which an estimate v^* can be found. The optimal match is then given by the integral equation

$$h^*(\mathbf{x}, 1) = \mathbf{x} - \int_0^1 \left(\left(\nabla_{\mathbf{x}}^T h^* \right)(\mathbf{x}, t) \right) v^*(\mathbf{x}, t) \, \mathrm{d}t.$$

Note that taking the minimum of the prior energy subject to $I_0\left(h(0)\right) = I_0$ and $I_0\left(h(1)\right) = I$ defines a distance on $\mathcal{I}$ (Trouvé 1998; Younes 1998). This enables us to compare images quantitatively, which is relevant for optimal estimation (see Section 4).

2.2 Data modelling

As the emphasis has moved towards real applications, the need for realistic modelling of how the data are formed has increased. Particularly in studies where interval estimates of image attributes are of interest, having a believable likelihood model may underpin the credibility of the results. Unfortunately, this is a rather neglected area in comparison with the effort that has gone into the development of new prior models (and the authors certainly would not claim to be an exception to this!). Some interesting work along these lines is found in microscopy (Qian *et al.* 1996; Higdon and Yamamoto 2001), impedance imaging (Nicholls and Fox 1998; Andersen *et al.* 2001), ultrasound imaging (Husby *et al.* 2001), and in much of the work in emission tomography, for example Green (1990), Weir (1997), and Higdon *et al.* (1997).

3 Treatment of parameters

This section discusses various approaches to the parameters of the models used; generally it is not these parameters that are of interest and so the issue may be whether to estimate them or to integrate them out in a fully Bayesian way.

3.1 Estimation techniques for low-level models

The most notable development in this area has been the use of MCMC maximum likelihood (Geyer 1991; Geyer and Thompson 1992; Tjelmeland and Besag 1998; Descombes *et al.* 1999a; Jalobeanu *et al.* 1999). This approach makes it possible to do (approximate) likelihood estimation of hyperparameters where MCMC is used to estimate (locally) the influence of the hyperparameters on the normalization constant. However, care is needed to tune the algorithm to make it work properly; suggestions are found in the above references. A comparison with pseudo-likelihood estimation is found in Tjelmeland (1996), Forbes and Raftery (1999), and Dryden *et al.* (2002), amongst others.

Although likelihood estimation is natural when the model is correct, it might not be so when trying to fit an approximate, computationally tractable model to a set of training data. Rue and Tjelmeland (2002) discuss this for fitting GMRFs, but more research along these lines is needed.

3.2 Estimation techniques for high-level models

Much of the recent work investigating the various deformable template models available has been quite theoretical and intertwined with the model formulation itself (Hobolth *et al.* 2002; Kent *et al.* 2000). Usually it is assumed that a training sample of data is available in the form of traced outlines, to which vertices are assigned according to some rule. An alternative approach is used in Cootes *et al.* (1995), who employ principal component analysis to estimate the probability density function for a class of shapes represented by landmarks. Extensions to non-linear methods can be found in, for example, Cootes and Taylor (1999). Finally, Hurn *et al.* (2001) assume that the training data are generated as noisy observations of deformed templates; this particular framework was intended to provide a black-box estimation procedure which could be driven by a purely graphical user-interface.

3.3 Fully Bayesian approaches

Another notable recent changes in image analysis has been the increasing use of interval estimates of image attributes. There are problems, in terms of the propagation of uncertainty through to the posterior, of fixing parameter values, as well as issues of model adequacy. Higdon *et al.* (1997), Weir (1997), and Higdon (1998) adopt fully Bayesian approaches in computed tomography using pairwise MRF prior models; these models are acknowledged to be somewhat unrealistic for this type of application, and likelihood or other estimation techniques that fix the parameter values do not allow for this. Hurn (1998) and Hansen *et al.* (2002) work in a high-level framework, where the prior models are rather more tailored to the application in hand; as techniques have developed in this area, it has again become more important to provide uncertainty estimates rather than just point values. In addition, in either case, there may well be parameters of the specified likelihood model which are very difficult to estimate accurately, for example Hurn (1998) considers a blind deconvolution problem.

One of the biggest impediments to a fully Bayesian approach has been that often it becomes necessary to evaluate complex normalizing functions. For example, suppose the Ising model is being used, so that

$$\pi(x|\beta) = \sum_{i \sim j} \exp(-\beta I[x_i \neq x_j])/Z(\beta),$$

where the normalizing constant $Z(\beta)$ is a function of the hyperparameter β. That $Z(\beta)$ is not known is not an issue when sampling for fixed β because it will cancel out in the acceptance ratio of a Metropolis–Hastings algorithm. However, as soon as β is allowed to vary, $Z(\beta)$ can no longer be ignored. Similar problems occur with the parameters of many other spatial prior models (although less so with parameters of the likelihood terms). Ratios of normalizing constants have to be estimated off-line by computationally intensive methods some of which require the generation of samples from the density at a grid of different values of the hyperparameter. A nice review of methods for estimating ratios of normalizing constants is given by the recent book Chen *et al.* (2000, Chapter 5).

4 Point estimation for images

As models and techniques have become more advanced and applicable to real data problems, attention has moved away from *maximum a posteriori* (MAP) and marginal posterior modes (MPM) point estimation. There has been a change in the choice of estimator itself. This is much needed in some cases, for example for polygonal templates, where the posterior 'mean' of the outline is not well-defined and so cannot be used as an estimator.

It had long been argued that neither MAP nor MPM were well-suited to image applications, one being insensitive to detail and the other too local. MAP corresponds to using a zero–one loss function, while MPM is the estimator for the loss function which counts the number of misclassified pixels. Unfortunately there are few other loss functions for which the estimator is known. However, Rue (1995) noted that for certain general forms of loss function $L(\mathbf{x}, \hat{\mathbf{x}})$ it is possible to approximate the corresponding estimate. Suppose, as a simple case, that $L(\mathbf{x}, \hat{\mathbf{x}})$ can be written in the form $L_1(\mathbf{x}) + L_2(\mathbf{x})L_3(\hat{\mathbf{x}}) + L_4(\hat{\mathbf{x}})$, then in finding the estimator

$$\hat{\mathbf{x}}^* = \arg\min_{\hat{\mathbf{x}}} \mathrm{E}_{\mathbf{x}|\mathbf{y}} L(\mathbf{x}, \hat{\mathbf{x}})$$

$$= \arg\min_{\hat{\mathbf{x}}} \left(\mathrm{E}_{\mathbf{x}|\mathbf{y}}(L_1(\mathbf{x})) + L_3(\hat{\mathbf{x}})\mathrm{E}_{\mathbf{x}|\mathbf{y}}(L_2(\mathbf{x})) + L_4(\hat{\mathbf{x}}) \right),$$

since the expectation is with respect to the posterior distribution of $\mathbf{x}$ under which $\hat{\mathbf{x}}$ behaves like a constant. The terms $\mathrm{E}_{\mathbf{x}|\mathbf{y}}(L_1(\mathbf{x}))$ and $\mathrm{E}_{\mathbf{x}|\mathbf{y}}(L_2(\mathbf{x}))$ can be estimated as ergodic averages by simulation from the posterior $\pi(\mathbf{x}|\mathbf{y})$. This leaves an estimated posterior loss that is minimized over $\hat{\mathbf{x}}$ using an appropriate optimization technique. The only restriction on the form of $L(\mathbf{x}, \hat{\mathbf{x}})$ for which this approach is computationally viable, is that L is separable into (a linear combination of) functions of $\mathbf{x}$ and $\hat{\mathbf{x}}$. In his initial work Rue (1995, 1997) built a spatial structure into the loss function by penalizing configurations of discrepancies between $\mathbf{x}$ and $\hat{\mathbf{x}}$. Higher-order tasks such as segmentation (Frigessi and Rue 1997) and object recognition (Rue and Syversveen 1998) are tackled with a function of Baddeley's Δ image metric as the loss function (Baddeley 1992; Rue 1999). This metric between two binary images $\mathbf{x}$ and $\mathbf{z}$ is defined as

$$\Delta(\mathbf{x}, \mathbf{z}) = \left(\frac{1}{|\Lambda|} \sum_{i \in \Lambda} |d(i, b(\mathbf{x})) - d(i, b(\mathbf{z}))|^2 \right)^{1/2} ,$$

where i indexes the pixels in grid Λ, $b(\mathbf{x})$ represents the set of foreground pixels in $\mathbf{x}$, and the distance $d(i, b(\mathbf{x}))$ measures the shortest distance between pixel i and the nearest foreground pixel. If $\tilde{\mathbf{x}}$ is used to denote the image $\mathbf{x}$ with foreground and background reversed, then the symmetrized loss function is defined as

$$L(\mathbf{x}, \mathbf{z}) = \Delta(\mathbf{x}, \mathbf{z})^2 + \Delta(\tilde{\mathbf{x}}, \tilde{\mathbf{z}})^2.$$

The terms $\sum_{i \in \Lambda} d(i, b(\mathbf{x}))^2$, $\sum_{i \in \Lambda} d(i, b(\mathbf{x}))$, $\sum_{i \in \Lambda} d(i, b(\tilde{\mathbf{x}}))^2$, and $\sum_{i \in \Lambda} d(i, b(\tilde{\mathbf{x}}))$ can be estimated by sampling from $\pi(\mathbf{x}|\mathbf{y})$; all other terms in the expected loss are a function only of $\mathbf{z}$. This leaves an estimated expected loss to be minimized over images $\mathbf{z}$ using simulated annealing. The idea can be generalized beyond binary images by considering an additional labelling term in the loss function in the style of the MPM loss function.

Similar ideas are now crossing over into areas other than image analysis using a number of different forms of loss function; for examples in mixture modelling see Celeux *et al.* (2000) and Hurn *et al.* (2002).

5   Example

In this example we will collect many of the ideas presented above, discussing semi-automatic contour detection in ultrasound images of the carotid artery. Such images are used in the detection of atherosclerosis using the fact that diseased arteries are less likely to dilate in response to an infusion of acetylcholine. Estimating the cross-sectional area of the artery before and after infusion is difficult because changes can be masked by blur and image artefacts (speckle) introduced in the imaging process. An automated procedure should be able to assess these problems properly, as well as quantifying the uncertainty of the given answer, for instance by means of an interval estimate. Thus there is a real need for Bayesian methods with emphasis on realistic modelling of both object and data. Previous attempts at high-level modelling in ultrasound images have not focused on the issue of data modelling, but there is a concern that the artefacts can seriously affect the performance of algorithms for contour detection, and we feel that proper modelling of the data formation can lessen the effect of blur and speckle. Previous work on contour detection in ultrasound images can be found in Glasbey (1998) and Hansen *et al.* (2002), but neither of these focus on the issue of data modelling.

5.1 A model for diffuse scattering in ultrasound imaging

The imaging model used in this example is presented in Husby *et al.* (2001) and Langø *et al.* (2001). In ultrasound imaging, a short pulse of ultra high frequency sound is sent into the body and the back-scattered echo is measured after a time delay corresponding to time taken to travel the distance from the transducer. The reflected signal consists of two parts, one resulting from the pulse being reflected at interfaces between different tissue types, and the other from locations in the tissue having spatial variations in acoustic impedance (Hokland 1995). These variations are called scatterers. We concentrate on this second part, called diffuse scattering. The resolution cell Ω_i corresponding to pixel i is assumed to consist of a large number N_i of uniformly distributed scatterers, and the received signal A_i is the sum of the reflections from all the scatterers:

$$A_i = \sum_{k=1}^{N_i} |a_k| e^{j\phi_k},$$

where a_k and ϕ_k are the amplitude and phase of the signal from scatterer k (Goodman 1975), and $j = \sqrt{-1}$. We assume that all amplitudes and phases are independent, and that the phases are uniformly distributed on $[0, 2\pi)$. Thus, by using the central limit theorem the resulting complex amplitude is assumed to be a complex Gaussian random variable with an independent and identically distributed real and complex part, both with zero mean and variance

$$\sigma_i^2 = \frac{1}{N_i} \sum_{k=1}^{N_i} \frac{E|a_k|^2}{2}$$

given by the mean square scattering amplitude of the scatterers within the resolution cell. The resulting true radio-frequency (RF) image $\mathbf{x}$ is obtained by taking the real value of the received signal. Thus, for all i on some lattice $\mathcal{I}$, $x_i \,|\, \sigma_i$ is Gaussian with mean zero and variance σ_i^2. Since the variance of the RF signal depends on the scattering properties of the underlying tissue and since those variances are believed to vary little within homogeneous tissue, we use a latent piecewise smooth log-variance field $\boldsymbol{\nu} = \log \boldsymbol{\sigma}$ to model the anatomical structure of the imaged tissue. For the reasons given in Section 2.1.1, a Geman–Reynolds model of the form of (2.1) is used as a prior $\pi(\boldsymbol{\nu})$.

The observed image is then modelled as resulting from a convolution of the imaging system point spread function h and the true RF image. We assume the point spread function to be spatially invariant; see, for example, Langø *et al.* (2001) for a discussion of this choice. Thus $y_i \,|\, x_i$ is Gaussian with mean $\sum_k h_k x_{i-k}$ and variance τ^2. Combining the above representations, we get the full conditional for the true RF signal and the log-variance field

$$\pi(\mathbf{x},\boldsymbol{\nu}\,|\,\ldots) \propto \prod_{i\in\mathcal{I}} \exp\left\{ -\frac{1}{2\tau^2}\left(y_i - \sum_k h_k x_{i-k}\right)^2 \right.$$

$$\left. -\frac{x_i^2}{2}e^{-2\nu_i} - \nu_i - \beta\sum_m \omega_m \phi\left(\mathcal{D}_i^{(m)}\boldsymbol{\nu}\right) \right\}. \quad (5.1)$$

5.2 Image restoration and MCMC algorithms

A point estimate $\hat{\mathbf{x}}$ of the true RF image $\mathbf{x}^*$ can be constructed using MCMC output, and a natural first choice is the posterior mean (Husby *et al.* 2001). Unfortunately, sampling from the posterior is not trivial, and the single-site algorithm of Husby *et al.* (2001) demonstrated slow convergence and poor mixing. This is not surprising given the strong spatial interactions in the model. A natural way out of this would be to consider block sampling. However, it is not straightforward to construct block updates from the posterior directly. Nevertheless, after augmenting the model with suitably chosen parameters, block sampling is indeed possible and can profit from the fast algorithm for sampling GMRFs (see Section 2.1.1). The augmentation is done using an idea from Geman and Yang (1995) as follows. We assume $\pi(\boldsymbol{\nu})$ to be the marginal of a model $\pi^*(\boldsymbol{\nu},\mathbf{b})$ augmented by M real-valued auxiliary fields $\mathbf{b}^{(1)},\ldots,\mathbf{b}^{(M)}$. This π^* is chosen so that the conditional distribution $\pi^*(\boldsymbol{\nu}\,|\,\mathbf{b})$ is a GMRF. The posterior density $\boldsymbol{\nu}$ under the dual model is

$$\pi^*(\boldsymbol{\nu}\,|\,\mathbf{x},\mathbf{b}) \propto \exp\left\{ -\beta\sum_m \omega_m \sum_i \left(\frac{1}{2}\left(\mathcal{D}_i^{(m)}\boldsymbol{\nu}\right)^2 - b_i^{(m)}\mathcal{D}_i^{(m)}\boldsymbol{\nu}\right) \right.$$

$$\left. -\frac{1}{2}\sum_i \left(x_i^2 e^{-2\nu_i} + 2\nu_i\right) \right\},$$

which can be block sampled using a Gaussian proposal distribution in a Metropolis–Hastings scheme; see Husby (2001) for details. Furthermore, the conditional distribution for the RF field $\mathbf{x}$ given all other variables is a GMRF, while the auxiliary fields $\mathbf{b}^{(m)}$ are conditionally independent. Hence, our MCMC algorithm block updates each of the three fields conditionally on the rest: $\mathbf{x}\,|\,\ldots,\,\mathbf{b}\,|\,\ldots$ and $\boldsymbol{\nu}\,|\,\ldots.$ This certainly improves mixing compared with single-site updating, but further improvements can be made, for instance by updating all variables jointly. As indicated in Knorr-Held and Rue (2002), the bottle-neck causing poor mixing can be high dependency between the three fields rather than within each field. Still, in our example the blocked sampler shows improved mixing over the single-site algorithm and is reasonably fast. Ways of improving the between-field mixing are currently being investigated.

Figure 3(b) shows an estimate of the posterior mean of x for fixed values of the parameters β, δ and τ^2. Note that the image looks less smooth than the data in

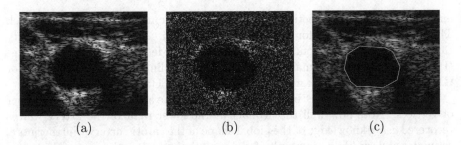

(a)　　　　　　　　(b)　　　　　　　　(c)

FIG. 3. (a) Observed 128×128 image of the carotid artery, (b) posterior mean estimate of the RF field $\mathbf{x}$, and (c) estimated artery outline using a circular template.

Fig. 3(a); this is a feature of the model since we assume the pixel values x_i to be a priori independent. This assumption is perhaps too simplistic, and can be relaxed by assuming a priori dependencies between the elements of $\mathbf{x}$. Compared with the filtering methods usually applied, our reconstruction method is successful at removing speckle and image artefacts. However, more work needs to be done on quantitatively comparing methods and on the estimation of hyperparameters.

5.3   Contour detection

We will now move to a high-level model for detecting the outline of the artery wall. A low-level procedure producing, for example, a segmentation of Fig. 3(a) would not be robust to image artefacts such as the missing edge at the lower right of the artery wall. The edge is missing because of strong reverberations at the parts orthogonal to the incoming ultrasound pulse, and an artefact of this kind is best dealt with using high-level models that constrain the solution to lie within some predefined set, in this case the set of smooth, closed, non-intersecting curves. Referring to Section 2.1.3, we model the artery outline $\mathbf{e}$ as the result of applying an edge transformation matrix $\mathbf{s}$ to a predefined circular template $\mathbf{e}^0$. To model the fact that data are missing along parts of the artery wall, we use a destructive deformation field acting on the template and indicating along which edges data are missing, see Rue and Husby (1998) for details. The deformation field is assumed a priori independent of $\mathbf{s}$, and given a simple Ising prior.

Having an explicit model for the artery wall, we no longer need the implicit edge model (5.1) but models for the two log-variance fields are still needed. We model these as two smooth Gaussian fields, ν_0 and ν_1, associated with back and foreground, respectively. This is somewhat akin to the approach taken in Qian *et al.* (1996). The fields are only observed within their respective regions; letting $\mathbf{d} = \{d_i \, ; \, i \in \mathcal{I}\}$ be an indicator variable having value 1 inside the template and 0 otherwise, the conditional distribution for the radio frequency signal x_i at site i is Gaussian with mean zero and variance $\exp{(2\nu_{d_i,i})}$. We assume the two log-variance fields have unknown spatially constant means, unknown variance, and an exponential correlation function with unknown range. We use GMRF approximations as discussed in Section 2.1.1 to derive a computationally efficient model. The fit is excellent even for a 5×5 neighbourhood. The full posterior density can be sampled using a combination of Gibbs and Metropolis–Hastings updates (Husby 2001), where we make use of the fast sampling algorithms available for GMRF to construct block updates of the variance fields.

Figure 3(c) shows an estimate of the outline of the artery with a fixed number of vertices. Note how well the artefact at the lower right of the artery wall is recovered using knowledge of the global shape of the carotid artery. Furthermore, estimates and credibility intervals of global statistics such as the area of the artery are directly available (we obtained $[2851, 2877]$ as the 95% credibility interval for this area). The results are quite stable for different starting values and choices of hyperparameters. More work is needed to investigate the properties of the model and the sampling algorithm, and their practical properties over a wider range of similar images.

Acknowledgements

We thank H. Tjelmeland for providing Fig. 1, Ø. Skare for providing Fig. 2(a), (b), and G. Nicholls for providing Fig. 2(c), (d). We would also like to thank the HSSS network for financial and scientific support.

References

Amit, Y., Grenander, U., and Piccioni, M. (1991). Structural image restoration through deformable templates. *Journal of the American Statistical Association*, **86**, 376–87.

Andersen, K. E., Brooks, S. P., and Hansen, M. B. (2001). A Bayesian approach to crack detection in electrically conducting media. *Inverse Problems*, **17**, 121–36.

Arak, T., Clifford, P., and Surgailis, D. (1993). Point based polygonal models for random graphs. *Advances in Applied Probability*, **25**, 348–72.

Baddeley, A. J. (1992). Errors in binary images and a L^p version of the Hausdorff metric. *Nieuw Archief voor Wiskunde*, **10**, 157–83.

Baddeley, A. J. and Van Lieshout, M. N. M. (1993). Stochastic geometry models in high–level vision. In *Statistics and Images*, Vol. 20 (eds K. V. Mardia and G. K. Kanji), pp. 235–56. Carfax Publishing, Abingdon.

Bakircioglu, M., Grenander, U., Khaneja, N., and Miller, M. I. (1998). Curve matching on brain surfaces using Frenet distance metrics. *Human Brain Mapping*, **6**, 329–32.

Besag, J. (1986). On the statistical analysis of dirty pictures (with discussion). *Journal of the Royal Statistical Society*, B, **48**, 259–302.

Besag, J. and Higdon, D. (1999). Bayesian analysis of agricultural field experiments (with discussion). *Journal of the Royal Statistical Society*, B, **61**, 691–746.

Besag, J. and Kooperberg, C. (1995). On conditional and intrinsic autoregressions. *Biometrika*, **82**, 733–46.

Besag, J., York, J., and Mollié, A. (1991). Bayesian image restoration with two applications in spatial statistics (with discussion). *Annals of the Institute of Statistical Mathematics*, **43**, 1–59.

Blake, A. and Isard, M. (1998). *Active Contours*. Springer-Verlag, Berlin.

Celeux, G., Hurn, M., and Robert, C. P. (2000). Computational and inferential difficulties with mixture posterior distributions. *Journal of the American Statistical Association*, **95**, 957–70.

Chen, M.-H., Shao, Q.-M., and Ibrahim, J. (2000). *Monte Carlo Methods in Bayesian Computation*. Springer-Verlag, New York.

Christensen, G. E., Rabbitt, R. D., and Miller, M. I. (1996). Deformable templates using large deformation kinematics. *IEEE Transactions on Image Processing*, **5**, 1435–47.

Clayton, D. G. and Bernardinelli, L. (1992). Bayesian methods for mapping disease risks. In *Small Area Studies in Geographical and Enviromental Epidemiology* (eds J. Cuzick and P. Elliot), pp. 205–20. Oxford University Press, UK.

Clifford, P. and Nicholls, G. (1994). A Metropolis sampler for polygonal image reconstruction. Technical Report, Department of Statistics, Oxford University, UK.

Cootes, T. F. and Taylor, C. J. (1999). A mixture model for representing shape variation. *Image and Vision Computing*, **17**, 567–73.

Cootes, T. F., Taylor, C. J., Cooper, D. H., and Graham, J. (1995). Active shape models — their training and application. *Computer Vision and Image Understanding*, **61**, 38–59.

Cressie, N. A. C. (1993). *Statistics for Spatial Data* (2nd edn). Wiley, New York.

Descombes, X., Morris, R. D., Zerubia, J., and Berthod, M. (1999a). Estimation of Markov random field prior parameters using Markov chain Monte Carlo maximum likelihood. *IEEE Transactions on Image Processing*, **8**, 954–63.

Descombes, X., Stoica, R., and Zerubia, J. (1999b). Two Markov point processes for simulating line networks. In *IEEE International Conference on Image Processing*. Kobe, Japan.

Dryden, I. L., Ippoliti, L., and Romagnoli, L. (2002). Adjusted maximum likelihood and pseudo-likelihood estimation for noisy Gaussian Markov random fields. *Journal of Computational and Graphical Statistics*, **11**, 370–88.

Dupuis, P., Grenander, U., and Miller, M. I. (1998). Variational problems on flows of diffeomorphisms for image matching. *Quarterly of Applied Mathematics*, **LVI**, 587–600.

Forbes, F. and Raftery, A. E. (1999). Bayesian morphology: fast unsupervised Bayesian image analysis. *Journal of the American Statistical Association*, **94**, 555–68.

Frigessi, A. and Rue, H. (1997). Bayesian image classification with Baddeley's delta loss. *Journal of Computational and Graphical Statistics*, **6**, 55–73.

Geman, S. and Geman, D. (1984). Stochastic relaxation, Gibbs distributions, and the Bayesian restoration of images. *IEEE Transactions on Pattern Analysis and Machine Intelligence*, **6**, 721–41.

Geman, D. and Reynolds, G. (1992). Constrained restoration and the recovery of discontinuities. *IEEE Transactions on Pattern Analysis and Machine Intelligence*, **14**, 367–83.

Geman, D. and Yang, C. (1995). Nonlinear image recovery with half-quadratic regularization. *IEEE Transactions on Image Processing*, **4**, 923–45.

Geyer, C. (1991). Markov chain Monte Carlo maximum likelihood. In *Computing Science and Statistics: Proceedings of the Twenty-third Symposium on the Interface* (ed. E. M. Keramidas), pp. 156–63. Interface Foundation of North America, Fairfax Station, VA.

Geyer, C. and Thompson, E. (1992). Constrained Monte Carlo maximum likelihood for dependent data. *Journal of the Royal Statistical Society*, B, **54**, 657–99.

Glasbey, C. A. (1998). Ultrasound image segmentation using stochastic templates. *Journal of Computing and Information Technology*, **6**, 107–16.

Glasbey, C. A. and Mardia, K. V. (1998). A review of image-warping methods. *Journal of Applied Statistics*, **25**, 155–71.

Glasbey, C. A. and Mardia, K. V. (2001). A penalised likelihood approach to image warping (with discussion). *Journal of the Royal Statistical Society*, B, **63**, 465–92.

Goodman, J. (1975). Statistical properties of laser speckle patterns. In *Laser Speckle and Related Phenomena* (ed. J. Dainty). Springer-Verlag, Berlin.

Green, P. J. (1990). Bayesian reconstruction from emission tomography data using a modified EM algorithm. *IEEE Transactions on Medical Imaging*, **9**, 84–93.

Green, P. J. (1995). Reversible jump MCMC computation and Bayesian model determination. *Biometrika*, **82**, 711–32.

Grenander, U. (1967). Toward a theory of patterns. In *Symposium on Probability Methods in Analysis*, Loutraki, Greece. Springer-Verlag, Berlin.

Grenander, U. (1976). *Lectures in Pattern Theory, Vol. 1: Pattern Synthesis.* Springer-Verlag, Berlin.

Grenander, U. (1978). *Lectures in Pattern Theory, Vol. 2: Pattern Analysis.*

Springer-Verlag, Berlin.

Grenander, U. (1981). *Lectures in Pattern Theory, Vol. 3: Regular Structures.* Springer-Verlag, Berlin.

Grenander, U. and Manbeck, K. M. (1993). A stochastic model for defect detection in potatoes. *Journal of Computational and Graphical Statistics*, **2**, 131–51.

Grenander, U. and Miller, M. I. (1994). Representations of knowledge in complex systems (with discussion). *Journal of the Royal Statistical Society*, B, **56**, 549–603.

Grenander, U. and Miller, M. I. (1998). Computational anatomy: an emerging discipline. *Quarterly of Applied Mathematics*, **LVI**, 617–94.

Grenander, U., Chow, Y., and Keenan, D. M. (1991). *Hands: A Pattern Theoretic Study of Biological Shapes*, Research Notes on Neural Computing. Springer-Verlag, Berlin.

Hansen, M. B., Møller, J., and Tøgersen, F. A. (2002). Bayesian contour detection in a time series of ultrasound images through dynamic deformable template models. *Biostatistics*, **3**, 213–28.

Heikkinen, J. and Arjas, E. (1998). Non-parametric Bayesian estimation of a spatial Poisson intensity. *Scandinavian Journal of Statistics*, **25**, 435–50.

Higdon, D. M. (1998). Auxiliary variable methods for Markov chain Monte Carlo with applications. *Journal of the American Statistical Association*, **93**, 585–95.

Higdon, D. and Yamamoto, S. (2001). Estimation of the head sensitivity function in scanning magnetoresistance microscopy. *Journal of the American Statistical Association*, **96**, 785–93.

Higdon, D. M., Bowsher, J. E., Johnson, V. E., Turkington, T. G., Gilland, D. R., and Jaszczak, R. J. (1997). Fully Bayesian estimation of Gibbs hyperparameters for emission computed tomography data. *IEEE Transactions on Medical Imaging*, **16**, 516–26.

Hobolth, A. and Jensen, E. (2000). Modelling stochastic changes in curve shape, with an application to cancer diagnostics. *Advances in Applied Probability (SGSA)*, **32**, 344–62.

Hobolth, A., Kent, J. T., and Dryden, I. L. (2002). On the relationship between edge and vertex modelling. *Scandinavian Journal of Statistics*, **29**, 355–74.

Hokland, J. (1995). *Speckle Reduction, Restoration, and Volume Visualization in Medical Ultrasonics*. PhD thesis, Department of Mathematics, University of Bergen, Norway.

Hurn, M. A. (1998). Confocal fluorescence microscopy of leaf cells: an application of Bayesian image analysis. *Journal of the Royal Statistical Society*, C, **47**, 361–77.

Hurn, M., Justel, A., and Robert, C. P. (2002). Estimating mixtures of regressions. *Journal of Computational and Graphical Statistics*. To appear.

Hurn, M., Steinsland, I., and Rue, H. (2001). Parameter estimation for a deformable template model. *Statistics and Computing*, **11**, 337–46.

Husby, O. (2001). Bayesian image analysis of medical ultrasound images. Technical Report, Department of Mathematical Sciences, Norwegian University of Science and Technology, Trondheim, Norway.

Husby, O., Lie, T., Langø, T., Hokland, J., and Rue, H. (2001). Bayesian 2-d deconvolution: a model for diffuse ultrasound scattering. *IEEE Transactions on Ultrasonics, Ferroelectronics and Frequency Control*, **48**, 121–30.

Jain, A. K., Zhong, Y., and Lakshmanan, S. (1996). Object matching using deformable templates. *IEEE Transactions on Pattern Analysis and Machine Intelligence*, **18**, 267–78.

Jalobeanu, A., Blanc-Feraud, L., and Zerubia, J. (1999). Hyperparameter estimation for satellite image restoration by a MCMCML method. *Proceedings of EMMCVPR 99*, York, UK.

Kent, J. T., Dryden, I. L., and Anderson, C. R. (2000). Using circulant symmetry to model featureless objects. *Biometrika*, **87**, 527–44.

Knorr-Held, L. and Rue, H. (2002). On block updating in Markov random field models for disease mapping. *Scandinavian Journal of Statistics*, **29**, 597–614.

Künsch, H. (1987). Intrinsic autoregressions and related models on the two-dimensional lattice. *Biometrika*, **74**, 517–24.

Langø, T., Lie, T., Husby, O., and Hokland, J. (2001). Bayesian 2-d deconvolution: effect of using spatially invariant point spread functions. *IEEE Transactions on Ultrasonics, Ferroelectronics and Frequency Control*, **48**, 131–41.

Møller, J. and Skare, Ø. (2001). Bayesian image analysis with coloured Voronoi tesselations and a view to applications in reservoir modelling. *Statistical Modelling*, **1**, 213–32.

Møller, J. and Waagepetersen, R. (1996). Markov connected component fields. *Advances in Applied Probability*, **28**, 340.

Nicholls, G. (1997). Coloured continuum triangulation models in the Bayesian analysis of two dimensional change point problems. Technical Report, Mathematics Department, The University of Auckland, NZ.

Nicholls, G. (1998). Bayesian image analysis with Markov chain Monte Carlo and coloured continuum triangulation models. *Journal of the Royal Statistical Society*, B, **60**, 643–59.

Nicholls, G. and Fox, C. (1998). Prior modelling and posterior sampling in impedance imaging. In *Bayesian Inference for Inverse Problems, Proceedings of SPIE*, Vol. 3459 (ed. A. Mohammad-Djafari), pp. 116–237. SPIE, Auckland, NZ.

Perrin, O. and Senoussi, R. (1999). Reducing non-stationary stochastic processes to stationarity by a time deformation. *Statistics and Probability Letters*, **43**, 393–407.

Qian, W., Titterington, D. M., and Chapman, J. N. (1996). An image analysis problem in electron microscopy. *Journal of the American Statistical Association*, **91**, 944–52.

Rue, H. (1995). New loss functions in Bayesian imaging. *Journal of the American Statistical Association*, **90**, 900–8.

Rue, H. (1997). A loss function model for the restoration of grey level images. *Scandinavian Journal of Statistics*, **24**, 103–14.

Rue, H. (1999). Baddeley's delta metric. In *Encyclopedia of Statistical Sciences*, Vol. 3 (eds S. Kotz, C. B. Read, and D. L. Banks), pp. 158–62. Wiley, New York.

Rue, H. (2001). Fast sampling of Gaussian Markov random fields with applications. *Journal of the Royal Statistical Society*, B, **63**, 325–38.

Rue, H. and Hurn, M. A. (1999). Bayesian object identification. *Biometrika*, **86**, 649–60.

Rue, H. and Husby, O. K. (1998). Identification of partly destroyed objects using deformable templates. *Statistics and Computing*, **8**, 221–8.

Rue, H. and Syversveen, A. R. (1998). Bayesian object recognition with Baddeley's delta loss. *Advances in Applied Probability (SGSA)*, **30**, 64–84.

Rue, H. and Tjelmeland, H. (2002). Fitting Gaussian Markov random fields to Gaussian fields. *Scandinavian Journal of Statistics*, **29**, 31–50.

Sampson, P. D. and Guttorp, P. (1992). Nonparametric estimation of non-stationary spatial covariance structure. *Journal of the American Statistical Association*, **87**, 108–19.

Stoica, R., Descombes, X., and Zerubia, J. (2000). A Markov point process for road extraction in remote sensed images. Technical Report 3923, INRIA.

Tjelmeland, H. (1996). *Stochastic Models in Reservoir Characterization and Markov Random Fields for Compact Objects*. PhD thesis, Norwegian University of Science and Technology, Trondheim, Norway.

Tjelmeland, H. and Besag, J. (1998). Markov random fields with higher order interactions. *Scandinavian Journal of Statistics*, **25**, 415–33.

Trouvé, A. (1998). Diffeormorphism groups and pattern matching in image analysis. *International Journal of Computer Vision*, **28**, 213–21.

Weir, I. S. (1997). Fully Bayesian reconstructions from single-photon emission computed tomography. *Journal of the American Statistical Association*, **92**, 49–60.

Younes, L. (1998). Computable elastic distances between shapes. *SIAM Journal on Applied Mathematics*, **58**, 565–86.

Zhu, S. C. and Mumford, D. (1997). Prior learning and Gibbs reaction-diffusion. *IEEE Transactions on Pattern Analysis and Machine Intelligence*, **19**, 1236–50.

Zhu, S. C., Wu, Y. N., and Mumford, D. (1998). Filters, random fields and maximum entropy (FRAME): towards an unifying theory for texture modelling. *International Journal of Computer Vision*, **27**, 107–26.

10A
Probabilistic image modelling

M. N. M. van Lieshout
Centrum voor Wiskunde en Informatica, Amsterdam, The Netherlands

1 Introduction

Since the pioneering work of Besag, Cross, and Jain, the Gemans, Ripley, and others (cf. the seminal papers reprinted in the monograph edited by Mardia and Kanji (1993)), Markov random fields have played a prominent role in statistical image analysis. At first, most attention was paid to tackling low-level problems like classification and segmentation with simple categorical or continuously valued pairwise interaction priors or Gaussian autoregression models with small neighbourhood systems. Gradually though it was realized that the above-mentioned models, despite their appealing simplicity and their relatively plausible local characteristics, which form the basic ingredient of most low-level algorithms, may have quite unrealistic global properties. Indeed, ensembles of connected geometrical shapes with smooth boundaries are unlikely to arise as realizations of such discrete random fields.

Accordingly, there has been a surge of interest in the construction of stochastic models with a more satisfying global behaviour. Hurn *et al.* mention several examples, notably the higher-order models of Tjelmeland and Besag (1998), and the Gaussian Markov random fields studied in Rue (2001) and Rue and Tjelmeland (2002). In Section 2 below, we present some more recent advances in this direction.

Another interesting development in the last decade has been the shift in emphasis towards high-level vision tasks that aim to understand and interpret an image in terms of the objects it contains. By their very nature, local Markov random fields are not well-suited as prior distributions for object scenes; far more natural are the Markov marked point processes suggested by Baddeley and Van Lieshout (1992); see also the papers collected in Mardia and Kanji (1993). In Section 3 we provide further details and point out some analogies with the region-based random fields discussed in Section 2.

2 Pixel-level modelling

Hurn *et al.* note that the Potts model and many other classical categorical models 'do not provide good prior models for real scenes, and estimates of attributes such as the number of connected components may be poor'. To tackle these problems, Møller and Waagepetersen (1998) introduce so-called Markov connected component fields, the probability density of which factorizes into terms associated with the connected components in the image rather than with cliques of neighbouring pixels. They prove that under an additional homogeneity and Markov condition with respect to horizontal and vertical neighbours, the probability density can be written as a product of two terms, one governing the area, the other the perimeter of the image components. For the second-order neighbourhood system, terms for the Euler–Poincaré characteristic and the numbers of corners and discontinuities must be added.

Another way to control the size and shape of regions is by means of morphological operators (Serra 1982). For instance, in the binary case, Chen and Kelly (1992) suggest a probabilistic model to favour images that are morphologically smooth in the sense that its foreground set does not feature narrow isthmuses, small islands, or sharp capes. A similar random field for the background pixels penalizes small holes. It can be shown that both models are Markov with respect to a neighbourhood system that depends on the underlying morphological operator. More generally, Sivakumar and Goutsias (1997) use a combination of operators to favour certain shapes and sizes of the foreground and background regions over others.

In summary, the higher-order models in Tjelmeland and Besag (1998), Markov connected component fields, and morphologically constrained random fields all offer some control over the global appearance of likely images, without sacrificing the local character of the full conditionals.

3　High-level modelling

As indicated in Section 1, the class of marked point processes provides a natural framework for object scenes. In such a setup (Baddeley and Van Lieshout 1992, 1993), a point represents the position of an object in the image, its mark captures other attributes. The latter may be a real-valued vector of a few size, shape, and texture parameters as in Descombes *et al.* (1999b); for more complex objects a deformable template mark may be more appropriate (Grenander and Miller 1994).

Hurn *et al.* discuss in detail a range of stochastic shape models, but pay little attention to the interactions between objects. As in low-level image analysis, it is highly desirable from a computational point of view that the conditional dependence structure be local. To quantify this notion, Baddeley and Van Lieshout (1992) assign to each marked point its 'silhouette' in the image, which may be thought of as the discrete representation of the actual object, and define two marked points to be neighbours if their silhouettes overlap. If the conditional intensity of finding an object at some given location with a given mark depends only on this object's neighbours, then the model is called a Markov overlapping object process. A simple, but nevertheless very useful, example is the hard core process (Baddeley and Van Lieshout 1993; Hurn 1998; Rue and Syversveen 1998) consisting of a Poisson number of independent objects conditioned on the event that all silhouettes are disjoint. Alternatively, the amount of overlap may be taken into account as in the area-interaction process (Baddeley and Van Lieshout

1995) defined by an unnormalized probability density that is exponential in the total area occupied by the silhouettes. Occlusion can be formalized by ordering scenes according to which objects lie on top of others (Mardia *et al.* 1997).

Various generalizations have been proposed recently. Considering the fact that images are discrete reflections of a continuous reality, it is especially gratifying to note that the models discussed in Section 2 have analogues in the class of Markov overlapping object processes. For instance, the quermass interaction processes (Mecke 1996; Kendall *et al.* 1999) form an exponential family in the plane with the area, perimeter, and Euler–Poincare characteristic as canonical sufficient statistics. In contrast to connected component fields (Møller and Waagepetersen 1998), in the continuous case some care has to be taken to ensure the model is well-defined (Kendall *et al.* 1999). The point process counterparts of morphologically constrained random fields (Chen and Kelly 1992; Sivakumar and Goutsias 1997) are studied in Van Lieshout (1999). Since the density of both models mentioned above depends on the silhouette image only, a corresponding random set is readily defined (Van Lieshout 2000).

4 Concluding remarks

In this note we have concentrated on models. However, no real progress would have been possible without the simultaneous development of efficient sampling and parameter estimation schemes. As remarked by Hurn *et al.*, perhaps the 'most notable development ... has been the use of Markov chain Monte Carlo maximum likelihood (MCMCML)' (Geyer and Thompson 1992) at the expense of the pseudo-likelihood method, Monte Carlo Newton–Raphson techniques, and stochastic approximation algorithms; see Geyer (1999) for a comprehensive overview. The MCMCML approach focuses on the full likelihood surface, thus allowing for the computation of statistics such as the Fisher information without extra computational effort. The method is easily embedded in a fully Bayesian inference scheme, and an asymptotic theory is available for the Monte Carlo error with respect to the 'true' maximum likelihood estimator. Recently, similar asymptotics have been developed for a combination of the EM algorithm and stochastic approximation (Delyon *et al.* 1999). However, careful tuning of the discount factors involved in the approximation part remains necessary.

Another exciting development is that of exact (or perfect) simulation, following the seminal paper by Propp and Wilson (1996). In contrast to classical MCMC techniques, an exact simulation algorithm outputs an unbiased sample from the target distribution, and neither requires a burn-in time nor sufficiently large time lags between sub-sampled states.

Finally, I would like to congratulate Merrilee Hurn, Oddvar Husby, and Håvard Rue on an interesting overview of recent developments in image analysis, and express the hope that their article will act as a stimulus to further progress.

10B
Prospects in Bayesian image analysis

Alain Trubuil
Institut National de la Recherche Agronomique, Jouy-en-Josas, France

1 Introduction

The ideas presented by Hurn *et al.* offer an excellent review of the impressive work carried out by the statisticial community on the Bayesian framework and its application to image analysis. Necessarily the presentation could not be exhaustive and with respect to applications, a consensus has guided the text with modelling illustrated mainly for restoration problems. My aim is to draw attention to detection, labelling, and eventually three-dimensional (3D) segmentation. This will be addressed in Section 3. The first part of the discussion gives a very short presentation of the work done by Zhu *et al.* (1997–2000) on texture modelling and estimation. In Section 2, I will illustrate the usefulness of intensity-level curves as very important clues for image interpretation.

2 Texture modelling

In relating some aspects of the work of Zhu *et al.* on texture modelling, my aim is to provide the reader with recent information on generic texture modelling and parameter estimation that complements Section 2.1 of the article.

Given a class of images (e.g. outdoor images) and a set of filters (e.g. Sobel filters and Gabor filters, the impulse response of which is a complex sinusoid centred at a frequency and modulated by a Gaussian envelope), the aim is to identify a probability distribution p, associated with elements of this class. An estimator of this distribution is obtained from a learning set of M images. Histogram statistics of filter responses are computed as

$$\mu_{\text{obs}}^l(z) = \frac{1}{M} \sum_{m=1}^{M} \#\{(x,y) \in \Lambda;\ F^l(I_m)(x,y) = z\},$$

where $(x,y) \in \Lambda$ denotes a pixel of the lattice Λ underlying the images; and $F^l(I_m)(x,y)$ is the filter response of the lth filter at position (x,y) when applied to the mth image of the learning set. If one looks for an estimator of maximum entropy with similar histogram statistics:

$$E_{\hat{p}}\left(\sum_{(x,y)} \mathbf{1}\{F^l(I)(x,y) = z\}\right) = \mu_{\text{obs}}^l(z), \quad l = 1,\ldots,L,\ z \in Z^l,$$

then $\hat{p}$ is a Gibbsian distribution: $\hat{p}(I) \propto \exp\{-\sum_l \sum_z \lambda^l(z) H^l(I)(z)\}$ with $H^l(I)(z) = \int \int \mathbf{1}\{F^l(I)(x,y) = z\} \, dx \, dy$.

Of course, the potential functions λ^l (analogous to Lagrange multipliers) depend upon the observed statistics μ^l_{obs}. First of all, two distinct potential functions are identified. At fine scales, V-shape potential functions with rather flat tails occur with derivative filters. At coarse scales, these curves turn 'upside down'. The Gibbs distribution learned from images has energy that one can attempt to minimize by gradient descent. This leads to a non-linear PDE equation of the diffusion–reaction type with two terms. Starting from an input image $I(x, y; 0)$, the first term diffuses the image whereas the second term forms patterns.

An interesting application of texture learning through this framework is clutter removal. Here the problem is eliminating or attenuating the presence of a complex background (e.g. trees, landscape) in an image displaying objects from a class of interest (e.g. buildings). This difficult problem is handled in three stages in Zhu *et al.* (1997):

(a) learning the background images class: $p(I^{\mathbf{b}}) \propto \exp\{-U^{\mathbf{b}}(I^{\mathbf{b}})\}$;

(b) learning the objects of interest images class: $p(I^{\mathbf{o}}) \propto \exp\{-U^{\mathbf{o}}(I^{\mathbf{o}})\}$;

(c) assuming that $I^{\mathrm{obs}} = I^{\mathbf{o}} + I^{\mathbf{b}}$, looking for a MAP estimator of $I^{\mathbf{o}}$ according to $p(I^{\mathbf{o}}|I^{\mathrm{obs}}) \propto \exp\{-U^{\mathbf{b}}(I^{\mathrm{obs}} - I^{\mathbf{o}}) + U^{\mathbf{o}}(I^{\mathbf{o}})\}$.

In order to gain accuracy and speed in the learning process, Zhu *et al.* realize that filter selection and parameter estimation can be separated. They address the selection problem in Zhu *et al.* (1999) and estimation in Zhu *et al.* (2000). In these papers, the reader can find new ideas for parameter estimation in Gibbsian models.

3 Level curves and segmentation

My aim in this section is to draw attention to level curves of filter response (mainly intensity filter) as a carrier of very useful information for image representation and segmentation. This is related to Section 2.2 devoted to intermediate-level modelling and illustrated with tessellations obtained through the simulation of a posterior distribution combining grey-level homogeneity and number or length of edges between regions. The supports for labels of regions are triangles. During the simulation process triangles are merged, split, or distorted with the help of different proposals of a Metropolis–Hastings algorithm. The posterior energy is related to the Mumford and Shah functional widely used in image segmentation. Triangles have a long history in computer science and approximation (e.g. in finite element techniques for partial differential equations) but in image representation other geometric primitives could be considered. Hence, it appears that in many cases, level curves of intensity function are good candidates for boundaries between objects. Actually, as pointed out by Kervrann *et al.* (2000), a subset of level curves is the optimal solution with respect to the energy:

$$U_\lambda(f, \Omega_1, \ldots, \Omega_P) = \sum_{i=1}^{P} \int_{\Omega_i} (f(x) - \bar{f}_{\Omega_i})^2 \, dx + \lambda \sum_{i=1}^{P-1} |\Omega_i|^q, \qquad (3.1)$$

where f denotes the image intensity function, $\bar{f}_{\Omega_i}$ the mean grey-level over object Ω_i with area $|\Omega_i|$, P the unknown number of objects including the background, and $q = -1, 0, 1, 2$. The first term of (3.1) controls homogeneity whereas the second term controls the number of objects when $q = 0$. It is not possible to penalize directly edge lengths while keeping solutions inside the set of solutions whose object boundaries are intensity level curves. However, it is clear that quite often the triangulation and the level curve approach should be able to provide representations similar to each other. An efficient algorithm has been designed for minimizing (3.1). It relies on a coarse discretization of the grey levels. An object-guided smoothing can be incorporated with the help of anisotropic diffusion to restore surfaces and boundaries of objects. The advantage of the approach, described by Nicholls (1997) over the level curve approach, lies in the direct inclusion of edge lengths which is not possible in the level curve framework, even if more or less controllable through penalization and the smoothing trick. The advantage of the level curve approach is mainly due to fast estimation and concordance in many situations of level lines of original or smoothed images with perceptible edges in images. Within the level curve selection approach described herein, different configurations of objects are obtained depending on the grey-level quantitation chosen. My feeling is that this family of configurations could be used by colouring schemes like the one described by Nicholls. This could make it possible to incorporate a priori some geometric on objects.

4   Applications

From a practical point of view, the relevance of the Bayesian framework has been demonstrated by Hurn *et al.* in restoration and object-based segmentation of complex ultrasound images. I would also like to draw attention to an important issue for imaging in biology: 3D microscopy of biological samples.

Let me consider the fine analysis of complex structures (e.g. antennal lobes) or tissues (e.g. intestinal crypts) for which the ultimate objective is comparing samples. Very often these structures or tissues are made of dozens to hundreds of smaller sub-structures (e.g. glomerulii, cells). The structures are partially observed and often very close to each other in a 3D domain, see Fig. 1 displaying an intestinal crypt. By formulating the problem in the right way, all the tricky algorithms designed for probability distributions may be used to obtain some hints on the spatial organization of sub-structures. Let me illustrate this on nuclei detection in a volume observed through an optical section in confocal microscopy. The nuclei detection process we have used can be divided into four stages:

(a) detection of seeds inside objects,

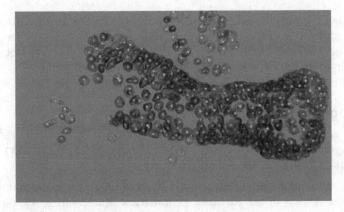

FIG. 1. Volume rendering of nuclei detection of an intestinal crypt.

(b) sub-sampling of seeds,

(c) seed aggregation,

(d) validation.

At the end of stage (b), we have a set $\mathcal{S} = \{s_i,\ i = 1, \ldots, N\}$ of seeds in 3D space. The objective is to associate a unique label, l_i, to each seed according to its membership to a nucleus. Here the number of objects (nuclei) is unknown but assumed bounded by N. Moreover, we expect there is at least one seed per object and few false seeds. We define a graph whose nodes are the seeds. An edge is created between two nodes when they are neighbours. In practice, the a priori knowledge about nucleus size is used to define neighbours, ν_i, of a seed, s_i, ν_i consisting of seeds that are no more than a given distance from the considered seed. The probability distribution of labels is chosen as

$$p(L = (l_1, \ldots, l_N)) \propto \exp\left\{ -\sum_{i=1}^{N} \sum_{j \in \nu_i} U_{i,j}(l_i, l_j) \right\},$$

with $U_{i,j}(l_i, l_j) = (\exp\{-Q(s_i, s_j)\} - e^{-1} \exp Q(s_i, s_j)) \mathbf{1}\{l_i = l_j\}$. The quantity $Q(s_i, s_j)$ is related to image content in a local region between s_i and s_j. If one knows that objects are convex, then a low value of Q is associated with the presence of a rupture in the grey-level intensity profile between s_i and s_j. So the potential Q expresses the relationship between parameters and data. This model is similar to an inhomogeneous Potts model on a graph. An estimate of a mode of p can be searched for by a relaxation algorithm or a more efficient algorithm like the one proposed by Boykov *et al.* (1999). Moreover, constraints can be introduced which may reduce energy computation towards plausible configurations. This has been done and is illustrated in Fig. 1 which depicts a volume rendering visualization of the crypt with incrustation of small spheres for each centre of

330 *Prospects in Bayesian image analysis*

inertia of homogeneous label subsets. Even if these results are encouraging, they are not completely satisfactory. Indeed, validation is a hard task when hundreds of labels have to be examined from 2D slices of a 3D volume. My feeling is that efficient simulation tools can provide us with relevant validation clues, detecting regions of interest in the data where visual inspection should be encouraged.

Additional references in discussion

Baddeley, A. J. and Van Lieshout, M. N. M. (1992). ICM for object recognition. In *Computational Statistics*, No. 2 (eds Y. Dodge and J. Whittaker), pp. 271–86. Physica/Springer, Heidelberg/New York.

Baddeley, A. J. and Van Lieshout, M. N. M. (1995). Area-interaction point processes. *Annals of the Institute of Statistical Mathematics*, **47**, 601–19.

Boykov, Y., Veksler, O., and Zabih, R. (1999). Fast approximate energy minimization via graph cuts. *Proceedings of the Seventh IEEE International Conference on Computer Vision*, Kerkyra, Greece.

Chen, F. and Kelly, P. A. (1992). Algorithms for generating and segmenting morphologically smooth binary images. *Proceedings of the Twenty-sixth Conference on Information Sciences*, Princeton.

Delyon, B., Lavielle, M., and Moulines, E. (1999). Convergence of a stochastic approximation version of the EM algorithm. *Annals of Statistics*, **27**, 94–128.

Geyer, C. J. (1999). Likelihood inference for spatial point processes. In *Proceedings Seminaire Européen de Statistique, Stochastic Geometry, Likelihood, and Computation* (eds O. Barndorff-Nielsen, W. S. Kendall, and M. N. M. Van Lieshout), pp. 79–140. CRC Press/Chapman & Hall, Boca Raton.

Kendall, W. S., Van Lieshout, M. N. M., and Baddeley, A. J. (1999). Quermass-interaction processes: conditions for stability. *Advances in Applied Probability (SGSA)*, **31**, 315–42.

Kervrann, C., Hoebeke, M., and Trubuil, A. (2000). Level lines as global minimizers of energy functionals in image segmentation. In *European Conference on Computer Vision*, Lecture Notes in Computer Science, No. 1843, pp. 241–56. IEEE Computer Society.

Van Lieshout, M. N. M. (1999). Size-biased random closed sets. *Pattern Recognition*, **32**, 1631–44.

Van Lieshout, M. N. M. (2000). *Markov Point Processes and their Applications*. Imperial College Press/World Scientific Publishing, London/Singapore.

Mardia, K. V. and Kanji, G. K. (eds) (1993). *Statistics and Images*, Vol. 1. Carfax Publishing, Abingdon.

Mardia, K. V., Qian, W., Shah, D., and de Souza, K. (1997). Deformable template recognition of multiple occluded objects. *IEEE Transactions on Pattern Analysis and Machine Intelligence*, **19**, 1036–42.

Mecke, K. R. (1996). Morphological model for complex fluids. *Journal of Physics: Condensed Matter*, **8**, 9663.

Møller, J. and Waagepetersen, R. (1998). Markov connected component fields. *Advances in Applied Probability (SGSA)*, **30**, 1–35.

Propp, J. G. and Wilson, D. B. (1996). Exact sampling with coupled Markov chains and applications to statistical mechanics. *Random Structures and Algorithms*, **9**, 223–52.

Serra, J. (1982). *Image Analysis and Mathematical Morphology*. Academic Press, London.

Sivakumar, K. and Goutsias, J. (1997). Morphologically constrained discrete random sets. In *Advances in Theory and Applications of Random Sets* (ed. D. Jeulin), pp. 49–66. World Scientific Publishing, Singapore.

Zhu, S., Liu, X., and Wu, Y. (1999). Statistics matching and model pursuit by efficient MCMC – A conclusion to Julesz's texture quest? *International Conference on Computer Vision*, Kerkyra, Greece.

Zhu, S. and Liu, X. (2000). Learning in Gibbsian fields: how accurate and how fast can it be? *Conference on Computer Vision and Pattern Recognition*, Hilton Head Island, South Carolina.

11

Preventing epidemics in heterogeneous environments

Niels G. Becker and Sergey Utev

Australian National University, Canberra, Australia

1 Introduction

Models that describe the transmission of infection have a number of uses. They underpin analyses of transmission data. They are also used to translate our understanding about transmission of infection between individuals to a description of transmission through the community. Of particular importance is how models are employed as an aid for predicting the likely consequences of different interventions in the transmission process. In most sciences interventions are compared by conducting planned experiments. It is usually not acceptable to conduct planned experiments involving outbreaks of an infectious disease, and therefore mathematical models become surrogates for planned experiments.

The most effective means of controlling some infectious diseases, such as measles and rubella, is by vaccinating community members. The assumptions made when using epidemic models to guide decisions about the control of transmission by vaccination are usually oversimplified, which makes public health professionals skeptical of insights gained from these models. In particular, it is often assumed that individuals are homogeneous and mix uniformly, and that conditions are constant over time. For the greater acceptance of insights gained from epidemic models, it is important to understand the consequences of weakening these assumptions. Previous work has allowed for some heterogeneity. For example, consideration has been given to the effect of different rates of mixing between different age groups and allowing the community to consist of a collection of households. Here we look at these and other types of heterogeneity, focusing on the effect of this heterogeneity on three parameters that are crucial to the problem of controlling transmission.

One parameter is the epidemic threshold parameter whose value indicates when herd immunity is achieved. An epidemic threshold parameter is a parameter such that the probability of a major epidemic is positive if, and only if, its value is greater than 1. Another parameter considered is the critical vaccination coverage, which is the proportion of the community that must be vaccinated to prevent epidemics or endemic transmission. In other words, what fraction of the community must be vaccinated to bring the threshold parameter below 1. The third parameter measures vaccine efficacy; that is, it measures how effectively the vaccine controls transmission. These are not distinct parameters. For example, the critical vaccination coverage depends on the epidemic threshold parameter and the vaccine efficacy.

Estimating parameters of transmission models is complicated by the fact that a relatively small part of the transmission process is observed. Expressions for the likelihood function typically involve many integrals over unobserved quantities, so that maximum likelihood estimation is often impractical unless unrealistic simplifying assumptions are made. This has led to the use of alternative methods of estimation, for example the use of estimating equations (Becker 1993). The fact that the likelihood generally admits an explicit expression when complete observation is possible suggests that methods of inference based on data augmentation can play an important role in this area. For example, applications of the expectation-maximization (EM) algorithm to facilitate maximum likelihood estimation of transmission and disease progression parameters are reviewed by Becker (1997). As conditional expectations needed for the EM algorithm are not always available, the search for methods of analysis under more realistic assumptions has recently focused on Bayesian analyses using Markov chain Monte Carlo (MCMC) methods, with unobserved data being treated as latent variables; see O'Neill and Roberts (1999), O'Neill *et al.* (2000) and O'Neill and Becker (2001). There is quite a lot of information about characteristics of common diseases from a variety of studies and, in view of the lack of detail in infectious disease data, it is important to use all available data. A Bayesian analysis brings with it the advantage of being able to incorporate prior information about parameters. In some situations this may estimate parameters more precisely than estimation based on the data set alone. Bayesian analysis also facilitates the combination of different sources of data that are sometimes available; see Farrington *et al.* (2001).

For our task of estimating parameters that are crucial for planning the control of infectious diseases, and associated modelling, it is useful to distinguish two types of situation with heterogeneity among individuals. One situation is when differences between individuals occur for *identifiable* types of individual. For example, we might need to distinguish between individuals from different age-groups because the mixing rates of school children differ from the mixing rates of other community members. For this form of heterogeneity it is useful to stratify individuals into types, since type-specific data can be observed. The other situation is when we cannot identify the types that differ. For example, individuals might have immune systems that differ in their capacity to control the invading infectious agent, resulting in differences in how susceptible they are and how infectious they become when infected. Stratification is less useful in this situation and, instead, we describe such heterogeneity by an unobservable random variable, whose distribution contains parameters that need to be estimated from available data.

Control programmes are generally planned for large communities. Accordingly, all models of transmission considered here are for epidemics in a large community. In Section 2, we describe a transmission model for a community of homogeneous individuals who mix uniformly. This model is the benchmark against which comparisons are made. In Section 3, individuals are allowed to differ in susceptibility and how infectious they become, covering the two cases where individuals who differ are distinguishable and where they are not. We also introduce differences arising from vaccination, which involve both distinguishable differences, namely being vaccinated or not vaccinated, and indistinguishable differences, for example responding to vaccination in unobserved ways. In Section 4 we consider heterogeneity arising from community structure, due to individuals being clustered into groups such as locations or households. In Section 5 we briefly look at the effects of networks of contacts and variations over time.

Dietz (1993) reviews some settings for which an epidemic threshold parameter is known and considers how it can be estimated in those settings. This article is in the same spirit, but we consider different settings. We also pay some attention to the critical vaccination coverage and vaccine efficacy, because these concepts are central to the problem of preventing epidemics.

2 The standard epidemic model

An infection occurs when a pathogen is passed from one individual to another and then reproduces within the infected individual. A transmission process is generated by infected hosts passing the pathogen on to other individuals. We assume that, initially, individuals are susceptible to infection. Over time they may become infected, whereupon their immune system reacts to the invading organism, permitting the infected individual to recover after some time. We assume that upon recovery, taken as the time when the infectious period ends, the individual acquires immunity from further infection for the remainder of the study period. This is the case for many infections, such as chickenpox, measles, and rubella. A recovered individual is said to be *removed*, to indicate that he no longer contributes to the transmission of infection. Note that individuals in the removed state continue to mix within the community. Models that assume Susceptible→Infected→Removed transitions are referred to as S–I–R models.

2.1 Model formulation

The extent to which an infective sheds a pathogen depends on the time since infection. To reflect this, we associate with each infected person a non-negative infectiousness function β_x, $x \geqslant 0$, where β_x measures how contagious an infected individual is x time units after being infected, see Becker (1989, Chapter 3). Our standard model assumes that each infected individual has exactly the same infectiousness function, although we make no specification about its shape. In other words, individuals are homogeneous as far as transmission of infection is concerned. The standard model assumes that individuals mix uniformly with each other. The infectious period is defined as $\{x\colon \beta_x > 0\}$, the time period during which the infected individual is capable of infecting others.

Choose the time when the first community member is infected, by an external contact, as the time origin and consider the evolution of the epidemic from that point on. Suppose that n, the size of the community, is large and that the population is closed. The probability that an individual, who was infected at time y, transmits the pathogen to any *given* susceptible individual in the small time increment $(t, t + dt)$ is assumed to be $n^{-1}\beta_{t-y}\,dt + o(dt)$. The term n^{-1} arises from the uniform mixing assumption, which means that the infection intensity is spread evenly over the n community members. Let S_t be the number of individuals who are still susceptible to infection at time t and N_t the total number of individuals infected in $[0, t]$. The number of susceptible individuals remaining at the end of the epidemic is denoted S_∞. The history of infections up to time t is denoted $\mathcal{H}_t = \sigma(N_x\,;\,0 \leqslant x \leqslant t)$. In terms of $dN_t = N_{t+dt} - N_t$, with $dN_{0-} = 1$ to include the initial infective, the infection process is defined by

$$\Pr(dN_t = 1 \mid \mathcal{H}_t) = 1 - \Pr(dN_t = 0 \mid \mathcal{H}_t) + o(dt) = \int_{0-}^{t} \beta_{t-y}\,dN_y\,\frac{S_t}{n}\,dt + o(dt).$$

2.2 A threshold parameter

A natural epidemic threshold parameter for this epidemic model is R_v, the initial reproduction number of infectives, defined as the mean number of individuals that an initial infective generates by direct contact if a fraction v of community members are initially immune and a fraction $s_0 = 1 - v$ are susceptible. The number of infections generated by an initial infective, in this large community, is a Poisson variate with mean $R_v = (1 - v)\int_0^\infty \beta_x\,dx$. The epidemic threshold result states that a major outbreak occurs with positive probability if, and only if, $R_v > 1$. Intuitively, this means that an epidemic can occur only if, on average, initial infectives more than replace themselves. More specifically, the eventual proportion of susceptible individuals,

$$\lim_{n \to \infty} \frac{S_\infty}{n} = s_\infty,$$

is strictly less than s_0 with positive probability if, and only if, $R_v > 1$ and is then given by

$$s_\infty = s_0 \exp[-R_0(s_0 - s_\infty)], \tag{2.1}$$

where $R_0 = \int_0^\infty \beta_x\,dx$, the *basic* reproduction number, is the initial reproduction number when everyone is initially susceptible. Equation (2.1) also arises from the corresponding deterministic epidemic model.

2.3 Critical immunity coverage

To illustrate how the threshold parameter is used to give insights into the control of epidemics, suppose that we can render a fraction v of the population immune by vaccination, say. Then, each time an initial infective makes a contact, close enough for transmission to occur, the probability that this contact is with a susceptible individual is $1 - v$. The initial reproductive number for infectives in this partially immunized community is $R_v = (1-v)R_0$. Epidemics are prevented if $R_v \leqslant 1$, that is when

$$v \geqslant (1 - 1/R_0)^+ = \text{CIC, the critical immunity coverage.} \qquad (2.2)$$

This is the minimum fraction that must be immunized to prevent epidemics and endemic transmission. The CIC is equal to the critical vaccination coverage (CVC) when each vaccination leads to 100% protection against infection.

2.4 Estimation

When the initial proportion of susceptible individuals is known and the eventual proportion of susceptible individuals is observed to be $\hat{s}_\infty$ for a major epidemic, we obtain from (2.1) the estimate of R_0 given by

$$\widehat{R_0} = \frac{\ln s_0 - \ln \hat{s}_\infty}{s_0 - \hat{s}_\infty}. \qquad (2.3)$$

Alternatively, from the martingale

$$M_t = N_t - n^{-1} \int_0^t S_u \int_0^u \beta(u - y) \, dN_y \, du \,,$$

we can derive a new martingale, using the martingale transform dM_t/S_t. By equating the resulting martingale to its mean, namely zero, we obtain an estimating equation. When evaluated at $t = \infty$, so that all the data are used, this estimating equation yields the estimate

$$\widehat{R_0} = \frac{n}{C_\infty} \sum_{i=S_\infty+1}^{S_0} \frac{1}{i} \simeq \frac{\ln s_0 - \ln \hat{s}_\infty}{s_0 - \hat{s}_\infty}, \qquad (2.4)$$

where C_∞ is the eventual total number of cases. Comparison with (2.3) shows that the estimate is similar, but the martingale approach enables us to deduce the standard error (Becker 1993)

$$\text{s.e.}(\widehat{R_0}) = \frac{n}{C_\infty} \left[\sum_{i=S_\infty+1}^{S_0} \frac{1}{i^2} \right]^{1/2} \simeq \left[n s_0 \hat{s}_\infty (s_0 - \hat{s}_\infty) \right]^{-1/2}. \qquad (2.5)$$

For large epidemics a central limit theorem applies and this provides a confidence interval for R_0. By the use of (2.2), and the delta method, we obtain

$$\widehat{\text{CIC}} = 1 - 1/\widehat{R_0} \quad \text{and} \quad \text{s.e.}(\widehat{\text{CIC}}) = \text{s.e.}(\widehat{R_0})/\widehat{R_0} \,,$$

which provides approximate confidence intervals for the critical immunity coverage when we have data on the eventual size of a major outbreak.

2.5   Dependence, heterogeneity, and structure

The standard model assumes that all individuals are identical, as far as transmission is concerned, and that community members mix uniformly with one another. Dependence arises due to the fact that the current force of infection, acting on each susceptible, is determined by the past infections, which are random. This results in some realized outbreaks having few cases and some having many cases. A transmission process is similar to a branching process, but with the complication that the depletion of susceptible individuals limits the growth of the number of infected individuals and thereby introduces additional dependence.

Below we consider ways in which individuals might differ, communities might differ, or conditions might vary over time. As a consequence, an individual's chance of being infected depends not only on how many have been infected, and when, but also on how the infected individuals are 'related' to this individual. This can affect results substantially, making it necessary to develop a methodology for these complex settings. For example, we can no longer simply ask: What is v, the fraction of individuals we need to immunize to prevent major outbreaks? To determine v we must first specify how vaccinees are to be selected from among the different types of individual and household.

3   Variation in infectivity and susceptibility

We begin by allowing some differences between individuals, starting with the situation where we have no knowledge about which individuals might be similar.

3.1   Unidentifiable types

A standard way to allow for unidentified variation is to represent this variation by a random variable.

3.1.1   *Variation in infectivity*

Suppose that the only heterogeneity between individuals consists of differences in their infectiousness function. There are a number of ways this function might vary. Many studies assume that the infectiousness function has the form

$$\beta_x = \left\{ \begin{array}{ll} \beta, & \text{if } 0 \leqslant x \leqslant D, \\ 0, & \text{otherwise,} \end{array} \right. \tag{3.1}$$

where D, the duration of the infectious period, is permitted to differ between individuals. The so-called stochastic general epidemic model is an example, in which D is assumed to have an exponential distribution and all infectious periods are independent. For our purpose we do not need the convenience of a Markovian model and can let D have an arbitrary distribution.

In this setting the basic reproduction number is given by $R_0 = \beta \, \mathrm{E}D$, the mean of the area under the infectiousness function. The martingale estimate

(2.4) for R_0 still applies, however now the standard error becomes

$$\text{s.e.}(\widehat{R_0}) = \frac{n}{C_\infty}\left[\sum_{i=S_t+1}^{S_0}\frac{1}{i^2} + \frac{C_\infty\widehat{\text{var}(\beta D)}}{n^2}\right]^{1/2}$$

$$\simeq \left\{\frac{1}{n(s_0-\hat{s}_\infty)}\left[\frac{1}{s_0\hat{s}_\infty}+\widehat{\text{var}(\beta D)}\right]\right\}^{1/2} \quad (3.2)$$

(Becker 1993). The extra term $\widehat{\text{var}(\beta D)}$ in (3.2), compared with (2.5), arises from variation in the area under the infectiousness function. The term (3.2) can also be obtained by applying the delta method to the central limit theorem of von Bahr and Martin-Löf (1980) for the eventual epidemic size.

Data on the initial number susceptible and eventual number infected are not adequate to estimate both R_0 and $\text{var}(\beta D)$ as separate parameters. If no additional data are available, we need to assume a relationship between the mean and variance of the area under the infectiousness function. For example, when $\text{var}(D) = (ED)^2$, as is the case for an exponentially distributed infectious period, we have $\text{var}(\beta D) = R_0^2$, so that $\widehat{\text{var}(\beta D)} = \widehat{R_0}^2$. One form of additional data that has been used to estimate the variance of the infectious period is data on the times between detection of cases in outbreaks occurring in small households. Data on transmission in small households seems best suited to estimation of $\text{var}(D)$, because they contain information about the source of the infection, as is pointed out by Auranen (discussion 11B in this volume). Estimates based on realistic distributions for the duration of the latent and infectious periods seem to require the flexibility of a Bayesian framework and MCMC methods. The potential of these methods is illustrated in O'Neill *et al.* (2000), who consider the times between cases of measles in households of size two. The variance of the infectious period has also been estimated by MCMC methods applied to data on the removal times of cases of an epidemics in a uniformly mixing community of homogeneous individuals; see O'Neill and Becker (2001).

The estimate (2.4) and the large-outbreak standard error (3.2) apply when the infectiousness function has forms other than (3.1). It seems sufficient to require that the mean and variance of $B_U = \int_0^\infty \beta_x \, \mathrm{d}x$, the total area under the infectiousness function, exist.

3.1.2 *Variation in susceptibility*

We now allow for variation in the susceptibility of individuals, considering first the case where every infective has the same infectiousness function β_x, $x \geqslant 0$. Every individual is furnished with a relative susceptibility, given by an independent realization on a non-negative random variable A. More specifically, x time units after being infected, an infective is assumed to exert a force of infection $a\beta_x/n$ on an individual whose realized value of susceptibility A is a.

The basic reproduction number for this setting is $R_0 = B_U EA$. When a proportion v of individuals are initially immune, the reproduction number of initial infectives is $R_v = (1 - v)R_0$ and therefore the critical immunity coverage still has the form (2.2). However, R_0 is now an epidemic threshold parameter for a setting with variable susceptibility.

The estimation of R_0 when there is variation in susceptibility is not as easy as estimation when only infectiousness varies. The difference stems from the fact that despite variation in infectiousness the mean area under the infectiousness function of those still uninfected does not change over time, while with varying susceptibility the mean susceptibility decreases over time because the more susceptible individuals tend to be infected earlier.

Conditional on a major outbreak, Scalia-Tomba (1990) shows that

$$\sqrt{n}\left(\frac{S_\infty}{n} - s_\infty\right) \xrightarrow{\mathcal{D}} \mathrm{N}(0, \sigma_s^2),\tag{3.3}$$

where

$$s_\infty = s_0 \, \mathrm{E}\exp[-AB_\mathrm{U}(s_0 - s_\infty)] \quad \text{and} \quad \sigma_s^2 = \frac{s_\infty(s_0 - s_\infty)}{s_0(1 - \rho)^2},\tag{3.4}$$

with $\rho = s_0 \, \mathrm{E}\{AB_\mathrm{U}\exp[-AB_\mathrm{U}(s_0 - s_\infty)]\}$. The law of large numbers equation, the first equation in (3.4), does not determine $R_0 = B_\mathrm{U}EA$, in general. However, applying a stochastic comparison argument, as in Becker and Utev (2001), we derive

$$\frac{\ln s_0 - \ln s_\infty}{s_0 - s_\infty} = \underline{R}_0 \leqslant R_0 \leqslant \overline{R}_0 = \frac{\ln s_0 - \ln(s_\infty - \varepsilon)}{s_0 - (s_\infty - \varepsilon)},$$

where $\varepsilon = (s_0 - s_\infty)\,\mathrm{var}(A)/\mathrm{E}^2 A$. These bounds are tight when ε is small. In particular, both bounds are equal to (2.4) when there is no variation in susceptibility. In the presence of variation in susceptibility the estimate (2.4) actually estimates $\underline{R}_0$, a lower bound for R_0. The central limit theorem (3.3) enables inferences to be made about $\underline{R}_0$ and this provides approximate inferences for R_0 when ε is small. A standard error for $\widehat{\underline{R}}_0$, the estimate of the lower bound is, using (3.4) and the delta method, found to be

$$\mathrm{s.e.}(\widehat{\underline{R}}_0) = \frac{1 - \hat{s}_\infty \widehat{\underline{R}}_0}{1 - \rho} \left[n s_0 \hat{s}_\infty(s_0 - \hat{s}_\infty)\right]^{-1/2},\tag{3.5}$$

where $\hat{s}_\infty$ is the observed proportion of eventual susceptible individuals. The expression (3.5) contains the parameter ρ, which is unknown and difficult to estimate. One way of dealing with the unknown ρ is to note that $g(x) = x - \mathrm{E}\exp[s_0 AB_\mathrm{U}(x - 1)]$ gives $1 - \rho = g'(s_\infty/s_0) \geqslant s_\infty(s_0 - s_\infty)/(2s_0^2)$, which may be substituted in (3.5) to give a conservative standard error. Inferences about $\overline{R}_0$ along these lines requires an assumption about $\mathrm{var}(A)/\mathrm{E}^2 A$, or an estimate for it from additional data.

Becker and Yip (1989) are motivated by a smallpox data set to consider parametric estimation for an epidemic model with variable susceptibility. The method relies on additional data, in the form of times of detection of cases, and some simplifying assumptions. They assume that the times of detection mark the ends of the infectious periods and that both the latent period and the infectious period are of constant duration. A gamma distribution is assumed for A. The focus is not on estimating R_0, but such estimation is immediate from the maximum likelihood estimates for $\mathrm{E}A$ and the duration of the infectious period.

When there is variation both in A, the relative susceptibility, and in the infectiousness function, but still uniform social mixing, the epidemic threshold parameter is $R_0 = \mathrm{E}(AB_{\mathrm{U}})$, where B_{U} is the area under the infectiousness function; see Becker and Starczak (1998). It is not possible to estimate R_0 from the eventual size of the epidemic alone. One possible way to estimate R_0 is to use the initial generation sizes. Generation 0 is comprised of the primary infectives, generation 1 consists of those individuals that are infected by direct contacts with the primary infectives, generation 2 consists of those infected by direct contacts with generation 1 infectives, and so forth. Let C_i denote the number of cases in generation i. If the disease has a fairly long latent period and a fairly short infectious period, then it is possible to observe $C_0, C_1, \ldots, C_k$, the initial generation sizes. Becker (1977) suggests $\widehat{R_0} = \sum_{i=1}^{k} C_i \big/ \sum_{j=0}^{k-1} C_j$, which is a consistent estimate of the reproduction number for a branching process approximation to the initial stages of a multitype epidemic. An approach with more practical appeal is that of O'Neill and Becker (2001), who consider parametric estimation for an epidemic model with both variable susceptibility and variable infectiousness, using data on the times of detection of cases. The generality in the model is made possible by basing inferences on MCMC methods. The inferences in O'Neill and Becker (2001) do not focus directly on estimating R_0, but the approach is clearly suited to such inferences.

3.2 Identifiable types

The description of a multitype epidemic is the same irrespective of whether types are identifiable or not. The distinction between identifiable and unidentifiable types becomes important for parameter estimation since data on the number at risk and cases can be provided separately for each type when they are distinguishable. This additional detail in the data provides scope for exploring the differences between types. In the above discussion of variation in infectiousness and susceptibility we assume uniform social mixing. A natural extension to the model when we have identifiable types is to allow different rates of mixing within and between types of individual.

Suppose we have k types, labeled $1, 2, \ldots, k$. Let π_i denote the proportion of all individuals that are of type i and $s_i(0)$ the proportion of all type-i individuals that are initially susceptible. Assume that a small number of infectives enter the large community and a major epidemic occurs that leaves a proportion $s_i(\infty)$ of all type-i individuals susceptible at the end of the epidemic. Let $m_{ij}\pi_j$ denote the mean number of individuals of type j that an infective of type i infects during the course of the infectious period assuming that every individual is susceptible. Suppose individuals mix uniformly apart from the differential rates of mixing implicit in (m_{ij}) and infectiousness functions are allocated to individuals independently. Under quite general assumptions about the type of infectiousness function, the eventual proportions of susceptible individuals converge in probability, as the population size n gets large, to a solution of the system of equations (Ball and Clancy 1993):

$$\sum_{i=1}^{k} \pi_i m_{in}[s_i(0) - s_i(\infty)] = \ln s_j(0) - \ln s_j(\infty), \qquad j = 1, 2, \ldots, k. \qquad (3.6)$$

The largest eigenvalue, R_v, of the mean matrix $(m_{ij}\pi_j s_j(0))$ is a threshold parameter for this multitype epidemic model. In a large community, the probability of a major epidemic is zero when $R_v < 1$. If individuals are selected for vaccination at random, irrespective of type, then the critical immunity coverage is again given by (2.2), with R_0 being the largest eigenvalue of $(m_{ij}\pi_j)$. This motivates the need to make inferences about R_0, but it should be noted that more effective vaccination strategies exist for multitype epidemics (Becker and Starczak 1997). The π_i are assumed known, but the m_{ij} are not. It is not feasible to estimate the k^2 values of m_{ij} from the k observed values $\hat{s}_i(\infty)$. Substituting the observed $\hat{s}_i(\infty)$ into (3.6) gives a set of k estimating equations and it is of interest to know how much information this provides about R_0. Using known results for the largest eigenvalue of a matrix subject to its elements satisfying certain linear equations, in our case the estimating equations (3.6), gives

$$\min_{1 \leqslant j \leqslant k} \frac{\ln s_j(0) - \ln s_j(\infty)}{s_j(0) - s_j(\infty)} \leqslant R_0 \leqslant \max_{1 \leqslant j \leqslant k} \frac{\ln s_j(0) - \ln s_j(\infty)}{s_j(0) - s_j(\infty)}. \qquad (3.7)$$

Awareness of such inequalities in the epidemic context is expressed in Greenhalgh and Dietz (1994) and Daley and Gani (1994).

We see that data on initial and eventual proportion susceptible, for each group, enables estimation of the lower and upper bounds of (3.7). The central limit theorem for the eventual proportion of each type infected (Scalia-Tomba 1986; Ball and Clancy 1993), can provide confidence limits for the estimates of the bounds in (3.7), which leads to an interval estimate for R_0 comprised of two parts, a part that reflects the limited identifiability of R_0 from the type of data available and a part that reflects sampling variation.

It is of interest to address this problem by Bayesian inference via MCMC methods, as it seems a natural way of incorporating both the limited identifiability and the uncertainty due to chance fluctuations. The flexibility of this approach allows us to incorporate two other desirable features. One of these is to add data from a second epidemic, with a similar mode of transmission. The idea is to double the amount of data but add only one new parameter, because it can be assumed that the mean matrix for the second epidemic is $\alpha(m_{ij}\pi_j)$, where α is a constant of proportionality. This idea is used by Farrington *et al.* (2001) in an approach to estimating R_0 from age-stratified seroprevalence data.

A second desirable feature, that helps to reduce the number of parameters, is to acknowledge that contacts made between individuals of type i and individuals of type j are, for some kinds of transmission, the same, irrespective of who is the infective. We then need to balance the number of contacts made between the two types of individual, giving the constraints $m_{ij}\pi_i = m_{ji}\pi_j$. This reduces the number of parameters considerably, although it does so at the cost of assuming that differences between types are entirely due to differences in social mixing. Inference about R_0 that incorporates these two features seems to be a most worthwhile project for future work.

Other assumptions can reduce the number of parameters. Indeed, the identifiability problem can be overcome if these assumptions reduce the number of parameters to k. For example, if the difference between types is entirely due to differences in susceptibility ($m_{ij} = m_{\bullet j}$ for all i) then we can solve (3.6) to give

$$\widehat{m}_{\bullet j} = [\ln s_j(0) - \ln \hat{s}_j(\infty)] \Big/ \sum_{i=1}^{k} \pi_i[s_i(0) - \hat{s}_i(\infty)].$$

3.3   Variable vaccine response

Many epidemiological studies are conducted to measure the effectiveness of a vaccine. It is important to have a measure of vaccine effectiveness, because vaccines differ in their ability to protect individuals from the disease. The concept of vaccine efficacy,

$$VE = 1 - \frac{\text{proportion of cases among vaccinated individuals}}{\text{proportion of cases among unvaccinated individuals}},$$

has played a central role for this purpose since Greenwood and Yule (1915). A weakness of this measure, as Smith *et al.* (1984) point out, is that VE takes values that depend substantially on the *way* a vaccine reduces susceptibility. It is also of concern that the measure VE takes no account of the fact that the disease is transmissible. An extreme example illustrates this. Consider a community whose members mix uniformly and are homogeneous, except that some are vaccinated. Suppose that vaccinated individuals have the same susceptibility as unvaccinated individuals, but the infectivity of vaccinees is substantially lower. Then the vaccine reduces the force of infection acting on community members, with the consequence that each individual has a lower chance of being infected and fewer community members are likely to be infected. The vaccine clearly provides a benefit, however vaccinated and unvaccinated individuals are equally likely to be infected so that VE estimates an 'effectiveness' of zero for such a vaccine.

For infectious diseases, we would like to know how well the vaccine protects an individual against infection and, when a vaccinated individual is infected, how much less infectious they are. For simplicity, suppose that unvaccinated individuals are homogeneous; however, vaccinated individuals may differ from each other because their response to vaccination differs. The effect of vaccination on susceptibility and infectiousness is modelled as in Section 3. Consider first the reduction in susceptibility. We follow the formulation given in Becker and Utev (2002). A force of infection λ_t acting on an unvaccinated individual at time t would be $A\lambda_t$ if that individual were vaccinated. Here A reflects the protective effect of the vaccination. The notation A, already used in Section 3.1.2, is chosen deliberately because the protective effect has been modelled as a relative susceptibility. It is random and expected to take values in $[0, 1]$. When A has the distribution $\Pr(A = 1) = f = 1 - \Pr(A = 0)$ the vaccination leads to either full protection or no protection against infection, with f being the proportion of vaccine failures. When A has the degenerate distribution $\Pr(A = a) = 1$, for some a, then every vaccinee has exactly the same response and is partially protected. These are the two responses considered by Smith *et al.* (1984). Our description captures a much wider set of responses. A description that seems to cover most applications is $\Pr(A = 0) = p_0$, $\Pr(A = 1) = p_1$ and with probability $1 - p_0 - p_1$ the variable A has a continuous distribution on the interval $[0, 1]$. The quantity $1 - \mathrm{E}A$ might be used to measure how well the vaccine protects against infection, i.e. a measure of protective vaccine efficacy.

Consider now the reduction in infectiousness arising from vaccination. Let B_U denote the area under the infectiousness function for an unvaccinated individual, and B_V the corresponding area for a vaccinated individual; $B = B_V/B_U < 1$ reflects a reduction in infectiousness. Note that B_V will be random when responses to vaccination vary, and then B is also random. The quantity $1 - \mathrm{E}B$ might be used to measure how much the vaccine reduces infectiousness.

It would be useful to have a measure that captures the effects on both susceptibility and infectiousness, and also measures the effect vaccination has on transmission in the community. One such measure is $1 - \mathrm{E}(AB)$, which is motivated by the fact that it is equal to CIC/CVC; see Becker and Starczak (1998).

Estimation of $\mathrm{E}A$, $\mathrm{E}B$, and $\mathrm{E}(AB)$ is of interest. Suppose we only observe the eventual number of cases for a large epidemic. That is, we observe $\hat{c}_V$ and $\hat{c}_U$ the eventual proportions of vaccinated and unvaccinated individuals infected. Becker and Utev (2002) show that, for large epidemics

$$\frac{\ln(1 - c_V)}{\ln(1 - c_U)} \leqslant \mathrm{E}A \leqslant \frac{c_V}{c_U}. \tag{3.8}$$

The two bounds of (3.8) can be estimated by substituting the observed values $\hat{c}_V$ and $\hat{c}_U$, and this provides inferences for $\mathrm{E}A$ without needing to specify a distribution for the relative susceptibility A. The lower bound of (3.8) is attained when $\Pr(A = a) = 1$ and then its estimate

$$\hat{a} = \frac{\ln(1 - \hat{c}_V)}{\ln(1 - \hat{c}_U)}$$

has the large-sample standard error given by

$$n \, [\mathrm{s.e.}(\hat{a})]^2 = \frac{1}{[\ln(1 - \hat{c}_U)]^2} \left[\frac{\hat{c}_V}{\pi_V(1 - \hat{c}_V)} + \hat{a}^2 \frac{\hat{c}_U}{\pi_U(1 - \hat{c}_U)} \right],$$

where π_{V} and π_{U} are the proportion of the community vaccinated and unvaccinated. The upper bound is attained when $\Pr(A = 1) = f = 1 - \Pr(A = 0)$ and then its estimate

$$\hat{f} = \frac{\hat{c}_{\mathrm{V}}}{\hat{c}_{\mathrm{U}}}$$

has the large-sample standard error given by

$$n \left[\mathrm{s.e.}(\hat{f})\right]^2 = \frac{\hat{f}(1-\hat{f})}{\pi_{\mathrm{V}}} + \frac{\hat{f}\,\hat{s}_\infty}{s_0 - \hat{s}_\infty}\left(\frac{1}{\pi_{\mathrm{V}}} + \frac{\hat{f}}{\pi_{\mathrm{U}}}\right).$$

Inference about variation in the infectiousness is more difficult, as is demonstrated by Britton (1998). To make inferences about EB and $E(AB)$ it seems necessary to obtain data on outbreaks in small groups, like households. Suppose, for example, that we observe outbreaks occurring over a time interval $[0, T]$ in pairs of 'susceptible' individuals. Let $Q = \exp\left(-\int_0^T \lambda_t \, dt\right)$ be the probability that an unvaccinated individual avoids being infected from outside the pair during the interval $[0, T]$, where λ_t is the community-based force of infection acting on an unvaccinated individual. The probability that an unvaccinated individual avoids being infected from an unvaccinated infected partner is $q = \exp(-B_{\mathrm{U}})$. For this illustration we assume that the vaccine reduces the susceptibility by a fixed factor a for every vaccinee and reduces the infectiousness by a fixed factor $b = B_{\mathrm{V}}/B_{\mathrm{U}}$ for every vaccinee. Then the probabilities for the various possible outcomes over the interval $[0, T]$ for a pair with neither member vaccinated are

Cases	0	1	2
Probability	Q^2	$2Q(1-Q)q$	balance

The probabilities of outcomes for a pair with both members vaccinated are

Cases	0	1	2
Probability	Q^{2a}	$2Q^a(1-Q^a)q^{ab}$	balance

The probabilities of outcomes for a pair with exactly one member vaccinated are

Cases	None	Unvac. member only	Vac. member only	Both
Probability	Q^{a+1}	$(1-Q)(qQ)^a$	$Q(1-Q^a)q^b$	balance

With observations on n_i pairs with exactly i members vaccinated, $i = 0, 1, 2$, our data consist of observations on three multinomial distributions in which each cell probability is expressed as a function of the parameters Q, q, a, and b. In principle, inferences for these parameters are straightforward.

To accommodate variation in vaccine response we must allow the relative susceptibility A and relative infectivity $B = B_{\mathrm{V}}/B_{\mathrm{U}}$ to be random. The unconditional probabilities of the various outcomes are then expectations, with respect to A and B, over probability terms like those shown above. Therefore, for likelihood inferences to be manageable we need to have explicit expressions for terms like $\mathrm{E}x^A$ and $\mathrm{E}(x^A y^B)$, where A and B are correlated. This requirement is a severe limitation and a more promising approach is to use MCMC methods in a Bayesian analysis, with A and B treated as latent variables. In view of the potential applications, the development of these methods is very worthwhile.

4   Community of households

The assumption of uniform mixing is violated in real societies by the fact that we mix more intensely within our network of friends and relatives. Here we consider two levels of mixing in a community consisting of a large number of households. While infectious, an infected individual exerts a force of infection β_{C}/n on each community member not in the same household and a force of infection β_{H} on each household member. Assume that every individual is a member of exactly one household. A detailed study of a model for transmission of infection through a community of households has been undertaken recently (Ball *et al.* 1997) but recognition of a threshold parameter for epidemics in such a community goes back to Bartoszyński (1972), who formulated an epidemic model based on a branching process. To emphasize that reproduction numbers are not unique, Becker and Dietz (1996) discuss four different threshold parameters for epidemics in a community of households, each with an interpretation as a reproduction number. As each reproduction number leads to the same conclusion about the critical vaccination coverage, it is natural to work with the reproduction number that is most manageable. This is the reproduction number for infected households introduced by Bartoszyński (1972). It may be expressed as

$$R_0 = \theta_{\mathrm{C}} \theta_{\mathrm{H}} \,.$$

To explain the meaning of θ_{C} and θ_{H} suppose that every community member is susceptible and one individual, selected at random from the community, is infected. Then θ_{C} is the mean number of individuals infected by this initial infective in the general community and θ_{H} is the mean size of the outbreak in the household of this initial infective, including the initial infective.

To see how vaccination affects the reproduction number it is useful to express θ_{H} in greater detail, to reflect the different possible household sizes. Let h_j be the proportion of households with exactly j members, $j = 1, 2, \ldots, k$, where $k \ll n$. The mean household size is $m_{\mathrm{H}} = \sum_{j=1}^{k} j h_j$. Then we can write

$$R_0 = \theta_{\mathrm{C}} \sum_{j=1}^{k} \frac{j h_j}{m_{\mathrm{H}}} \mu_j \,, \tag{4.1}$$

where μ_j is the mean number of eventual cases in a household with j members, all initially susceptible, when exactly one of them is infected and the outbreak evolves without further infection from outside. In particular, $\mu_1 = 1$.

Consider now R_v the initial reproduction number when a fraction v of individuals in the community are initially immune. Let ω_{js} denote the proportion of households of size j with s susceptible members. When we assume that the mean outbreak size in a household with s susceptible members does not depend on the number of immune individuals it contains, we can write

$$R_v = \frac{\theta_C}{m_H} \sum_{j=1}^{k} h_j \sum_{s=1}^{j} s\, \omega_{js}\, \mu_s \,, \tag{4.2}$$

where the immunity coverage is $v = \sum_{j=1}^{k} \sum_{s=0}^{j-1} (j-s) h_j \omega_{js}/m_H$ (Becker and Starczak 1997). Equation (4.2) reveals that the effect of vaccination depends on the way immunity is distributed over households.

From expressions (4.1) and (4.2) we see that to estimate R_0 or CIC we need estimates of $\mu_2, \mu_3, \ldots, \mu_k$ and θ_C. The values of the h_j, and therefore m_H, can be assumed known from census data. Britton and Becker (2000) consider this estimation problem on the basis of data on household outbreaks, where all members of a sample of households are serologically tested before the epidemic season and again after the epidemic season, to see who was initially susceptible and who was eventually infected. They estimate the critical immunity coverage by maximum likelihood methods applied to the model of Addy *et al.* (1991) for transmission of infection within households, and illustrate the approach with data on influenza outbreaks. Bayesian inference for the parameters of the model of Addy *et al.*, by MCMC methods, is demonstrated by O'Neill *et al.* (2000).

As such estimation is not a trivial matter, one is tempted to look for simpler approximate methods. Becker (1995) proposes an estimator for θ_C derived from a martingale for the number of infected households and suggests sample means of household outbreak sizes as estimates for the μ_j. Sample means tend to overestimate the μ_j because they ignore the infection from outside the household. This means that R_0 and CIC would be overestimated, which is acceptable when the goal is to prevent epidemics. An advantage of this method is that we do not need to specify a model for transmission within households.

An alternative method of making approximate inferences for diseases that are highly infectious within households is now suggested. Let p be the probability that a given susceptible household member is infected by a single initial infective. Then $\mu_j \leqslant 1 + EX_j$, where X_j is a Binomial$(j-1, p)$ variable. This leads to the inequality

$$\theta_C(1-p) + pR_H \leqslant R_0 \leqslant R_H \,, \tag{4.3}$$

where $R_H = \theta_C(m_H + \sigma_H^2/m_H)$ is the basic reproduction number under the assumption that all household members eventually become infected whenever one household member is infected. The mean m_H and variance σ_H^2 of household sizes can be assumed known and so only estimates of θ_C and p are required. Estimation of these parameters seems relatively simple. For example, an estimate for θ_C can be obtained from data on the number of households infected, as in Becker (1995), and p can be estimated from data on outbreaks in households of size two and three, say.

Expressions for the reproduction number of infected households are also available when there are distinguishable types of individual, see Becker and Hall (1996), but parameter estimation for this setting needs further work. A start has been made by Addy *et al.* (1991) and Britton and Becker (2000).

5 Contact networks and other models

In this section we mention briefly some other modelling approaches by which the effect of heterogeneity should be studied.

5.1 Disease transmission via contact sets

The standard model described in Section 2 is able to trace the progress of an epidemic over calendar time. Our parameters of interest only require this when there is variation over time, as in Section 5.2. In the absence of such variation we can consider transmission in terms of the number, or likely number, of contacts individuals make while infectious. This is used above when considering epidemics in a community of households and is useful for a variety of other contact networks. We mention a couple of these.

Label the individuals of the community 1, 2, ..., n and let $C_i \subseteq \{1, 2, \ldots, n\}$ be the contact set for individual i, which is the set of individuals that i infects should he become infected. An epidemic progresses by the infection being transmitted from each infective to every member of their contact set, who in turn infect every member of their contact set, and so forth. There is no reference to calendar time.

Different settings are created by different assumptions about the choice of the contact sets C_i. A standard setting is when the C_i are independent and identically distributed sets. Let $C_n = |C_i|$, the number in a randomly chosen contact set. If the limit $C_n \to C$ exists, then a basic reproduction number is given by $R_0 = \mathrm{E}\,C$, which is of standard form. A basic reproduction number of non-standard form arises under the alternative assumption that the contact sets C_i are exchangeable and commutative, so that $i \in C_j$ if and only if $j \in C_i$. If the random sets C_i are approximately independent and the limit $C_n \to C$ exists, then a basic reproduction number is given by $R_0 = \mathrm{E}\,C - 1 + \mathrm{var}(C)/\mathrm{E}\,C$, see Andersson (1998).

The idea of viewing transmission in terms of contact sets is also applicable to models of the percolation type, which are characterized by a focus on transmission to 'nearby' neighbours. For example, each infective might choose, independently and at random, some of the nearest neighbours for his contact set. The new element that these models bring is the addition of a direction of transmission. An extreme example is a one-dimensional percolation, where the epidemic progresses only along a line. In this situation, there will only be minor outbreaks if $\lim_{x \to \infty} x \Pr(C > x) = 0$ and the proportion of immune individuals is positive.

Parameter estimation for models based on contact sets presents a challenge, but a start has been made by Britton and O'Neill (2002), who apply Bayesian inferences by MCMC methods to the case where the social structure of the community is determined by a Bernoulli random graph.

5.2 Variation over time

Little attention has been paid to the fact that mixing rates of individuals vary over time. It is of interest to study the circumstances under which this variation has a significant effect on our parameters of interest. An example given in Becker *et al.* (1995, Section 3) shows that it sometimes is an important factor.

It is not possible to accommodate a completely arbitrary dependence of the transmission rate on time. Consider a systematic pattern in the mixing rate. Assume that during the course of each day, say, the mixing rate of each individual is described by a periodic piecewise continuous function g_x, so that the mean number of transmissions in a day is $m = \int_0^1 g_x \, dx$, when the individual is infectious for the entire day. For example, the simple form

$$g_x = \begin{cases} \beta & \text{for } x \in [0, \tau], \\ 0 & \text{for } x \in [\tau, 1], \end{cases} \tag{5.1}$$

reflects a situation where an individual mixes at a constant rate during the active period and does not mix at all during the inactive period. The infectiousness function is determined by g_x and the duration of the infectious period.

The particular case where D, the duration of the infectious period, has a negative exponential distribution with mean μ_D gives the familiar result $R_0 = m\mu_D$, derived by Kendall (1948). Time variation only affects the threshold parameter through a minor change in the interpretation of the parameter m. This is a consequence of the memoryless property of the exponential distribution. The following example demonstrates that variation over time can have a substantial effect on the form of R_0.

Suppose that each individual has a daily mixing rate of the form given by (5.1), with $\tau \leqslant 1/2$. Let d be an integer and assume that an infective remains infectious for D days, where $\Pr(d + \tau \leqslant D \leqslant d + 1 - \tau) = 1$. This means that an infective can infect others during the active periods on the day he is infected and on each of the next d days. Therefore the potential to infect others depends on the time of day an individual is infected. Using an approximation by a multitype branching process during the early stages of the epidemic, we find R_0, the largest eigenvalue of the mean matrix, to be

$$R_0 = \frac{\beta\tau}{\ln(1 + 1/d)}$$

(Becker *et al.* 1995). This reproduction number does not have the familiar form, demonstrating that time variation can have an appreciable effect on control issues. More specifically, suppose that $d = 0$, so that the infective can infect individuals only in the remaining part of the active period on the day he is infected. Then $R_0 = 0$, suggesting that no epidemics can occur. However, since β may be arbitrarily large, a huge number of infections may occur on day 1. This illustrates the fact that threshold parameters are meant to indicate whether an epidemic can take off, but we tend to determine this by exploring the eventual behaviour of a branching process that approximates the early stages of the epidemic. This approach ignores any short-term explosion of infectives and may therefore be unsatisfactory for practical use in some circumstances.

5.3 Other models

Individuals are allowed to have a systematic contact pattern over time in Section 5.2. However, many social events occur haphazardly and one way to accommodate this is to allow changes in the contact process of an individual at random times (Ball and Clancy 1993; Becker *et al.* 1995).

For some diseases the rate of transmission in the community depends on how many individuals are alive, or active. This is the case when individuals are physically removed, or are no longer part of the community interaction for some other reason. Examples where such models are needed include the spread of myxomatosis in a rabbit population (Saunders 1980), and some highly dangerous sexually transmitted diseases (O'Neill 1995).

For many diseases, an individual recovers and acquires immunity for the remainder of the epidemic period at the end his infectious period. For some diseases, such as pneumococcal infections and gonorrhoea, re-infection is possible. Models that allow re-infection are often referred to as S–I–S models. The reproduction number for many of S–I–S epidemic models is the same as for their S–I–R counterpart. A difference arises for a community of households, or similar social setting. Formula (4.1) still applies, but the interpretation of μ_k must be modified by counting each infection of the same individual as a separate case.

Acknowledgements

Support from the Australian Research Council is gratefully acknowledged. The inference problems in Sections 3.2 and 3.3 were discussed at an HSSS research kitchen on Statistical Inference using Stochastic Epidemic Models, held 10–13 July 2000.

References

Addy, C. L., Longini, I. M., and Haber, M. (1991). A generalized stochastic model for the analysis of infectious disease final size data. *Biometrics*, **47**, 961–74.

Andersson, H. (1998). Limit theorems for a random graph epidemic model. *Annals of Applied Probability*, **8**, 1331–49.

von Bahr, B. and Martin-Löf, A. (1980). Threshold limit theorems for some epidemic processes. *Advances in Applied Probability*, **12**, 319–49.

Ball, F. G. and Clancy, D. (1993). The final size and severity of a generalized stochastic multitype epidemic model. *Advances in Applied Probability*, **25**,

721–36.

Ball, F. G., Mollison, D., and Scalia-Tomba, G. (1997). Epidemics in populations with two levels of mixing. *Annals of Applied Probability*, **7**, 46–89.

Bartoszyński, R. (1972). On a certain model of an epidemic. *Applicationes Mathematicae*, **13**, 139–51.

Becker, N. G. (1977). Estimation for discrete time branching processes with application to epidemics. *Biometrics*, **33**, 515–22.

Becker, N. G. (1989). *Analysis of Infectious Disease Data*. Chapman & Hall, London.

Becker, N. G. (1993). Martingale methods for the analysis of epidemic data. *Statistical Methods in Medical Research*, **2**, 93–112.

Becker, N. G. (1995). Estimation of parameters relevant for vaccination strategies. *Bulletin of the International Statistical Institute*, **56**, Book 2, 1279–89.

Becker, N. G. (1997). Uses of the EM algorithm in the analysis of data on HIV/AIDS and other infectious diseases. *Statistical Methods in Medical Research*, **6**, 24–38.

Becker, N. G. and Dietz, K. (1996). Reproduction numbers and critical immunity levels for epidemics in a community of households. In *Athens Conference on Applied Probability and Time Series, Volume 1: Applied Probability*, Lecture Notes in Statistics, No. 114 (eds C. C. Heyde, Yu. V. Prohorov, R. Pyke and S. T. Rachev), pp. 267–76. Springer-Verlag, New York.

Becker, N. G. and Hall, R. (1996). Immunisation levels for preventing epidemics in a community of households made up of individuals of different types. *Mathematical Biosciences*, **132**, 205–16.

Becker, N. G. and Starczak, D. N. (1997). Optimal vaccination strategies for a community of households. *Mathematical Biosciences*, **139**, 117–32.

Becker, N. G. and Starczak, D. N. (1998). The effect of random vaccine response to prevent epidemics. *Mathematical Biosciences*, **154**, 117–35.

Becker, N. G. and Utev, S. (2001). A distribution functional arising in epidemic control. In *Probability and Statistical Models with Applications* (eds Ch. A. Charalambides, M. V. Koutras, and N. Balakrishnan), pp. 573–82. Chapman & Hall, London.

Becker, N. G. and Utev, S. (2002). Protective vaccine efficacy when vaccine response is random. *Biometrical Journal*, **44**, 29–42.

Becker, N. G. and Yip, P. (1989). Analysis of variations in infection rates. *Australian Journal of Statistics*, **31**, 42–52.

Becker, N. G., Bahrampour, A., and Dietz, K. (1995). Threshold parameters for epidemics in different community settings. *Mathematical Biosciences*, **129**, 189–208.

Britton, T. (1998). Estimation in multitype epidemics. *Journal of the Royal Statistical Society*, B, **60**, 665–79.

Britton, T. and Becker, N. G. (2000). Estimating the immunity coverage required to prevent epidemics in a community of households. *Biostatistics*, **1**, 389–402.

Britton, T. and O'Neill, P. D. (2002). Bayesian inference for stochastic epidemics in populations with random social structure. *Scandinavian Journal of Statistics*, **29**, 375–90.

Daley, D. J. and Gani, J. (1994). A deterministic general epidemic model in a stratified population. In *Probability, Statistics and Optimisation* (ed. F. P. Kelly), pp. 117–32. Wiley, New York.

Dietz, K. (1993). The estimation of the basic reproduction number for infectious diseases. *Statistical Methods in Medical Research*, **2**, 23–41.

Farrington, C. P., Kanaan, M. N., and Gay, N. J. (2001). Estimation of the basic reproduction number for an infectious disease from age-stratified serological survey data. *Applied Statistics*, **50**, 251–92.

Greenhalgh, D. and Dietz, K. (1994). Some bounds on estimates for reproductive ratios derived from the age-specific force of infection. *Mathematical Biosciences*, **124**, 9–57.

Greenwood, M. and Yule, U. G. (1915). The statistics of anti-typhoid and anti-cholera inoculations, and the interpretations of such statistics in general. *Proceedings of the Royal Society of Medicine*, **8**, 113–94.

Kendall, D. G. (1948). On the generalized 'birth-and-death' process. *Annals of Mathematical Statistics*, **19**, 1–15.

O'Neill, P. D. (1995). Epidemic models featuring behaviour change. *Advances in Applied Probability*, **27**, 960–79.

O'Neill, P. D. and Becker, N. G. (2001). Inference for an epidemic when susceptibility varies. *Biostatistics*, **2**, 99–108.

O'Neill, P. D. and Roberts, G. O. (1999). Bayesian inference for partially observed stochastic epidemics. *Journal of the Royal Statistical Society*, A, **162**, 121–9.

O'Neill, P. D., Balding, D. J., Becker, N. G., Eerola, M., and Mollison, D. (2000). Analyses of infectious disease data from household outbreaks by Markov chain Monte Carlo methods. *Journal of the Royal Statistical Society*, C, **49**, 517–42.

Saunders, I. W. (1980). A model for myxomatosis. *Mathematical Biosciences*, **48**, 1–15.

Scalia-Tomba, G. (1986). Asymptotic final size distribution of the multitype Reed–Frost process. *Journal of Applied Probability*, **23**, 563–84.

Scalia-Tomba, G. (1990). On the asymptotic final size distribution of epidemics in heterogeneous populations. In *Stochastic Processes and Epidemic Theory*,

Lecture Notes in Biomathematics, No. 86, (eds J.-P. Gabriel, C. Lefevre, and P. Picard), pp. 189–96. Springer-Verlag, New York.

Smith, P. G., Rodrigues, L. C., and Fine, P. E. M. (1984). Assessment of the protective efficacy of vaccines against common diseases using case–control and cohort studies. *International Journal of Epidemiology*, **13**, 87–96.

11A
MCMC methods for stochastic epidemic models

Philip D. O'Neill
University of Nottingham, UK

1   Introduction

The previous article highlighted some of the challenges involved in performing statistical inference using stochastic models for infectious diseases. These challenges arise due to both the inherent features of infectious disease outbreak data, for instance a high level of dependency in such data and a lack of complete observational information, and also the relative mathematical intractability of even very simple stochastic epidemic models. In what follows we shall describe the use of Markov chain Monte Carlo (MCMC) methods as a tool for inference, listing advantages and drawbacks, and illustrating the methodology with some examples.

1.1   Advantages and challenges of MCMC

Three key advantages of using MCMC in this area as follows.

- *Model flexibility* Arguably the greatest benefit that MCMC brings to the area of inference for epidemics is that it allows for a high degree of modelling flexibility. In particular, it becomes possible to consider models with more realistic features as opposed to models that contain mathematically convenient but unrealistic simplifying assumptions.

- *Missing data* When considering temporal data on an epidemic outbreak, information concerning the times at which infections occurred is invariably unavailable. In an MCMC setting such missing data can be included as additional model parameters in a straightforward manner. Other useful kinds of missing data, for instance who infected who, can be similarly included.

- *Bayesian framework* MCMC naturally caters for a Bayesian framework in which posterior densities of interest can be explored. As described in the previous article, the Bayesian setting seems a natural one in certain situations in epidemic modelling. Moreover, credible interval information can often be obtained in situations where corresponding classical confidence intervals are not available.

The major obstacle to using MCMC is that of implementation; in other words, how algorithms actually perform in practice. There are two notable difficulties that can arise, as follows.

- *Correlations and convergence* It is well known (Hills and Smith 1992) that strong correlations in the target density can be a source of slow convergence and poor mixing of MCMC algorithms. Algorithms for epidemic models often contain highly correlated parameters, partly due to inherent structures (e.g. unknown infection times are highly interdependent) and partly due to problems of identifiability (e.g. if a combination of parameters may be estimable with high precision then estimates of individual parameters will be correlated accordingly). Overcoming these difficulties is often a problem-specific task, although generally speaking, careful re-parameterization of the epidemic model is beneficial.

- *Calculating final size distributions* It is not uncommon for available data to consist only of final size information, i.e. the numbers ultimately infected by an epidemic. In this case it is important, both for MCMC methods and other approaches, to be able to compute the likelihood of the observed data for a given set of parameter values. For a wide range of epidemic models, including multitype models, a triangular system of equations can be written down and solved recursively to find the probabilities of all different outcomes, in terms of numbers infected, at the end of the epidemic (see, for example, Ball and Clancy (1993)). However, such systems have two drawbacks. First, there are often problems of numerical instability, essentially due to the presence of probabilities very close to zero, with the consequence that the calculations can become computationally prohibitive. This phenomenon can occur even for very small population sizes (e.g. less than 20) for certain parameter values. It is sometimes possible to overcome this problem by using software routines to increase the effective machine precision, although this can lead to very slow programs. Secondly, for non-homogeneous-population epidemic models, the triangular system of equations can be of an enormous size, which again creates computational difficulties.

2 Examples

We now outline three different scenarios involving different kinds of model and data set.

2.1 Temporal data, general stochastic epidemic

Recall from the previous article, Section 3.1.1, that an S–I–R model in which the infectiousness function β_x equals β throughout an exponentially distributed infectious period D, and equals zero otherwise, is known as the general stochastic epidemic. Suppose that the mean of D is γ^{-1}. O'Neill and Roberts (1999) give details of an MCMC algorithm for performing Bayesian inference for β and γ given data consisting of the times at which removals occur. The key features of this algorithm are as follows.

First, suppose that removals are observed to occur at times $r_1 = 0 \leqslant r_2 \leqslant \cdots \leqslant r_m = \tau$ among a population consisting initially of one infective and n susceptibles. Let infections occur at (unobserved) times $i_1, i_2, \ldots, i_p,$

where $m \leqslant p \leqslant n + 1$ and $i_p \leqslant \tau$, and denote by S_t and I_t respectively the numbers of susceptibles and infectives at time t. Writing $\mathbf{i} = (i_2, \ldots, i_p)$ and $\mathbf{r} = (r_1, \ldots, r_m)$, the likelihood for β and γ is given by

$$L(\beta, \gamma) = \pi(\mathbf{i}, \mathbf{r}|\beta, \gamma, i_1) \propto \beta^{p-1} \gamma^m \exp\left\{-\int_{i_1}^{\tau} (n^{-1}\beta S_t I_t + \gamma I_t)\, dt\right\}. \quad (2.1)$$

Next, parameters can be updated as follows. If each β and γ are assigned gamma-distributed prior densities, then by (2.1) their full conditional densities will also be gamma-distributed. If the initial infection time i_1 has an exponential prior density $\theta \exp(\theta y)$ for $y < 0$ (it must be the case that $i_1 < r_1 = 0$), then the full conditional of y is also of exponential form on the interval $(-\infty, i_2)$. Finally, the unknown infection times $i_2, \ldots, i_p$ can be updated by using a Hastings algorithm with three possible moves, namely (i) an infection time is moved; (ii) a new infection time is added; and (iii) an infection time is deleted. In each case the corresponding acceptance probability is straightforward to calculate.

2.2   Temporal data, network model

As described in the previous article, one approach to modelling heterogeneity is the use of contact sets which list those individuals that a specified individual infects if they themselves become infectious. A simple model involving random contact sets is described in Britton and O'Neill (2002), and is defined as follows. A population consists of N individuals, each of whom has social contact with any other given individual with probability p, with different pairs assumed independent. The social structure is thus represented by a Bernoulli random graph. For a given realization of the random graph, a general stochastic epidemic model can be defined. Specifically, each infectious individual remains so for an exponentially distributed period of time with mean γ^{-1}, during which time infectious contacts occur with each neighbour in the graph according to independent Poisson processes, each of rate β.

An MCMC algorithm for this network model can be developed in a similar manner to that described above for the general stochastic epidemic, again using data on removal times, as described in Britton and O'Neill (2002). The network model has three parameters, and joint posterior densities for these indicate clear correlations, essentially corresponding to different ways in which the data could have arisen under the assumptions of the model. For example, β and p are found to be negatively correlated, essentially since removal data alone make it hard to distinguish between a highly-infectious disease on a sparse network and a less infectious disease on a more dense network. It seems likely that larger data sets would lessen this identifiability problem. The correlations also affect the MCMC algorithm, so that an extra step to improve mixing was found to be necessary.

2.3   Final size data, households model

Section 4 of the previous article describes estimation methods for models that involve a community of households. Such a model, described in Addy *et al.*

(1991), is considered using MCMC methods in O'Neill *et al.* (2000). This model has three parameters: q_h, the probability that a given infective fails to infect a given susceptible in the same household; q_c, the probability that a susceptible household member avoids infection from the community at large; and v, the probability of an individual being immune to the disease. This last parameter is included in the analysis to account for some degree of heterogeneity in the population.

The model is used to consider data from outbreaks of influenza. These data are of the form $\mathcal{D} = \{n_{ij}\}$, where n_{ij} is the number of households containing $i \geqslant 1$ individuals in which $0 \leqslant j \leqslant i$ become infected during the epidemic. Households are assumed to be independent of one another, and thus the posterior density of (q_c, q_h, v) given $\mathcal{D}$ satisfies

$$\pi(q_c, q_h, v | \mathcal{D}) \propto \pi(q_c, q_h, v) \prod_{i,j} [P(T_i = j)]^{n_{ij}}, \qquad (2.2)$$

where T_i is the total number ultimately infected in a household containing i individuals and $\pi(q_c, q_h, v)$ is the prior on (q_c, q_h, v). Calculating the probability mass function of T_i is relatively straightforward, and can be achieved in several ways, as described in O'Neill *et al.* (2000). A simple Metropolis–Hastings algorithm for approximate sampling from $\pi(q_c, q_h, v | \mathcal{D})$ is easily defined; the three parameters can be updated either individually, or in a single block. The acceptance probability is calculated using (2.2). As described in O'Neill *et al.* (2000), this algorithm (with suitable choices of prior and proposal density) was found to perform well in practice, with no convergence difficulties. As for the previous example, notable posterior parameter correlations were seen when the algorithm was applied using data taken from influenza outbreaks.

2.4 Future directions

In this short review, we have indicated that MCMC methods appear to be suitable tools for analysing infectious disease outbreak data using stochastic epidemic models. There are two ways in which this area of work can develop. First, as described in the previous article, there are a number of problem areas for which it seems natural to apply MCMC methods. Secondly, and more generically, more work is needed on the construction of efficient algorithms, ideally developing techniques that are not simply problem-specific. This includes methodology both for small or medium sized data sets, and also outbreaks in much larger populations. Regarding larger data sets, there are additional challenges due to the potential number of latent variables, although this need not be prohibitive. In particular, a careful choice of imputed variables can lead to efficient algorithms, even for very large data sets.

11B
Towards Bayesian inference in epidemic models

Kari Auranen
National Public Health Institute, Finland
Rolf Nevanlinna Institute, University of Helsinki, Finland

Estimation in epidemic models involving unidentifiable types of heterogeneous individuals requires observations of appropriate contrasts. It is well understood that under variability in susceptibility, the more frail individuals tend to contract infection earlier during an epidemic or earlier in life in endemic situations. Assuming such heterogeneity, it is appropriate to characterize the observed trend, i.e. a contrast over time in the rate of infection by a hierarchical (frailty) model with regard to susceptibility (Becker and Utev, Chapter 11 in this volume; Halloran *et al.* 1996). Assessing heterogeneity in infectiousness is more difficult because its direct estimation requires knowledge of contacts between susceptibles and infectives (Halloran 1998). Often only exposure (opportunity) to infection can be defined within an epidemiologic study (Rhodes *et al.* 1996). Moreover, periods of infectiousness are scattered over the course of the epidemic, and the estimation of variability in infectiousness may only depend on stochastic fluctuations of the proportions of infectives in different strata (Britton 1998). For recurrent infections, with alternating periods of susceptibility and infectiousness, estimation under heterogeneities is obviously even more difficult.

Lacking an appropriate temporal contrast in the eventual proportion of susceptibles ('final size' of epidemic), the mere estimation of a threshold parameter (R_0) in a heterogeneous population may seem dubious. Except for the simplest of epidemic models, it is not even possible without recourse to additional knowledge. This was clearly exposed by Becker and Utev who used various methods within the frequentist paradigm to derive estimates and bounds for R_0. Nevertheless, comparing inferences from different frequentist approaches is not clearcut. In particular, studies of influence of varying amounts of information on parameter uncertainty should benefit from a coherent framework. It would be interesting to see their ambitious programme carried out using likelihood or Bayesian methods, with proper conditioning on the body of knowledge in each particular case.

In the general epidemic model, likelihood-based inference of R_0 can make use of the marginal likelihood for the set of infection events (acquisitions and clearances of infection) during the epidemic. The data augmentation algorithm of Gibson (1997) can be modified to estimate R_0 in this setting. Although Bayesian modelling *per se* does not resolve the unidentifiability problem in more heterogeneous situations, it provides straightforward means to incorporate the required prior knowledge or more structured data. As an example of the latter possibility, O'Neill *et al.* (2000) studied the usefulness of final size information from a large number of families in the assessment of heterogeneity in susceptibility. In any case, whether final size, current status, or incomplete time to event data are available, likelihood calculations soon become cumbersome, and different data augmentation schemes are required in likelihood or Bayesian approaches.

In addition to the efficient use of prior knowledge, there are other reasons to adopt a fully Bayesian analysis of transmission data. Combining data from different sources is straightforward in the Bayesian framework. Auranen *et al.* (1996) considered transmission of a recurrent infection. Rates of acquiring and clearing infection were identifiable when two sets of current status data were analysed jointly, one bearing information on the endemic and the other on the epidemic phase of transmission within families. A full account of uncertainties about different types of unknown model quantities, including unobserved heterogeneities, can be dealt with in a single framework using hierarchical Bayesian models: (population level) parameters, unobserved (individual) events in the epidemic processes, and unobservable individual or group-specific frailties are treated in the same way with regard to the hierarchical structure and computing (Auranen *et al.* 2000).

The interpretation and estimation of threshold parameters often requires data from a major epidemic in a large population. Obviously, the use of transmission models in data analysis should not be restricted to such situations. An often neglected alternative to studies of threshold behaviour is predictive inference of the evolution of epidemics in time. This emphasizes the importance of collecting genuinely temporal data and of using predictive models. In a model of clinical disease and underlying spread of an asymptomatic infection, Ranta *et al.* (1999) used Bayesian methods to incorporate informative prior knowledge and to quantify uncertainty in predictions of the future course of a specific meningococcal outbreak. Bayesian modelling is a natural tool to analyse data of infections where transmission occurs via asymptomatic infection and only a fraction of infectious contacts lead to clinical disease.

Data augmentation often involves explicit imputation of events in continuous time. Although data would be available on an aggregate level, augmented (multivariate point) processes are usually described on the individual level to obtain tractable likelihood expressions, and suitable 'mutations' are applied through Markov chain Monte Carlo (MCMC) to realizations of these processes. As long as these are compatible with the observed data, the choice of mutations is free in the usual sense of choosing updating steps in MCMC. For example, when current status data are available, the likelihood term for the augmented processes reduces to an indicator function, corresponding to agreement with the observations (Auranen *et al.* 2000). In practice, efficient algorithms are difficult to devise in the case of large aggregated data sets which may introduce strong dependencies between the latent individual processes.

If the number of infection events to be augmented can be retrieved from the data, then versions of the Metropolis–Hastings algorithm are applicable. This is the case when incomplete observation of the epidemic process arises from some type of censoring. Although only removals (times of clearance of infection) were observed in an S–I–R type epidemic, the occurrence of infection can be inferred in those initially susceptible for whom a removal was recorded during the study. If it can be assumed that there are no infectious individuals present at the end of the study (completed epidemic), then the total number of infections equals that of removals; only the times of infection remain latent (O'Neill *et al.* 2000). MCMC updating schemes involve perturbing the latent times.

The number of infection events to be augmented cannot always be deduced from the data. If an S–I–R epidemic is not completed, the number of infections during the observation period cannot be fixed (Gibson and Renshaw 1998; O'Neill *et al.* 2000). For a recurrent infection, the numbers of acquisitions and clearances cannot be determined from sequential current status data, that is, when the times of transition are not directly observed (Auranen *et al.* 2000). Owing to this type of filtering, the epidemic process model has an unknown number of occurrences which requires the use of reversible jump MCMC (Green, Chapter 6 in this volume). In addition to perturbing times of infection, now also infection events, or sets of events, have to be added to and removed from the realizations.

Identifiability is again of interest. Reversible jump MCMC updating schemes were first used in epidemic models in which only aggregate-level information was available on the total number of individuals in different states of infection (Gibson and Renshaw 1998). However, using only the total number of individuals in the population as the data in their examples, the models were not identifiable without informative prior knowledge (e.g. the rate of removals). Using MCMC, O'Neill and Roberts (1999) estimated acquisition and removal rates from data on times of removal, subject to filtering in the sense described above. Halloran *et al.* (1996) considered estimability in models with unidentifiable heterogeneity in susceptibility and vaccine effects. They assumed longitudinal, aggregated data. In general, it seems optimal to have longitudinal, individual-level data to fit continuous-time models with an unknown number of occurrences. An application of reversible jump MCMC to a recurrent infection was presented in Auranen *et al.* (2000). The data recorded the presence/absence of sub-clinical pneumococcal infection in all members of some 100 families. Intuitively, the identifiability of acquisition and clearance rates relied on the stability of observed individual patterns of presence/absence of infection across consecutive observations.

An important use of epidemic models in infectious disease epidemiology is to help deal with dependencies in the data, induced by the underlying transmission. Epidemic models with the help of data augmentation also support parameters of disease incidence or intervention effects that are conditional on exposure to infection (cf. Halloran 1998). For example, in discrete time models, in which contacts are assumed to occur at discrete time points, data augmentation may allow for the estimation of transmission probabilities (conditional upon a contact), despite incomplete observation of the contacts. Parameters that are conditional on exposure may provide more natural measures of incidence or intervention effects. More importantly, they are instrumental in deriving predictions of effects of various interventions. Investigations of direct and indirect effects of vaccination, possibly with heterogeneous response to vaccine (Halloran 1998; Becker and Utev, Chapter 11 in this volume) are central in many practical applications. It seems certainly worth developing Bayesian methods to design and analyse such studies.

Additional references in discussion

Auranen, K., Ranta, J., Takala, A. K., and Arjas, E. (1996). A statistical model of transmission of Hib bacteria in a family. *Statistics in Medicine*, **15**, 2235–52.

Auranen, K., Arjas, E., Leino, T., and Takala, A. K. (2000). Transmission of pneumococcal carriage in families: a latent Markov process model for binary data. *Journal of the American Statistical Association*, **95**, 1044–53.

Gibson, G. J. (1997). Markov chain Monte Carlo methods for fitting spatiotemporal stochastic models in plant epidemiology. *Journal of the Royal Statistical Society*, C, **46**, 215–33.

Gibson, G. J. and Renshaw E. (1998). Estimating parameters in stochastic compartmental models using Markov chain methods. *IMA Journal of Mathematics Applied in Medicine and Biology*, **15**, 19–40.

Halloran, M. E. (1998). Concepts in infectious disease epidemiology. In *Modern Epidemiology* (eds K. J. Rothman and S. Greeland). Lippincott-Raven, Philadelphia.

Halloran, M. E., Longini, I. M., and Struchiner, C. J. (1996). Estimability and interpretation of vaccine efficacy using frailty mixing models. *American Journal of Epidemiology*, **144**, 83–7.

Hills, S. E. and Smith, A. F. M. (1992). Paramaterization issues in Bayesian inference (with discussion). In *Bayesian Statistics 4* (eds Bernardo, J. M., Berger, J. O., Dawid, A. P., and Smith, A. F. M.), pp. 641–9. Oxford University Press, UK.

Ranta, J., Mäkelä, P. H., Takala, A. K., and Arjas, E. (1999). Predicting the course of meningococcal disease outbreaks in closed subpopulations. *Statistics in Medicine*, **123**, 359–71.

Rhodes, P. H., Halloran, M. E., and Longini, I. M. (1996). Counting process models for infectious disease data: distinguishing exposure to infection from susceptibility. *Journal of the Royal Statistical Society*, B, **58**, 751–62.

12
Genetic linkage analysis using Markov chain Monte Carlo techniques

Simon C. Heath

Memorial Sloan–Kettering Cancer Center, New York, USA

1 Introduction

Linkage analysis is a general technique for mapping genes relative to each other. It has proven very useful for localizing genes affecting a wide range of diseases, for example cystic fibrosis (Kerem *et al.* 1989), some forms of breast cancer (Hall *et al.* 1990), and Alzheimer's disease (Schellenberg *et al.* 1992; Levy-Lahad *et al.* 1995). Localization of the disease gene is a first step towards identifying the gene, discovering what it does and, hopefully, developing therapies to prevent or reduce the effects of the disease. To understand linkage analysis it is necessary to begin with some definitions.

Genes occur at positions or *loci* on chromosomes. Each gene can exist in several variants called *alleles*. Each chromosome is present in two copies (ignoring the sex chromosomes), one from each parent. Each individual therefore has two alleles at each locus which form the individual's *genotype*; if the two alleles are the same, then the individual is *homozygotic*, otherwise they are *heterozygotic*. The simple pedigree in Fig. 1 shows the genotypes of pedigree members at two loci. Both loci have four alleles, the first locus (on the left) has alleles A, B, C, and D, and the second locus has alleles 1, 2, 3, and 4. In keeping with normal

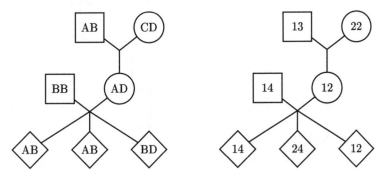

FIG. 1. A simple pedigree, with two loci.

pedigree drawing conventions, a square represents a father, a circle a mother, and a diamond a child with no descendants. From Fig. 1, the *segregation* of alleles from parents to offspring can be observed. At a given locus, a child gets one allele with equal probability from the two available from each parent. When multiple loci are considered, the segregations of alleles may be correlated if the loci are located close together on the same chromosomes; this phenomenon is called linkage.

During gamete formation, in a process called *meiosis*, the maternal and paternal copies of each chromosome pair up. The chromosomes can then break at several random positions, allowing segments of a chromosome to be exchanged in a process called *crossing over*. The resulting chromosomes that form the gametes can therefore contain a mixture of maternal and paternal chromosome segments. If loci are physically close, then it is unlikely that a cross-over will occur between them – this produces the correlation in segregations between linked loci.

Considering the parent–offspring trio at the top of Fig. 1, it can be seen that at the first locus the daughter must get allele A from her father and D from her mother, and at the second locus she gets allele 1 from her father and 2 from her mother. Considering the two loci together, she gets *haplotype* A1 from the father, and D2 from the mother. If a recombination occurs between the two loci during gamete formation in the daughter, then the recombinant haplotypes A2 and D1 are formed. This process can be seen in Fig. 2. Part (a) of the figure shows the chromosomes she received from her parents with the maternal chromosome shaded. A recombination between the two loci will result in the situation shown in part (b) of the figure, where there are two recombinant chromosomes which are a mixture of the original parental chromosomes.

When two alleles at different loci have the same parental origins (i.e. A and 1 in this example) they are said to be *in phase*, and a genotype that specifies the phase relationships is called an *ordered* genotype. If the two loci are on separated

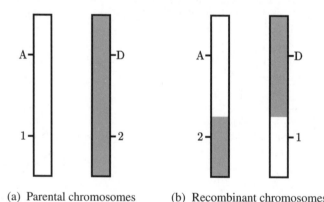

(a) Parental chromosomes   (b) Recombinant chromosomes

FIG. 2. Illustration of recombination.

chromosomes, or at widely separate positions on the same chromosome, then all four possible allele combinations should be found in the offspring with equal probability (all other factors being equal). However, if the loci are physically close, then the parental types A1 and D2 should occur with greater frequency than the *recombinant* types A2 and D1. If a recombination is seen, it means that an odd number of cross-overs has occurred. Estimation of the distances between loci can therefore be performed by estimating the recombination frequency r between them from the frequency of recombinant haplotypes. A simple model for cross-overs is that they occur with a uniform probability along a chromosome, with multiple cross-overs occurring independently. This simple model is incorrect because cross-overs on the same chromosome can *interfere* so that they are not independent; however, use of the no interference model does not lead to a large reduction in power for linkage analysis, and so is widely used.

1.1   Markers and trait loci

There are typically two functional types of loci involved in linkage analysis, namely marker loci and trait loci. Trait loci have a direct effect on the trait measured, with the relationship between genotypes at a trait locus and the trait being given by a penetrance function $p(y|g,\theta)$, the probability of trait data y given a genotype g and model parameters θ. Marker loci do not normally have any direct effect on the trait; instead, their role is to act as a map against which the trait loci are located. A linkage analysis is usually performed on the marker loci alone first to produce the map; the trait loci are then located on this map in a separate analysis. In human studies it is rarely necessary to map the marker loci as most available markers have already been placed onto publicly available maps; this is not necessarily true with other species.

The majority of marker loci used currently differ from trait loci in that it is possible to observe the genotypes directly, although there is a chance of error, so a penetrance function allowing for errors should be used. In practice, however, obvious errors (those causing inconsistencies between marker and pedigree data) are removed prior to the analysis with the remaining genotypes being assumed to be error free. This simplifies the analysis in several important ways (discussed in detail later), but the possible existence of undetected errors has implications for the validity of any results.

1.2   The computational difficulties of linkage analysis

Linkage analysis can be presented as a missing data problem – given complete information on genotypes and phase, it is simple to estimate recombination frequencies. A powerful way of performing linkage analysis is to design crosses between well categorized genetic lines of experimental organisms, which enable unambiguous determination of phase. It is often desirable, however, to perform linkage analysis in non-experimental organisms where planned crosses are not possible. Two examples of this are human linkage analysis and commercial animal production where producing large crosses is expensive. In these cases linkage analysis must be performed using the pedigree structures that arise 'naturally'. There is often a large amount of missing phase and genotype information (for example, it may not be possible to genotype ancestors in a pedigree) so it is necessary to sum over all combinations of this 'missing data' which are consistent with the observed data. This is complicated because genotypes of related individuals are correlated, so the summation cannot be carried out individually for each family member. In addition, there are correlations between the segregations of alleles at linked loci (which determine phase). These correlations are, however, local in nature, and this fact can be exploited to make the summation tractable in many cases.

1.3 Relationship of linkage to linkage disequilibrium analysis

Linkage analysis estimates the distance between loci from the number of recombinations occurring between the loci within the pedigrees being studied. This means that the pedigree size and structure places a limit on the resolution that can be obtained; to get good estimation of the probability of recombination when the true value is 0.01 would require a data set with several hundred informative meioses, which is large for human studies. However, a probability of recombination of 0.01 corresponds in humans to $\approx 10^6$ base pairs of DNA which, given that the average gene size in eukaryotes is 3×10^5 base pairs, can contain many genes. Note that the relationship between genetic and physical distance is not precise and varies between males and females and even between different chromosomal locations within the same individual.

If a trait is monogenic and is not influenced by environmental factors, then it is often possible to precisely determine whether a recombination has occurred between a marker locus and a trait locus. This is because, in this case, the observed trait data can precisely determine the trait locus genotype. With traits controlled by multiple genetic and environmental factors, this is no longer possible as the trait data cannot determine the trait locus genotype. For human studies of multifactorial traits, a practical limit on the resolution of linkage analysis is closer to a recombination probability of 0.1 corresponding to $\approx 10^7$ base pairs. For this reason, while linkage analysis is useful for determining the general location of trait loci, it cannot be used to determine exactly which gene is the trait gene. For this, a method with a finer resolution is required.

Linkage disequilibrium (LD) analysis works on a similar principle to linkage analysis, but rather than considering recombinations occurring within a pedigree, it uses all recombinations between a marker and a mutation at a trait locus since that mutation entered the population. Good descriptions of LD analysis can be found in many population genetics textbooks, but the important point for this discussion is that LD analysis can have a resolution on the order of a few thousand base pairs. Conversely, LD analysis typically works over a much smaller range than linkage analysis. With linkage analysis it is possible to detect linkage between a marker and a trait locus over a range of $\approx 3 \times 10^7$ base pairs in humans, whereas with LD analysis the range over which a signal can be detected could be as small as 3000 bases. LD studies, therefore, typically require a denser map of markers than linkage analysis studies. The resolution and range of LD analysis is highly population and trait specific as it depends on many factors, such as the age of the population, the age of the trait locus mutation, and the demographic history of the population. Linkage and LD analysis therefore tend to complement each other, and hierarchical approaches using linkage analysis to identify chromosome regions, followed by LD analysis for fine mapping of these regions are popular. It should be noted that the dichotomy between linkage and LD analysis is somewhat artificial in that they both are based on inferring recombination events, and it should be possible to combine information from pedigree and population sources when both are available. A way this can be done is to use techniques from population genetics (such as coalescent models, see Griffiths and Tavaré, Chapter 13 in this volume) to model the distribution of haplotypes in the population the pedigree founders are drawn from. This would be an improvement over standard linkage analysis where it is assumed that the founder population is in linkage and Hardy–Weinberg equilibrium.

2 Exact probability calculations

For linkage analysis, it is necessary to calculate $p(\mathbf{y}|\theta)$, where $\mathbf{y}$ denotes the observed data (both trait data $\mathbf{y}_t$ and marker data $\mathbf{y}_m$) and θ is a vector of model parameters such as the recombination frequencies between marker loci. As stated above, given the ordered genotypes at all loci for all individuals, this calculation is relatively simple. However, in the normal case where complete genotype information is not available, it is necessary to perform a summation over all possible ordered genotype configurations, which is complicated because of the correlations between the genotypes of related individuals caused by the fact that alleles segregate from parents to offspring. Assuming that no other methods of transferring genetic information can occur, all direct correlations are between parents and offspring. The correlation structure of a pedigree can be drawn as a directed acyclic graph by representing the multilocus ordered (i.e. with phase) genotypes of pedigree members as nodes, and adding arrows from every parent to each of their offspring showing the transmission of alleles. This can be moralized to form an undirected graph by converting the arrows to edges and adding an edge between the parental nodes (Fig. 3). The important feature of the graph is that all direct correlations are local – the only direct connections to an individual are to their parents, spouse(s), and children (if present). This can be seen in Fig. 3 which shows in grey the local neighbourhood of the individual in black.

2.1 The peeling algorithm

An algorithm (the peeling algorithm) has been described (Elston and Stewart 1971) that exploits the local nature of the dependencies to allow fast summation over all genotype combinations for simple unlooped pedigrees. An unlooped pedigree is one in which no parents are related (either by blood or by marriage).

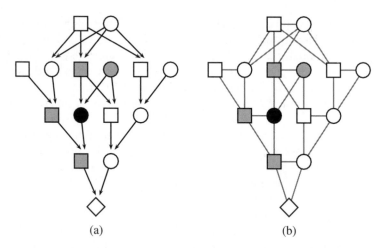

(a)   (b)

FIG. 3. Example pedigree drawn as a directed acyclic graph (DAG) showing the transmission of genes from parents to offspring (a) and as an undirected 'dependency' graph (b). The individuals in grey are the local neighbourhood of the individual in black.

The method was later extended to general (i.e. looped) pedigrees (Cannings *et al.* 1978). The idea behind peeling is to progressively simplify the dependency graph in Fig. 3 by removing nodes and expressing their contribution to the quantity being calculated as a joint function (an R-function in the terminology of Cannings *et al.*) of their immediate neighbours in the graph (i.e. those nodes that are directly connected to the selected node). To represent the joint function on the graph in Fig. 3, edges are drawn between the neighbouring nodes if they were not there before. For example, peeling (removing) the node in black from Fig. 3 (not the most efficient choice) would give the graph in Fig. 4(a). Subsequent removal of the mother of the individual in black, would then give the graph in Fig. 4(b).

It is clear from the way that peeling has been described here that the algorithm is a special case of the method of Lauritzen and Spiegelhalter (1988) for calculating probabilities on graphical structures. In common with all problems of this type, the efficiency of peeling depends on the initial complexity of the graph structure, the order in which nodes are removed from the graph, and the number of possible combinations at each node.

The initial complexity of the graph depends on the size and, more particularly, the structure of the pedigree. A pedigree with many loops will result in a highly connected graph that will be harder to peel. The order of removing nodes has a very large effect on the computational requirements. It can be seen from Fig. 4 that removing nodes can add extra edges to the graph, thereby increasing the complexity. The challenge is therefore to find an order that minimizes

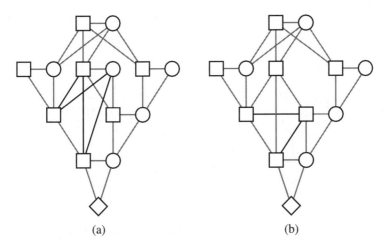

(a)   (b)

FIG. 4. Dependency graph for example pedigree after peeling the individual originally in black in Fig. 3 (a), and then the mother of that individual (b). New dependencies induced by each operation are shown in black.

this increase in complexity. Finding the optimal order is only feasible for small problems (Yannakakis 1981), but several approaches (developed for the ordering of the factorization of sparse matrices) are available which can quickly produce orders that are close to the best order (George and Liu 1989; Amestoy *et al.* 1996).

The number of possible combinations at each node depends on the number of alleles and, if multiple loci are being considered, the number of loci. The dependence on the number of loci is roughly exponential because the total number of combinations at a node is the product of the number of possible alleles at each locus. This makes the peeling algorithm unsuitable for large numbers of loci, and an alternative algorithm was developed (Lander and Green 1987), which behaves better in this case.

2.2   The Lander–Green algorithm

At first glance the Lander–Green (LG) algorithm appears quite different to the peeling algorithm. Rather than summing over genotypes, the LG algorithm performs summation over *segregation indicators* (SIs). We define *founders* as pedigree members who have no parents in the pedigree, and *founder alleles* as the alleles possessed by the founders. SIs are binary variables that describe, for each non-founder allele, the grandparental origin of that allele. For example, in Fig. 5(a), there are two non-founders, each with two SIs. For each non-founder, the left SI describes the segregation of the paternal allele, and the right SI describes the maternal allele. An SI of 0 denotes a grandmaternal origin for the corresponding allele, and 1 denotes a grandpaternal origin. The pattern of segregation that corresponds to this example is shown in Fig. 5(b). In this

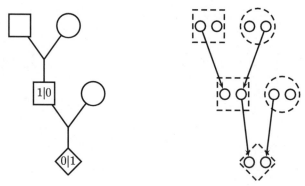

(a) Segregation indicators   (b) Corresponding segregation pattern

FIG. 5. Use of segregation indicators and patterns.

diagram the paternal and maternal alleles for each individual are on the left and right, respectively. Note that the four SIs in the figure completely determine the founder allele origins of each non-founder allele.

If the segregation indicators for a given meiosis at two loci differ, then a recombination has occurred. The SIs for all non-founders at one locus can be described by a binary string of length $2n$, where n is the number of non-founders. If the strings from two loci are compared, then the number of bits that differ gives the number of recombinations occurring between the loci. Assuming a no interference model for recombination events, dependencies between the SIs at different loci only occur between adjacent loci, in the sense that the dependency structure of this system can be represented as an undirected graph with one node per locus, where each node represents the joint states of all the SIs at that locus and edges connect adjacent loci. Calculating the probability of this graph can then be performed using the Lauritzen and Spiegelhalter (1988) approach. This calculation is relatively simple compared with pedigree peeling in theory because of the linear nature of the dependency graph, but is not practical for all but small pedigree structures due to the large number of possible states at each node (2^n).

Pedigree peeling and the LG algorithm can both be considered as special cases of the Lauritzen and Spiegelhalter approach, but using different latent variables (genotypes and SIs, respectively) and with the calculation proceeding in a different direction; pedigree peeling works by moving through the pedigree while LG works along the chromosome. Pedigree peeling ignores the local dependency structure along the chromosome, and the LG algorithm ignores the local dependency structure within the pedigree. For both algorithms, an important factor is the number of possible states at each node. With pedigree peeling this is primarily determined by the number of loci, whereas for chromosome peeling it is determined by the number of non-founders in the pedigree. In both cases, the number of states is exponential in the corresponding factor. For this reason, pedigree peeling is only suitable for a few loci (typically 3–6), and chromosome peeling is restricted to small pedigrees with 16–20 non-founders.

If all individuals could have all possible genotypes and all inheritance vectors were possible at all loci (i.e. the observed data did not place any restriction on which states were possible), then the best strategy would be to pick peeling over LG if the number of possible multilocus genotypes was less than the number of possible segregation patterns, which would depend on the size and structure of the pedigree and the number of loci. In many cases, however, the observed data place constraints on the possible genotype configurations or inheritance vectors. Typically in a linkage analysis, as has been previously described, many marker loci and relatively few trait loci are modelled. With marker loci, assuming no errors, observed individuals have at most two possible ordered genotypes, and in many cases the order is also determined by the genotypes of relatives. The number of possible genotypes for unobserved individuals may also be small if they have observed relatives. Similarly, certain SIs can be fixed by the observed data, reducing the number of possible segregation patterns. Returning to the dependency graphs for pedigree and chromosome peeling, when analysing marker loci the data can act to reduce the number of possible states at each node, thus simplifying the calculation. In most situations, this has a large effect in reducing the computational cost, and many modern programs for linkage analysis perform some type of pre-analysis whereby most impossible states are identified so they do not need to be further considered (Lange and Goradia 1987; Lange and Weeks 1989).

For pedigree peeling, further reduction in the number of states at each node can be achieved by allele recoding. The idea behind this is simple: linkage studies are often carried out on samples consisting of many separate pedigrees. For a particular locus, there may be many different allelic types when the whole sample is considered, but in a particular pedigree there may only be a subset of alleles present. It is then possible to reduce the number of possible states that have to be considered by lumping together the alleles which are not observed in a pedigree. A more sophisticated version of recoding can also be performed (O'Connel and Weeks 1995) where the recoding is done on an individual rather than pedigree basis. For each unobserved individual, the possible alleles that they could transmit to their descendants are noted, with all other alleles being lumped together and treated as one. The procedure is called set recoding or 'fuzzy inheritance' because non-transmitted alleles are treated as a set rather than as individual alleles. This procedure can be very effective at reducing the number of states that must be considered.

If any of the nodes have only one possible state due to the observed data, then any non-local dependencies mediated through that node disappear. This simplifies the dependency graph making the calculation easier. With multilocus data, using either peeling or the LG algorithm, it is rare that a node is completely determined. However, by modifying the dependency graph so that multilocus genotypes are split into multiple allele and SI nodes, it is possible to take more advantage of the constraints placed by the data.

To do this requires a more complicated representation of an individual's state at a single locus, using four nodes for each non-founder and two nodes for each founder. For a founder the two nodes are the maternal and paternal alleles. A non-founder has the same two allelic nodes plus two nodes representing the maternal and paternal SIs for that individual. A dependency graph showing the connections between the members of a parent–offspring trio at a single locus is shown in Fig. 6. If multiple linked loci are considered, then there will be a similar graph for each locus, with adjacent loci being connected only at corresponding SIs (Fig. 7). As before, this assumes no interference, as otherwise longer range dependencies between loci can occur which, while they can be accommodated, complicate the discussion needlessly.

If a SI is determined by the data, then the dependencies between adjacent loci for the corresponding alleles are broken. If alleles are fixed by the data known, then this can have a similar simplifying effect on the dependency graph. An important point to note is that there are no direct dependencies between different SIs at the same locus; this means that if some alleles are fixed, then

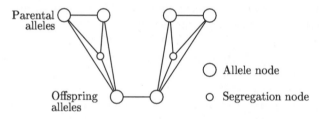

FIG. 6. Single locus dependency graph for pedigree trio.

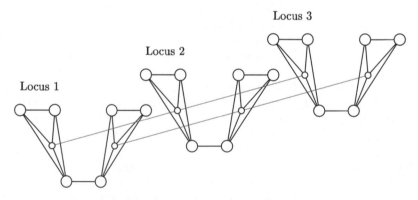

FIG. 7. Multilocus dependency graph for trio.

groups of disconnected SIs may be formed. Considering the crude approach of the LG algorithm, instead of having to consider 2^n states when moving from one locus to another, it may be possible to consider $\sum_i 2^{n_i}$ states, where n_i is the number of SIs in the ith disconnected group. The effect of this would be to greatly increase the size of problems for which exact computation is possible. A similar approach, effectively combining the LG and peeling algorithms making use of the observed data to simplify the calculations, was proposed by Lange (1997). The advantage of the graphical approach presented here over the scheme proposed by Lange is that the effect of the data on simplifying the calculations is more explicit and easier to exploit using the general Lauritzen and Spiegelhalter (1988) approach.

3   Markov chain Monte Carlo sampling schemes

The techniques described previously allow linkage calculations to be carried out on moderately complicated examples. However, as the number of loci, the number of untyped individuals, or the number of pedigree loops increase, the point is rapidly reached where exact calculations are not possible. Monte Carlo techniques are simple to use for small examples. Alleles for founders can be sampled independently, and then can be 'dropped' down the pedigree using Mendelian segregation rules, with the complete samples being weighted by their posterior probability. This strategy is only workable for very small examples otherwise the vast majority of samples are inconsistent with the observed data resulting in low efficiency. A large improvement in efficiency can be made using Markov chain Monte Carlo (MCMC) techniques (Metropolis *et al.* 1953; Hastings 1970). The use of these techniques for pedigree data was first suggested by (Sheehan *et al.* 1989), and many papers have been presented on the topic since then, for example Lange and Sobel (1991), Thompson (1991), Kong (1991a), Sheehan and Thomas (1993), Geyer and Thompson (1995), Lin (1995), Thompson (1996), Thomas and Gauderman (1996), Heath (1997), and Thompson and Heath (1999).

The application of MCMC to pedigree data is very natural because of the local dependency structure that was previously described in the context of exact calculations. Starting with the dependency graph for multilocus data given in Fig. 7, a general MCMC sampling scheme would be to fix a subset of the nodes, sample the remainder conditional on the fixed set, and then repeat fixing a different set of nodes. The set of nodes to fix is selected to make the task of sampling the remainder simple.

3.1   Single locus genotype samplers

The first pedigree sampler proposed was concerned with only a single locus, and proposed a Gibbs (Geman and Geman 1984) sampling step where the genotype of a single individual is updated conditional on the genotypes of the other pedigree members (Sheehan *et al.* 1989). Calculation of the conditional probabilities required for this only depends on the immediate neighbours in the dependency graph in Fig. 3, which are the genotypes of parents, spouses, and offspring. As this neighbourhood never gets very complicated, even for very large pedigrees, the update step is always straightforward.

It has been shown that for most biologically interesting models of inheritance, this sampling scheme is irreducible as long as there are just two alleles at the locus. When there are more alleles it is possible that the sampler is reducible (Sheehan and Thomas 1993); this is because there can be many configurations that have zero probability given the observed data, and the sampler cannot move through these invalid states. Several solutions have been proposed to the irreducibility problem, including relaxing model parameters so that all states have a non-zero probability and then re-weighting the resulting samples (Sheehan and Thomas 1993), and identifying non-communicating sets of configurations and explicitly proposing moves between them (Lin 1995). In practice, however, the irreducibility problem is less important than that of achieving a well *mixing* sampler. A sampler can be irreducible in theory, but be *practically* reducible because of the presence of subsets of configurations, where the probability of moving between subsets, while non-zero, is so low that it almost never happens. An obvious way to improve the mixing is to perform block sampling, updating genotypes for multiple (related) individuals simultaneously. For example, jointly updating all members of a nuclear family simultaneously can eliminate mixing/reducibility problems which are due to tight correlation between the genotypes of the family members. In practice, sampling nuclear families is not the best choice of strategy because with a little extra effort it is possible to sample much larger blocks of the pedigree simultaneously, which can be done using the peeling algorithm (Ploughman and Boehnke 1989). The danger with this approach is that it is difficult to prove that a particular sampling scheme is irreducible without being able to peel the entire pedigree.

3.2 Multilocus genotype samplers

Single locus samplers are not normally of direct use for linkage analysis because for this it is generally necessary to consider systems with at least two loci, a trait locus and a marker locus. More information can be obtained by considering multiple marker loci, so analyses may be carried out with 5, 10, or more marker loci. When dealing with multiple loci, as long as the probability of recombination between adjacent loci is greater than 0, the transition probability of moving from any configuration at one locus to any other configuration at the adjacent locus is also greater than 0. It is, therefore, *possible* to use the single locus samplers for multilocus data by performing the sampling conditional on the genotypes at neighbouring loci as well as on the genotypes of pedigree neighbours. If the mixing was bad for a single locus, however, using the same scheme for multiple loci will result in even worse mixing.

For pedigrees where it is possible to perform a single locus peeling calculation, the block samplers discussed earlier can be extended to sample the entire pedigree at one locus conditional on the SIs at the neighbouring loci (Kong 1991a). The resulting sampler is called the *locus sampler* as it updates one locus at a time.

The main drawbacks with the Kong approach are that it requires each locus to be peelable, and that it can have mixing problems due to strong correlations between the segregation patterns at adjacent loci.

The mixing of the sampler can be improved by conditioning on the ordered genotypes at neighbouring loci rather than on the SIs directly. For example, if a mother has a heterozygous sampled genotype for a locus, then it is possible to precisely determine the maternal SIs for her children. If, however, she has a homozygous sampled genotype, then the ordered genotypes provide no information about her children's maternal SIs at that locus. We denote such undetermined SIs as *non-informative*. Modifying the locus sampler to condition on ordered genotypes rather than on SIs therefore involves conditioning on the two closest informative SIs at each meiosis (in each direction along the chromosome), rather than on the SIs at the flanking loci.

3.3 Sampling segregation indicators

A different sampling framework has been proposed where single SIs rather than genotypes are sampled (Thompson 1994a,b) using a Metropolis update step. This was extended by Thompson and Heath (1999) to a multisite sampler; a meiosis from a parent to a child is chosen, and SIs at all loci for that meiosis are sampled simultaneously using a Gibbs step. The sampling step is a modification of the LG algorithm to allow sampling of a single meiosis conditional on observed data and the segregation indicators at other meioses. The original sampler described updated a single meiosis at a time, but there is no reason why multiple meioses cannot be updated simultaneously, except that the time and memory requirements for the sampler are proportional to 2^i, where i is the number of meioses being updated.

This *meiosis sampler* is reducible in some cases (Sobel and Lange 1996; Thompson and Heath 1999), and can also have mixing problems when there are large amounts of genetic data. The sampler was originally proposed for problems with large pedigrees that had data available only on a few people. In these cases, the locus sampler can be slow because the large amount of missing data makes peeling difficult. Where data are available on large proportions of the pedigree, however, the meiosis sampler tends to be slow and mix poorly when compared with the locus sampler (Szydlowski and Heath 2000).

3.4 General genetic samplers

The locus and meiosis samplers use modifications of the peeling and LG algorithms, respectively, to perform the update steps. In the same way that improvements in efficiency for exact calculations can be made by working with a more general graph structure where alleles and SIs are handled separately, the efficiency of samplers could be improved by working with the multilocus dependency graph (Fig. 7) to sample multilocus pedigree blocks rather than a single locus or a single meiosis.

A possible sampling scheme would be to fix enough of the nodes on the graph so that an exact sample is possible, and then repeat changing the fixed nodes each time. This has the advantage that exact sampling (i.e. independent samples from the posterior) will be used on simple pedigrees, with the sampler switching to updating relatively smaller blocks as pedigree size and complexity increase.

For example, with a data set containing a large number of loci it may be possible to perform an exact calculation on two or three loci together, but for more loci than that the calculation is no longer possible. This is because for a large number of loci the calculation has to progress more or less in a direction along the chromosome. After a certain number of loci have been processed, large numbers of the SI nodes become linked together by edges. To move from one locus to the next it is necessary to consider all the linked SI nodes jointly. As the sizes of the linked SI groups become larger, the number of possible states for each group rapidly becomes unmanageable. It is necessary, therefore, to prevent this joining together of the SI nodes by judicious fixing of nodes. It may be possible for this example that fixing of a relatively small number of allele or SI nodes at every second or third locus would allow the calculation to proceed along the chromosome. By switching the loci at which nodes are fixed at each sampling iteration, a valid sampler could be simply achieved. The main challenge to this approach is in finding an effective algorithm for identifying the nodes to fix at each sampling iteration for all pedigree types.

3.5   Samplers for imperfect data

Much of this article has been concerned with using the restrictions in the state space imposed by marker data. These restrictions only arise if it assumed that the marker data are perfect observations of the marker genotypes. This is not the case as the rate of genotyping errors (when an allele is mistakenly read as another allele) is on the order of 1%. A common approach to handling errors is to remove them prior to the linkage analysis, and this can be done through a variety of laboratory based (i.e. repeated observations) and statistical techniques (i.e. checking for consistency between the pedigree and marker data). These processes are not perfect; not all errors can be detected, and correctly observed alleles may be flagged as errors.

It is simple to develop an MCMC sampler that allows for errors in the marker data. However, as discussed earlier, to achieve an effectively mixing sampler it is necessary to implement some form of block sampler where multiple variables are updated together, and block sampling is computationally much more efficient when it can be assumed that marker data are error free. This problem may be avoided in the following way. Let y be the observed marker data, which may be inconsistent with the pedigree, and let z be a sample of 'corrected' marker data containing no errors. An MCMC scheme to sample z from $p(z|y)$ is developed; this scheme may not be computationally efficient because it has to allow for the presence of errors. Conditional on z, a more efficient sampling scheme can be employed, such as those discussed previously, to sample the missing genotypes and phase information necessary for the linkage calculations. By defining a complete sampling scheme where the updating of z is performed relatively rarely, the computational time required per iteration could be kept low. More work is required to establish how well this type of sampler would work in practice.

4 Sampling trait locus number

Some traits are largely controlled by a single gene, with environmental factors and other loci contributing little. For such traits, the relationship between the gene and the trait may be clearly defined by the pattern of inheritance of the trait. For many traits, however, the picture is less clear as they can be controlled by multiple interacting genetic and environmental factors. In these appropriately named *complex traits*, the pattern of inheritance gives little information about how the genes and environmental factors interact to affect the trait. Linkage analysis was traditionally focused on simple traits under single gene control, and conventional analytical techniques assume a single trait locus. Modelling multiple trait loci adds greatly to the computational difficulty of the calculation of linkage probabilities, although the use of MCMC schemes make such analyses possible. For complex traits, the number of trait loci is unknown prior to the analysis, which causes difficulties for conventional approaches. However, this problem can also be handled in the MCMC framework by making use of reversible jump MCMC (Green 1995; Richardson and Green 1997).

Several applications of reversible jump MCMC to linkage analysis have been proposed (Heath 1997; Thomas *et al.* 1997; Sillanpää and Arjas 1998, 1999; Stephens and Fisch 1998; Lee and Thomas 2000). These applications have concentrated on line crosses or nuclear pedigrees, with the exception of Heath (1997), who presented a method for much larger pedigrees based on the locus sampler described earlier. The difficulty of using reversible jump MCMC for genetic data is in achieving good joint proposals for the genotypes and effects of the new trait loci. For simple pedigree structures it is relatively straightforward to adjust the model parameters when changing the number of trait loci in a similar fashion to that described by Green (Chapter 6 in this volume). For general pedigrees, the problem of proposing genotypes *de novo* for new trait loci is a difficult problem in itself, and complicated joint proposals have to be formulated to raise the acceptance probability of the reversible jump steps to acceptable levels.

Despite the difficulties, the use of reversible jump MCMC for general pedigrees has shown itself to be effective at analysing several complex traits, and the estimates of the number of trait loci have proven surprisingly accurate (Daw *et al.* 1997; Heath 1997; Heath *et al.* 1997) with both simulated and real data sets.

5 Discussion

This article has covered linkage analysis and the difficulty of calculating linkage probabilities in general pedigrees with many marker loci. It has also covered how representing the problem as a graphical model makes it clear how to utilize local dependencies to simplify the calculations to a greater extent than achieved by existing algorithms. As pedigree size and complexity are increased, however, a limit is reached where exact calculation is infeasible, and a range of MCMC sampling schemes can be developed for these cases. As opposed to existing MCMC samplers for genetic data, the sampling schemes proposed here condition on the minimum number of variables necessary to allow efficient updating of the remaining variables. One benefit of this approach is that it should be more efficient when applied to a range of different pedigree types.

Current MCMC samplers for genetic data have generally been designed for particular types of data sets. For example, the meiosis sampler (Thompson and Heath 1999) was designed, and is most advantageous for, complex pedigrees with relatively few observed individuals. The locus sampler (Kong 1991a; Heath 1997), however, works well on general pedigrees with many people observed, but can be inefficient on small pedigrees, and may not work at all on complex pedigrees with few observed individuals. A recent study on nuclear families, which compared different reversible jump MCMC samplers for genetic data, concluded, not surprisingly, that the approaches developed specifically for nuclear families were most efficient in this case (Lee and Thomas 2000). This emphasizes the need for more general schemes that are efficient across different pedigree structures. The approach outlined in this article should avoid many of these problems; for small pedigrees it will perform exact probability calculations, while for large pedigrees it will use block sampling where the blocks are as large as possible.

This article may have given the impression that MCMC should be used only rarely in linkage analysis, as for many situations exact calculation is possible. In practice this is not true because linkage analysis of complex traits involves many unknown variables in addition to the missing genotype data, such as allele frequencies, effects of the trait locus, effects of environmental factors and, of course, the number of trait loci. Even if the pedigree is small enough to allow exact calculation of linkage probabilities, an MCMC scheme that samples the genetic data conditional on the other model parameters and vice versa will still be necessary in other than very simple situations. The use of MCMC techniques in this way can produce very flexible tools for linkage analysis of complex traits which are potentially more efficient and more widely applicable than current approaches to linkage analysis. The use of such tools could allow finer mapping and greater power when using linkage mapping by allowing larger and more complex data sets to be analysed in their entirety. However, the resolution of linkage analysis will always be too coarse to precisely determine the gene involved, and higher resolution techniques, such as LD analysis, will be needed to perform the final identification of the genes involved.

References

Amestoy, P. R., Davis, T. A., and Duff, I. S. (1996). An approximate minimum degree ordering algorithm. *SIAM Journal of Matrix Analysis and Applications*, **17**, 886–905.

Cannings, C., Thompson, E. A., and Skolnick, M. H. (1978). Probability functions on complex pedigrees. *Advances in Applied Probability*, **10**, 26–61.

Daw, E. W., Heath, S. C., and Wijsman, E. M. (1997). Multipoint oligogenic analysis of age of onset data with applications to large Alzheimer's disease pedigrees. *American Journal of Human Genetics*, **61**, A273.

Elston, R. C. and Stewart, J. (1971). A general model for the genetic analysis of pedigree data. *Human Heredity*, **21**, 523–42.

Geman, S. and Geman, D. (1984). Stochastic relaxation, Gibbs distributions and the Bayesian restoration of images. *IEEE Transactions on Pattern Analysis and Machine Intelligence*, **6**, 721–41.

George, A. and Liu, J. W. H. (1989). The evolution of the minimum degree ordering algorithm. *SIAM Review*, **31**, 1–19.

Geyer, C. J. and Thompson, E. A. (1995). Annealing Markov chain Monte Carlo with applications to ancestral inference. *Journal of the American Statistical Association*, **90**, 909–20.

Green, P. J. (1995). Reversible jump Markov chain Monte Carlo computation and Bayesian model determination. *Biometrika*, **82**, 711–32.

Hall, J. M., Lee, M. K., Newman, B., Morrow, J. E., Anderson, L. A., and King, B. H. M. (1990). Linkage of early-onset familial breast cancer of chromosome 17q21. *Science*, **250**, 1684–9.

Hastings, W. K. (1970). Monte Carlo sampling methods using Markov chains and their applications. *Biometrika*, **57**, 97–109.

Heath, S. C. (1997). Markov chain Monte Carlo segregation and linkage analysis for oligogenic models. *American Journal of Human Genetics*, **61**, 748–60.

Heath, S. C., Snow, G. L., Thompson, E. A., Tseng, C., and Wijsman, E. M. (1997). MCMC segregation and linkage analysis. *Genetic Epidemiology*, **14**, 1011–16.

Kerem, B., Rommens, J. M., Buchanan, J. A., Markiewicz, D., Cox, T. K., Chakravarti, A., Buchwald, M., and Tsui, L. (1989). Identification of the cystic fibrosis gene: genetic analysis. *Science*, **245**, 1073–80.

Kong, A. (1991a). Analysis of pedigree data using methods combining peeling and Gibbs sampling. In *Computer Science and Statistics: Proceedings of the Twenty-third Symposium on the Interface*, pp. 379–85. Interface Foundation of North America, Fairfax Station, VA.

Lander, E. S. and Green, P. (1987). Construction of multilocus genetic maps in humans. *Proceedings of the National Academy of Science of the U.S.A.*, **84**, 2363–7.

Lange, K. (1997). *Mathematical and Statistical Methods for Genetic Analysis*. Springer-Verlag, New York.

Lange, K. and Goradia, T. M. (1987). An algorithm for automatic genotype elimination. *American Journal of Human Genetics*, **40**, 250–6.

Lange, K. and Sobel, E. (1991). A random walk method for computing genetic location scores. *American Journal of Human Genetics*, **49**, 1320–34.

Lange, K. and Weeks, D. E. (1989). Efficient computation of lod scores: genotype elimination, genotype redefinition, and hybrid maximum likelihood algorithms. *Annals of Human Genetics*, **53**, 67–83.

Lauritzen, S. L. and Spiegelhalter, D. J. (1988). Local computations with probabilities on graphical structures and their application to expert systems. *Journal of the Royal Statistical Society*, B, **50**, 157–224.

Lee, J. K. and Thomas, D. C. (2000). Performance of Markov chain–Monte Carlo approaches for mapping genes in oligogenic models with an unknown number of loci. *American Journal of Human Genetics*, **67**, 1232–50.

Levy-Lahad, E., Wijsman, E. M., Nemans, E., Anderson, L., Goddard, K., Weber, J. L., Bird, T. D., and Schellenberg, G. D. (1995). A familial Alzheimer's disease locus on chromosome 1. *Science*, **269**, 970–3.

Lin, S. (1995). A scheme for constructing an irreducible Markov chain for pedigree data. *Biometrics*, **51**, 318–22.

Metropolis, N., Rosenbluth, A. W., Rosenbluth, M. N., Teller, A. H., and Teller, E. (1953). Equations of state calculations by fast computing machines. *The Journal of Chemical Physics*, **21**, 1087–91.

O'Connell, J. R. and Weeks, D. E. (1995). The VITESSE algorithm for rapid exact multilocus linkage analysis via genotype set-recoding and fuzzy inheritance. *Nature Genetics*, **11**, 402–8.

Ploughman, L. M. and Boehnke, M. (1989). Estimating the power of a proposed linkage study for a complex genetic trait. *American Journal of Human Genetics*, **44**, 543–51.

Richardson, S. and Green, P. J. (1997). On Bayesian analysis of mixtures with an unknown number of components. *Journal of the Royal Statistical Society*, B, **59**, 731–92.

Schellenberg, G. D., Bird, T. D., Wijsman, E. M., Orr, H. T., Anderson, L., Nemans, E., White, J. A., Bonnycastle, L., Weber, J. L., Alonso, M. E., Potter, H., Heston, L. L., and Martin, G. M. (1992). Genetic linkage evidence for a familial Alzheimer's disease locus on chromosome 14. *Science*, **258**, 668–71.

Sheehan, N., Possolo, A., and Thompson, E. A. (1989). Image processing procedures applied to the estimation of genotypes on pedigrees. *American Journal of Human Genetics*, **45** (Suppl.), A248.

Sheehan, N. and Thomas, A. (1993). On the irreducibility of a Markov chain defined on a space of genotype configurations by a sampling scheme. *Biometrics*, **49**, 163–75.

Sillanpää, M. J. and Arjas, E. (1998). Bayesian mapping of multiple quantitative trait loci from incomplete inbred line crosses. *Genetics*, **148**, 1373–88.

Sillanpää, M. J. and Arjas, E. (1999). Bayesian mapping of multiple quantitative trait loci from incomplete outbred offspring data. *Genetics*, **151**, 1605–19.

Sobel, E. and Lange, K. (1996). Descent graphs in pedigree analysis: applications to haplotyping, location scores and marker-sharing statistics. *American Journal of Human Genetics*, **58**, 1323–37.

Stephens, D. A. and Fisch, R. D. (1998). Bayesian analysis of quantitative trait locus data using reversible jump Markov chain Monte Carlo. *Biometrics*, **54**, 1334–47.

Szydlowski, M. and Heath, S. C. (2000). Performance of hybrid MCMC samplers on a large complex pedigree [abstract]. *American Journal of Human Genetics*, **67** (Suppl. 2), 207.

Thomas, D. C. and Gauderman, W. J. (1996). Gibbs sampling methods in genetics. In *Markov chain Monte Carlo in Practice* (eds W. Gilks, S. Richardson, and D. Spiegelhalter). Chapman & Hall, London.

Thomas, D. C., Richardson, S., Gauderman, J., and Pitkäniemi, J. (1997). A Bayesian approach to multipoint mapping in nuclear families. *Genetic Epidemiology*, **14**, 903–8.

Thompson, E. A. (1991). Probabilities on complex pedigrees; the Gibbs sampler approach. In *Computer Science and Statistics: Proceedings of the Twenty-third Symposium on the Interface*, pp. 371–8. Interface Foundation of North America, Fairfax Station, VA.

Thompson, E. A. (1994a). Monte Carlo estimation of multilocus autozygosity probabilities. In *Proceedings of the 1994 Interface Conference* (eds J. Sall and A. Lehman), pp. 498–506. Interface Foundation of North America, Fairfax Station, VA.

Thompson, E. A. (1994b). Monte Carlo likelihood in genetic mapping. *Statistical Science*, **9**, 355–66.

Thompson, E. A. (1996). Likelihood and linkage: from Fisher to the future. *Annals of Statistics*, **24**, 449–65.

Thompson, E. A. and Heath, S. C. (1999). Estimation of conditional multilocus gene identity among relatives. In *Statistics in Molecular Biology and Genetics*, IMS Lecture Notes–Monograph Series, 33, pp. 95–113. Institute of Mathematical Statistics, Hayward, CA.

Yannakakis, M. (1981). Computing the minimum fill-in is NP-complete. *SIAM Journal of Algebraic Discrete Methods*, **2**, 77–9.

12A
Graphical models for mapping continuous traits

Nuala A. Sheehan
University of Leicester, UK

Daniel A. Sorensen
Danish Institute of Agricultural Sciences, Foulum, Denmark

1 Introduction

As discussed in Heath's article, probability and likelihood calculations routinely required for linkage analyses can pose enormous computational problems when general pedigrees and many marker loci are involved. They are completely intractable, even for a single locus model, on the large complex pedigrees that frequently arise in animal populations. Markov chain Monte Carlo (MCMC) methods have not really been tested extensively on these large problems and tend to be viewed with some suspicion in practice, due to the unreliability of the resulting estimates. Consequently, pedigree information is often discarded and the power to detect linkage reduced. A modelling framework that facilitates the development and testing of different MCMC sampling schemes is crucial in these applications. Here, we continue the discussion on dependency graphs introduced by Heath and argue that *graphical models* (Lauritzen 1996), to use the more general term, provide such a framework.

The use of graphs is not new in genetics, dating back to the path analysis diagrams of Wright (1934). Indeed, the standard representation of a pedigree (see Fig. 2(a)) is a graphical model that ignores the extra probabilistic dependencies imposed by the genetic model. Although featuring explicitly in specific applications in genetics (Kong 1991b; Jensen and Kong 1999; Lund and Jensen 1999), the general applicability of a graphical model approach to solving complex problems in genetics due to the natural modular structure of these problems, has not been widely appreciated. The ideas are illustrated here with an application to the question of mapping a quantitative trait locus (QTL) from marker data on a simple design.

2 QTL mapping on a granddaughter design

Quantitative traits, such as height, weight, etc. exhibit variation without natural discontinuities. This is a consequence of the simultaneous segregation of many genes (*polygenic* variation) superimposed, perhaps, by some truly non-genetic continuous variation. We can think of a QTL as a segment of chromosome affecting a quantitative trait and is essentially a 'gene' with a sizeable effect on the trait of interest but which cannot be detected with Mendelian methods (Falconer and Mackay 1996). We assume that the locus is diallelic with alleles Q and q.

Consider the simplest possible scenario where there are only two marker loci with alleles M, m and N, n, respectively, and known allele frequencies p_M, $1 - p_M$, p_N, and $1 - p_N$. The question of interest is whether there is a section of chromosome between these two markers that codes for the trait. Differences, within families, in mean phenotypes among the genotype classes of a marker would support this hypothesis. The map distance is known so the probability of recombination between the two markers, r, is given. The QTL position is not known so the recombination fractions, r_m and r_n, between each marker and the QTL as shown in Fig. 1, have to be estimated. Under the assumption of no genetic interference, r_m and r_n are independent so only one of the two is necessary.

Standard approaches including analysis of variance (ANOVA) and likelihood methods are restricted to crossed genetic lines or very simple designs extracted from a much larger pedigree structure, such as the granddaughter design which we consider here. Bayesian methods tend to be restricted in practice to pedigrees that can be peeled for one locus in order to avoid the potential reducibility of the single-site Gibbs sampler; see Hoeschele *et al.* (1997) for an overview.

Consider the QTL detection problem of Fig. 1 where the trait of interest is milk yield in dairy cows, for example. In a typical granddaughter design, 10 to 15 bulls are chosen (the grandsires) each of which has from 50 to 100 sons (the sires) who each in turn has about 50 daughters. The grandsires and sires are typed at the two marker loci. The daughters have a phenotypic record (i.e. milk yield) giving information on the QTL. All information on the dams (the mothers of the sires and the daughters) is ignored and in the absence of such knowledge, they are all assumed to be different. The basic structure of the design illustrating one grandsire, one sire, and two daughters is shown in Fig. 2(a).

To represent this problem as a graphical model the following nodes and their interrelating probabilities are defined. For each grandsire, maternally (indexed by 0) and paternally (indexed by 1) inherited genes at each of the three loci are randomly assigned from a Bernoulli distribution with the relevant allele frequencies: M_{G0}, M_{G1} at the 'M-locus', Q_{G0}, Q_{G1} at the QTL, and N_{G0}, N_{G1} at the

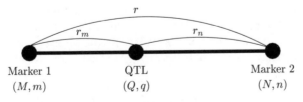

FIG. 1. QTL mapping.

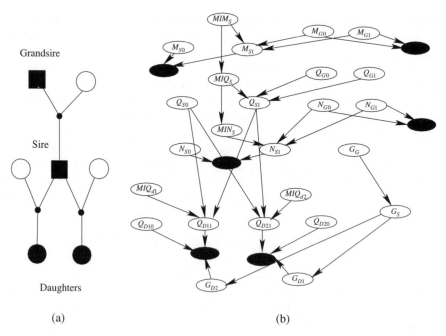

Grandsire

Sire

Daughters

(a)   (b)

FIG. 2. A simple granddaughter design and corresponding graphical model for the QTL mapping problem. The black nodes in (a) indicate individuals for whom we have genetic data. The black nodes in (b) represent the observed genotypes at both marker loci for the grandsire and sire and the milk yield records, P_{D1} and P_{D2}, for both daughters.

'N-locus'. There is a further node for each observed marker genotype with values determined by the assigned genes and an additional node, G_G, to represent the unlinked polygenic effect that defines the covariance structure between related animals. It is assumed that $G_G \sim \mathrm{N}(0, \sigma_a^2)$, where σ_a^2 is the polygenic variance.

For the sires, phase and linkage must be accounted for, so *meiosis* (or *segregation*) indicators are required for each locus with values 0 and 1 representing maternal and paternal inheritance, respectively (see Fig. 5 in Heath's article). At the M-locus, say, the sire's meiosis indicator (MIM_S) is randomly assigned from a Bernoulli distribution with probability 0.5, i.e. inheritance is Mendelian. Whether or not the maternal or paternal allele of the grandsire is passed to the sire at the QTL now depends on the value of this indicator and the recombination fraction, r_m. Specifically, for any given value of r_m, the sire's meiosis indicator at the QTL is given by

$$MIQ_S \sim \begin{cases} B(1, r_m) & \text{if } MIM_S = 0, \\ B(1, 1 - r_m) & \text{if } MIM_S = 1. \end{cases}$$

The meiosis indicator for the second marker locus, MIN_S, depends on MIQ_S and r_n and is defined analogously. Nodes representing the sire's marker and QTL genes are now assigned. The maternal genes, M_{S0}, Q_{S0}, and N_{S0} are randomly drawn from the general population with relevant allele frequencies and the paternal gene at each locus will be a copy of the paternal gene in the grandsire if the corresponding meiosis indicator has a value of 1 and will be a copy of the grandsire's maternal gene, otherwise. These are denoted by M_{S1}, Q_{S1}, and N_{S1}, respectively. Note that this part of the model is essentially the modified dependency graph for three linked loci shown in Fig. 7 in Heath's article without the maternal segregations. For each marker, a further node represents the sire's observed genotype and depends on the assigned alleles. Finally, the unlinked polygenic effect inherited by the sire from the grandsire, G_S, is included with conditional distribution $G_S \sim \mathrm{N}(\frac{1}{2}G_G, \frac{3}{4}\sigma_a^2)$, since an offspring shares half its genes with its sire.

For the ith daughter, in the absence of marker information, the QTL meiosis indicator, MIQ_{di}, is a Bernoulli random variable with probability 0.5. The maternal QTL gene, Q_{Di0}, is drawn randomly from the general population, as before, and the paternal gene inherited from the sire, Q_{Di1}, depends on the value of MIQ_{di}. The unlinked polygenic effect has the distribution $G_{Di} \sim \mathrm{N}(\frac{1}{2}G_S, \frac{3}{4}\sigma_a^2)$ and the observed phenotype is also assumed to have a normal distribution which depends on the polygenic effect and the QTL genotype: $P_{Di} \sim \mathrm{N}(G_{Di} + \alpha_j, \frac{3}{4}\sigma_a^2 + \sigma_e^2)$, where $\alpha_j, j = 1, 2, 3$ is the QTL genotype effect and σ_e^2 is the environmental variance. Figure 2(b) shows this graphical model for the design of Fig. 2(a). The graph for a typical design would comprise over 200 000 nodes.

Clearly there are several faults with the design and model: high selectivity amongst breeding bulls leads to an obvious violation of the normality assumptions on grandsire effects, maternal inheritance is completely ignored even when data on dams are available, known complex pedigree links between all these animals are also ignored, and a model for multiple polymorphic markers and multiple QTLs would be more realistic. In addition, the genetic map may be incorrect and marker data are frequently incomplete due to typing errors. Despite its simplicity, however, exact calculation for the above model is already a problem for any reasonably sized granddaughter design and MCMC methods are required.

Model extensions and extra pedigree links can easily be incorporated into the graphical model by careful addition of extra nodes and links, and any program that computes probabilities on general graphical models or *Bayesian Networks* (Jensen 1996) such as HUGIN (Andersen *et al.* 1989), can be adapted to carry out these analyses. A pedigree is a very special type of Bayesian network and some modifications to a general program are required in order to perform these calculations efficiently. In particular, it is crucial to be able to exploit the often considerable reductions in complexity imposed by the data. Whole sections of the pedigree might be uninformative for a given data set and should be discarded before a peeling sequence is sought. This process of *clipping* is equivalent to the requirement that data be entered prior to triangulation of a Bayesian network and is a standard component in the pre-analysis routines for reducing computational complexity, as discussed more fully by Heath.

3 Conclusion

The main advantage of this approach is that many complex problems in genetics can be studied with the same program and commercially available software can be used. The computational issues discussed by Heath are still pertinent but calculations are simplified by exploiting local dependencies to a greater extent than existing algorithms, which also tend to be more problem-specific. In particular, the graphical model does not distinguish between pedigree peeling and chromosome peeling as there is no clear differentiation between the various nodes and links, and the best features of both can be combined. The representation of a complex problem via its simplest components allows for greater flexibility and changes to the model or pedigree structure are much easier to incorporate. These features are very appealing in an area where extensive testing and development of MCMC sampling schemes is required.

Acknowledgements

We acknowledge support from Wellcome Biomedical Research Collaboration Grant 056266/Z/98/Z and support for Nuala Sheehan from the TVW Telethon Institute for Child Health Research, Perth, Western Australia.

12B
Statistical approaches to genetic mapping

David A. Stephens
Imperial College, London, UK

The analysis of genetic data, specifically in relation to the modelling of population-level variation in disease, is one of the most challenging areas for statisticians. Advances in molecular technology have now facilitated routine, large-scale genetic screening, and such data will provide key information in the study of diseases such as hypertension and cardiovascular disease, helping to quantify purely genetic and gene–environment components of variation and risk.

Traditionally, in classical statistical genetics, emphasis has centred on inference for genetically related individuals, usually in relatively small families or pedigrees. As Heath's article clearly demonstrates, probabilistically-based inference for such data can be extremely complicated, requiring sophisticated models and algorithms, even in such small problems. As pedigree size increases (to include up to tens of thousands of individuals) and model complexity increases (multi-locus, quantitative traits, etc.), complete and accurate statistical inference becomes a demanding task.

The developments and extensions to pedigree-based analysis, the primary emphasis of this article, that will become increasingly important in a medical or public health context relate to the following issues:

- the population under study – increasing emphasis will be placed on general population or case–control studies on apparently unrelated individuals, rather than on directly related family members;

- the form of the data – for example *fine-scale mapping* (via high-resolution genetic data at the marker or even nucleotide level), augmented by data from other sources also used for genetic analysis (molecular methods and sequence data, single nucleotide polymorphisms, gene function-related data), to be incorporated in a coherent fashion into the linkage-based analysis;

- the specifics of the statistical inference – discrete traits further extended to quantitative traits, models of *epistasis* (co-dependence amongst the genetic effects/contributions of different genes) and environmental factors, etc.

The first of these points is perhaps the most relevant to practical modern medical genetics, and the fundamental issues are discussed by Clayton (2001); the problems that are present in most population-level or epidemiological studies are also inevitable in population-level genetic analysis. Chapter 12 does not include population genetics ideas; they are discussed elsewhere in this volume (see the article in Chapter 13 by Griffiths and Tavaré). However, it is clear that issues in population genetics (population modelling and dynamics, evolutionary processes, population heterogeneity) will play more central roles in practical statistical genetics analysis.

The distinction between traditional pedigree-based analysis and population statistical genetic approaches reduces to how the joint distribution for the genetic information present in the data is represented. As both approaches rely on ancestral inference of some kind, the real distinction between these two types of analysis is merely the degree to which missing genotype information is to be imputed; in pedigree analysis, the missing data (on the pedigree and of pedigree founders) are relatively few, and can be imputed probabilistically using the Mendelian laws of inheritance, whereas in population statistical genetics, the missing data (in the form of an ancestral tree, genotypes, mutation, and recombination events), are many, and the stochastic process models used to define population development are somewhat different. As outlined in Heath's article, simulation methods have provided a natural inferential tool for pedigree analysis. They are also beginning to become routine in population statistical genetics; see, for example, Wilson and Balding (1998) and Stephens and Donnelly (2000). The computational problems associated with the two approaches are apparently somewhat different. In pedigree analysis, concern typically centres on convergence of a Markov chain Monte Carlo (MCMC) algorithm, and irreducibility of the Markov chain itself, alongside efficiency, whereas for ancestral inference issues of traversing a large parameter (ancestral tree) space are encountered. Despite such difficulties, MCMC methods are proving valuable in practical population statistical genetics; see, for example, Cooper *et al.* (1999) and Nielsen (2000). The impact of these computational methods on techniques for the mapping of disease genes is yet fully to be felt, but will be profound.

1   Association mapping and genome-wide maps

Classical statistical genetics analysis has centred on data from individuals who are related through some family or pedigree structure; for linkage analysis, of course, such pedigree information is essential so that recombination events can be counted. An alternative approach to the localization of disease genes is *allelic association* or *linkage disequilibrium (LD)* mapping. There are many related methods for assessing the level of association statistically; one method is based on *transmission disequilibrium (TD)*, that is, the relative frequency with which each allele is transmitted/not transmitted from parent to affected child; any serious imbalance may indicate significant association.

Pure association mapping is often feasible, but is more profitably used in conjunction with linkage mapping. Theoretical reasoning suggests that the most effective mapping strategy is to use linkage analysis to obtain a general localization, and then to use association methods to provide more accurate estimates of locus position; in any case, it is widely suggested that informative imbalances in association only extends over relatively short genomic ranges. An example of the use of association mapping in conjunction with linkage analysis is given by Copeman *et al.* (1995), re-analysed by Curnow *et al.* (1998); localizations of a disease locus determining susceptibility to insulin dependent diabetes mellitus (IDDM) in the vicinity of marker D2S152 on chromosome 2 is required. D2S152 is a multiallelic marker; TD tests were used to discover which if any of the alleles is most associated with the disease phenotype.

In a real mapping exercise, the genetic data may become available from different types of study. Conventionally, data from affected nuclear families permit linkage and association analysis, but perhaps more and more typically, data arising from prospective studies on unrelated individuals, or from case–control studies again on affected/unaffected but unrelated individuals, will be required. There are many potential pitfalls of such population-level analysis; in particular, assumptions of population heterogeneity may lead to seriously misleading results. A discussion of the advantages and disadvantages of different types of study, and alternatives to disequilibrium-based tests is given by Pritchard *et al.* (2000).

In *fine scale mapping*, the linkage/association mapping problem is computationally demanding. The details and requirements of formal statistical inference begin to become prohibitive – the likelihood is constructed in the usual way, but contains many more terms. Morris and Whittaker (2000) give a detailed description and discussion of likelihood construction and inference for fine scale association mapping of loci for Huntington's disease and Cystic Fibrosis.

As a final extension, mapping on a genomic scale has been successful in uncovering evidence of genes responsible for disease susceptibility. The IDDM case again proves illustrative; genome-wide mapping studies have provided evidence for several susceptibility loci involved in type 1 diabetes (see, for example, Davies *et al.* 1994; Owerbach and Gabbay 1996; Owerbach 2000). In these genome wide studies, however, genes are localized to a very broad region on a specific chromosome, and therefore fine scale mapping is the final stage of a gradual localization process of increasing accuracy, from upwards of tens of Mb initially, down to a few kb at the last stage of analysis. A multistage mapping approach is inevitable, and there is a requirement for coherence between the different stages of the analysis.

Many authors have attempted comparisons of the various available techniques, and provided evidence (for linkage over LD, or vice versa) in specific genetic contexts. The approach described above (broad scale via linkage, fine scale via allelic association or disequilibrium) seems to be generally effective, but a systematic comparison of these approaches in a sufficiently wide variety of real data problems, in a formal model validation setting, is yet to be completed. Bayesian methods are becoming more and more prominent, as indicated by the material and references in this chapter, and discussion on the use of MCMC approaches to population inference (see Griffiths and Tavaré, Chapter 13 in this volume). Data used in different types of analysis, but derived from the same genetic location, are highly dependent. It is not immediately clear which approach provides the most accurate means by which the required region is identified, and how accuracy trades off with efficiency and cost. A hierarchical modelling strategy resolves any issues of lack of coherence in multistage analyses; provided we formulate the different components of the model correctly. Furthermore, as advances in molecular technology lead to the availability of more and more refined data at the sequence level, it is legitimate to assess the future use of genetic mapping methods over high-resolution analysis.

2 Future developments: data mining approaches to genetic mapping

The generic inference problem in relation to explaining variation in genetic disease is a familiar statistical one; given a large amount of predictor variable information (multilocus marker genotype data for individuals and possibly their family members, other patient-specific covariate information, data on environmental factors) and a possibly multivariate response variable (disease status, quantitative trait values), we seek a statistical model in the predictors that explains the variation in the response. The genetic aspect (that informs us about which statistical models are more appropriate, by specifying rules that govern how the predictor information can/must be related) is important, but sometimes clouds our impression of how we might most effectively go about inference.

High-throughput statistical data analysis, including *data mining* techniques, have proved useful in drawing inferences in large and complex data sets. Although often informal and unprincipled, data mining techniques can be used as an exploratory tool, a precursor to formal statistical analysis. However, principled data mining approaches, based on legitimate statistical principles but utilizing approximation methods, and algorithmic or computational devices, can be regarded as potential competitors to exact approaches.

One aspect of data mining that has wide potential use is the detection of significant or anomalous patterns in the data. In a genetic context, for example, we may wish to relate a coded marker genotype vector pattern to the incidence of a disease, or the level of some quantitative trait. Recently, such techniques have begun to make an impact in statistical genetics. For example, Toivonen *et al.* (2000) report on the use of pattern detection techniques in the association mapping of a gene again related to type I diabetes. In essence their analysis (referred to as Haplotype Pattern Mining, or HPM) proceeds as follows: genotype data, comprising 25 polymorphic microsatellite marker loci from a large (14 Mb) region of chromosome 6, spanning two candidate IDDM susceptibility loci (HLA-DQB1 and HLA-DRB1), were obtained from 385 affected sib-pairs (see Bain *et al.* (1990) for full details). Thus, for each individual, a vecto observation of genotype (in fact, haplotype) data (25 integers) was used to predict/model the disease status of the individual. The HPM approach searches the data for haplotype patterns that correspond to (or predict, or have 'close' association with) haplotypes of the disease-associated chromosomes; exact and approximate matches are weighted appropriately. This approach takes into account mutation and recombination (in the way that test patterns are formulated and matched) but does not incorporate them explicitly into a likelihood model. Toivonen *et al.* (2000) use extensive simulation studies (for microsatellite and SNP data), and the real HLA data, to verify that their HPM approach produces comparable localization accuracy to other association mapping techniques, and indicate that their approach is generalizable to more complex mapping problems.

This general approach, where the genetic nature of the data analysis is sublimated, opens up a range of intriguing possibilities, and poses some interesting questions. If the HPM approach is competitive for large data sets, and Toivonen *et al.* (2000) claim large power even in small simulated studies of 100 affected individuals, then to what extent can the genetic nature of the data be ignored in analysis? Can other complicating factors (epistasis, environmental effects, the influence of other covariates) be readily incorporated into the HPM analysis? From a purely statistical viewpoint, if the genetic context is temporarily ignored, then the inferential issues are immediately evident, and there are many potential methods of inference – we wish to relate a response to a covariate vector, and so standard generalized linear, additive, or flexible modelling techniques can be used.

As the amount of relevant genetic and ancillary information increases, it is possible that the complexity of formal genetic modelling of the data necessitates a modelling and/or computational compromise; in any case, such approaches deserve attention. As a parallel, it may be that generic classification/discrimination and pattern detection methods will also prove to be extremely useful in a genetic mapping context. These methods are already being used in the analysis of gene expression profiles (for example, to identify and characterize types of tumour, but in such way that the genetic context is being largely ignored) and in biological sequence analysis for nucleotide and protein sequences, but less so in the mapping environment. A new challenge for statisticians is to develop and utilize these techniques as a partial solution to the localization of disease genes.

Additional references in discussion

Andersen, S. K., Olesen, K. G., Jensen, F. V., and Jensen, F. (1989). HUGIN — a shell for building Bayesian belief universes for expert systems. In *Proceedings of the Eleventh International Joint Conference on Artificial Intelligence*, pp. 1080–5. Morgan Kaufmann, San Mateo, CA.

Bain, S., Todd, J., and Barnett, A. (1990). The British Diabetic Association – Warren repository. *Autoimmunity*, **7**, 83–5.

Clayton, D. (2001). Population association. In *Handbook of Statistical Genetics* (eds D. J. Balding, M. Bishop, and C. Cannings). Wiley, Chichester.

Cooper, G., Burroughs, N. J., Rand, D. A., Rubinsztein, D. C., and Amos, W. (1999). Markov chain Monte Carlo analysis of human Y-chromosome microsatellites provides evidence of biased mutation. *Proceedings of the National Academy of Science*, **96**, 11916–21.

Copeman, J. B., Cucca, F., Hearne, C. M., Cornall, R. J., Reed, P. W., Ronningen K. S., Undlien, D. E., Nistico, L., Buzzetti, R., and Tosi, R. (1995). Linkage disequilibrium mapping of a type 1 diabetes susceptibility gene (IDDM7) to chromosome 2q31-q33. *Nature Genetics*, **9**, 80–5.

Curnow, R. N., Morris, A. P., and Whittaker, J. C. (1998). Locating genes involved in human diseases. *Applied Statistics*, **47**, 63–76.

Davies, J. L., Kawaguchi, Y., Bennett, S. T., Copeman, J. B., Cordell, H. J., Pritchard, L. E., Reed, P. W., Gough, S. C. L., Jenkins, S. C., Palmer, S. M., Balfour, K. M., Rowe, B. R., Farrall, M., Barnett, A. H., Bain, S. C., and Todd, J. A. (1994). A genome-wide search for human type 1 diabetes susceptibility genes. *Nature*, **371**, 130–6.

Falconer, D. S. and Mackay, F. C. (1996). *Introduction to Quantitative Genetics* (4th edn). Longman, Harlow, UK.

Hoeschele, I., Uimari, P., Grignola, F. E., Zhang, Q., and Gage, F. M. (1997). Advances in statistical methods to map quantitative trait loci in outbred populations. *Genetics*, **147**, 1445–57.

Jensen, F. V. (1996). *An Introduction to Bayesian Networks*. UCL Press, London.

Jensen, C. S. and Kong, A. (1999). Blocking Gibbs sampling for linkage analysis in large pedigrees with many loops. *American Journal of Human Genetics*, **65**, 885–901.

Kong, A. (1991b). Efficient methods for computing linkage likelihoods of recessive diseases in inbred pedigrees. *Genetic Epidemiology*, **8**, 81–103.

Lauritzen, S. L. (1996). *Graphical Models*. Oxford University Press, UK.

Lund, M. S. and Jensen, C. S. (1999). Blocking Gibbs sampling in the mixed inheritance model using graph theory. *Genetics, Selection and Evolution*, **31**, 3–24.

Morris, A. P. and Whittaker, J. C. (2000). Fine scale association mapping of disease loci using simplex families. *Annals of Human Genetics*, **64**, 223–37.

Nielsen, R. (2000). Estimation of population parameters and recombination rates from single nucleotide polymorphisms. *Genetics*, **154**, 931–42.

Owerbach, D. (2000). Physical and genetic mapping of IDDM8 on chromosome 6q27. *Diabetes*, **49**, 508–12.

Owerbach, D. and Gabbay, K. H. (1996). The search for type 1 diabetes susceptibility genes: the next generation. *Diabetes*, **45**, 544–51.

Pritchard, J. K., Stephens, M., Rosenberg, N. A., and Donnelly, P. (2000). Association mapping in structured populations. *American Journal of Human Genetics*, **67**, 170–81.

Stephens, M. and Donnelly, P. (2000). Inference in molecular population genetics. *Journal of the Royal Statistical Society*, B, **62**, 605–55.

Toivonen, H. T. T., Onkamo, P., Vasko, K., Ollikainen, V., Sevon, P., Mannila, H., Herr, M., and Kere, J. (2000). Data mining applied to linkage disequilibrium mapping. *American Journal of Human Genetics*, **67**, 133–45.

Wilson, I. J. and Balding, D. J. (1998). Genealogical inference from microsatellite data. *Genetics*, **150**, 499–510.

Wright, S. (1934). The method of path coefficients. *Annals of Mathematical Statistics*, **5**, 161–215.

13

The genealogy of a neutral mutation

R. C. Griffiths
University of Oxford, UK

Simon Tavaré
University of Southern California, USA

1   Introduction

Technological advances in molecular biology have made it possible to survey
DNA sequence variation in natural populations. These data include restriction
fragment length polymorphisms, microsatellite repeats, single nucleotide polymorphisms, and complete DNA sequences of many loci; for more details, we
refer to Chapter 2 of Hartl and Jones (2001). The analysis and interpretation of
the patterns of variation seen in such data are complicated by the fact that the
sampled chromosomes share a common ancestry, thus making the data highly
dependent; for example, a mutation appearing in an ancestor is carried by all
descendants of that ancestor. To make matters worse, the nature of this common ancestry is not known precisely and therefore needs to be modelled. Since
the pioneering work of Kingman (1982a), Tajima (1983), and Hudson (1983),
population geneticists have used *coalescent models* as a stochastic description of
the ancestry of a sample of chromosomes, and there is now an extensive literature on theory and inference for such models. See, for example, Hudson (1990),
Donnelly and Tavaré (1995), Nordborg (2001), and Stephens (2001).

One aspect of the theory that has received a lot of attention concerns the
age of mutations. For historical overviews, see Watterson (1996) and Slatkin
and Rannala (2000). The ages of mutations are of interest to human geneticists
trying to map disease mutations using linkage disequilibrium methods. More
on this aspect can be found in Nordborg and Tavaré (2002). Kimura and Ohta
(1973) studied the age of a mutation known to have a certain frequency in a
population using diffusion methods, and these results were recast in the coalescent framework by Griffiths and Tavaré (1998, 1999), Wiuf and Donnelly (1999),
Stephens (2000), and Wiuf (discussion 13B in this volume). These authors also
studied the age of a mutation observed to have a given frequency in a sample of
chromosomes. Slatkin and Rannala (1997) addressed the problem of estimating
the age of a mutation when given not just its frequency in a sample but also an
estimate of the number of mutations occurring in a completely linked region of

DNA. Their approach models the age of the mutation as a parameter, and so differs from the coalescent-based approach in which the age is an unobservable random variable; the natural quantity to report is the conditional distribution of the age, given the available data.

In this article we study aspects of the age of a mutation from the coalescent perspective. After a brief introduction to the general coalescent tree (in which coalescence times can have any continuous distribution), we describe in Section 2 an urn model that can be used to study the combinatorics of coalescent trees having a mutation of a given frequency. Section 3 describes the infinitely-many-sites model of mutation. Sections 4–6 give various properties of the age of a mutation having a given frequency in a sample. Section 7 discusses simulation algorithms and illustrates them by studying Slatkin and Rannala's (1997) problem as well as the distribution of Tajima's D in a subtree. Section 8 studies the coalescent subtree of that part of the population known to carry a given mutation, and Section 9 exploits these results to study the age of a mutation in a sample taken at random from chromosomes carrying a given mutation (the disease registry model), or together with data from chromosomes not carrying that mutation (the case–control model).

Figure 1 illustrates a coalescent tree of a sample of 10 genes with mutations occurring in the ancestry of the sample. Our interest focuses on the descendants of one single given mutation. For example, the mutation on the far right of the tree subtends five descendants.

1.1   Coalescent trees

In the absence of recombination, the ancestry of a sample of n genes from a large population can be described by a coalescent tree (Kingman 1982a). Let $T_n, T_{n-1}, \ldots, T_2$ denote the lengths of time for which the sample has $n, n-1, \ldots, 2$

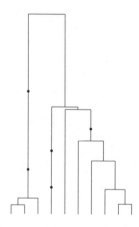

FIG. 1. Coalescent tree with mutations (symbolized by dots).

distinct ancestors back in time to its most recent common ancestor (MRCA). In the usual coalescent process, corresponding to a constant population size, the number of distinct ancestors $A_n(t)$ time t ago is a time-homogeneous death process with death rate μ_j from state j given by

$$\mu_j = \binom{j}{2}, \qquad j = n, n-1, \ldots, 2. \tag{1.1}$$

The times T_j are therefore distributed as independent exponential random variables with means μ_j^{-1}. This article discusses results about the structure of coalescent trees under a general joint distribution for the times $(T_n, \ldots, T_2)$. As in Griffiths and Tavaré (1998) we assume that:

(A1) $T_n, \ldots, T_2$ are continuous random variables.

(A2) The ancestral tree is binary, and such that when there are k ancestral lines each pair has probability $\binom{k}{2}^{-1}$ of being the next pair to coalesce.

In this article we use coalescent time units. In the population genetics setting, these may be converted to generations by letting one coalescent time unit correspond to $2N$ generations, appropriate for the coalescent approximation of a population of size N diploid individuals at the time of sampling.

1.2 Variable population size

The motivation for considering a general tree comes from a coalescent model with variable population size, and from other models such as the birth-and-death process generated forward in time.

The coalescent for a population undergoing deterministic population size fluctuations is described in Slatkin and Hudson (1991) and Griffiths and Tavaré (1994a). For the Wright–Fisher model, let $\lambda(t)$ denote the ratio of the population size time t ago and the population size at the time of sampling. Let $\{A_n(t), t \geqslant 0\}$ be the death process described by (1.1), and let $\{A_n^\lambda(t), t \geqslant 0\}$ denote the corresponding process in the variable population size case. Then

$$A_n^\lambda(t) = A_n\left(\int_0^t \lambda(u)^{-1}\,\mathrm{d}u\right), \qquad t \geqslant 0. \tag{1.2}$$

A formula for the distribution of $A_n(t)$ in the constant population size case is well known (Tavaré 1984; Griffiths 1980), and it follows from (1.2) that

$$\Pr(A_n^\lambda(t) = k) = \sum_{j=k}^n \rho_j(t)\frac{(-1)^{j-k}(2j-1)k_{(j-1)}n_{[j]}}{k!(j-k)!n_{(j)}}, \qquad k = 1, \ldots, n, \tag{1.3}$$

where $\rho_j(t) = \exp\left(-\binom{j}{2}\int_0^t \lambda(u)^{-1}\,\mathrm{d}u\right)$ and

$$a_{(0)} = 1, \qquad a_{(j)} = a(a+1)\cdots(a+j-1), \qquad j \geqslant 1; \tag{1.4}$$
$$a_{[0]} = 1, \qquad a_{[j]} = a(a-1)\cdots(a-j+1), \qquad j \geqslant 1. \tag{1.5}$$

The mean waiting time in state j is given by

$$\mathrm{E}(T_j) = \int_0^\infty \mathrm{Pr}(A_n^\lambda(t) = j)\,\mathrm{d}t, \qquad j = 2,\ldots,n.$$

2 Relationship between coalescent trees and urn models

In the classical Pólya urn model, an urn contains balls of k distinct colours. At discrete time instants, a ball is chosen at random from the urn and replaced with an additional ball of the same colour. If the initial configuration of colours is $\mathbf{c} = (c_1,\ldots,c_k)$, then after r draws the probability of a configuration $\mathbf{r} + \mathbf{c} = (r_1 + c_1,\ldots,r_k + c_k)$ is

$$\mathrm{Pr}(\mathbf{r} + \mathbf{c}) = \binom{r}{\mathbf{r}} \frac{(c_1)_{(r_1)} \cdots (c_k)_{(r_k)}}{(c_1 + \cdots + c_k)_{(r)}}$$

$$= \binom{c_1 + \cdots + c_k + r - 1}{r}^{-1} \prod_{j=1}^k \binom{c_j + r_j - 1}{r_j}, \qquad (2.1)$$

where $r = r_1 + \cdots + r_k$ and $c_{(r)}$ is defined by (1.4); cf. Feller (1968, Chapter V). This distribution is the Multinomial–Dirichlet distribution:

$$\mathrm{Pr}(\mathbf{r} + \mathbf{c}) = \int \binom{r}{\mathbf{r}} x_1^{r_1} \ldots x_k^{r_k} f(x_1,\ldots,x_k)\,\mathrm{d}\mathbf{x}, \qquad (2.2)$$

where $f(x_1,\ldots,x_k)$ is a Dirichlet$(c_1,\ldots,c_k)$ density defined by

$$f(x_1,\ldots,x_k) = \frac{\Gamma(c_1 + \cdots + c_k)}{\Gamma(c_1)\ldots\Gamma(c_k)} x_1^{c_1-1} \ldots x_k^{c_k-1}, \qquad (x_1,\ldots,x_k) \in \Delta,$$

with $\Delta = \{(x_1,\ldots,x_k) \in \mathbb{R}_+^k : x_1 + \cdots + x_k = 1\}$. As $n \to \infty$, the limit distribution of the relative proportions of the balls has this Dirichlet distribution. For $j \geqslant 1$, let $\mathbf{u}_j$ be an indicator vector whose lth component is 1 if the jth draw is colour l. Then $\mathbf{u}_j, j \geqslant 1$ are exchangeable random vectors, and the density f above is de Finetti's representing measure of the sequence; cf. Feller (1971, Chapter VII).

A coalescent tree can be generated either forward in time or backward in time. In forward time, an edge of the tree is chosen at random to branch to increase the number of descendants, corresponding to coalescence decreasing the number of ancestors backwards in time. The descendants of edges in a general coalescent tree generated forward in time can be identified with this classical urn model. Consider a cross-section of a coalescent tree at a particular time in the past when there are k edges, and give each a distinct colour. Then at each branch point in forward time an additional edge is added, analogous to the urn model with $c_1 = \cdots = c_k = 1$. The probability of getting $n_1,\ldots,n_k$ descendants of edges $1, 2, \ldots, k$ in a sample of size n is

$$\binom{n-1}{k-1}^{-1} \tag{2.3}$$

for $n_1 + \cdots + n_k = n$. This follows from (2.1) by setting $r_i = n_i - 1$, $1 \leqslant i \leqslant k$, and $r = n - k$, and shows that the distribution is uniform on the collection of ordered non-empty k-subsets of a set with n elements. This result is derived in a different way in Kingman (1982a).

2.1 The number of descendants of an edge

The probability $p_{nk}(b)$ that a particular edge among k ancestors subtends b particular descendants in a sample of n also follows from (2.1) by setting $k = 2$, $c_1 = 1$, $c_2 = k - 1$, $r_1 = b - 1$, and $r_2 = n - b - k + 1$:

$$p_{nk}(b) = \binom{n-b-1}{k-2}\binom{n-1}{k-1}^{-1}, \qquad 1 \leqslant b \leqslant n - k + 1, \tag{2.4}$$

the number of ways the $n - b$ descendants can be assigned to $k - 1$ ancestors, divided by the total number of ways the n descendants can be assigned to their k ancestors.

2.2 Coalescence times in a subtree

A b-subtree of an n-tree is the subtree formed by the ancestral tree of a particular b genes from the sample of n. In an exchangeable model, all b-subtrees are identically distributed. The coalescence times in a b-subtree are

$$T_i^{\#} = T_{M_i} + \cdots + T_{M_{i-1}+1},$$

where for $i = 1, 2, \ldots, b - 1$, M_i is the number of edges in the n-tree at the time the b-subtree first has i edges (and we define $M_b \equiv n$). From Theorem 2 of Saunders *et al.* (1984), we have

$$
\begin{aligned}
\phi&(k, i; n, b) \\
&:= \Pr(b\text{-subtree has } i \text{ edges when the } n\text{-tree has } k \text{ edges}) \\
&= \frac{(n-b)!(n-k)!b!(b-1)!k!(k-1)!(n+i-1)!}{(b-i)!(k-i)!n!(n-1)!i!(i-1)!(k+b-1)!(n+i-k-b)!}.
\end{aligned} \tag{2.5}
$$

The limit as $n \to \infty$ is

$$\phi(k, i; \infty, b) = \frac{\binom{k}{i}\binom{b-1}{i-1}}{\binom{k+b-1}{k-1}}. \tag{2.6}$$

Since T_k is included in the sum defining $T_i^{\#}$ if, and only if, there are i edges in the b-subtree when the n-tree has k edges, it follows that, for n finite or infinite,

$$\mathrm{E}\left(T_i^{\#}\right) = \sum_{k=i}^{n} \phi(k, i; n, b)\mathrm{E}(T_k). \tag{2.7}$$

The reverse Markov chain $\{M_i, i = b-1, \ldots, 1\}$ can be simulated from $M_b = n$ by noting that the transition probabilities are, for $i - 1 \leqslant \ell < k$,

$$\Pr(M_{i-1} = \ell \mid M_i = k) = \left(1 - \frac{i(i-1)}{k(k-1)}\right) \cdots \left(1 - \frac{i(i-1)}{(\ell+2)(\ell+1)}\right) \cdot \frac{i(i-1)}{(\ell+1)\ell}. \tag{2.8}$$

As $n \to \infty$, the limiting marginal distribution of M_i is, for $k \geqslant i$,

$$\Pr(M_i = k) = \frac{b!(b-1)!k!(k-1)!}{(b-i-1)!(k-i)!i!(i-1)!(b+k)!}$$

$$= \binom{k-1}{i-1} \mathrm{E}\left\{\rho^i(1-\rho)^{k-i}\right\}, \tag{2.9}$$

where ρ has a Beta$(b-i, i+1)$ distribution. Thus M_i has a Negative Binomial–Beta mixture distribution. This is useful for simulating the $\{M_i\}$: generate M_{b-1} with a Negative Binomial–Beta$(1, b)$ mixture, then use the transition probabilities (2.8) to simulate $\{M_i, b-1 > i \geqslant 1\}$. See Saunders *et al.* (1984) for details of (2.8) and (2.9).

2.3 Branches in the subtree below a mutation

Later in the chapter we study the subtree of a general coalescent tree formed by considering the individuals who carry a particular mutation that arises just once in the history of the sample. If there are b copies of the mutation in the sample, and the mutation arose while there were $J_0 = j$ ancestors, then the subtree describing the ancestry of the individuals carrying the mutation has $b - 1$ coalescence events. The first of these results in $J_{b-1} \leqslant n - 1$ ancestors of the sample, the second in J_{b-2} ancestors, and so on until the most recent common ancestor of the individuals carrying the mutation occurs when the sample has $J_1 \geqslant j$ ancestors. Note that individuals not carrying the mutation cannot share common ancestors with those that do until the mutant individuals have coalesced to their most recent common ancestor. Wiuf and Donnelly (1999) study properties of the coalescent conditional on having this property. Here we use the urn model to derive the basic results we need.

Consider the urn model starting from $j - 1$ red balls and one black ball. For $k = j, \ldots, n - 1$, let U_k be the indicator function of the event that a black ball is added when there are k balls in the urn. According to the de Finetti urn representation, the $\{U_k\}$ are conditionally independent Bernoulli trials, given success probability Z having a Beta$(1, j - 1)$ distribution with density $(j - 1)(1-z)^{j-2}, 0 < z < 1$. There will be b black balls in the urn when there are n balls altogether if $U_j + \cdots + U_{n-1} = b-1$. It follows that conditioning the urn on having b black balls out of n is equivalent to conditioning on $U_j + \cdots + U_{n-1} = b - 1$.

For $i = 1, 2, \ldots, b - 1$ let J_i be the number of balls already in the urn when the ith additional black ball is added. Given that there are b black balls in the urn containing n balls, we obtain the joint distribution of $(J_1, \ldots, J_{b-1})$ by exchangeability:

$$\Pr(J_1 = j_1, \ldots, J_{b-1} = j_{b-1} \mid J_0 = j_0)$$

$$= \Pr\left(U_{j_1} = 1, \ldots, U_{j_{b-1}} = 1 \,\middle|\, \sum_{l=j}^{n-1} U_l = b - 1\right) \qquad (2.10)$$

$$= \binom{n-j}{b-1}^{-1}, \qquad (j_1, \ldots, j_{b-1}) \in \mathbf{I}(j, n),$$

where $\mathbf{I}(j, n) = \{(j_1, \ldots, j_{b-1}) : j \leqslant j_1 < j_2 \cdots < j_{b-1} \leqslant n - 1\}$.

We can identify J_i as the number of ancestors of the sample at the time the subtree of size b has i distinct ancestors, for $i = b - 1, b - 2, \ldots, 1$. The result in (2.10) says that, conditional on $J_0 = j_0$, $J_1, \ldots, J_{b-1}$ are uniformly distributed over $\mathbf{I}(j, n)$.

3 Mutations in the tree

We are interested in the effects of mutation on general coalescent trees. We assume that for some $\theta \in (0, \infty)$ the following holds:

(A3) Conditional on the edge lengths of the tree, mutations occur according to independent Poisson processes of rate $\theta/2$ along the edges of the tree.

In the population genetics setting, the compound parameter θ is given by $\theta = 4Nu$, where u is the mutation rate per sequence per generation. To model the effects of each mutation, we use the *infinitely-many-sites model* (cf. Watterson 1975), under which a new mutation in the population is assumed to occur at a site in an infinitely-long DNA sequence where there has never previously been a mutation. Thus in this model the number of mutations in the ancestral tree of n sample genes is the number of *segregating sites* in these n sequences, that is, sites that contain two distinct types of base. A mutation on an edge of the tree at a site occurs at that site in all leaves subtended by that edge, while other leaves contain the ancestral base. Note that each mutation that has arisen in the history of the sample back to its most recent common ancestor is represented in that sample. In a finite-sites model of mutation, there is a fixed number of sites in a gene and mutation at the same site more than once is possible. The infinitely-many-sites model can be obtained as a limit from the finite-sites model as the number of sites tends to infinity and the mutation rate per gene is kept fixed.

It is often convenient to think of the DNA sequences as being represented by unit intervals, and to label the locations of new mutations that arise in the sample using a sequence of independent and identically distributed random variables having an arbitrary distribution on (0,1). Thus for any set $M \subset (0,1)$, mutations with locations in M arise in a branch of the coalescent tree at rate $\theta\lambda(M)/2$, where $\lambda(M)$ is the probability of M under the mutation distribution.

When mutation is uniform in (0,1) $\lambda(M) = |M|$, and non-uniform choices for λ correspond to mutational hotspots.

3.1 The distribution of the number of segregating sites

Let S_n be the number of segregating sites in a sample of n sequences under the infinitely-many-sites model. Since mutations occur as a Poisson process on the edges of the tree, S_n has a compound Poisson distribution with mean $\theta \sum_{j=2}^n jT_j/2$, and it follows immediately that

$$E(S_n) = \frac{\theta}{2}\sum_{j=2}^n jE(T_j), \quad \text{var}(S_n) = \frac{\theta^2}{4}\text{var}\left(\sum_{j=2}^n jT_j\right) + \frac{\theta}{2}\sum_{j=2}^n jE(T_j).$$

Watterson (1975) showed that in the standard coalescent process $E(S_n) = \theta\sum_{j=1}^{n-1} j^{-1}$, $\text{var}(S_n) = \sum_{j=1}^{n-1}\left(j^{-2}\theta^2 + j^{-1}\theta\right)$, and suggested the now commonly used moment estimator of θ given by $S_n/\sum_{j=1}^{n-1} j^{-1}$.

3.2 The distribution of pairwise differences

Let Π_n be the average number of pairwise differences between the $\binom{n}{2}$ sequences in a sample of n sequences in a general coalescent tree. If the coalescence time for two randomly chosen genes is T'_{n2}, then $E(\Pi_n) = \theta E(T'_{n2})$. Letting η_n be the number of ancestral lines in the coalescent tree when the two genes coalesce, we see that

$$\Pr(\eta_n \leqslant k) = \prod_{j=k+1}^n \left(1 - \binom{j}{2}^{-1}\right) = \frac{(n+1)(k-1)}{(n-1)(k+1)}.$$

Since

$$T'_{n2} = \sum_{j=\eta_n}^n T_j,$$

it follows that

$$E(T'_{n2}) = \sum_{k=2}^n \Pr(\eta_n \leqslant k)E(T_k) = \frac{(n+1)}{(n-1)}\sum_{k=2}^n \frac{(k-1)}{(k+1)}E(T_k). \quad (3.1)$$

Note that if there is a set of coalescent trees for $n = 2, 3, \ldots$ with coalescence times $\{T_{nj}, j = n, \ldots, 2\}$ on the same probability space such that, for $2 \leqslant k \leqslant n$, $\{T_{nj}, j = k, \ldots, 2\}$ is distributed as $\{T_{kj}, j = k, \ldots, 2\}$, then T'_{n2} is distributed as T_{22}. This is true for a coalescent process modelling constant or varying population size, but not necessarily true for a general coalescent tree.

3.3 Tajima's D

An important problem in the interpretation of genomic polymorphism data is the detection of regions of a chromosome that have undergone selection. One of the statistics most widely used for this purpose is Tajima's D, a standardized version of $S_{\text{scaled}} - \Pi_{\text{scaled}}$ (Tajima 1989), where S_n and Π_n are scaled to be unbiased estimates of θ. Departures from the assumptions of the usual neutral coalescent model of Kingman can be detected using the null distribution of D.

In the general coalescent tree

$$S_{\text{scaled}} = S_n \Big/ \sum_{j=2}^{n} \tfrac{1}{2} j \mathrm{E}(T_j), \qquad \Pi_{\text{scaled}} = \Pi_n / \mathrm{E}(T'_{n2}). \tag{3.2}$$

In the usual coalescent process Π_n is an unbiased estimate of θ because $\mathrm{E}(T'_{n2}) = \mathrm{E}(T_2) = 1$, so $\Pi_{\text{scaled}} = \Pi_n$. The analogue of D in the general coalescent tree, derived from eqn (3.2), is $S_{\text{scaled}} - \Pi_{\text{scaled}}$. The distribution of D is discussed later in the article.

4 Frequency spectra

The distribution of the number of mutant genes arising from a single mutation in an ancestor is of considerable interest. We assume that the mutation is segregating in the sample, so that it arose between the present and the time of the most recent common ancestor of the sample. Here we derive this distribution under the general conditions (A1), (A2), and (A3).

Let $\mathbf{T}$ denote the sequence of waiting times $T_2, \ldots, T_n$ in the coalescent tree of the sample. Consider a mutation arising at rate $\mu/2$, and let C denote the event that this mutation arises just once. Let $C_b \subseteq C$ denote the event that this mutation has b copies in the sample, and let I_k denote the event that the mutation arises when the sample has k ancestors. First we calculate $\Pr(C_b \cap I_k)$ using the Poisson nature of the mutation process:

$$\Pr(C_b \cap I_k | \mathbf{T}) = p_{n,k}(b) \left(k T_k \frac{\mu}{2} e^{-k T_k \mu/2} \times e^{-(L_n - k T_k)\mu/2} \right),$$

where $L_n = \sum_j j T_j$ is the length of the tree. Averaging over the distribution of $\mathbf{T}$ gives

$$\Pr(C_b \cap I_k) = k p_{nk}(b) \mathrm{E} \left(T_k \frac{\mu}{2} e^{-L_n \mu/2} \right). \tag{4.1}$$

Summing (4.1) over $b = 1, \ldots, n-1$ and $k = 2, \ldots, n-b+1$ gives

$$\Pr(C) = \sum_{k=2}^{n} k \mathrm{E} \left(T_k \frac{\mu}{2} e^{-L_n \mu/2} \right). \tag{4.2}$$

Dividing (4.1) by (4.2) shows that

$$\Pr(C_b \cap I_k \mid C) =: q_{n,b;k} = \frac{k p_{nk}(b) \mathrm{E} \left(T_k e^{-L_n \mu/2} \right)}{\sum_{j=2}^{n} j \mathrm{E} \left(T_j e^{-L_n \mu/2} \right)}, \tag{4.3}$$

for $0 < b < n$, $k = 2, \ldots, n-b+1$. Letting $\mu \to 0$ we obtain

$$q_{n,b;k} = \frac{k p_{nk}(b) \mathrm{E}(T_k)}{\sum_{j=2}^{n} j \mathrm{E}(T_j)}, \qquad 0 < b < n, \ k = 2, \ldots, n-b+1. \tag{4.4}$$

The *frequency spectrum* is the probability distribution $q_{n,b}, b = 1, \ldots, n-1$, of the number of times the mutation is represented in the sample. Since $q_{n,b} = \sum_{k=2}^{n-b+1} q_{n,b;k}$, we see that

$$q_{n,b} = \frac{\sum_{k=2}^{n-b+1} k p_{nk}(b) \mathrm{E}(T_k)}{\sum_{k=2}^{n} k \mathrm{E}(T_k)}, \qquad 0 < b < n, \tag{4.5}$$

as given in Griffiths and Tavaré (1998). Equation (4.4) provides the frequency spectrum for a *particular* segregating site in the infinitely-many-sites model. In the case of a constant population size, we see that $\sum_{k=2}^{n-b+1} k p_{nk}(b) \mathrm{E}(T_k) = 2/b$, so that

$$q_{n,b} = \frac{1}{b} \left(\sum_{k=1}^{n-1} \frac{1}{k} \right)^{-1}.$$

Stephens (2000) derived the result analogous to (4.5) for arbitrary μ. We note that this is a particular case of (4.5) obtained by modifying the distribution of $\mathbf{T}$ to $\mathbf{T}'$ such that the Laplace transform of $\mathbf{T}'$ is

$$\mathrm{E}\left(e^{-s_n T_n' - \cdots - s_2 T_2'} \right) = \frac{\mathrm{E}\left(e^{-s_n T_n - \cdots - s_2 T_2} e^{-\frac{1}{2}\mu L_n} \right)}{\mathrm{E}\left(e^{-\frac{1}{2}\mu L_n} \right)}.$$

Therefore

$$\mathrm{E}\left(T_k' \right) = \frac{\mathrm{E}\left(T_k e^{-\frac{1}{2}\mu L_n} \right)}{\mathrm{E}\left(e^{-\frac{1}{2}\mu L_n} \right)},$$

which evaluates to

$$\mathrm{E}\left(T_k' \right) = \frac{2}{k(k-1+\mu)}$$

in the coalescent process with constant population size.

5 Distribution of the age of a mutation

Let $\xi_{n,b}$ denote the age of a mutant having b copies in a sample of size n, for $0 < b < n$. Griffiths and Tavaré (1998) showed that the density of $\xi_{n,b}$ is given by

$$g_{n,b}(t) = \frac{\sum_{k=2}^{n} k p_{nk}(b) \Pr(A_n(t) = k)}{\sum_{k=2}^{n} k p_{nk}(b) \mathrm{E}(T_k)}, \qquad t > 0, \tag{5.1}$$

where $A_n(t)$ denotes the number of ancestors of the sample of n time t ago. Furthermore, the moments of $\xi_{n,b}$ are given by

$$\mathrm{E}(\xi_{n,b}^j) = \frac{\sum_{k=2}^{n} k(k-1)\binom{n-k}{b-1}\frac{1}{j+1}\mathrm{E}\left(W_k^{j+1} - W_{k+1}^{j+1} \right)}{\sum_{k=2}^{n} k(k-1)\binom{n-k}{b-1}\mathrm{E}(T_k)}, \qquad j = 1, 2 \ldots, \tag{5.2}$$

where for $k = n, n-1, \ldots, 2$,

$$W_k = T_n + \cdots + T_k \tag{5.3}$$

is the time taken to reach state $k-1$, with $W_{n+1} \equiv 0$. The mean and variance of $\xi_{n,b}$ can be derived from (5.2).

The population versions of (5.1) and (5.2) were also studied in Griffiths and Tavaré (1998). If we assume that $\{A_n(t), t \geqslant 0\}$ converges in distribution to a process $\{A(t), t \geqslant 0\}$ as $n \to \infty$, and that the time taken for $A(\cdot)$ to reach 1 is finite with probability one, then as $n \to \infty$, and $b/n \to x$, $0 < x < 1$, we obtain the density of the age ξ_x as

$$g_x(t) = \frac{\sum_{k=2}^{\infty} k(k-1)(1-x)^{k-2}\Pr(A(t)=k)}{\sum_{k=2}^{\infty} k(k-1)(1-x)^{k-2}\mathrm{E}(T_k)}, \tag{5.4}$$

and the moments of ξ_x are given by

$$\mathrm{E}(\xi_x^j) = \frac{\sum_{k=2}^{\infty} k(k-1)(1-x)^{k-2}\frac{1}{j+1}\mathrm{E}\left(W_k^{j+1} - W_{k+1}^{j+1}\right)}{\sum_{k=2}^{\infty} k(k-1)(1-x)^{k-2}\mathrm{E}(T_k)}, \qquad j = 1, 2 \ldots. \tag{5.5}$$

Related results appear in Wiuf and Donnelly (1999).

6   Coalescence times in a subtree

In this section we study properties of the subtree that relate to a sample of chromosomal regions carrying a particular mutation. In the b-subtree under a mutation in an n-tree, the coalescence times are

$$T_i' = T_{J_{i-1}+1} + \cdots + T_{J_i} = W_{J_{i-1}+1} - W_{J_i+1}, \qquad i = 2, \ldots, b, \tag{6.1}$$

where W_k is defined in (5.3) and we define $J_b \equiv n$. Note that $W_i' := T_b' + \cdots + T_i' = W_{J_{i-1}+1}$.

We can use the results of Section 4 to find the distribution of J_0, the number of ancestors at the time the mutation arose. From (4.4) and (4.5) we have

$$\Pr(J_0 = j) = \frac{q_{n,b;j}}{q_{n,b}} = \frac{jp_{nj}(b)\mathrm{E}(T_j)}{\sum_{l=2}^{n-b+1} lp_{nl}(b)\mathrm{E}(T_l)}, \qquad j = 2, \ldots, n-b+1. \tag{6.2}$$

Conditional on $J_0 = j$, we saw in (2.10) that $J_1, \ldots, J_{b-1}$ are uniform in $\mathbf{I}(j,n) = \{j \leqslant j_1 < j_2 < \ldots < j_{b-1} \leqslant n-1\}$. By considering the $k-j$ available indices less than k for $J_1, \ldots, J_{i-1}$ and the $n-k-1$ available indices for $J_{i+1}, \ldots, J_{b-1}$, it follows that

$$\Pr(J_i = k \mid J_0 = j) = \frac{\binom{k-j}{i-1}\binom{n-k-1}{b-i-1}}{\binom{n-j}{b-1}}, \qquad k = j+i-1, \ldots, n-b+i. \tag{6.3}$$

The unconditional mean waiting times may be computed from the formula

$$E(W_i') = \frac{\sum_{j=2}^{n-b+1} jp_{nj}(b) \sum_{k=j+i-1}^{n-b+i} E(W_k T_j) \Pr\left(J_{i-1} + 1 = k \mid J_0 = j\right)}{\sum_{j=2}^{n-b+1} jp_{nj}(b) E(T_j)} \tag{6.4}$$

and $E(T_i') = E(W_i') - E(W_{i+1}')$.

The length of the subtree is $L_{nb} = \sum_{l=2}^{b} l T_l'$, and its mean can be computed from (6.4) and the fact that

$$E\sum_{l=2}^{b} l T_l' = \sum_{l=2}^{b} l E(W_l' - W_{l+1}') = EW_2' + \sum_{l=2}^{b} EW_l'.$$

However, we can also exploit the urn representation from Section 2.3.

Suppose the mutation occurs when there are $J_0 = j$ ancestors of the sample. Let $U_k, k = j, \ldots, n-1$, be the indicator of the event that the subtree branches while the sample has k ancestors. We saw earlier that conditioning on b descendants in the subtree is equivalent to conditioning on $\sum_{i=j}^{n-1} U_i = b - 1$, and we define J_1 to be the number of ancestors of the sample when the subtree reaches its most recent common ancestor. In terms of $\{U_i\}$, we have

$$L_{nb} = \sum_{k=J_1+1}^{n} \left(1 + \sum_{i=J_1}^{k-1} U_i\right) T_k,$$

with $U_i = 0$, $j \leqslant i \leqslant J_1 - 1$, and $U_{J_1} = 1$. Consider the conditional distribution of L_{nb} given $J_1 = j_1, \sum_{i=j_1+1}^{n-1} U_i = b - 2$. By exchangeability, the conditional distribution of $\{U_i, j_1 + 1 \leqslant i \leqslant n-1\}$ is uniform on $\binom{n-j_1-1}{b-2}$ positions for which $b - 2$ of the indicator variables are 1.

It follows from this approach after some algebra that the mean edge length, conditional on a mutation subtending b descendants, is

$$E(L_{nb}) = \frac{\sum_{j=2}^{n-b+1} jp_{nj}(b) \sum_{k=j+1}^{n} c_{jk} E(T_j T_k)}{\sum_{j=2}^{n-b+1} jp_{nj}(b) E(T_j)}, \tag{6.5}$$

where

$$c_{jk} = b - (b-1)\frac{n-k}{n-j} - \frac{(n-k)!(n-j-b+1)!}{(n-j)!(n-k-b+1)!}.$$

In the usual constant-size coalescent, (6.2) reduces to

$$\Pr(J_0 = j) = \frac{\binom{n-j}{b-1}}{\binom{n-1}{b}}, \tag{6.6}$$

and it follows that

$$E(L_{nb}) = \binom{n-1}{b}^{-1} \sum_{j=2}^{n-b+1} \binom{n-j}{b-1} \sum_{k=j+1}^{n} \frac{2}{k(k-1)} c_{jk}. \tag{6.7}$$

7 Further mutations in subtrees

In the remainder of this article, we discuss the theoretical issues relating to frequency spectra and properties of ages of mutations. To this end, recall that the number of additional mutations falling in the subtree determined by the given mutation has, under an infinitely-many-sites assumption for these additional mutations, a Compound Poisson($\theta L_{nb}/2$) distribution, where L_{nb} is the total edge length of the subtree up to its MRCA and θ is the mutation parameter appropriate for the additional mutations. In particular, the expected number of segregating sites in the subtree is just $\theta E(L_{nb})/2$, which can be found from (6.5).

7.1 A simulation algorithm

While a number of explicit results are available for properties of subtrees, it is useful to have a simulation algorithm that produces them. Here we focus on subtrees arising below a mutation having frequency b in a sample of size n. One approach is provided by Wiuf and Donnelly (1999). Another method is:

(B1) Choose j_0 according to the distribution of J_0 in (6.2).

(B2) Choose $j_1 < \cdots < j_{b-1}$ from the conditional distribution of $J_1, \ldots, J_{b-1}$ given $J_0 = j_0$; this is uniform over $\mathbf{I}(j_0, n)$, as in (2.10).

(B3) Join edges at random to form the subtree.

If, in addition, coalescence times T_i', $i = b, b-1, \ldots, 2$, in the subtree are required, we need only add:

(B4) Simulate an observation from the joint distribution of $T_n, \ldots, T_{j_0+1}$.

(B5) Compute the times T_i' via (6.1).

This allows us to calculate summary statistics about the subtree, such as its height (that is, the time to the MRCA of the subtree), and its length L_{nb}.

7.2 The age of the mutation

To simulate from the age of the mutation, we add:

(B6) Conditional on the results of (B4), simulate from the random variable Z having the size-biased distribution of T_{j_0} and set $T^* = UZ$, where U is an independent U(0,1) random variable.

The time $A = T_n + \cdots + T_{j_0+1} + T^*$ is the required age of the mutation. Note that the random variable Z has density proportional to $xf(x)$, where f is the density of T_{j_0}. Hence if T_{j_0} has an exponential distribution with parameter τ, then so too does UZ.

If one wants to simulate observations from the posterior distribution of trees and times conditional on the number k of segregating sites appearing *in the b individuals carrying the mutation* in a region completely linked to the mutation, then one can add a rejection step (cf. Tavaré *et al.* 1997):

(B7) Accept the results of (B1)–(B6) with probability $\text{Po}(\theta L_{nb}/2)\{k\}/\text{Po}(k)\{k\}$, where θ is the mutation parameter appropriate for the linked region, L_{nb} is the subtree length, and we use the notation $\text{Po}(\mu)\{k\} = \mu^k e^{-\mu}/k!$. Otherwise, go to (B1).

Note that in this case step (B3) is not needed.

Slatkin and Rannala (1997) discussed the problem of estimating the age of a mutation given its frequency in the sample together with (an estimate of) the number of mutations that had arisen in a completely linked region among the chromosomes carrying the mutation. Under an infinitely-many-sites model for these extra mutations, the algorithm in (B7) provides one approach to Slatkin and Rannala's problem in the coalescent setting. An example appears in the next section. When the additional data are complete DNA sequences from the linked region, this algorithm no longer works, essentially because the acceptance probability is far too small. In this case, a Markov chain Monte Carlo approach can be implemented, as in Markovtsova *et al.* (2000).

7.3   An example of simulation of ages

The distribution of the age of a mutation and the height of the subtree were simulated using 50 000 runs of algorithm (B7) for the case $n = 200$, $b = 30$, $\theta = 4.0$, and five segregating sites. The mean age was 1.01 with standard deviation 0.91, while the mean subtree height was 0.40 with a standard deviation of 0.25. Percentiles of the distributions are given below. The estimated densities are given in Figs. 2 and 3.

	2.5%	25%	50%	75%	97.5%
Age	0.156	0.412	0.721	1.289	3.544
Subtree height	0.099	0.218	0.334	0.514	1.056

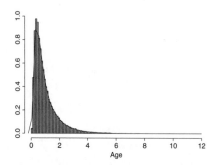
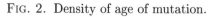

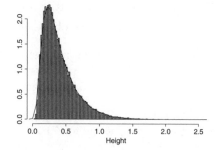

FIG. 2. Density of age of mutation.   FIG. 3. Density of height of subtree.

7.4 Simulation of Tajima's D in a subtree

Formula (6.5) can be used together with (3.1) and (3.2) to find the expected number of segregating sites and the expected pairwise difference in a subtree, and to study the analogue of Tajima's D in the subtree. This last provides a way to test for neutrality of the region around the mutation of interest; see Innan and Tajima (1997) for related material. It is straightforward to use the simulation algorithm outlined above to simulate from the distribution of $S_{\text{scaled}} - \Pi_{\text{scaled}}$: use steps (B1)–(B3) to produce the subtree, and then simulate mutations (with parameter θ) on that tree. Once done, the observed values of S_n and Π_n can be recovered.

We used this approach to simulate 50 000 observations for the case $n = 200$, $b = 30$, and $\theta = 4.0$. The mean of $S_{\text{scaled}} = 4.0143$ and the mean of $\Pi_{\text{scaled}} = 4.0243$. The percentage points of $S_{\text{scaled}} - \Pi_{\text{scaled}}$ are given in the following table:

2.5%	5.0%	10.0%	25.0%	50.0%	75.0%	90.0%	95.0%	97.5%
−5.3	−4.0	−2.4	−0.8	0.0	1.4	2.3	2.9	3.8

In a situation such as this the percentage points can be used to test for departures from neutrality for a subsample of b genes under a mutation. For example, if genes with the mutation were under positive selection $S_{\text{scaled}} - \Pi_{\text{scaled}}$ would have a heavier distribution on the negative side.

8 Sampling under a mutation in the population

Consider a mutation that has frequency $x \in (0,1)$ in the population. Our interest is in characteristics of a sample taken from this proportion of the population. To obtain the distribution of coalescence times under the mutation suppose, as previously in the article, that a sample of n has b copies of a mutation and then let $n \to \infty$, $b \to \infty$ in such a way that $b/n \to x$. Recall that conditional on $J_0 = j_0$, $J_1, \ldots, J_{b-1}$ have a uniform distribution on $\mathbf{I}(j_0, n)$. The joint conditional distribution of $J_1 < J_2 < \cdots < J_i$ is thus

$$\Pr(j_1, \ldots, j_i) = \frac{\binom{n-1-j_i}{b-i-1}}{\binom{n-j_0}{b-1}}, \qquad j_1 < \cdots < j_i. \tag{8.1}$$

The limit distribution of (8.1) as $n \to \infty$, $b/n \to x$ is

$$\Pr(j_1, \ldots, j_i) = x^i (1-x)^{j_i - i - j_0 + 1}, \qquad j_1 < \cdots < j_i. \tag{8.2}$$

We denote the limit random variables by $J_0^x, J_1^x, \ldots$. Informally, in this limit, when the whole tree branches there is probability x that the branch is in the subtree, since the subtree has a total proportion x of the branching points. $J_1^x < J_2^x < \cdots$ are thus distributed, from (8.2), as success epochs in a sequence of Bernoulli trials, shifted by J_0. If G is a geometric random variable with distribution

$$x(1-x)^g, \qquad g = 0, 1, \ldots,$$

then $Q_i = J_i^x - J_{i-1}^x$, $i = 1, 2, \ldots$, are independent random variables with Q_1 distributed as G, and Q_i, $i > 1$ distributed as $G + 1$. $L_i = J_i^x - J_0^x - i + 1$ has a negative binomial distribution

$$\binom{\ell + i - 1}{i - 1} x^i (1 - x)^\ell, \qquad \ell = 0, 1, \ldots.$$

The distribution of J_0^x depends on the coalescence times in the population. Assuming convergence of $\{T_{nj}\}$ to a proper collection of coalescence times $\{T_j\}$, with the means also converging, the distribution of J_0^x converges to

$$\frac{j(j-1)(1-x)^{j-2}\mathrm{E}(T_j)}{\sum_{i=2}^{\infty} i(i-1)(1-x)^{i-2}\mathrm{E}(T_i)}, \qquad j \geqslant 2, \tag{8.3}$$

and the population then has coalescence times

$$T_i^x = T_{J_{i-1}^x + 1} + \cdots + T_{J_i^x}, \qquad i = 2, 3, \ldots. \tag{8.4}$$

The collection of coalescence times $\{T_i^x\}$ can then be used in formulae for characteristics of samples under the mutation in the population for general coalescent trees. It is straightforward to simulate $\{T_i^x\}$ by simulation of J_0^x from the distribution (8.3), $\{J_i^x; i \geqslant 1\}$ from their geometric structure, and $\{T_i^x\}$ from (8.4).

8.1   Results for the standard coalescent

In the usual coalescent process, it is possible to find explicit formulae for the means and second moments of the coalescence times $\{T_j^x\}$ and then use these in applications. In particular,

$$\Pr(J_0^x = j) = x(1-x)^{j-2}, \qquad j \geqslant 2.$$

It follows that

$$\mathrm{E}(T_i^x) = 2 \int_0^1 z^{-1} f^i \left(1 - f \right) \mathrm{d}z, \tag{8.5}$$

and

$$\mathrm{E}(W_i^x) = 2 \int_0^1 z^{-1} f^i \, \mathrm{d}z,$$
$$\mathrm{E}((W_i^x)^2) = 8 \int_0^1 z^{-1} f^i \left((-\log z)(1-z)^{-1} - 1 \right) \mathrm{d}z, \tag{8.6}$$

where

$$f(z; x) = zx(1 - z(1 - x))^{-1}$$

is the probability generating function of a geometric random variable shifted by 1. It follows from (8.6) that the mean time to coalescence of the subtree under the mutation is

$$\mathrm{E}(W_2^x) = 2x(1-x)^{-2}(x \log(x) - x + 1). \tag{8.7}$$

Substituting into (5.5) with $j = 1$ and simplifying, we see that the mean age of another mutation occurring in the subtree under the mutation which subtends a frequency $0 < y < 1$ (relative to x) is given by

$$A(x,y) = -\frac{4xy\big((1+xy)\log(xy) - 2xy + 2\big)}{2xy\log(xy) - x^2y^2 + 1}.$$ (8.8)

Note that $A(x,y)$ is a function of xy, the proportion of the second mutant in the total population.

Alternative forms for integrals in this section can be found by changing the variable of integration from z to f, noting that $0 \leqslant f \leqslant 1$, and that

$$\frac{dz}{df} = x^{-1}\left(1 + \frac{1-x}{x}f\right)^{-2}.$$

9 Other sampling schemes

In this section we develop the theory required for studying two different sampling schemes. Motivated by the problem of sampling from a disease registry, we study the genealogy of a random sample from the population carrying a particular mutation known to have frequency x in the whole population. For related material, see Wiuf (2000). The second scheme we address is motivated by the case–control design, in which one considers in addition a sample of the same size from that part of the population not carrying the mutation. We note that the results (9.2), (9.4), (9.5), and (9.7) apply to the standard coalescent.

9.1 Sampling from the disease population

Suppose then that a random sample of n genes is taken from a mutant class whose frequency is x in the population. The coalescence times in the sample are distributed as $\{T_i^\#\}$ of Section 2.2, where

$$T_i^\# = T_{M_i}^x + \cdots + T_{M_{i-1}+1}^x, \qquad i = n, \ldots, 2.$$ (9.1)

Thus the subpopulation of mutant genes of frequency x plays the role of the population, with $\{T_i^\#\}$ the population coalescence times. Formula (2.7) for $E(T_i^\#)$ holds with n there set equal to ∞, and substituting we obtain

$$E\left(T_i^\#\right) = \sum_{k=i}^{\infty} \frac{\binom{k}{i}\binom{n-1}{i-1}}{\binom{k+n-1}{k-1}} E\left(T_k^x\right) =$$
$$2n\binom{n-1}{i-1}\int_0^1\int_0^1 z^{-1}f^i w^{i-1}(1-fw)^{-(i+1)}(1-w)^{n-1}(1-f)\,dw\,dz,$$ (9.2)

where $f \equiv f(z;x)$.

The site frequency spectrum in a random sample of size n under a mutation of frequency x in the population is, from (4.5),

$$q_{n,j} = \frac{(n-j-1)!(j-1)!\sum_{k=2}^{n} k(k-1)\binom{n-k}{j-1}E(T_k^\#)}{(n-1)!\sum_{k=2}^{n} kE(T_k^\#)}.$$ (9.3)

The properties of Tajima's D in this setting may be studied using the following results. From Section 3.1, the mean number of mutations in a sample of n is

$$\mathrm{E}(S_n) = \frac{\theta}{2} \sum_{i=2}^{n} i \mathrm{E}\left(T_i^{\#}\right)$$

$$= n\theta \int_0^1 \int_0^1 z^{-1}(1-w)^{n-1} f(1-f)(1-fw)^{-2}$$

$$\times \left((1+\phi)^{n-1} + (n-1)\phi(1+\phi)^{n-2} - 1\right) dw\, dz, \quad (9.4)$$

where $\phi = fw/(1 - fw)$. From Section 3.2, the expected number of pairwise differences between two of the n sequences is

$$\mathrm{E}(\Pi_n) = \theta \mathrm{E}(T_2^{\#}) = \theta \sum_{k=2}^{\infty} \frac{k-1}{k+1} \mathrm{E}(T_k^x)$$

$$= 2\theta \int_0^1 x(x + (1-x)f)^{-3} \left(2 - f + \frac{2(1-f)\log(1-f)}{f}\right) df, \quad (9.5)$$

because a random sample of two genes from a random sample of n is distributed as a random sample of two from under the subtree. Tabulated below are mean coalescence times for a sample of two from the subtree:

x	0.0	0.1	0.2	0.3	0.4	
$\mathrm{E}(T_2^{\#})$	0.00	0.13	0.20	0.25	0.28	
x	0.5	0.6	0.7	0.8	0.9	1.0
$\mathrm{E}(T_2^{\#})$	0.31	0.34	0.36	0.39	0.40	0.42

Further properties are perhaps most simply studied by a simulation approach. First, simulate $\{T_i^{\#}\}$ and the coalescence pattern in the tree, and add mutations along the edges according to a Poisson process of rate $\theta/2$. Simulation of $\{T_i^{\#}\}$ involves simulation of $\{J_i^x\}$, $\{T_j^x\}$, $\{M_i\}$, and $\{T_i^{\#}\}$ in order. It is of interest to consider simulated percentage points of Tajima's D analogue $S_{\text{scaled}} - \Pi_{\text{scaled}}$ for a test of the standard neutral model under a mutation, against various alternatives such as selection or growth.

9.2 Case–control sampling

In a case–control model, a sample is taken from chromosomes carrying a particular mutation and another sample taken from genes without that mutation. Denote the coalescence times in the population of genes without the mutation as $\{T_k^{1-x}\}$ and the number of ancestors of the population when the subtree branches as $\{J_i^{1-x}\}$. If $J_0^x = j$, then the two coalescent trees under and not under the mutation are coupled, with branching occurring in the respective trees with probabilities x and $1-x$ while there are greater than or equal to j ancestors of the total population. Let H be a geometric random variable with distribution $(1 - x)x^h, h = 0, 1, \ldots$. Then $\{J_{j+r}^{1-x} - J_{j+r-1}^{1-x}, r = 0, 2, \ldots\}$ are independent with $J_j^{1-x} - j$ distributed as H, and $J_{j+r}^{1-x} - J_{j+r-1}^{1-x}$ distributed as $H + 1, r \geqslant 1$. Coalescence times are

$$T_k^{1-x} = \begin{cases} T_k, & k < j - 1, \\ T_{J_{k-1}^{1-x}+1} + \cdots + T_{J_k^{1-x}}, & k \geqslant j - 1. \end{cases} \tag{9.6}$$

It can be shown that the mean coalescence time in the subtree is

$$E(W_2^{1-x}) = (1 - x)^{-3}(1 - 2x)^{-1}\Big(2x^2(x^4 - 6x^3 + 15x^2 - 12x + 3)\log(x)$$

$$- 2x(1 - x)^5\log(1 - x) + (1 - x)(1 - 2x)(x^4 - x^3 - 4x^2 + 2)\Big). \tag{9.7}$$

The table below gives the mean coalescence times in the two subtrees by numerically evaluating (8.7) and (9.7):

x	0.0	0.1	0.2	0.3	0.4	
$E(W_2^x)$	0.00	0.17	0.30	0.41	0.52	
$E(W_2^{1-x})$	2.00	2.29	2.48	2.63	2.77	
x	0.5	0.6	0.7	0.8	0.9	1.0
$E(W_2^x)$	0.61	0.70	0.78	0.86	0.93	1.00
$E(W_2^{1-x})$	2.90	3.05	3.20	3.36	3.52	3.67

10 Conclusion

There is no doubt that coalescent methods have revolutionized the way in which molecular variation data are studied. They provide a way to model the ancestral relationships among gene regions, which in turn provides the basis for inference and estimation for such data. There have been two basic applications of the coalescent approach: *'forward' methods* that are used to study the properties of typical samples, and *'backward' methods* that is used to infer the features of a coalescent consistent with a given set of data. Ancestral inference, an example of the second type, has provided some challenging statistical problems, some of which are discussed here. As more genome-wide data are collected, these inference questions become more challenging. For example, it should be possible to use genome-wide data to understand better the fluctuations that have occurred in population sizes through time. Weiss provides an illustrative example in the subsequent discussion.

Acknowledgements

Simon Tavaré was supported in part by NSF grant DBI95-04393 and NIH grant GM58897.

References

Donnelly, P. and Tavaré, S. (1995). Coalescents and genealogical structure under neutrality. *Annual Review of Genetics*, **29**, 410–21.

Feller, W. (1968). *An Introduction to Probability Theory and its Applications*, Vol. 1 (3rd edn). Wiley, New York.

Feller, W. (1971). *An Introduction to Probability Theory and its Applications*, Vol. 2 (2nd edn). Wiley, New York.

Griffiths, R. C. (1980). Lines of descent in the diffusion approximation of neutral Wright–Fisher models. *Theoretical Population Biology*, **17**, 37–50.

Griffiths, R. C. and Tavaré, S. (1994a). Sampling theory for neutral alleles in a varying environment. *Proceedings of the Royal Society of London*, B, **344**, 403–10.

Griffiths, R. C. and Tavaré, S. (1998). The age of a mutation in a general coalescent tree. *Stochastic Models*, **14**, 273–95.

Griffiths, R. C. and Tavaré, S. (1999). The ages of mutations in gene trees. *Annals of Applied Probability*, **9**, 567–90.

Hartl, D. L. and Jones, E. W. (2001). *Genetics. Analysis of Genes and Genomes* (5th edn). Jones and Bartlett, Sudbury, Mass.

Hudson, R. R. (1983). Properties of a neutral allele model with intragenic recombination. *Theoretical Population Biology*, **23**, 183–201.

Hudson, R. R. (1990). Gene genealogies and the coalescent process. In *Oxford Surveys in Evolutionary Biology*, No. 7 (eds D. Futuyma and J. Antonovics), pp. 1–44. Oxford University Press, UK.

Innan, H. and Tajima, F. (1997). The amounts of nucleotide variation within and between allelic classes and the reconstruction of the common ancestral sequence in a population. *Genetics*, **147**, 1431–44.

Kimura, M. and Ohta, T. (1973). The age of a neutral mutant persisting in a finite population. *Genetics*, **75**, 199–212.

Kingman, J. F. C. (1982a). On the genealogy of large populations. *Journal of Applied Probability*, **19A**, 27–43.

Markovtsova, L., Marjoram, P., and Tavaré, S. (2000). The age of a unique event polymorphism. *Genetics*, **156**, 401–9.

Nordborg, M. (2001). Coalescent theory. In *Handbook of Statistical Genetics* (eds D. J. Balding, M. Bishop, and C. Cannings), pp. 179–208. Wiley, Chichester.

Nordborg, M. and Tavaré, S. (2002). Linkage disequilibrium: what history has to tell us. *Trends in Genetics*, **18**, 83–90.

Saunders, I. W., Tavaré, S., and Watterson G. A. (1984). On the genealogy of nested subsamples from a haploid population. *Advances in Applied Probability*,

16, 471–91.

Slatkin, M. W. and Hudson, R. R. (1991). Pairwise comparisons of mitochondrial DNA sequences in stable and exponentially growing populations. *Genetics*, **129**, 555–62.

Slatkin, M. W. and Rannala, B. (1997). Estimating the age of alleles by use of intra-allelic variability. *American Journal of Human Genetics*, **60**, 447–58.

Slatkin, M. W. and Rannala, B. (2000). Estimating allele age. *Annual Review of Genomics and Human Genetics*, **1**, 225–49.

Stephens, M. (2000). Times on trees and the age of an allele. *Theoretical Population Biology*, **57**, 109–19.

Stephens, M. (2001). Inference under the coalescent. In *Handbook of Statistical Genetics* (eds D. J. Balding, M. Bishop, and C. Cannings), pp. 213–38. Wiley, Chichester.

Tajima, F. (1983). Evolutionary relationship of DNA sequences in finite populations. *Genetics*, **105**, 437–60.

Tajima, F. (1989). Statistical methods for testing the neutral mutation hypothesis by DNA polymorphism. *Genetics*, **123**, 585–95.

Tavaré, S. (1984). Line-of-descent and genealogical processes, and their application in population genetics models. *Theoretical Population Biology*, **26**, 119–64.

Tavaré, S., Balding, D. J., Griffiths, R. C., and Donnelly, P. (1997). Inferring coalescence times from DNA sequence data. *Genetics*, **145**, 505–18.

Watterson, G. A. (1975). On the number of segregating sites in genetical models without recombination. *Theoretical Population Biology*, **7**, 256–76.

Watterson, G. A. (1996). Motoo Kimura's use of diffusion theory in population genetics. *Theoretical Population Biology*, **49**, 154–88.

Wiuf, C. (2000). On the genealogy of a sample of neutral rare alleles. *Theoretical Population Biology*, **58**, 61–75.

Wiuf, C. and Donnelly, P. (1999). Conditional genealogies and the age of a neutral mutant. *Theoretical Population Biology*, **56**, 183–201.

13A
Linked versus unlinked DNA data – a comparison based on ancestral inference

Gunter Weiss
Max Planck Institute for Evolutionary Anthropology, Leipzig, Germany

Coalescent theory provides a conceptually simple – but very general – way to describe the ancestral relationship of individuals randomly selected from a large population. The application of coalescent ideas in population genetics has mainly focused on the exploration of demographic scenarios and genetic forces that shaped the intraspecific DNA samples we observe today, as well as the dating of specific genealogical events.

The main sources for this inference are DNA sequence data. These data consist of stretches of DNA sampled from, say, n individuals. The region studied is usually chosen in such a way that selective forces can be neglected and recombination does not dilute the signals of demographic events. With recombination being absent, the sequence is inherited as a whole so that the base positions are completely linked. The evolution of the sampled sequences occurs on a single genealogy, which is conveniently described by a coalescent tree.

However, new technology provides a novel sort of data potentially available to population genetic analysis. These data consist of single base polymorphisms (SNPs) spread over the whole genome with a formerly unexpected high density of about 1 SNP per 300–1000 base positions. Here, the data consist of a set of, say, m biallelic loci sampled from n individuals. If we choose these loci such that they are far apart on a chromosome, or even on different chromosomes, then the loci can be regarded as being mutually independent.

We study two extreme cases of molecular data: DNA sequence data from a single region are treated and referred to as completely linked loci. In contrast, SNP data are regarded as completely unlinked loci.

In what follows we will compare the potentials of these two kinds of data for ancestral inference by their efficiency in estimating two basic coalescent quantities, namely a scaled mutation rate and a growth rate parameter.

1 Estimation of a scaled mutation rate parameter

Measuring coalescent times in units of twice the population size, let θ be the usual scaled mutation rate parameter. We assume that the data evolved according to an infinitely-many-sites model (see Section 3 in Griffiths and Tavaré's article) and was sampled from a population of constant size. Then, θ is the single parameter to be estimated in this model. If the data D consist of m completely unlinked

SNP loci with d_i the frequency of the derived allele, then the likelihood function $\Pr(D|\theta) = \prod_{i=1}^{m} \Pr(d_i|\theta)$ can be computed explicitly. To calculate $\Pr(d_i|\theta)$, we condition on the event 'the mutation occurred when there were k individuals in the coalescent tree', which occurs with probability $\binom{n-d_i-1}{k-2}/\binom{n-1}{k-1}^{-1}$ (see formula (2.3) in Griffiths and Tavaré's article). With currently k individuals in the coalescent tree, the probability of coalescence and mutation is given by $(k-1)/(k-1+\theta)$ and $\theta/(k-1+\theta)$, respectively. Thus we have

$$\Pr(d_i|\theta) = \sum_{k=2}^{n-d_i+1} \frac{1}{1+\theta} \cdots \frac{k-1}{k-1+\theta} \frac{\theta}{k-1+\theta} \frac{k}{k+\theta} \cdots \frac{n-1}{n-1+\theta} \frac{\binom{n-d_i-1}{k-2}}{\binom{n-1}{k-1}}.$$

For m completely linked loci, the likelihood function takes the form

$$\Pr(D|\theta) = \frac{(n-1)!}{\prod_{j=2}^{n}(j-1+\theta)} \sum_{k_1,\ldots,k_m=2}^{n} \frac{\theta^m}{\prod_{j=1}^{m}(k_j-1+\theta)} \Pr(G_{(k_1,\ldots,k_m)} \equiv D),$$

where $\Pr(G_{(k_1,\ldots,k_m)} \equiv D)$ is the probability that a genealogy G with mutation events when there are $k_1, \ldots, k_m$ individuals were present, is compatible with the data D. Solutions for general m can be obtained by the simulation scheme of Griffiths and Tavaré (1994b).

2 Estimation of a growth rate parameter

Now let the sample be from a population that grew to its present size exponentially with rate λ. For m completely unlinked loci, we follow the derivation by Nielsen (2000) of the likelihood function for λ in the limiting case of low mutation rate with $\theta \to 0$ (see also Griffiths and Tavaré (1998)). Let T_{ij} denote the length of branch b_{ij} of the gene genealogy of locus i, $T_i = \sum_j T_{ij}$ the total tree length of this genealogy, t_i the sum of the branches in which a mutation could have caused the data d_i, and S_i the number of mutations at locus i. We have $\Pr(D|\lambda,\theta) = \prod_{i=1}^{m} \Pr(d_i|\lambda,\theta,S_i > 0)$. By conditioning on the genealogy G and summing over the set of branches B_i that could have caused the data d_i, we get

$$\Pr(d_i|\lambda,\theta,S_i > 0) = \frac{1}{\Pr(S_i > 0|\lambda,\theta)} \int \Pr(d_i|\theta,G)\,dF(G|\lambda)$$

$$= \frac{\int \sum_{j:b_{ij}\in B_i}(1-e^{-\theta T_{ij}/2})e^{-\theta(T_i-T_{ij})/2}\,dF(G|\lambda)}{\int (1-e^{-\theta T_i/2})\,dF(G|\lambda)}.$$

Taking the limit $\theta \to 0$ leads to

$$\mathrm{lik}(\lambda|d_i) = \lim_{\theta \to 0} \Pr(d_i|\lambda,\theta,S_i > 0) = \frac{\int t_i\,dF(G|\lambda)}{\int T_i\,dF(G|\lambda)} = \frac{\mathrm{E}(t_i|\lambda)}{\mathrm{E}(T_i|\lambda)}.$$

The expectations in the last formula can easily be obtained by simulations.

Changing again to the case of m completely linked loci, let $\theta_1, \ldots, \theta_m$ be the respective mutation parameters. Similar computations as in the unlinked data case yield in the limit of $\theta_1, \ldots, \theta_m \to 0$ the following representation for the likelihood function for λ:

$$\text{lik}(\lambda|D) = \lim_{\theta_1,\ldots,\theta_m \to 0} \Pr(D|\lambda, \theta_1, \ldots, \theta_m, S_1 > 0, \ldots, S_m > 0)$$

$$= \frac{\int t_1 \ldots t_m \, dF(G|\lambda)}{\int T^m \, dF(G|\lambda)} = \frac{\text{E}(t_1 \ldots t_m|\lambda)}{\text{E}(T^m|\lambda)}.$$

Contrary to the unlinked data case, this formula does not allow for a general evaluation by a simple simulation method. Here, the approximation of the expectations via simulations needs the search of a large space of genealogies. Therefore, we use the results of the complex simulation scheme by Griffiths and Tavaré (1994a) for comparison with the case of unlinked loci.

3   An illustrative example

We apply the four different estimation methods to a data set of a 10 kb part of the human X chromosome. Kaessmann *et al.* (1999) sampled 69 individuals representing all major linguistic groups and observed a total of 33 variable positions. The region sequenced is known to be of low recombination such that the assumption of full linkage is a reasonable approximation. However, we will (mis)treat the data also as if it consists of unlinked loci so that we can compare the performance of the methods in these two extreme cases.

The results of the estimation are depicted in Fig. 1 where the solid curves correspond to the unlinked loci case and the dashed curves to the linked loci case. Figure 1 contains also approximate 95% confidence regions if the usual χ^2-approximation for the distribution of likelihood ratios applies. The result for the

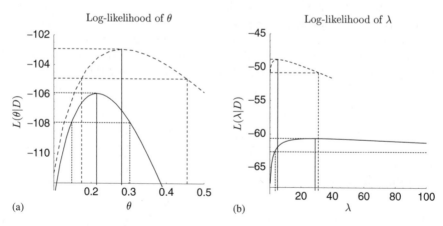

FIG. 1. Log-likelihood function of the parameters θ (a) and λ (b) in the unlinked (solid curve) and linked (dashed curve) cases using data from Kaessmann *et al.* (1999).

mutation rate θ is shown on the left side of Fig. 1. The independence assumption connected to the unlinked loci data reduces the uncertainty in the estimate compared with the intrinsically large variability of the linked case estimate. If we use the length of the confidence interval scaled by the estimate as a crude measure of uncertainty of an estimation procedure, then this scaled length for the unlinked loci case is 72% of that for the linked loci data.

The efficiency of the estimate of the growth rate λ (right part of Fig. 1) is known to be poor in the linked data case. In this example the length of an approximate confidence region is about six times as large as the estimate itself. However, the performance of the estimator with unlinked loci is worse. The interval is not even finite. This was a frequent observation when we conducted simulation studies with simulated data sets of comparable size. The other typical outcome from the analysis of a simulated data set was a monotonically decreasing likelihood function with a maximum at zero, the lower boundary of the parameter space (see also a real data example in Nielsen (2000), and its interpretation). This unattractive performance of the estimation procedure is due to the minimal genealogical information contained in frequency data. Theoretically, one can overcome this problem by drastically increasing the number of independent loci. Then, however, the independence assumption can hardly be satisfied in real data.

The estimation of the scaled mutation rate shows that the independence structure in SNP data is advantageous in terms of variance reduction. However, the estimation of a growth rate parameter already shows that this extreme case of single, independent positions is not suitable to study ancestral demography which is based on genealogical information. Thus, a set of independent loci, each containing a handful of completely linked informative sites, gives the best opportunity to successfully conduct ancestral inference. A different approach is to take recombination explicitly into account. However, this means not just adding an additional parameter to the model, but changing from a tree-valued to a graph-valued process.

13B
The age of a rare mutation

Carsten Wiuf
University of Oxford, UK

1 Introduction

Within the last 10 years coalescent theory has become an integrated part of population genetics and the advances in computational statistics have made coalescent theory a standard component in the analysis of genetic variation data. The coalescent and its modifications, discussed by Griffiths, Tavaré and Weiss, are highly structured stochastic systems and offer a framework in which many problems in population genetics have been – and can be – cast and solved.

This discussion deals with one such problem: the problem of dating the age of a mutation, in particular that of a rare mutation. This is not a new problem and dates back to before the emergence of the standard coalescent process (Kingman 1982b), to papers by Kimura and Ohta (1973) and Thompson (1976). Here, I will discuss two solutions to the problem, one that springs from the work of Thompson (1976) and treats the age as a fixed parameter in a likelihood framework, and one based on coalescent theory where the age is treated as a random variable. It will be argued that the latter provides the most satisfactory solution.

The age of a mutation is the time since it arose in the population. Recent interest in estimating mutation age stems from the extensive DNA sequencing and marker typing being performed in order to map mutations that cause genetic diseases. Estimating mutation age is partly done out of curiosity and partly to make further use of data that have been gathered for other purposes, though it relates to other interesting and important problems. To get a sense of this, imagine we have a sample from a population in which we find k copies of the mutation (Fig. 1). One approach to estimate the age would be to trace the ancestry of the k genes that bear the mutation and estimate when the mutation happened prior to the event of a most recent common ancestor of the k mutant

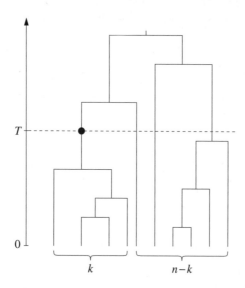

FIG. 1. Genealogy of a sample of size n with k mutants. The mutants find a most recent common ancestor before sharing an ancestor with any other gene in the sample. The mutation arose at time T ago.

genes; that is, the genealogy that relates the k mutants provides information about the age. In the article by Griffiths and Tavaré the genealogical structure of a sample of both mutants and non-mutants is discussed more generally.

However, this genealogy is useful in other contexts. It will be shaped by the demographic history of the population and by forces of selection; as such, inference on the age relates closely to inference on selection and demography. In the context of gene mapping, modelling of the genealogy of a mutation also provides insight into the genealogy and the genetic variation at linked loci, insight that can be used in development of gene mapping methods.

Examples of such mutations include mutations in the CF locus causing cystic fibrosis and mutations in the BRCA1 and BRCA2 loci causing breast cancer. These mutations are all found in low frequencies, $q = \lim_n k/n < 2\%$.

2 Defining the problem

It is assumed that the mutation is unique in the history of the entire population, otherwise the age cannot be unambiguously defined. On the other hand, a stationary stochastic evolution process is required for the age to be defined properly as a stochastic variable (Watterson 1976). If only one mutation is allowed, then this cannot be achieved.

Several solutions have been suggested. Kimura and Ohta (1973) were first to consider the relationship between the frequency of a mutant gene and the age, T, of the mutant. It was assumed that the mutant gene arose initially at frequency $1/2N$ (where $2N$ is the number of genes in a population of size N), and that its subsequent evolution to its present frequency, q, was a diffusion process. In the absence of selection, they found that T, conditional on q, has expectation

$$\mathrm{E}(T \mid q) = -\frac{2q \log(q)}{1 - q} \tag{2.1}$$

(T is measured in units of $2N$ generations). The mutation process giving rise to the mutant gene was not explained.

We will later see that (2.1) can be justified if the rate, θ, by which new mutants arise is low. In contrast, Thompson (1976) found it more natural to consider T to be a fixed parameter, thereby avoiding the issue of defining the age distribution.

3 The likelihood approach

Thompson's (1976) work is closely related to work by Slatkin and Rannala (1997) and it suffices to consider their work. They formulate the problem in the following way. Assume the mutation arose time T ago, that is at time T there is one single copy of the mutation. Owing to drift and selection, this copy is multiplied over the years and reaches frequency q today.

The number of copies present at any time between T and the present is modelled by a branching process. Slatkin and Rannala (1997) use a continuous birth–death process, whereas Thompson (1976) uses its discrete counterpart, a branching process with a modified geometric offspring distribution. Both these processes are commonly used in genetics and appear as approximations to the evolution of mutations in low frequency. The birth–death process allows certain realistic demographic features to be introduced in the model, e.g. population growth.

The form of the likelihood function of the data (here q) allows the maximum likelihood estimator, $\hat{T}$, of T to be derived analytically:

$$\hat{T} = \frac{1}{\xi} \log \{2\xi q + 1\} \quad \text{for } \xi > 0 \quad \text{and} \quad \hat{T} = 2q \quad \text{for } \xi > 0. \tag{3.1}$$

Here, ξ is a combined measure of population growth and selection. In the absence of selection, $\xi = 2Nr$, where r is growth rate per generation. This approach has a number of drawbacks:

- The mutation process is not modelled.
- T is a parameter, all other times in the genealogy are stochastic.
- Copies of the mutation could have existed before time T; it is only assumed that there is one at time T.

4   The coalescent approach

The genealogy, G, of the sample or the entire population is modelled by the coalescent process. On top of the genealogy, mutations are imposed according to a Poisson process with intensity $\theta t/2$, where t is the length of a branch (see also Griffiths and Tavaré's article). The genealogy we seek is G conditional on exactly one mutation in the sample's (population's) history. Furthermore, we assume $\theta \approx 0$ (formally, we consider $\theta \to 0$), so that mutations are rare.

In this setting, T is an unobserved random variable and it is natural to report the posterior density of T given the data. As such, the coalescent approach is Bayesian. Furthermore, we see that all three points of concern mentioned above have been taken into account.

The general form of the age distribution, $f(t \mid q)$, conditional on its current frequency, is mathematically untractable, but simulations from the distribution can, in general, be performed (Griffiths and Tavaré 1998; Wiuf 2000). If the total population size is constant over time and the mutation is neutral, then the expectation is given by (2.1). If we compare (2.1) with $\hat{T} = 2q$ we find that $\mathrm{E}(T \mid q)/\hat{T} \to \infty$ for $q \to 0$, that is the ratio between the two quantities becomes arbitrary large for small (and realistic) q. For example, if $q = 1\%$, then $\mathrm{E}(T \mid q) = 0.09$ and $\hat{T} = 0.02$.

If the frequency is less than, say, $q < 10\%$, then very accurate and simple approximations exist (Wiuf 2000); for $\xi > 0$,

$$f(t \mid q) \propto \frac{1}{(e^{\xi t} - 1)^2} \exp \left\{ -\frac{2\xi q}{e^{\xi t} - 1} \right\}, \tag{4.1}$$

where $\xi = 2Nr$ measures population growth. If ξ is large ($\xi > 10$), then $f(t|q)$ is centred around the mode, T_{mode}, and

$$\mathrm{E}(T|q) \approx T_{\mathrm{mode}} = \frac{1}{\xi} \log(\xi q + 1).$$ (4.2)

Considering T_{mode} as a point estimate of the age, we find $T_{\mathrm{mode}}/\hat{T} \to 1$ for $\xi \to \infty$ and the two approaches give estimates of the same order of magnitude.

5 Conclusion

It has been demonstrated that the likelihood approach has a number of features that make it inappropriate. The coalescent approach is natural in that it considers the genealogy to be wholly stochastic. For mutations in low frequency, the difference between $\hat{T}$ and $\mathrm{E}(T|q)$ can be substantial.

Additional references in discussion

Griffiths, R. C. and Tavaré, S. (1994b). Simulating probability distributions in the coalescent. *Theoretical Population Biology*, **46**, 131–59.

Kaessmann, H., Heissig, F., von Haeseler, A., and Pääbo, S. (1999). DNA sequence variation in a non-coding region of low recombination on the human X chromosome. *Nature Genetics*, **22**, 78–81.

Kingman, J. F. C. (1982b). The coalescent. *Stochastic Processes and their Applications*, **13**, 235–48.

Nielsen, R. (2000). Estimation of population parameters and recombination rates from single nucleotide polymorphisms. *Genetics*, **154**, 931–42.

Thompson, E. A. (1976). Estimation of age and rate of increase of rare variants. *American Journal of Human Genetics*, **28**, 442–52.

Watterson, G. A. (1976). Reversibility and the age of an allele. I. Moran's infinitely many neutral alleles model. *Theoretical Population Biology*, **10**, 239–53.

14

HSSS model criticism

Anthony O'Hagan
University of Sheffield, UK

1 Introduction

1.1 The role of model criticism

There has been an enormous surge in practical applications of Bayesian statistics in the last decade. The primary reason for this has been the development of Markov chain Monte Carlo (MCMC) tools for computing posterior inferences. HSSS has played a major part in that process, through helping to extend and consolidate the computational tools themselves, and through enriching the ways that we build models. Modern applications tackle subtle problems with complex data structures, and the modelling and computational tools employed are commensurately complex and subtle. There is a feeling that we can model just about any practical statistical problem, and no matter how complex that model might be, we can almost always compute relevant posterior inferences. It is tempting to think that all the technical problems for applied Bayesian statistics have been solved. That temptation needs to be resisted.

Given a model, and given a specification of which summaries we wish to derive from the posterior distribution for inference purposes, it is true that we can now generally apply Bayes' theorem algorithmically and more or less automatically to compute those summaries. There remain, however, several serious practical and technical questions about how we specify the model and the required summaries.

This article concerns one of those practical issues. Model criticism is one technique that can aid in effective specification of the model. It is worth emphasizing that in the Bayesian framework the model comprises both a representation of the data in terms of unknown parameters and a prior distribution for those parameters.

Model criticism may be defined briefly as the process of checking the model assumptions, using the observed data, without explicit reference to alternative models or assumptions. It is intended as an open-minded phase of investigation to identify any problems with the model. Formulation of explicit alternatives comes after the model criticism phase has identified some problems.

This definition clearly separates model criticism from model comparison or model choice, since those activities require two or more models to have been formulated. It is helpful to think of model criticism as occupying a different place in the modelling cycle. We might, for instance, represent that cycle as having three steps.

(1) *Model criticism.* Given a current model, we attempt to evaluate its adequacy as a representation of the data, without reference to explicit alternative models. It is recognized that alternatives are at least loosely implicit in any choice of model criticism tools, but the role of model criticism is essentially exploratory.

(2) *Model extension.* Any potential model shortcomings exposed in the model criticism stage are now considered, and appropriate extensions of the current model are formulated. The result is a proposal of a range of models containing the current model.

(3) *Model comparison/choice.* The range of models proposed in the model extension stage are now formally compared. The result may be to choose one model as the new current model. It may also be to select a mixture of some or all of the models as a new current model (in a model averaging sense). It may even be to confirm that the original model is adequate.

There are of course many other formulations of the modelling cycle that have been proposed in the literature, and the above is certainly inadequate for the purposes of discussing all aspects of modelling. It has, for example, the implication that the current model will typically become more complex in each cycle, and does not allow for model simplification. Nevertheless, it serves to identify model criticism as an important activity in its own right.

The terms *model validation* and *model checking* are often used in a similar sense, and in practice are almost synonymous with *model criticism*. I prefer *model criticism* because it seems to describe the activity most accurately. We do not in practice validate models, and since the usually accepted view is that all models are strictly wrong, it is clear that we cannot hope to validate a model in the sense of proving it true. By failing to find any sufficiently strong criticism with which to demonstrate a particular aspect of the model that is invalid, we can hope to attain some confidence that a model is adequate, but this is a long way from validating it. I have less objection to *model checking*, only that it conveys to me a sense of a rather passive activity that does not expect to uncover any problems, while *model criticism* conveys a more vigorous and open-minded process.

Before attempting to define model criticism more precisely, it is helpful to examine some other concepts.

1.2 Model adequacy and sensitivity analysis

The statement that all models are wrong is usually continued by asserting that some models are useful. No model is strictly correct, but no model is ever intended to be. Any model is a simplification of a reality that is more complex than we can (or need to) represent in practice. The test of a model is whether it is adequate for the purpose for which it was created.

The purpose of a statistical model is to derive relevant posterior inferences or summaries. In a formal decision-making context, the relevant posterior inferences are dictated by the decision problem. In the more informal 'scientific' context of simply wishing to understand whatever mechanisms or processes underlie the data, the relevant posterior summaries will describe the important features of the posterior distribution of some of the unknown parameters.

In sensitivity analysis, we consider the effect on a posterior inference or summary of varying one or more aspects of the model. For instance, we might examine sensitivity to the prior distribution by considering the effect of changing the prior distribution. If a posterior inference or summary changes substantially when we vary the model, then we say that it is sensitive to that variation in the model.

Sensitivity in itself does not imply that the model is inadequate. Suppose that we have sensitivity in the sense that an alternative model would lead to a substantially different posterior inference. This should be of no concern if the alternative model is not credible.

If we admit the possibility of both models, giving each a prior probability according to its plausibility a priori, the posterior inference in question would now be computed by model averaging. Because of the sensitivity, the inferences under the two separate models, which are to be averaged, are substantially different. If the posterior probability of the alternative model is near zero, however (and this is what is intended by saying that 'the alternative model is not credible' in the preceding paragraph), then the result of averaging will be a posterior inference almost the same as that which would have been obtained from the original model. It is only if the posterior probability of the alternative model is appreciable that the potential problem identified by the sensitivity becomes a real problem.

A proposed model is therefore inadequate for its purpose of deriving relevant posterior inferences or summaries if there exists at least one alternative model that

(a) is credible in the sense that it has sufficiently high prior probability and fit the data sufficiently well, so that it would have an appreciable posterior probability if admitted alongside the proposed model, and

(b) leads to substantially different values for one or more of the relevant posterior inferences.

1.3 Model criticism

Model criticism cannot address whether a model is adequate, since we exclude consideration of specific alternatives. Its role rather is to identify what kinds of alternative model should be constructed (model elaboration) in order to test formally the adequacy of the model (model comparison). This is the position that I first expressed in my discussion (O'Hagan 1980), of Box (1980).

To further clarify the scope of model criticism, as I understand and define it, it may be useful to think of a specific model. Consider the simple hierarchical model for several normal samples (one-way analysis of variance). Observations y_{ij} for $i = 1, 2, \ldots, k$ and $j = 1, 2, \ldots, n_i$ are available. At the first model stage, these are expressed as normally distributed with means λ_i and common variance σ^2. At the next stage (the first stage of the prior distribution) we represent the λ_is as normally distributed with common mean μ and variance τ^2. The model is completed by a prior distribution for σ^2, τ^2, and μ. Thus the model has the form,

$$y_{ij} \mid \lambda, \sigma^2 \sim \mathrm{N}(\lambda_i, \sigma^2), \quad i = 1, 2, \ldots, k;\ j = 1, 2, \ldots, n_i \quad \text{(independent)},$$
$$\lambda_i \mid \mu, \tau^2 \sim \mathrm{N}(\mu, \tau^2), \quad i = 1, 2, \ldots, k \quad \text{(independent)}, \tag{1.1}$$
$$p(\sigma^2, \tau^2, \mu).$$

We will suppose that the primary interest is in estimating the λ_is, and so relevant summaries will be posterior means (or some other suitable location summaries) of these parameters. There are obviously a number of specific assumptions in this model that might be criticized.

The assumption of normality may be a problem if we were to get substantially different posterior means for the λ_is under a credible non-normal model structure at either stage of the hierarchy. This might very well be the case if the data include outlying observations, and in particular if we have outlying groups of observations. So one method of model criticism might be to look for outliers or other evidence of non-normality in the data. If problems of this kind are identified, then at the model elaboration stage we should propose alternative distributions in place of normal distributions. Then at the model comparison stage we check whether the use of such models does demonstrate inadequacy of the original model, and if necessary select a new current model.

Another aspect to check in this model is the degree of shrinkage. The shrinkage will be affected, and hence the posterior means of the λ_is will change, if we alter the posterior distribution of the ratio of σ^2 to τ^2. We therefore expect that the shrinkage may be sensitive to the prior distributions of σ^2 and τ^2. Model criticism might then consist of looking at whether the data suggest different values for these parameters than their prior distributions. If this is the case, then model elaboration would consider how far we might plausibly vary these prior distributions. The result might be to extend the original model by allowing the prior variances of either σ^2 or τ^2 to be extra unknown parameters. Model comparison would explore the consequences of working using this larger model, and whether it should be adopted in place of the original model.

The model (1.1) will be used extensively to illustrate the various model criticism tools described in subsequent sections, both theoretically and in the context of some data presented in Section 3.3.

Any tool that is used to criticize a model may be called a *diagnostic*, and it is clear that to criticize a model in practice it will be necessary to deploy a variety of diagnostics, checking different aspects of the model.

1.4   The origins and structure of this article

The genesis of this work lies in my growing unease that the power of HSSS modelling with MCMC computational tools was tempting us to build models that we did not know how to criticize effectively.

The essence of HSSS modelling is that we construct models using simple local structures, each local component of which may be well understood separately, but such that the global behaviour of the models can be very complex and far less well understood. Such models are powerful and highly adaptive, but their complexity makes it difficult to predict the consequences of apparently small local changes. The fact that there are typically many local components and high-dimensional parameter spaces makes it difficult to criticize such models in any kind of systematic way.

Sharing these concerns with colleagues in HSSS led to a 'Research Kitchen' on model criticism being held in March 2000 with HSSS funding. The 'kitchen' was organized by Arnoldo Frigessi with a little help from me, and held in Oslo. Other participants were Roger Cooke (Delft), Antti Penttinen (Jyväskylä), Sylvia Richardson (then Paris, but now London), and David Spiegelhalter (Cambridge). The sharing of knowledge and genesis of ideas over the three days of the 'kitchen' was extremely stimulating, and this article owes a great deal to my fellow 'chefs' in the kitchen.

Although some elements of my description of the nature and role of model criticism in the preceding discussion is based on discussions at the 'kitchen', I have presented very much my own view. I am aware that other 'chefs' might disagree with some aspects of my presentation. However, there are strong parallels between the view presented here and that advocated by Gelman *et al.* (1995, pp. 161-2).

In Section 2 and the beginning of Section 3 I discuss some existing methods of model criticism. It is not my intention to provide in any sense a thorough review of the literature on Bayesian model criticism. That would be an enormous task. A bibliography of some related work is given by Bernardo and Smith (1994, pp. 418-20) and many more references can be found in Gelfand *et al.* (1992), Bayarri and Berger (1999), and other references cited here. Instead, my review of previous work is limited to providing motivation for the methods proposed in Sections 3 and 4.

Much of the previous literature on model criticism has been based on predictive distributions for observables, and this work is reviewed in Section 2. However, the key motivation for the Research Kitchen and for this article is that these methods only rather indirectly address the structure of the kind of complex hierarchical and graphical models that are routinely built in HSSS applications. In Section 3, I first review ideas of Bayesian residual analysis. Although the idea of examining residuals arises in the same context of checking the observations themselves, Chaloner (1994) provided the important step of looking at residuals at other levels of a hierarchical model. This idea is extended further in Section 3 to suggest some general tools for examining fit at any node of a complex model.

All of the techniques discussed in Sections 2 and 3 concern diagnosing lack of fit in some form. A quite different idea is outlined in Section 4, where the objective is to identify potentially important sensitivities directly.

I have tried to consider the value of each available tool in the context of complex HSSS models, where computation is typically by MCMC sampling. To be effective tools of HSSS model criticism, they should be applicable with minimal effort in large, typically hierarchical, models and capable of identifying a wide range of problems wherever they might manifest themselves within the model structure.

2 Predictive diagnostic techniques

2.1 Comparing data with model predictions

Traditionally, Bayesian model criticism has been based largely on comparing data with the predictions made of those data by the model.

The simplest way to do this is to reserve some of the data purely for the purpose of model criticism. Thus, the data $\mathbf{y}$ are divided into two parts, $\mathbf{y} = (\mathbf{y}_f, \mathbf{y}_c)$. The data $\mathbf{y}_f$ are used to 'fit' the model, i.e. to derive the posterior distribution of the underlying parameters. From this posterior distribution, we obtain the predictive distribution $p(\mathbf{y}_c \mid \mathbf{y}_f)$ for the remaining data $\mathbf{y}_c$. We then compare the observed $\mathbf{y}_c$ with its predictive distribution. The way in which this comparison is made will depend on the structure of the model, but in general involves deriving the predictive distributions of various univariate functions of $\mathbf{y}_c$, an n_c-dimensional vector. If, for instance, the predictive distribution of $\mathbf{y}_c$ is multivariate normal, so that $\mathbf{y}_c \mid \mathbf{y}_f \sim \mathrm{N}(\mathbf{m}_c, \mathbf{V}_c)$, then one such function could be the (squared) Mahalanobis distance $M = (\mathbf{y}_c - \mathbf{m}_c)^{\mathrm{T}} \mathbf{V}_c^{-1} (\mathbf{y}_c - \mathbf{m}_c)$. We might then regard a value of M that is large relative to its predictive $\chi^2_{n_c}$ distribution as casting some doubt on the model.

This *external data* technique provides a check of the predictive ability of the model. If our objective is only to use the model for prediction, then it is not actually important whether any other aspect of the model might be doubtful, as long as its predictions are valid. The disadvantage of such an approach lies in the separation of the data in two parts. First, the choice of how to split the data is essentially arbitrary, and we therefore have the unattractive feature that splitting the data a different way would lead to different diagnostics. Furthermore, we would ideally wish to use as much data as possible to fit the model and obtain the posterior distributions, and we would also ideally wish to use as much data as possible for predictive checking.

At one extreme of this general approach is the case where the model fit data $\mathbf{y}_f$ are empty, and the model criticism data comprise the whole dataset, $\mathbf{y}_c = \mathbf{y}$. Then the predictive distribution in question is $p(\mathbf{y})$, conditioned only on the prior information. This has been called the *prior predictive* approach by Bayarri and Berger (1999), and is advocated in the pioneering paper by Box (1980). The principal disadvantage with this as a general approach may be sensitivity of the diagnostics to the prior information. If prior information is generally weak, then $p(\mathbf{y})$ will be diffuse, and may give little diagnostic value.

At the other extreme, we can let the model criticism data comprise just a single observation y_i. The *cross-validation* method extends this idea, and attempts to achieve effectively larger samples for both model fit and model criticism, by splitting the data in different ways. We consider the split where $\mathbf{y}_c = y_i$ comprises simply the ith observation, and therefore $\mathbf{y}_f = \mathbf{y}_{-i}$ is the remainder of the data. This leads to a comparison of y_i with its predictive distribution $p(y_i \mid \mathbf{y}_{-i})$. Furthermore, this is repeated for each i. Cross-validation uses almost the entire sample to fit the model in each case. Although only a single observation is predicted each time, the fact that the prediction is repeated for each i means that in effect we predict all the data. Note, however, that predictions are made only of individual y_is, not jointly. The cross-validation approach is good at identifying outliers in the data (but even then suffers from the usual problems of masking when there are multiple outliers). It may not be so effective in examining overall fit or for devising diagnostics to target specific model assumptions.

There is also a computational issue with this approach, since we need to calculate predictive distributions based on different datasets $\mathbf{y}_{-i}$ as i varies. Gelfand *et al.* (1992) suggest, however, that this may be done within a single MCMC run. The MCMC sampling produces draws from $p(\theta \mid \mathbf{y})$, where θ is the parameter vector, whereas to construct the predictive distribution $p(y_i \mid \mathbf{y}_{-i})$ we require samples from $p(\theta \mid \mathbf{y}_{-i})$. The idea is to regard the MCMC sample as generated by importance sampling. Importance resampling can be used to generate samples from the predictive distribution of interest, or the usual importance sampling formula can be used to compute predictive expectations of relevant functions of y_i.

Despite this computational device, it is clearly simpler to compute, or sample from, not $p(y_i \mid \mathbf{y}_{-i})$ but $p(y_i \mid \mathbf{y})$. Formally, the latter is the predictive distribution of a future *replicate* observation made under the same conditions (same covariates, etc.) as the ith sample observation. We then compare the observed y_i with the predictive distribution $p(y_i^r \mid \mathbf{y})$ of its replicate y_i^r. Under these conditions, we no longer need to consider the observations singly but can derive or sample from the predictive distribution $p(\mathbf{y}^r \mid \mathbf{y})$ of a complete replicate sample $\mathbf{y}^r$, and compare the observed y with this distribution.

This *predictive replicate* method is the one advocated by Gelman *et al.* (1995), also termed *posterior predictive*. Diagnostics produced by this approach will clearly need more careful interpretation, since the observed data $\mathbf{y}$ have been used in devising the predictive distribution. Having fitted the model to $\mathbf{y}$, we must expect those data to agree more closely with the predictive replicate distribution than would be the case in the external data or cross-validation methods. Bayarri and Berger (1999) criticize the predictive replicate method on these grounds, both theoretically and through examples.

2.2 Predictive diagnostics

Although the general predictive approach admits a number of variations as above, there is a much greater variety of particular predictive diagnostics available in the literature. I will describe just some of these here, to illustrate the general principles.

Consider first the prediction of a single observation y_i. We have a predictive distribution in which y_i is regarded as a random variable, say Y_i. Gelfand *et al.* (1992) define a general diagnostic d which is the expectation with respect to this predictive distribution of a function $g(Y_i; y_i)$. They propose this general diagnostic in the context of the cross-validation method where the predictive distribution is $p(y_i \,|\, y_{-i})$, but it could equally be a prior predictive distribution $p(y_i)$ or the replicate predictive distribution $p(y_i^r \,|\, \mathbf{y})$. Particular choices for the function g that they suggest are as follows.

- $g_1(Y_i; y_i) = y_i - Y_i$. The diagnostic d_1 is then the difference between y_i and its predictive expectation.

- $g_2(Y_i; y_i) = 1$ if $Y_i \leqslant y_i$, otherwise 0. The diagnostic d_2 is just the probability that $Y_i \leqslant y_i$.

- $g_3(Y_i; y_i) = 1$ if the predictive density at Y_i is less than or equal to that at y_i. Then d_3 is the tail-area predictive probability that a random Y_i would have density lower than that of the observed y_i.

- $g_4(Y_i; y_i)$ defined so that d_4 is just the predictive density at the observed y_i.

The first diagnostic d_1 is related to residuals, discussed in Section 3, while d_4 is termed the predictive ordinate by Geisser and Eddy (1979) and Pettit and Smith (1985).

This general formulation of single observation diagnostics generalizes further to deal with diagnostics based on predictive distributions for multiple observations. The predictive distribution might be obtained by the external data or predictive replicate methods, and again we can consider a diagnostic d obtained as the predictive expectation of a function $g(\mathbf{Y}; \mathbf{y})$ comparing the observed $\mathbf{y}$ with its random replicate $\mathbf{Y}$. The analogue of g_1 might, for instance, be the sum of squared differences $(\mathbf{Y} - \mathbf{y})^{\mathrm{T}}(\mathbf{Y} - \mathbf{y})$. The measures g_2, g_3, and g_4 clearly generalize very easily to the multivariate case. The tail-area prior predictive probability d_3 for the whole data $\mathbf{y}$ was proposed as a simple overall model check by Box (1980).

We might alternatively consider a general function $g(T(\mathbf{Y}); T(\mathbf{y}))$, where T is some scalar function of the data, and then consider functions g such as g_1 to g_4. For instance, Bayarri and Berger (1999) use $T(\mathbf{y}) = |\bar{y}|$ in one of their examples and the range $y_{(n)} - y_{(1)}$ in another. The studentized range

$$T(\mathbf{y}) = \frac{y_{(n)} - y_{(1)}}{\left(\sum (y_i - \bar{y})^2 / n\right)^{1/2}} \tag{2.1}$$

has also been used.

Gelman *et al.* (1995) allow T to be a function also of the parameters θ. Their approach is generally that of the predictive replicate, and the relevant distribution against which the measure g is then evaluated is the joint posterior distribution of θ and a random replicate $\mathbf{Y}$. For instance, they call

$$T(\mathbf{y}, \theta) = \sum \frac{(y_i - \mathrm{E}(y_i \mid \theta))^2}{\mathrm{var}(y_i \mid \theta)} \tag{2.2}$$

the χ^2 discrepancy.

The idea of using tail-area probabilities (such as arise in the case of g_3), which are usually referred to as p-values, sits uncomfortably in a Bayesian analysis. Bayesians are generally firmly opposed to the frequentist notion of a p-value in the analysis of data. Nevertheless, p-values of various kinds have been advocated in the context of Bayesian model criticism: Box (1980) proposed the prior predictive p-value, Gelman *et al.* (1996) the posterior predictive p-value, and Bayarri and Berger (1999) the conditional predictive and partial posterior predictive p-values.

When using measures such as (2.1) and (2.2) that are functions of all the data it is clearly not possible to employ a cross-validation approach. These measures can be adopted in the context of external data methods, where y is the model criticism part of the data, y_c. Although advocated for use in the predictive replicate approach, the double use of the data (both for model fit and model criticism) raises concerns, in response to which Bayarri and Berger (1999) proposed their conditional predictive and partial posterior predictive methods. Broadly, these methods correspond to thinking of $\mathbf{y}_c$ as comprising the function T of interest, while $\mathbf{y}_f$ is made up of other functions of $\mathbf{y}$. So we consider predicting T based on simply observing those other functions of the data. The basic idea is that $\mathbf{y}_f$ should contain those functions most relevant to learning about the parameters of the model, while $\mathbf{y}_c = T$ should be a function related to diagnosing a departure from the model, and hence relating to a potential parameter in a larger model. Whilst that interpretation can be useful in straightforward models, it is hard to see how it might be applied in complex HSSS modelling!

This seems to be an active area of research still. In particular, an important new contribution by Marshall and Spiegelhalter (2000) is another direct result of the HSSS Research Kitchen in Oslo. They develop diagnostics of divergent behaviour in models that are similar to our measures of conflict in Section 3.2.

3 Node-based methods

In this section I will introduce some tools for model criticism that can be applied at each node of a complex hierarchical or graphical model, with a view to diagnosing problems of model fit at any point in the model structure. In general, the model will be supposed to be expressed as a directed acyclic graph (DAG), but it is possible that the methods proposed here may be generalized to other kinds of graphical models. The tools will first be introduced in a heuristic way and illustrated through a numerical example. The important question of how to interpret the various measures will then be addressed in Section 3.3.

3.1 Residuals

Much of frequentist model testing, at least in the context of linear models, is based on analysis of residuals. In a linear model $\mathbf{y} = \mathbf{X}\beta + \varepsilon$, the residuals $\mathbf{e} = \mathbf{y} - \mathbf{X}\hat{\beta}$ are regarded as estimates of the observation errors ε, and residual analysis is based on the sampling distribution of these and related random variables. The natural Bayesian analogue of frequentist residual analysis would be based on the posterior distribution of the error vector ε. This was first proposed by Zellner (1975) and further developed more recently by Chaloner (1994).

In the context of the model (1.1), the basic data residuals are the values $y_{ij} - \lambda_i$. According to the model, they are distributed independently as $\mathrm{N}(0, \sigma^2)$. Chaloner (1994) denotes these by $\varepsilon^*_{ij} = \sigma\varepsilon_{ij}$, so that in her notation the ε_{ij}s are standardized residuals

$$\varepsilon_{ij} = \frac{y_{ij} - \lambda_i}{\sigma} \sim \mathrm{N}(0, 1) \,. \tag{3.1}$$

This prior distribution is then compared with the posterior distribution $p(\varepsilon_{ij} \mid \mathbf{y})$. In particular, Chaloner defines the observation y_{ij} to be an outlier if $|\varepsilon_{ij}|$ is sufficiently large, and then proposes computing the posterior probability of this proposition. She chooses the criterion of 'sufficiently large' so that there is a high prior probability, such as 0.95, that *none* of the observations will be an outlier. Then, if the posterior probability that any individual $|\varepsilon_{ij}|$ exceeds this threshold is appreciable, it is declared an outlier and this is regarded as a criticism of the validity of the underlying model.

Note that this approach is clearly related to the discussion in Section 2.2, and in particular ε_{ij} is, in this normal model, just one of the standardized values $(y_i - \mathrm{E}(y_i \mid \theta))/(\mathrm{var}(y_i \mid \theta))^{1/2}$ in (2.2). As in the predictive replicate method, there is no element of cross-validation in this calculation. The posterior distribution of (3.1) is calculated based on all the data $\mathbf{y}$, rather than on the reduced data $\mathbf{y}_{-(i,j)}$ leaving out y_{ij}. In general, it seems clear that a more sensitive device for model criticism would result from looking at $p(\varepsilon_{ij} \mid \mathbf{y}_{-(i,j)})$.

An important distinguishing feature of this approach is that the comparison is between the posterior distribution of ε_{ij} and its prior distribution. It therefore has a different basis from the predictive methods of Section 2. It is this feature that allows aspects of the model to be checked that are less directly related to its predictive properties.

Specifically, Chaloner (1994) advocates extending the comparison to higher-level residuals in hierarchical models. In the context of the one-way model (1.1), she considers the standardized parameter residuals (that she calls 'between-group residuals')

$$\varepsilon_i = \frac{\lambda_i - \mu}{\tau} \,,$$

which again are independent N(0, 1) according to the model. Setting a threshold such that there is a high prior probability that none of the $|\varepsilon_i|$ values will exceed it, she again regards an appreciable posterior probability of an individual $|\varepsilon_i|$ exceeding the threshold as a criticism of the model.

Chaloner (1994) derives posterior distributions for the residuals in the case of known σ^2 and τ^2, and employs the Laplace method to approximate their distributions in the case of unknown variances. In the context of HSSS models, it is trivial to compute the distributions of residuals as by-products of the MCMC simulation of the full joint posterior distribution. Indeed, it is clear that the general approach can be extended to any level of a hierarchical model and to more general graphical models.

Consider a single node in such a model, where a random variable θ is modelled as having a density $p(\theta\,|\,\beta)$ dependent on the parameters β at its parent nodes. If this is a normal distribution, then the standardized form

$$\varepsilon_\theta = \frac{\theta - \mathrm{E}(\theta\,|\,\beta)}{(\mathrm{var}(\theta\,|\,\beta))^{1/2}}$$

has, according to the model, the N(0, 1) distribution. We can readily compute the posterior distribution of ε_θ from an MCMC sample and define it to be an 'outlier' according to a criterion such as Chaloner's.

The consideration of residuals in this way is strongly motivated by the notion of additive errors in normal models, and in general of course $p(\theta\,|\,\beta)$ will not be normal. To generalize the approach, note that ε_θ is a pivot, i.e. a function of both θ and β whose distribution, given β, does not depend on β. Similarly, the event that $|\varepsilon_\theta| > Z_{\alpha/2}$, where $Z_{\alpha/2}$ is some suitable upper percentage point of the standard normal distribution, is a pivotal event, i.e. one whose probability, given β, is fixed (equal to α in this case). In general, we can consider any pivotal event $h(\theta,\beta)$ whose probability α is fixed independent of β. Then we can compute the posterior probability $\mathrm{Pr}(h(\theta,\beta)\,|\,\mathbf{y})$. If this is sufficiently large compared with the corresponding (small) prior probability α, then we can regard this as indicative of a possible problem with the model specification at the θ node.

For instance, we can analogously define $h(\theta,\beta)$ to be the event that θ lies within the $100(1-\alpha)\%$ central probability interval of $p(\theta\,|\,\beta)$. (We can also invert this to obtain yet another kind of Bayesian p-value.)

Chaloner's approach only considers the marginal distributions of individual residuals, and looks at a particular criterion of whether each exceeds in absolute value some threshold. She uses the fact that there are many residuals being considered when setting the threshold, and this can be applied in the general case when there are a number of identical θ_i nodes in the model. We can set a threshold for ε_θ, or in general define the pivotal events $h(\theta_i,\beta)$, so as to fix the prior probability that none of the events occurs. More generally still, we can define pivotals across the group of θ_is to produce other measures for model criticism. With MCMC it would be simple, for instance, to obtain the posterior distribution of the number of $h(\theta_i,\beta)$s that occur, or the posterior probability of a pivotal event based on the prior chi-square distribution of the sum of squares of normal residuals.

Note that these diagnostics are based on the full posterior distribution, analogous to using the posterior predictive distribution. It is therefore open to the same kind of criticism as Bayarri and Berger (1999) make of the predictive replicate approach. Bayarri's discussion in this chapter shows that the diagnostics discussed here and in Section 3.2 can be seriously weakened by their 're-use' of the data. Nevertheless, I believe that the ability to explore model fit at any node of a complex model, via simple MCMC computations, makes this approach worthy of serious study.

An alternative solution in the case of hierarchical models is provided by Dey *et al.* (1998). Their approach avoids the re-use of data but is highly computational. Gelfand gives details in his discussion in this chapter.

3.2 Conflict

The above generalization of residuals allows for consideration of diagnostics at each node of a complex model. However, it makes use of only one aspect of the structure of that node, namely the distribution specified in the model for that node conditional on its parents. At the node for a variable θ, we can in general think of there being several sources of information about θ, and it is then of interest to consider the possibility of conflicts between these information sources.

In the one-way model (1.1), consider the node for parameter λ_i. In addition to its parents μ and τ^2, to which it is linked through the model component $\lambda_i \,|\, \mu, \tau^2 \sim \mathrm{N}(\mu, \tau^2)$, it is linked to each of the observations $y_{i1}, y_{i2}, \ldots, y_{in_i}$ through their distributions $y_{ij} \,|\, \lambda_i, \sigma^2 \sim \mathrm{N}(\lambda_i, \sigma^2)$. The parameter λ_i is a parent of each of these y_{ij}s and they are its child nodes. Together, these distributions provide the full conditional posterior distribution of λ_i:

$$p(\lambda_i \,|\, \mathbf{y}, \lambda_{-i}, \mu, \sigma^2, \tau^2) \propto p(\lambda_i \,|\, \mu, \tau^2) \prod_{j=1}^{n} p(y_{ij} \,|\, \lambda_i, \sigma^2).$$

This shows clearly how each distribution can be considered as a source of information about λ_i. In this case we have $n+1$ sources of information. When we are considering the possibility of conflict at the λ_i node, we must thus consider each of these contributing distributions as functions of λ_i. In general, a child node contributes information in the form of a likelihood, while the node's parents contribute information in the form of a prior distribution.

The simplest case of conflict between information sources is an outlier in a sample. A single outlier is an observation that conflicts with the other members of the sample. In the one-way model, suppose for the moment that σ^2 is known. Then the information provided by y_{ij} is to suggest that λ_i would probably lie in the range $y_{ij} \pm 2\sigma$. An outlying y_{ij} in the ith group provides information about λ_i that conflicts with the information given by the other observations in that group.

Formally, we can identify an outlier by contrasting the density $p(y_{ij} \mid \lambda_i, \sigma^2)$ for the outlying observation with the distribution $\prod_{f \neq j} p(y_{if} \mid \lambda_i, \sigma^2)$ obtained by pooling the other $n - 1$ data information sources at that node (considering each as a function of λ_i). The two distributions conflict if they give most support to quite different ranges of λ_i values. We can think of this as a data–data conflict. We can also have a data–prior conflict at this node if the prior distribution of λ_i conflicts with the information in the data. This would be seen through contrasting $p(\lambda_i \mid \mu, \tau^2)$ with $\prod_{j=1}^{n} p(y_{ij} \mid \lambda_i, \sigma^2)$.

If we now look at the μ node, its full conditional derives from the $N(m, v)$ prior distribution of μ and the $N(\mu, \tau^2)$ prior densities of each of the λ_is. Any conflict between these sources of information is a prior–prior conflict.

In a complex hierarchical or graphical model there are many possible sources of information and many ways in which conflict can arise. In general, the full conditional distribution at any node is proportional to the product of component distributions contributed by each neighbouring node in the graph. We consider each of the neighbouring nodes as a source of information about the parameter at the current node, and the information in each case is represented by the component distribution. Conflict can arise between any of the information sources at any node in the graph, and any such conflict is indicative of a problem with some part of the model.

A number of practical issues need to be addressed if we wish to implement this idea.

1. When we consider the information sources at a given node, we must condition on the values of the neighbouring nodes. If these are data nodes, then their values are fixed, and their distributions are therefore fixed. At the other extreme of a hierarchical model, the distributions of final-stage hyperparameters are also fixed. However, in general the distributions representing the different sources of information at a node will depend on values at neighbouring nodes that are not fixed.

   In a Gibbs sampler, we typically cycle through the random nodes drawing from the full conditional distribution of the parameter at that node. For a Gibbs sampler or any other kind of single site updating MCMC sampler, at each visit to a given node the values of all the neighbouring nodes are fixed, and so it would be possible to routinely examine the sources of information at each node in such a model, to check for conflicts.

2. We need to identify meaningful conflicts to look for.

   I do not have general rules for generating interesting conflict checks. When there are many sources of information at a node, we will often find that most come from groups of nodes of identical structure, such as the y_{ij} nodes bearing on a given λ_i in the one-way model, or the λ_i nodes bearing on the μ node. In such cases, it seems sensible to look for within-group conflicts by considering pairs of nodes within a group, but there may be many pairs to consider. The alternative is to compare each with the aggregate information of the remaining members of the group, as in the above proposed comparison of $p(y_{ij} \mid \lambda_i, \sigma^2)$ with $\prod_{f \neq j} p(y_{if} \mid \lambda_i, \sigma^2)$, although here we can expect problems of masking that are familiar in outlier detection methods. We can also compare the aggregate information from a group with that of other individual nodes, or with other groups, as in the above suggested comparison of $p(\lambda_i \mid \mu, \tau^2)$ with $\prod_{j=1}^{n} p(y_{ij} \mid \lambda_i, \sigma^2)$.

3. We need a measure of conflict between two distributions/likelihoods.

In general, the measure of conflict should be smaller the more the two curves overlap. Consider two normal densities, say the $N(m_1, v_1)$ density and the $N(m_2, v_2)$ density. Intuitively, the measure of conflict should compare $|m_1 - m_2|$ with some measure of spread obtained from v_1 and v_2. It is less easy to see how to proceed in general. The following general measure is proposed for comparing unimodal densities/likelihoods. (A more general form and some justification will be given in a paper currently in preparation, but it is clear that many other measures might be considered.)

First normalize both densities to have unit maximum height. We now consider the height of both curves at the point where they cross between the two modes. If this height is z, then the conflict measure is $c = -2 \ln z$. It can be shown that for the above case of two normal densities we have

$$c = \left(\frac{m_1 - m_2}{\sqrt{v_1} + \sqrt{v_2}} \right)^2 , \qquad (3.2)$$

and the measure therefore agrees with intuition for this case.

As discussed, the conflict measures will in practice be computed at each iteration of a Gibbs sampler or other MCMC algorithm. So for the comparison between any two specific sources of information at any given node we will have not a single value for the conflict but a sample of values. The median c value (recognizing that c will typically have a skew distribution) is perhaps the most obvious single measure of conflict, but others might be suggested. A value less than 1 should be thought of as indicative of no conflict, whereas values of 4 or more would certainly indicate a conflict to be taken seriously.

3.3   Example

Chaloner (1994) applies her methods to some artificial data previously analysed by Sharples (1990). These data were generated from a one-way model (1.1), but with a number of individual observations being outlying, and with one group mean λ_i also being substantially different from the others.

We present here another artificial data set, in which there is a clearly outlying group. There are $n = 6$ observations in each of $k = 5$ groups, and the data are as follows:

Group 1	2.73, 0.56, 0.87, 0.90, 2.27, 0.82	$\bar{y}_1 = 1.3583$
Group 2	1.60, 2.17, 1.78, 1.84, 1.83, 0.80	$\bar{y}_2 = 1.67$
Group 3	1.62, 0.19, 4.10, 0.65, 1.98, 0.86	$\bar{y}_3 = 1.5667$
Group 4	0.96, 1.92, 0.96, 1.83, 0.94, 1.42	$\bar{y}_4 = 1.3383$
Group 5	6.32, 3.66, 4.51, 3.29, 5.61, 3.27	$\bar{y}_5 = 4.4433$

The data were analysed according to the model (1.1), with the following independent prior distributions for the hyperparameters:

$$\mu \sim \mathrm{N}(2, 10), \qquad \sigma^2 \sim 22\chi_{20}^{-2}, \qquad \tau^2 \sim 6\chi_{20}^{-2}.$$

Considering first the residual methods of Chaloner (1994), most of the ε_{ij}s had posterior means close to zero, and the largest residuals were indicated by $\mathrm{E}(\varepsilon_{32} \,|\, \mathbf{y}) = -1.456$, $\mathrm{E}(\varepsilon_{33} \,|\, \mathbf{y}) = 2.269$, $\mathrm{E}(\varepsilon_{51} \,|\, \mathbf{y}) = 2.473$, and $\mathrm{E}(\varepsilon_{55} \,|\, \mathbf{y}) = 1.796$. Using Chaloner's outlier criterion a threshold of 3.14 was set such that the prior probability that all residuals would be below the criterion (in absolute value) was 0.95. The only non-zero estimated posterior probabilities of outliers from an MCMC sample of 10 000 were $\mathrm{Pr}(|\varepsilon_{33}| > 3.14 \,|\, \mathbf{y}) = 0.0285$, $\mathrm{Pr}(|\varepsilon_{51}| > 3.14 \,|\, \mathbf{y}) = 0.0563$, and $\mathrm{E}(|\varepsilon_{55}| > 3.14 \,|\, \mathbf{y}) = 0.0003$. The prior probability that an individual residual would be an outlier is 0.0017, so Chaloner's method identifies y_{33} and y_{51} as quite clear outliers. However, no outliers were seeded into this simulated data set and none is seen on visual inspection of the data.

The parameter-level residuals ε_i were also examined. It was found that $\mathrm{E}(\varepsilon_5 \,|\, \mathbf{y}) = 2.492$, and $\mathrm{Pr}(|\varepsilon_5| > 2.57 \,|\, \mathbf{y}) = 0.453$, compared with a prior probability of 0.01 for an individual ε_i to exceed this threshold. The method therefore correctly identifies the outlying fifth group.

We now consider the conflict measures applied to these data. We examine conflict at the λ_i and μ nodes. Since the contributing information sources are all normal, (3.2) was applied in each case. At the λ_i node, we first consider conflict between each y_{ij} and the remaining observations in that group. The posterior expectations of these conflict measures were all less than 1, the largest being $\mathrm{E}(c_{33} \,|\, \mathbf{y}) = 0.811$. Therefore there is no evidence of conflict within the group data, and no outlying observations are indicated. Still at the λ_i node, we also measure conflict between $\bar{y}_i$ and the prior distribution for λ_i. The expectation of the corresponding c_is are all below 1 except $\mathrm{E}(c_5 \,|\, \mathbf{y}) = 4.927$. There is therefore clear evidence that the fifth group is outlying.

At the μ node, we can consider conflict within the λ_is, which would be another diagnostic for outlying groups. Although the fifth group again emerges as in some conflict with the remaining groups, the conflict measure now has expectation only slightly larger than 1. It is therefore less clear-cut in identifying this group as different from the others. Finally, we can compare the information at the μ node provided by the mean of the λ_is and by the prior distribution of μ. The conflict measure has mean 0.004, and so gives no indication of any conflict at this level.

Both the residual diagnostics and conflict diagnostics correctly flag a problem with the fifth group. The residual methods also indicate the likely presence of outlying individual observations, although there are in reality no such outliers in the data and the conflict measures therefore seem more reliable in this case.

More work is needed to explore these measures, and in particular to apply them to non-normal distributions in complex models. It is planned to provide some theoretical results and practical examples in a forthcoming article.

3.4   Interpretation of conflict

Conflict measures play a similar role in model criticism to that of residual-based diagnostics, in the sense that those diagnostics look for conflict between the prior and posterior distributions of a residual or other diagnostic measure. Of course, we do not expect the posterior distribution of any quantity to be the same as its prior distribution. However, the prior distribution should represent our prior belief about where the posterior distribution will be centred, and we do not expect the posterior distribution to be so different from the prior distribution that they conflict in the sense of Section 3.2. If the posterior distribution is surprisingly different from the prior, then this suggests that there is some information, ultimately coming from information in the data, that is unexpected if the model is correct.

At the node in question, that information is coming via its child nodes and their distributions, which act as likelihoods. This analysis shows a link with the conflict measures. The residual analysis will typically identify an outlier if the conflict measure indicates a conflict between the information in child nodes and that in the prior distribution, since this situation typically leads to a posterior distribution very different from the prior distribution. In general, we may expect the residual and conflict measures to respond to similar kinds of possible problems with the model. I think that measures of conflict are likely to be more sensitive because they address the underlying problem more directly, but this will only be confirmed or contradicted by practical experience with the methods.

To put this abstract argument in a more practical context, consider again the example in Section 3.3. The data in group 5 are flagged by the conflict measure as providing information about λ_5 that is in conflict with its prior distribution. The data therefore exert an influence to pull the posterior distribution of λ_5 away from its prior distribution, and this leads to the discrepancy between prior and posterior being flagged as an outlier by the residual analysis. This problem is identified equally clearly by the conflict measure c_5 and by the residual diagnostic ε_5. However, looking for conflict between the observations in a group more correctly concludes that there is no evidence of outliers, illustrating the flexibility of the conflict analysis.

It is important to consider carefully the interpretation of these diagnostics, and in particular what alternative models they indicate should be considered at the model extension stage.

The residual-based diagnostics examine particular distributional components of the model. When an outlier or some other problem is indicated, then the natural response should be to consider models in which these distributions are varied. For instance, if outliers are indicated, at any level of the model, then it would be appropriate to consider models in which the possibility of outlying or aberrant observations is explicitly or implicitly included. Explicit allowance for outliers is usually done by introducing mixture models. Thus, the indications of outliers in the example data could be responded to by allowing each y_{ij} to be either $N(\lambda_i, \sigma^2)$ distributed with probability $1 - \alpha$ or else with probability α to follow some other distribution such as the mean-shifted $N(\lambda_i + \delta, \sigma^2)$. Implicit allowance for outliers could be by changing from normal distributions to some more heavy-tailed distributions such as t distributions; see O'Hagan (1979, 1990).

However, while it is certainly possible that the cause of an apparent outlier is a misspecification of the distribution under examination, it is not the only possible explanation. When the posterior distribution differs from the prior, it is equally possible for this to indicate an error in the 'likelihood' that is combining with the prior. This duality is clear in the conflict analysis, where a conflict clearly indicates only that one of the sources of information might be formulated wrongly, without favouring either as a more likely explanation.

O'Hagan (1994, Sections 7.19–7.21) links the existence of conflict to sensitivity of inferences to changes in the model. In general, the posterior inferences may be changed substantially if we change the model such that the variance or tail thickness of one or more of the conflicting distributions will change. Again, there is an indication that we might consider mixtures or heavy-tailed models.

Implementation of these methods in the context of MCMC sample-based inference is straightforward, since the computation of sample values of a relevant diagnostic can be carried out whenever a node or group of nodes is updated. Suitable diagnostics could be generated automatically within a program such as WinBUGS, and the user warned whenever they indicate a potential problem with the model.

I hope to discuss these methods in greater detail, with more complex examples, in a forthcoming paper.

3.5 Some related issues

It should be noted that conflict has long been recognized as important in the context of graphical models. For instance, Jensen *et al.* (1991) refer to ideas going back to Habbema (1976), who proposed a diagnostic similar to Box's prior predictive tail area. Jensen *et al.* suggest that instead of using the tail area one could normalize the probability of joint occurrence of a number of data items by the product of their marginal probabilities. Conflict then arises when we observe a combination of things that are relatively unlikely to have occurred together. This work intrinsically relates to conflict between nodes in the graph, rather than between sources of information bearing on a single node. There may be scope for adapting this idea to Bayesian hierarchical models.

It is often considered that 'non-conflict' is interesting, in the sense that data may fit a model 'too well'. Consider the predictive diagnostics of Section 2. A single y_i might be argued to be surprising if it is very close to the centre of its predictive distribution, but this argument only really has force when we consider several observations. For instance, it is suggested in Section 2 that in the case of data following a multivariate normal predictive distribution, a large value of the squared Mahalanobis distance M would cast doubt on the model, but a small value would also be of interest. Such a value would indicate that the data fit their predictive distribution too well. Indeed, exactly this idea is used to criticize models based on expert judgements in O'Hagan (1997). Similar ideas might be considered as worth applying to the diagnostics of outliers and conflict developed here (looking for 'inliers' and 'non-conflict'). However, in general any kind of extreme lack of conflict does not indicate a likely sensitivity of the posterior inferences to small changes in the distributions and relationships assumed in the model.

Finally, there is of course an enormous body of frequentist literature on goodness of fit, and a smaller amount of Bayesian work with the same focus. I have not attempted to examine this work here, for the same reason that the link between lack of fit and sensitivity of posterior inferences is usually weak.

4   Sensitivity to prior distributions

4.1   Diagnosing potential sensitivity

All the diagnostics considered so far have been targeted at identifying possible problems with the model. It is intuitively sensible to suppose that if problems are identified, then the posterior inferences of interest would be sensitive to a change in the model that 'corrects' the problem. A quite different approach is to target diagnostics directly at identifying potential sensitivities.

Consider the following proposal for a simple diagnostic of potential sensitivity. Let ϕ be some parameter (or function of parameters) of interest, and consider how its posterior might be sensitive to the prior distribution of some parameter θ_i. There are two stages:

1. Compare the prior and posterior distributions of θ_i. If $p(\theta_i \,|\, \mathbf{y})$ is similar to $p(\theta_i)$ in some sense, then it suggests that the data are not modifying this prior. It is clear that any change to the prior $p(\theta_i)$ would also change the posterior. In other words, θ_i's own posterior is sensitive to its prior.

2. Look to see how dependent ϕ and θ_i are in the posterior. If there is strong dependence, then the change in the posterior of θ_i would be reflected in a change of some kind to the posterior of ϕ.

Given both of these features, the posterior $p(\phi \,|\, \mathbf{y})$ of interest can be expected to be sensitive to $p(\theta_i)$.

In order to turn this idea into a useful tool, we must determine how to apply it in practice. For step 2 it might be adequate to compute the posterior correlation between ϕ and θ_i, although in general a rank correlation coefficient might be better.

For step 1, the task is superficially similar to the residual-based methods;

however, the objective is rather different. We are not looking for conflict between prior and posterior but simply to see to what extent they are different. Furthermore, we now regard a difference as *not* indicative of possible sensitivity, because we are trying to identify a different source of sensitivity.

A natural general measure (but by no means the only one) of how far two distributions differ is the Kullback–Leibler distance between them. We want to evaluate

$$D = \int \ln \frac{p(\theta_i \mid \mathbf{y})}{p(\theta_i)} p(\theta_i \mid \mathbf{y}) \, d\theta_i \,.$$

Note that in this context we require a general measure of discrepancy between the two distributions, whereas for the purpose of measuring conflict we needed to quantify their lack of overlap, which led to the more sophisticated measure (3.2).

Given a MCMC sample from $p(\theta_i \mid \mathbf{y})$, we can calculate D in the usual way of calculating posterior expectations if we can evaluate the log-ratio at each sample point θ_i^j. Now in practice we will not have an explicit formula for $p(\theta_i \mid \mathbf{y})$ (since otherwise we would not need MCMC to compute it). I will, however, assume that $p(\theta_i)$ is known. This approach is primarily for use in examining sensitivity to the prior distributions at the final stage of a hierarchical model (or at the originator nodes of a more general graphical model).

I am not aware of any existing method for estimating D under these conditions. The following is based on an original idea of Roger Cooke, and a paper giving more details is in preparation. In effect, we use the MCMC sample to make a crude density estimate for $p(\theta_i \mid \mathbf{y})$. Let $\theta_i^{(j)}$ be the *ordered* MCMC sample values of θ_i, $j = 1, 2, \ldots, N$. Then the density at $\theta_i^{(j)}$ is estimated by putting a histogram block of probability content N^{-1} on the part of the line between $(\theta_i^{(j-1)} + \theta_i^{(j)})/2$ and $(\theta_i^{(j)} + \theta_i^{(j+1)})/2$, which therefore has height $\hat{p}(\theta_i^{(j)} \mid \mathbf{y}) = N^{-1} 2 / (\theta_i^{(j+1)} - \theta_i^{(j-1)})$.

This leads to the following estimate of the Kullback–Leibler distance:

$$N^{-1} \sum_j \ln \frac{\hat{p}(\theta_i^{(j)} \mid \mathbf{y})}{p(\theta_i^{(j)})} = \ln(2/N) - N^{-1} \sum_j \ln(\theta_i^{(j+1)} - \theta_i^{(j-1)}) - N^{-1} \sum_j \ln p(\theta_i^{(j)}) \,.$$

Strictly, this formula needs $\theta_i^{(0)}$ and $\theta_i^{(N+1)}$ to be defined; in practice we simply set $\theta_i^{(2)} - \theta_i^{(0)} = 2(\theta_i^{(2)} - \theta_i^{(1)})$ and $\theta_i^{(N+1)} - \theta_i^{(N-1)} = 2(\theta_i^{(N)} - \theta_i^{(N-1)})$.

However, it is possible to show that this estimator is biased, because randomness in the data causes $\hat{p}(\theta_i^{(j)} \mid \mathbf{y})$ to be rougher than the true density, and this results in a positive bias in the Kullback–Leibler estimate. The bias persists even in large samples, and can be shown to be asymptotically equal to $\gamma - 1 + \ln 2$, where $\gamma = 0.577245\ldots$ is Euler's constant. Hence the proposed estimator is

$$\hat{D} = 1 - \gamma - \ln N - N^{-1} \sum_j \ln(\theta_i^{(j+1)} - \theta_i^{(j-1)}) - N^{-1} \sum_j \ln p(\theta_i^{(j)}) \,. \quad (4.1)$$

Here $\hat{D}$ is actually a member of a class of estimators obtained by using varying degrees of smoothing of the crude density estimate, with appropriate changes to the bias correction, but (4.1) seems to work well in the few applications that we have considered.

It is clear that this kind of diagnostic could also be routinely implemented in a program such as WinBUGS. The Kullback–Leibler distance estimate $\hat{D}$ could be computed for each θ_i. For each θ_i for which $\hat{D}$ is suitably small (see the example below for a discussion of how to determine this) the correlations between θ_i and each node being monitored in the MCMC run can be calculated and reported.

4.2   Example

To illustrate this sensitivity diagnostic method, consider the one-way model (1.1) with the same data as in Section 3.3. However, we employ a different prior structure in which I have deliberately introduced an underidentified parameter that will be influential on the amount of shrinkage of the λ_is in the posterior distribution. The prior distribution for μ, σ^2 and τ^2 is specified by

$$\mu \sim N(2, 10)\,, \qquad \sigma^2\,|\,\gamma \sim 22\gamma\,\chi_{20}^{-2}\,, \qquad \tau^2\,|\,\gamma \sim 6\gamma^{-1}\chi_{20}^{-2}\,, \qquad \gamma \sim 38^{-1}\chi_{40}^2\,.$$

The parameter γ will be poorly identified by the data, but influences the shrinkage because it will strongly influence the ratio σ^2/τ^2. Thus, γ plays the role of the parameter θ_i in Section 4.1, whose posterior distribution may be expected to be sensitive to its own prior.

The posterior distribution was simulated in WinBUGS, and the estimated Kullback–Leibler distance between the prior and posterior distributions of γ was 3.16. Is this small? There is no natural scale for the Kullback–Leibler distance between two density functions, so we must judge the magnitude of this figure in relative terms. One way to construct a comparator is to think of other prior distributions that would not be implausibly different from the actual assumed prior. I computed the Kullback–Leibler distance between the $38^{-1}\chi_{40}^2$ prior for γ and the $48^{-1}\chi_{40}^2$ distribution. This distribution is clearly different but overlaps substantially with the assumed prior and it might not be implausible as an alternative in practice. The distance in this case is 3.52. On this basis, it seems that the prior distribution has, as expected, been changed rather little by the data.

We therefore suspect that the prior information is not dominated by the data, and that the posterior distribution of γ would be sensitive to its prior. We now consider the correlations between γ and other parameters of interest, specifically the λ_is, which therefore play the role of ϕ in Section 4.1. The correlations with λ_1 to λ_4 were all of the order of 0.1, but the correlation with λ_5 is -0.31. It therefore seems likely that the posterior distribution of λ_5 might be sensitive to the prior distribution of γ. Again, this is not surprising in this example because it is λ_5 that will experience the greatest shrinkage.

5 Discussion

The growing complexity of modelling that has been stimulated by developments in HSSS makes Bayesian model criticism increasingly important and difficult. This article has focused on techniques of model criticism that are appropriate in the context of HSSS models and inferences computed by MCMC, beginning with a selective review of this active research area. I have suggested some new diagnostics that try to criticize complex models locally rather than globally, and which can be implemented readily within the MCMC computations. More work is needed to assess their practical value, and in particular it seems likely that the measures of conflict will be over-conservative. I hope that this work will serve to stimulate more research into tools for criticizing complex, highly-structured Bayesian models.

Acknowledgements

Finally, I should reiterate my thanks to those who contributed to the HSSS Research Kitchen on model criticism, and add my appreciation of the comments from my two discussants, from a referee, and from the editors. All of these people have materially improved this article by their contributions. All remaining errors and flaws are entirely my own responsibility.

References

Bayarri, M. J. and Berger, J. O. (1999). Quantifying surprise in the data and model verification (with discussion). In *Bayesian Statistics 6* (eds J. M. Bernardo, J. O. Berger, A. P. Dawid, and A. F. M. Smith), pp. 53–82. Oxford University Press, UK.

Bernardo, J. M. and Smith, A. F. M. (1994). *Bayesian Theory.* Wiley, Chichester.

Box, G. E. P. (1980). Sampling and Bayes' inference in scientific modelling and robustness (with discussion). *Journal of the Royal Statistical Society*, A, **143**, 383–430.

Chaloner, K. (1994). Residual analysis and outliers in Bayesian hierarchical models. In *Aspects of Uncertainty: a Tribute to D. V. Lindley* (eds P. R. Freeman and A. F. M. Smith), pp. 149–57. Wiley, Chichester.

Dey, D. K., Gelfand, A. E., Swartz, T. B., and Vlachos, P. K. (1998). A simulation-intensive approach for checking hierarchical models. *Test*, **7**, 325–46.

Geisser, S. and Eddy, W. (1979). A predictive approach to model selection. *Journal of the American Statistical Association*, **74**, 153–60.

Gelfand, A. E., Dey, D. K., and Chang, H. (1992). Model determination using predictive distributions, with implementation via sampling-based methods (with discussion). In *Bayesian Statistics 4* (eds J. M. Bernardo *et al.*), pp. 147–67. Oxford University Press, UK.

Gelman, A., Carlin, J. B., Stern, H. S., and Rubin, D. B. (1995). *Bayesian Data Analysis*. Chapman & Hall, London.

Gelman, A., Meng, X. L., and Stern, H. S. (1996). Posterior predictive assessment of model fitness via realized discrepancies (with discussion). *Statistica Sinica*, **6**, 733–807.

Habbema, J. D. F. (1976). Models for diagnosis and detection of combinations of diseases. In *Decision Making and Medical Care* (eds de Dombal *et al.*), 399–411. North-Holland, Amsterdam.

Jensen, V. J., Chamberlain, B., Nordahl, T., and Jensen, F. (1991). Analysis in HUGIN of data conflict. In *Uncertainty in Artificial Intelligence 6* (eds P. P. Bonissone *et al.*), pp. 519–28. Elsevier Science, New York.

Marshall, E. C. and Spiegelhalter, D. J. (2000). Simulation-based tests for divergent behaviour in hierarchical models. In submission. Available at http://www.med.ic.ac.uk/divisions/60/biostat/criticism.ps.

O'Hagan, A. (1979). On outlier rejection phenomena in Bayes inference. *Journal of the Royal Statistical Society*, B, **41**, 358–67.

O'Hagan, A. (1980). Discussion of Box (1980). *Journal of the Royal Statistical Society*, A, **143**, 408.

O'Hagan, A. (1990). On outliers and credence for location parameter inference. *Journal of the American Statistical Association*, **85**, 172–6.

O'Hagan, A. (1994). *Kendall's Advanced Theory of Statistics Volume 2B, Bayesian Inference*. Edward Arnold, London.

O'Hagan, A. (1997). The ABLE story: Bayesian asset management in the water industry. In *The Practice of Bayesian Analysis* (eds S. French and J. Q. Smith), pp. 173–98. Edward Arnold, London.

Pettit, L. I. and Smith, A. F. M. (1985). Outliers and influential observations in linear models (with discussion). In *Bayesian Statistics 2* (eds J. M. Bernardo *et al.*), pp. 473–94. North-Holland, Amsterdam.

Sharples, L. D. (1990). Identification and accommodation of outliers in general hierarchical models. *Biometrika*, **77**, 445–53.

Zellner, A. (1975). Bayesian analysis of regression error terms. *Journal of the American Statistical Association*, **70**, 138–44.

14A
Which 'base' distribution for model criticism?

M. J. Bayarri
Universitat de València, Spain

This is a very interesting article. It addresses a difficult question for which good answers are sorely needed. I fully agree with the author in his perception of the role of model criticism, and believe that he has provided very useful insights. I, however, disagree with some aspects of the proposed methodology. I focus on the one I perceive as most important, namely the reuse of the data to both compute the realized value of the diagnostic statistic and to locate this value in *posterior predictive distributions*. I put my concerns into perspective in the next section. The general developments underlying this discussion are taken from Bayarri and Berger (1999, 2000).

1 Basic issues in model criticism

In the spirit of O'Hagan's article, assume that we are considering a (null) model $\mathbf{Y}|\theta \sim p(\mathbf{y}|\theta)$, that we observe $\mathbf{Y} = \mathbf{y}_{\text{obs}}$, and that we want to investigate the question of whether or not $\mathbf{y}_{\text{obs}}$ is compatible with $p(\mathbf{y}|\theta)$. Many common frameworks for model criticism correspond to particular choices of:

(1) A diagnostic statistic $T = T(\mathbf{Y})$ to measure incompatibility of the realized value $t_{\text{obs}} = T(\mathbf{y}_{\text{obs}})$ with the model.

(2) A 'base', completely specified, distribution of t values, $p(t)$ (under the 'null') in which to 'locate' the observed t_{obs}.

(3) A way to measure conflict between t_{obs} and the base, null distribution $p(t)$. Popular choices include tail areas (as in p-values) and relative height of the density $p(t)$ at t_{obs}.

The scenario in the article is more general in that it considers *discrepancy* measures in which T is allowed to also depend on the parameters. Also, conflict is assessed between distributions, and not between a value and a distribution. The scenario above is easier to understand, and will be adopted in this discussion, but virtually identical concerns, questions, and potential undesirable behaviour arise in the general framework of the article.

All three choices are very important and determine the power of the selected model criticism procedure. I concentrate only on (2), and discuss the choice of the appropriate 'base' distribution $p(t)$ that we consider optimal for any choice of the diagnostic T and any way to measure conflict.

2   Distributions for the diagnostic measure

The basic Bayesian tool for model criticism is the prior predictive distribution $p(\mathbf{y}) = \int p(\mathbf{y}|\theta)p(\theta)\,\mathrm{d}\theta$. Sometimes, however, the use of $p(\mathbf{y})$ is unsuitable, for example when the prior is improper (or too weak). (Note that improper priors are most common at the exploratory stages at which model criticism occurs.) Instead of abandoning the use of $p(\mathbf{y})$ in these circumstances, we propose to use some (suitable) conditional distributions from $p(\mathbf{y})$. That is, choose a *conditioning* statistic $U = U(\mathbf{Y})$, and use as the relevant 'base' distribution for model criticism the *U-conditional predictive distribution*:

$$p(t|u_{\mathrm{obs}}) = \int p(t|u_{\mathrm{obs}}, \theta)p(\theta|u_{\mathrm{obs}})\,\mathrm{d}\theta \ . \tag{2.1}$$

A very popular choice (mentioned in the article) partitions the data and considers $T = \mathbf{Y_c}$ (the part to 'criticize' the model) and $U = \mathbf{Y_f}$, (the part to 'fit' the model) so that $p(t|u) = p(\mathbf{y_c}|\mathbf{y_f})$. Another popular choice among frequentists takes U to be sufficient for θ (so that $p(t|u, \theta)$ does not depend on θ, and therefore $p(t|u)$ is the same irrespective of the distribution of θ). Other choices for U are discussed in Bayarri and Berger (1997). Some comments are in order:

1. Because (2.1) is a valid probability computation, data do not get used twice.

2. 'Predictive replicate methods' (use of the posterior predictive distribution $p(t\,|\,\mathbf{y}_{\mathrm{obs}})$ as a 'base' distribution) cannot be obtained as a conditional distribution from the prior predictive. Indeed, if we take $U = \mathbf{Y}$, $p(t|u_{\mathrm{obs}})$ is degenerate.

3. Once T and U have been chosen, any of the measures of conflict can be used to asses incompatibility of t_{obs} with $p(t|u)$.

Of course, not every choice of $p(t|u)$ is optimal: if we condition too much there might be very little randomness left to assess conflict. If, on the other hand, we condition too little, there might not be enough 'learning' about the parameters. In Bayarri and Berger (1999) we argue that an optimal choice for any given T is to use U, the conditional maximum likelihood estimator (MLE), from $p(\mathbf{y}|t_{\mathrm{obs}}, \theta)$, as the conditioning statistic. This, however (as the author notes), could give rise to burdensome computations, especially in complicated models, such as those typical in HSSS. A very nice approximation (which avoids explicit identification of U) consists of integrating θ out from $p(t|\theta)$ w.r.t. the *partial posterior* distribution, defined as

$$p(\theta|\,\mathbf{y}_{\mathrm{obs}} \setminus t_{\mathrm{obs}}) \propto p(\mathbf{y}_{\mathrm{obs}}|t_{\mathrm{obs}}, \theta)p(\theta) \propto \frac{p(\mathbf{y}_{\mathrm{obs}}|\theta)}{p(t_{\mathrm{obs}}|\theta)}p(\theta) \ .$$

The notation was chosen to emphasize the interpretation of $p(\theta|\,\mathbf{y}_{\mathrm{obs}} \setminus t_{\mathrm{obs}})$ as a posterior given the information in y_{obs} not in t_{obs} (Bayarri and Berger 2000). Integrating θ out w.r.t. $p(\theta|\,\mathbf{y}_{\mathrm{obs}} \setminus t_{\mathrm{obs}})$ results in the *partial posterior predictive distribution*

$$p(\mathbf{y}|\mathbf{y}_{\text{obs}} \setminus t_{\text{obs}}) = \int p(\mathbf{y}|\theta)p(\theta|\mathbf{y}_{\text{obs}} \setminus t_{\text{obs}}) \, d\theta . \tag{2.2}$$

Computations of incompatibility measures with (2.2) are much easier than with (2.1), especially in cases for which $f(t|\theta)$ is available in closed form (in fact, work is in progress with M.E. Castellanos applying (2.2) to a criticism of hierarchical models). Moreover, they tend to be very similar (if not identical; see Bayarri and Berger (1999, 2000)). Posterior replicate methods use $p(\mathbf{y}|\mathbf{y}_{\text{obs}}) = \int p(\mathbf{y}|\theta)p(\theta|\mathbf{y}_{\text{obs}}) \, d\theta$, which usually results in even easier computations. Because of this (very compelling) reason, $p(\mathbf{y}|\mathbf{y}_{\text{obs}})$ is precisely the 'base' distribution adopted in O'Hagan's article for model criticism. However, the double use of the data involved often results in severe conservatism, so severe that clearly unsuitable models are still found compatible with the data, thus rendering the model criticism exercise useless. We present an example in the next section (from Bayarri and Morales (2003)), and many others can be found in previously mentioned references. In fact, when ease of computation is a severe constraint (thus precluding the use of (2.2) as the base distribution), we recommend the use of $p(t|\hat{\theta})$ instead ($\hat{\theta}$ could be the MLE, or the posterior mean obtained by Markov chain Monte Carlo (MCMC) sampling), which has proven to produce more satisfactory results than $p(\mathbf{y}|\mathbf{y}_{\text{obs}})$ at a sensibly smaller computational burden. (See also the results in Robins *et al.* (2000)).

3   An example (Bayarri and Morales 2003)

As a simple, intuitive example for which the model is clearly inadequate, consider checking for outliers taking as a 'diagnostic' statistic $T = Y_{(1)} = \min\{Y_1, \dots, Y_n\}$. The 'null' model is that observations are independent and identically distributed normal, that is θ here is the unknown mean and variance of the normal. We simulate 10 observations from a $N(0,1)$ distribution and substitute -8 for the minimum value. The ordered resulting sample is: -8, -1.27, -1.059, -0.986, -0.874, -0.204, 0.315, 0.42, 0.49, 2.457. We use two different measures of conflict (to exemplify that the problem is not with the particular measure of conflict, but rather with the choice of the posterior predictive as the 'base' distribution); specifically, for a given 'base' distribution, $p(t)$, we compute: (i) the p-value: $\Pr(T \leqslant t_{\text{obs}})$, and (ii) the relative height of $p(t)$ at t_{obs}, which we called the 'relative predictive surprise', RPS $= p(t_{\text{obs}})/\sup p(t)$. We use three different possibilities for $p(t)$: (i) the *partial posterior predictive* distribution (ppp): $p(t|\mathbf{y}_{\text{obs}} \setminus t_{\text{obs}}) = \int p(t|\theta)p(\theta|\mathbf{y}_{\text{obs}} \setminus t_{\text{obs}}) \, d\theta$, (ii) the *posterior predictive* distribution (post): $p(t|\mathbf{y}_{\text{obs}}) = \int p(t|\theta)p(\theta|\mathbf{y}_{\text{obs}}) \, d\theta$, and (iii) the *plug-in* distribution (plug): $p(t|\hat{\theta})$ (with $\hat{\theta}$ the MLE). The results are shown in Table 1. Because the outlier was 8 standard deviations away from the null mean, it seems rather clear that the only measures of conflict that adequately represent this incompatibility are those corresponding to ppp. Note how useless the measures based on the posterior predictive are, with a p-value of 0.133 (and a RPS of 0.341) even in this very extreme example.

TABLE 1. Measures of conflict derived from three different 'base' distributions for the 'diagnostic' statistic.

	p-value	RPS
ppp	1.59×10^{-3}	1.87×10^{-3}
post	0.133	0.341
plug-in	0.030	0.128

4   Concluding remarks

The approach presented in O'Hagan's article is a nice generalization of previous approaches, in principle suitably tailored to check models arising from HSSS. However, this nice attempt might fail due to the extreme conservativeness induced by the use of posterior replicate measures. We do agree that computations with such measures can usually be easily incorporated into the regular MCMC routine, but we feel it very dangerous to use such an extremely conservative approach for model checking, since it gives the user a false sense of confidence on a possibly disastrous model. The example presented in the article does find one of the means incompatible with the assumption of exchangeability of all five means, but it should be noted that the 'offending' (sample) mean ($\bar{y}_5 = 4.44$) is 17 (sample) standard deviations away from the group formed by the other 4 means (with mean 1.5 and standard deviation 0.17);

Although the potential for extreme conservatism is the most worrying, we also have some other minor concerns with the approach taken in the article. Namely: (i) it is based on the values of densities at particular points, which is known not to be easily computable by MCMC algorithms in general; (ii) it might be difficult to calibrate for all models, all dimensions, etc.; (iii) it needs a very carefully specified (and moderately informative) prior distribution – a weak prior will not do; (iv) as a consequence of (iii), the method cannot distinguish misspecification of prior opinions from inadequate structure.

In summary, we believe that this article is a most valuable contribution to a very challenging problem. However, in its present form, it should be used with extreme care: a lack of incompatibility should not be taken as an indication that the model is appropriate, but rather that a more powerful checking procedure should be used.

14B
Some comments on model criticism

Alan E. Gelfand
University of Connecticut, USA

As increasingly complex hierarchical models are brought to the analysis of data arising in a wide variety of disciplines, the need for effective model criticism becomes evident. As O'Hagan observes in Chapter 14, the terms model criticism, model checking, and model validation are essentially synonymous in practice (though the last might be more appropriately called model invalidation). But then, O'Hagan distinguishes model criticism from model adequacy, defining the former as 'the process of checking model assumptions without reference to explicit alternative models or assumptions'. The implication is that model criticism is an 'absolute' notion. With regard to the latter, he asserts that 'model criticism cannot address whether a model is adequate since we exclude consideration of specific alternatives'. Again, this may be a semantic issue since it would be generally agreed that, for example, the model $Y \sim N(0, 1)$ is inadequate if the observed value of Y is 100, without reference to an alternative model.

Still, I believe O'Hagan is correct in claiming that 'the role [of model criticism] rather is to seek to identify what kinds of alternative model should be constructed (model elaboration)'. That is, if we can identify a model failure we can suggest alternative elaborations to remedy this failure. However, I am not comfortable with his continuation, 'in order to test formally the adequacy of the model (model comparison)'. In my view, model adequacy is not model comparison. Assessing model adequacy is an informal activity; on the other hand, model comparison, say in pairs, is a more precisely defined problem suggesting more formal approaches.

In fact, accepting that, in all but the simplest cases, 'all models are wrong', for a given set of models, does it make sense to assign to each 'a prior probability according to its plausibility a priori'? Furthermore, does it make sense to use these probabilities to render one model inadequate relative to another? That is, in the above example, the alternative model $Y \sim N(0, 100)$ renders the $N(0, 1)$ model inadequate without reference to prior probability assignments. Would it not be best to conclude that many models can be adequate, but deferring to a model choice criterion, the selection of a particular one?

Continuing in this spirit, with hierarchical models, specifications at stages farther removed from the data are often intentionally less precise. How can we think of prior probabilities reflecting the plausibility of such models? Indeed, if any prior specification is improper, then the resulting model could not possibly have yielded the observed data. In the case of multi-level, over-parameterized, weak-hyperprior models, are criticism and adequacy even meaningful issues to raise? I think not. These concerns should be addressed to that *middle ground* of models that adopt rather informative priors and, in addition, are not so parsimonious as to preclude useful hierarchical modelling but not so high dimensional as to leave adequacy a non-issue.

Lastly, in this regard, model checking means checking the entire specification. Model failures can occur at each hierarchical stage. Such failures include outliers, mean misspecification, dispersion misspecification, and inappropriate exchangeabilities. Recent work by Dey *et al.* (1998) presents a general methodology for informally assessing such failures with the foregoing middle ground of models. It also provides some clarification with regard to whether a model permits stagewise checking. Such capability is a property of the model specification and has nothing to do with the data to which the model might be fitted. As an aside, taking the formal view that only marginal models (those that arise from integrating out all model unknowns) are unique, i.e. that conceptually, many hierarchical specifications yield the same marginal model, how can we claim that the data are really criticizing any stage of the specification? A practical counter is that, if a model is to be used for explanation rather than purely for prediction, we wish to be able to criticize a particular hierarchical specification.

To provide some supplementary historical perspective, Box (1980) proposed as a formal Bayesian model adequacy criterion, the marginal density of the data evaluated at the observations. Computing this criterion for the models we envisage is challenging enough; calibrating it seems hopeless. Chaloner and Brant (1988), Chaloner (1994), and Weiss (1995), focusing on outlier detection, suggest posterior–prior comparison. In general, if the entire model specification is correct, then such comparisons will be successful on average but will fail to appreciate the variability in the posterior across data sets realized under the model. Without such appreciation it is difficult to assess how large a posterior–prior discrepancy can be attributed to chance under the model.

Model expansion or elaboration nests the model of interest within a larger (full) model chosen to accommodate a possible model failure in the (reduced) model of interest; e.g. Pettit and Smith (1985) or Albert and Chib (1997). Model choice procedures then provide the checking of adequacy of the reduced model. A concern here is that, for a particular failure, which of many possible elaborations should be chosen? It seems more attractive to use the given model and the observed data to directly reveal the presence of a failure.

A posterior predictive strategy is proposed in Gelman *et al.* (1996) which O'Hagan describes and seems comfortable with. However, this approach has been criticized for using the data twice and thus for being conservative. As the example below shows, this strategy will find it very difficult to criticize a model. The problem is that the observed data through the posterior suggest parameter values that tend to be likely under the model. Then, to assess adequacy, the observed data are checked against data generated using such parameter values.

Evidently it will be hard for a discrepancy measure to find inadequacy.

The proposal of Dey *et al.* (1998) is also to study the posterior distributions of various discrepancy measures given the observed data. However, for a particular measure, comparison is made with what is expected under the model rather than what is expected under the model and observed data. Indeed, by simulating data replicates under the model and thus, replicating a posterior of interest under the model using these replicates, we can 'see' the extent of variability in such a posterior. Then we can compare the posterior obtained from the observed data with this medley of posterior replicates to ascertain whether the former is in agreement with them and, accordingly, whether it is plausible that the observed data came from the proposed model.

With a two-stage hierarchical model, Dey *et al.* suggest using first-stage discrepancy functions, second-stage discrepancy functions, marginal checking functions (associated with partly marginalized models), and overall checking functions (associated with model prediction). Hence, for a particular application, many discrepancy functions will be of interest. In turn this means numerous replicated datasets each requiring Markov chain Monte Carlo fitting and then, for each fitted model, the creation of many posteriors and, subsequently, many comparisons. The approach, though straightforward, is computationally intensive requiring care in storage and book-keeping. Further details are provided in Dey *et al.*

We conclude with an illustrative example. The data in Table 1 are abstracted from the United Network for Organ Sharing (UNOS) public-use database. The overall database consists of 3688 transplants at 131 transplant centres. We confine ourselves to the 10 centres with the largest number of heart transplants. Subjects were classified into 10-year age groups with binary patient response indicating whether or not short-term problems leading to organ rejection or death occurred.

TABLE 1. For UNOS heart transplant data, for 10 centres and five age groups, number of patients developing problems leading to short-term rejection of organ or death/number of transplant patients.

Centre	Age group					Total
	25	35	45	55	65	
1	0/3	0/11	9/27	16/36	0/2	25/79
2	3/7	1/8	4/15	10/31	7/15	25/76
3	6/15	3/11	2/14	7/30	0/5	18/75
4	1/14	1/6	2/23	10/32	5/16	19/91
5	3/14	1/6	4/19	5/28	0/7	13/74
6	2/11	1/12	3/23	12/51	3/9	21/106
7	5/16	6/19	5/34	9/39	11/26	36/134
8	1/9	2/7	3/28	4/25	0/4	10/73
9	6/33	7/17	12/42	11/42	4/18	40/152
10	13/31	2/7	5/28	8/34	0/1	28/101

The hierarchical model sets as the first stage $Y_{ij} \sim \text{Bin}(n_{ij}, p_{ij})$, where $i = 1, \ldots, 10$ denotes the transplant centre and $j = 1, \ldots, 5$ the age group. Then, with X_j denoting the centred age of the jth age group,

$$\log \frac{p_{ij}}{1 - p_{ij}} = \alpha_i + \beta_i X_j,$$

with

$$\begin{pmatrix} \alpha_i \\ \beta_i \end{pmatrix} \sim \text{N}\left(\begin{pmatrix} \mu_\alpha \\ \mu_\beta \end{pmatrix}, \begin{pmatrix} \sigma_\alpha^2 & 0 \\ 0 & \sigma_\beta^2 \end{pmatrix} \right)$$

and $\mu_\alpha \sim \text{N}(-0.9, a^2)$, $a = 0.2$ or 0.13, $\mu_\beta \sim \text{N}(0.17, (0.05)^2)$. These prior means arise, after a bit of algebra, from the information that roughly 30% of transplant patients develop problems and that the oldest patients have roughly twice the chance of the youngest ones. The looser prior variance on μ_α keeps $p_{ij} \in (0.18, 0.42)$, the tighter one keeps $p_{ij} \in (0.21, 0.39)$. The variance of μ_β encourages $\mu_\beta > 0$. σ_α^2 and σ_β^2 receive inverse gamma priors with infinite variances.

We consider the 50 first-stage discrepancies, $y_{ij} - n_{ij} p_{ij}$, the 10 second-stage intercept discrepancies, $\alpha_i - \mu_\alpha$, the 10 second-stage slope discrepancies $\beta_i - \mu_\beta$, and finally the 50 marginal discrepancies $y_{ij} - n_{ij} \tilde{p}_{ij}$ where, for convenience,

$$\log \frac{\tilde{p}_{ij}}{1 - \tilde{p}_{ij}} = \mu_\alpha + \mu_\beta X_j.$$

In this context, Gelman *et al.* (1996) compute $\Pr(y_{ij,\text{new}} \geqslant y_{ij,\text{obs}} | \mathbf{Y}_{\text{obs}})$. Posteriors were obtained by fitting the model to replicated data sets generated under the model using the priors. Monte Carlo tests are then used to compare, for any discrepancy, the posterior for the observed data with those arising under the replication.

At the 5% level under the loose prior we find 4 first-stage failures out of 50 comparisons, 1 of 10 'α failures', 2 of 10 'β failures', and 0 of 50 marginal failures. Under the tighter prior we still find 4 first-stage failures but now 10 of 10 α failures, 3 of 10 β failures, and 25 of 50 marginal failures. Our approach reveals that the tighter prior on μ_α is too tight to provide an adequate second-stage model. The approach of Gelman *et al.* finds 0 failures out of 50 in both cases and thus does not criticize the tighter specification.

Additional references in discussion

Albert, J. H. and Chib, S. (1997). Bayesian tests and model diagnostics in conditionally independent hierarchical models. *Journal of the American Statistical Association*, **92**, 916–25.

Bayarri, M. J. and Berger, J. O. (1997). Measures of surprise in Bayesian analysis. *ISDS Discussion Paper* 97–46, Duke University.

Bayarri, M. J. and Berger, J. O. (2000). *P*-values for composite null models. *Journal of the American Statistical Association*, **95**, 1127–42.

Bayarri, M. J. and Morales, J. (2003). Bayesian measures of surprise for outlier detection. *Journal of Statistical Planning and Inference,* **111**, 3–22.

Chaloner, K. and Brant, R. (1988). A Bayesian approach to outlier detection and residual analysis. *Biometrika,* **75**, 651–9.

Robins, J. M., van der Vaart, A., and Ventura, V. (2000). The asymptotic distribution of P-values in composite null models. *Journal of the American Statistical Association,* **95**, 1143–57.

Weiss, R. E. (1995). Residuals and outliers in repeated measures random effects models. Technical Report, Department of Biostatistics, UCLA.

15

Topics in non-parametric Bayesian statistics

Nils Lid Hjort
University of Oslo, Norway

1 Introduction and summary

The intersection set of Bayesian and non-parametric statistics was almost empty until about 1973, but now seems to be growing at a healthy rate. This article gives an overview of various theoretical and applied research themes in this field, partly complementing and extending recent reviews of Dey *et al.* (1998) and Walker *et al.* (1999). The intention is not to be complete or exhaustive, but rather to touch on research areas of interest, partly by example.

1.1 What is it, and why?

In this article we do not use a very precise definition of what constitute 'non-parametric Bayesian methods', and might err on the liberal side. Specifically, examples are included of statistical modelling and inference situations where placing a distribution over large sets of probability distributions is one of the ingredients. Some of these situations do not have to be intrinsically Bayesian *per se*.

One of the goals of non-parametric Bayesian statistics is to ease up on traditional 'hard' model assumptions, without essential loss of inference power. Pure finite-parametric models can never be fully correct, whereas non-parametric Bayes constructions may succeed in having most conceivable true data generating mechanisms inside its prior scope, i.e. its support. If the setup is satisfactory, and the data quality reasonable, then one often finds that the data themselves help dictate to what extent solutions are close or not close to what they would have been under simpler assumptions. Such findings, along with easily available software tools that most statisticians can learn to use, make good selling points. These 'harder' model assumptions of traditional statistics, both frequentist and Bayesian, might include both the error and the signal structure of models. Thus, one might soften up the linear normal regression textbook methods by using a nearly linear mean function, nearly Gaussian errors, nearly constant variance level across covariates, and if relevant some dependence structure. This also serves to illustrate that, by necessity, there is a broad range of possible non-parametric constructions, where one should not anticipate clear winners.

The essence of the 'non-parametric' word is that what is being modelled is not seen as well enough described by a fixed (and perhaps low) number of parameters. Otherwise the term in current usage is not very strictly interpreted. It might allude to broad flexibility (many shapes of the underlying curve or surface being possible under the model), which can be achieved in several ways. Some constructions use a growing number of parameters, or perhaps a growing range of candidate models to choose from or average over, e.g. involving mixtures. Here the operative word is flexibility, there being no clear division between parametric and non-parametric; see also Green and Richardson (2001). The non-parametric term might also allude to certain mathematical aspects of performance as the data volume grows, such as consistency or optimality of precision.

Statistics has witnessed the first three decades of non-parametric Bayesian life, which arguably has passed through its infancy and early youth. The first period was primarily a mathematical or probabilistic one, by necessity concerned with setting up the right probability structures on the proper spaces and deducing, when possible, relevant aspects of the posterior distributions. The second period has been more explorative, making different approaches more flexible and amenable to practical analysis in a growing list of applications. This has also, through serendipitous timing, been aided by broadly enhanced computing abilities and methodology, including software and bigger toolboxes for stochastic simulation, in particular Markov chain Monte Carlo methods (see, for example, Chapters 5 and 6 in this volume, with discussions), along with more frequent use of numerical analysis software. Of course both 'periods' are in a sense never-ending stories.

In spite of broad impressive developments, many non-parametric Bayes setups will continue to pose challenges of construction, deduction, interpretation, and computation. Given these complexities, related also to probability calculus over infinite-dimensional spaces, it is not surprising that a fair portion of published work in this area has been in the 'one can do' spirit. This is also true for many applications to real data. For the envisaged upcoming third period in the life of non-parametric Bayes one might predict a further broadening and maturing of the field, leading with experience to more finesse and possibly a higher degree of scientific relevance in old and new segments of substantial statistical application. At the same time more theory and a broader range of models will be developed. It is also likely that more hybrid constructions will evolve, perhaps mixing together not only parametric and non-parametric ingredients for given problems, but also by perhaps pragmatic frequentist-inspired solutions to aid constructions that at the outset are meant as pure Bayesian. There will be challenges of combining different data sources of different quality and complexity, where non-parametric Bayes might play a role, but along with other elements. Efron (2003) predicts a wave of empirical Bayes statistics for the twenty-first century, for example in connection with problems of microarrays and data mining for big databases. This wave should also encompass empirical non-parametric Bayes methods.

1.2 On the present article

In Section 2, the Dirichlet process is proved to be a distributional limit of certain simpler processes with symmetrically distributed probability weights. This suggests suitable generalizations of the Dirichlet for use as priors in Bayesian inference, and also proves useful in connection with transform identities for distributions of random means, as shown in Section 3. Identities are obtained there which partly generalize earlier results of Cifarelli and Regazzini (1990) and Diaconis and Kemperman (1996). Section 2 also provides another generalization of the Dirichlet, starting with the infinite series representation due to Sethuraman and Tiwari (1982). Section 4 deals with quantile inference, first based on the Dirichlet process and then using a more general non-parametric prior quantile process, which is constructed in a pyramidal fashion. These quantile trees aim to be for quantiles what Pólya trees are for cumulative distribution functions. A natural quantile function estimator is seen to lead to an attractive Bayesian density estimator, which does not require any smoothing parameters. Some interpretational and consistency issues are then discussed for Bayesian density estimation in general.

Section 6 shows how elements of non-parametric Bayesian modelling may be used for a different purpose than merely analysing data, namely to derive functional forms of statistical functions in regression contexts. It is shown there how a broad class of Lévy-type cumulative damage processes, viewed as frailty processes for individuals, influence their survival distributions in a way that leads to the multiplicative hazard regression structure. Then in Section 7 we briefly discuss the use of Beta processes to a linear hazard regression model, before going on in Section 8 to a broad class of Bayesian extensions of the by now traditional way of carrying out non-parametric regression, namely that of local polynomial modelling. In Section 9, a use is found for smoothed Dirichlet processes as a modelling tool for random shapes, and in Section 10, non-parametric envelopes around parametric models are studied. Finally, Section 11 offers some concluding remarks.

2 The Dirichlet process prior and some extensions

The Dirichlet process prior was introduced in Ferguson (1973, 1974) and remains a cornerstone in Bayesian non-parametric statistics. It is also a favourite special case of various generalizations that have been worked with, including neutral to the right and tailfree processes (Doksum 1974; Ferguson 1974), Pólya trees (Kraft 1964; Ferguson 1974; Lavine 1992), Beta processes (Hjort 1990), and mixtures of Dirichlets (Lo 1984; Escobar and West 1995). Below we establish some notation, review the Dirichlet, and briefly discuss two useful extensions.

2.1 The Dirichlet process

To define the Dirichlet process on a sample space Ω, let P_0 be a probability measure thereon, interpreted as the prior guess distribution for data, and let b be positive. Then P is a Dirichlet process with parameters (b, P_0), for which we write $P \sim \text{Dir}(b, P_0)$, if for each partition $A_1, \ldots, A_k$,

$$\big(P(A_1), \ldots, P(A_k)\big) \sim \text{Dir}\big(bP_0(A_1), \ldots, bP_0(A_k)\big). \tag{2.1}$$

We may refer to bP_0 as the total measure. In particular, for each set A, $P(A) \sim$ Beta$\{bP_0(A), b(1-P_0(A))\}$ with mean $P_0(A)$ and variance $P_0(A)(1-P_0(A))/(1+b)$. Perhaps the most attractive property of the Dirichlet prior is the ease with which it is updated with incoming data; if $x_1, \ldots, x_n$ are observations having arisen as an independent n-sample from the random P, then P given these is another Dirichlet, with total measure $bP_0 + n\widehat{P}_n$. Here $\widehat{P}_n$ is the empirical distribution giving mass $1/n$ to each data point.

One often refers to the limiting case $b \to 0$, where P is concentrated at the data points with probabilities given by a flat Dirichlet $(1, \ldots, 1)$, as corresponding to using a non-informative prior for P. There are many cases where Bayesian inference using this posterior gives natural parallels to perhaps canonical frequentist methods; cases in point include the empirical distribution $\widehat{P}_n$ as a limiting Bayes estimate, and the Bayesian bootstrap developed by Rubin (1981) and others. See also Sections 4.1 and 4.2 below. The notion of non-informativeness is debatable here, however, as the behaviour of the prior process P is peculiar when b is small. In the limit, it is concentrated at a single value, chosen from P_0.

2.2   The Dirichlet as a limit

Hjort and Ongaro (2002) give a new constructive definition of the Dirichlet process as a limit of simpler processes. Let

$$P_m = \sum_{j=1}^{m} \beta_j \delta(\xi_j), \quad \text{where } \beta = (\beta_1, \ldots, \beta_m) \sim \text{Dir}(b/m, \ldots, b/m), \tag{2.2}$$

where $\xi_1, \xi_2, \ldots$ are independent from P_0 and independent of β. Here $\delta(\xi)$ denotes unit point mass at position ξ. For a set A, and conditionally on the ξ_js, $P_m(A)$ is a Beta with parameters $\{b\widehat{P}_m(A), b(1 - \widehat{P}_m(A))\}$, where $\widehat{P}_m$ is the empirical distribution of $\xi_1, \ldots, \xi_m$. Hence $P_m(A)$ is distributed as a binomial mixture over such Beta distributions. Since $\widehat{P}_m(A)$ goes to $P_0(A)$ as $m \to \infty$ and the Beta is continuous in its parameters, the limit distribution of $P_m(A)$ is the Beta distribution of $P(A)$ when $P \sim \text{Dir}(b, P_0)$. An extension of this argument shows that all finite-dimensional distributions of P_m converge to those given in (2.1). An interesting connection is that $\Pr\{P_m(A) \in C\}$, for any set C, can be seen to be the Bernshteĭn polynomial approximation (see, for example, Billingsley (1995)) to the function $h(p) = \Pr\{\text{Beta}(bp, b(1-p)) \in C\}$ at the point $p = P_0(A)$. Proving convergence of P_m can also be done via results on the so-called Poisson–Dirichlet distribution (see Kingman (1975)), and which also has connections to representation (2.3) below.

The limit construction (2.2) is different from, but shares some of the ingredients of, the infinite series representation considered below. Among its advantages is the simplicity of the symmetric Dirichlet for the weights. In addition to being useful for deriving facts about the Dirichlet, as indicated in the next section, it also invites suitable bona fide generalizations of the Dirichlet process prior. A simple construction of interest is to let $P = \sum_{j=1}^{M} \beta_j \delta(\xi_j)$, where M has any distribution on the integers with $\Pr\{M > m\}$ positive for all m. One may develop methods for Bayesian inference using this non-parametric prior. The Dirichlet is the limiting case where M tends to infinity.

2.3 An extension via an infinite sum representation

Consider independent $B_1, B_2, \ldots$ drawn from the same distribution H on $(0, 1)$. These generate random probabilities $\gamma_1 = B_1$, $\gamma_2 = \overline{B}_1 B_2$, $\gamma_3 = \overline{B}_1 \overline{B}_2 B_3$, and so on, where $\overline{B}_j = 1 - B_j$; these sum a.s. to 1 since $1 - \sum_{j=1}^{n} \gamma_j = \overline{B}_1 \ldots \overline{B}_n$. Accordingly, we may define a random probability measure by

$$P = \sum_{j=1}^{\infty} \gamma_j \delta(\xi_j) \quad \text{with } \xi_1, \xi_2, \ldots \text{i.i.d.} \sim P_0. \tag{2.3}$$

Sethuraman and Tiwari (1982) showed that the Dirichlet process (b, P_0) can be represented in this form for a particular choice of the Beta$(1, b)$ distribution for the B_js; see also Sethuraman (1994). Ishwaran and Zarepour (2000) and Hjort (2000) have independently studied the extension to a general distribution H for these. A fruitful family of priors emerges by letting $H = \text{Beta}(a, b)$, creating a generalized Dirichlet process with parameters (a, b, P_0). Ishwaran and Zarepour (2000) develop computational algorithms using simulation, while more explicit theoretical results about estimators and performance are given in Hjort (2000). Note that the Dirichlet corresponds to $a = 1$, an inner point in the parameter space of its generalization. This extension allows more modelling flexibility regarding the skewness, kurtosis, and so on of random means. Explicit formulae are available for posterior means and variances of random mean parameters. One finding of general importance is that the speed with which the data wash out aspects of the prior is of the order $O(n^{-a})$, which can be slower or faster than the ordinary rate n^{-1} found for nearly all parametric problems as well as for the Dirichlet prior.

3 Random Dirichlet means

For P a Dirichlet process (b, P_0) on a sample space Ω, consider a random mean $\theta = \mathrm{E}_P g(X) = \int g \, dP$. There are many uses of such constructions besides the most immediate one where it is a focus parameter for Bayesian inference. Recently much attention has been given to the study of the distribution of such a θ; a partial list is Diaconis and Kemperman (1996), Guglielmi and Tweedie (2000), Regazzini *et al.* (2000), Tsilevich *et al.* (2000), and Hjort and Ongaro (2002).

3.1 Transform identities

The task is to derive aspects of the distribution of θ using information about $Y = g(\xi)$, where $\xi \sim P_0$. Assume for simplicity of presentation that g is non-negative. Cifarelli and Regazzini (1990, 1994) were the first to show an identity linking the so-called Hilbert transform of θ to a transform of Y. This connection may be written

$$\mathrm{E}\exp\left\{-b\log\left(1+u\int g\,\mathrm{d}P\right)\right\} = \exp\left[-b\int\log\{1+ug(\xi)\}\,\mathrm{d}P_0(\xi)\right]. \quad (3.1)$$

Cifarelli and Regazzini gave a rather long proof of (3.1) and some of its variations, and used integration in the complex plane to indicate how the transform may be inverted to find the distribution of θ numerically.

A quite straightforward derivation of (3.1), and without unnecessary side conditions, are among the consequences of construction (2.2) discussed in Hjort and Ongaro (2002). One may write $\beta_j = G_j/S_m$ in terms of independent Gamma$(b/m, 1)$ variables $G_1, \ldots, G_m$ and their sum S_m. Write therefore $\theta_m = \theta(P_m) = R_m/S_m$, with $R_m = \sum_{j=1}^m G_j Y_j$ being a random mixture of many small Gammas; here, $Y_j = g(\xi_j)$. First, exploit the independence between θ_m and S_m to derive

$$\mathrm{E}\exp(-uR_m) = \mathrm{E}[\mathrm{E}\exp(-u\theta_m S_m) \,|\, S_m] = \mathrm{E}\exp\{-b\log(1+u\theta_m)\},$$

using the fact that S_m has a Laplace transform $(1+u)^{-b}$. Next, use the Laplace transform $(1+u)^{-b/m}$ for the G_js to obtain

$$\mathrm{E}[\exp(-uR_m) \,|\, \xi_1, \ldots, \xi_m] = \prod_{j=1}^m (1+uY_j)^{-b/m} = \exp\left\{-b\frac{1}{m}\sum_{j=1}^m \log(1+uY_j)\right\}.$$

Taking the limit, and supplying some extra arguments, one proves (3.1); both sides are equal to the Laplace transform of the variable R to which R_m converges in distribution. Hjort and Ongaro also give a multivariate version of this in the form of a formula for $\mathrm{E}\exp\{-b\log(1+\sum_{j=1}^k u_j\theta_j)\}$, where $\theta_j = \int g_j\,\mathrm{d}P$. Such results were explicitly mentioned as missing in the literature by Diaconis and Kemperman (1996). See also Kerov and Tsilevich (1998).

Let G be a gamma process on the sample space with parameter bP_0; it has independent contributions for disjoint sets and $G(A)$ is Gamma$(bP_0(A), 1)$ for each A. The arguments used above are connected to the representation of a Dirichlet process as a normalized gamma process, namely $P(\cdot) = G(\cdot)/G(\Omega)$, where, in addition, one may demonstrate that $P(\cdot)$ is independent of $G(\Omega)$. Given the simplicity of these arguments, it is perhaps not surprising that several authors recently and independently have come up with somewhat different but related proofs of (3.1) and its relatives; in addition to Hjort and Ongaro (2002), see Regazzini *et al.* (2000) and Tsilevich *et al.* (2000). One may in fact trace the roots of identity (3.1) back to Markov (1896).

3.2 Stochastic equations and the full moment sequence

The mean of $\int g\,\mathrm{d}P$ is $\theta_0 = \int g\,\mathrm{d}P_0$, and Ferguson (1973) gave a formula for the variance. Among the uses of (2.2) and (3.1) is the possibility of deriving formulae for further moments; see Hjort and Ongaro (2002) for a list of the first ten centralized moments $E(\theta - \theta_0)^p$. One may also derive a stochastic equation for the distribution of θ, as follows. Use representation (2.3) to write $\theta = \sum_{j=1}^{\infty} \gamma_j Y_j$ in the form $B_1 Y_1 + \overline{B}_1 (B_2 Y_2 + \overline{B}_2 B_3 Y_3 + \cdots)$, from which it is apparent that

$$\theta =_d BY + \overline{B}\theta. \tag{3.2}$$

On the right-hand side, $B \sim H$, $Y = g(\xi)$ with $\xi \sim P_0$ and θ independent, and '$=_d$' indicates equality in distribution. This stochastic equation determines the distribution of θ uniquely. Only rarely can this distribution be exhibited in closed form, however. The equation at least gives a simple recursive method of finding all moments, via

$$E(\theta - \theta_0)^p = \sum_{j=0}^{p} \binom{p}{j} E_0 (Y - \theta_0)^{p-j} \, E B^{p-j} \overline{B}^j \, E(\theta - \theta_0)^j \quad \text{for } p \geqslant 2. \tag{3.3}$$

Here 'E_0' indicates the expected value when Y has its null distribution $Q_0 = P_0 g^{-1}$. Note that this gives a recipe for finding all moments not only for the Dirichlet case, where $B \sim \text{Beta}(1, b)$, but also for the generalized process of (2.3), where B has an arbitrary distribution on $(0, 1)$.

It is not difficult to use eqn (3.2) to construct a Markov chain with the distribution of θ as its equilibrium. Such simulation output can be further repaired to give improved accuracy via knowledge of the exact moments, as demonstrated in Hjort and Ongaro (2002).

4 Quantile pyramid processes

Let $Q(y) = F^{-1}(y)$ be the quantile function for a distribution on the real line, and suppose data are observed from this distribution. One may attempt to carry out quantile inference via a given non-parametric prior for F, and this is done below for the Dirichlet case. It is also worthwhile to place priors directly on the set of quantile functions, leading to direct Bayes estimators of Q and related functions.

4.1 Quantile inference with the Dirichlet process

Let F be the cumulative function of a $\text{Dir}(b, F_0)$ process, where F_0 is a suitable prior guess distribution with density f_0, and define more formally

$$Q(y) = F^{-1}(y) = \inf\{t \colon F(t) \geqslant y\} \quad \text{for } y \in (0, 1). \tag{4.1}$$

For this left-continuous inverse of the right-continuous F it holds that $Q(y) \leqslant x$ if and only if $y \leqslant F(x)$. It follows that the distribution of $Q(y)$, prior to data, is given by

$$\Pr\{Q(y) \leqslant x\} = 1 - G(y; bF_0(x), b\overline{F}_0(x)) = G(1 - y; b\overline{F}_0(x), bF_0(x)), \tag{4.2}$$

where $G(y; a, c)$ is the distribution function for a $\text{Beta}(a, c)$ and $\overline{F}_0 = 1 - F_0$. Note that (4.2) may be written $J_b(F_0(x))$, where $J_b(x) = G(1 - y; b(1 - x), bx)$ is the distribution of a random y-quantile for the special case of F_0 being uniform on $(0,1)$. It also follows that $Q(y)$ has a density of the form $j_b(F_0(x))f_0(x)$, where $j_b = J'_b$ is the density under the uniform prior measure.

The above leads immediately to results for the posterior distributions of quantiles given a set of data $x_1, \ldots, x_n$, in view of the updating mechanism for the Dirichlet. Assume for simplicity of presentation that data points are distinct; rank them as $x_{(1)} < \cdots < x_{(n)}$, and add on $x_{(0)} = -\infty$ and $x_{(n+1)} = \infty$. Then, for $x_{(i)} \leqslant x < x_{(i+1)}$,

$$H_{n,b}(x) = \Pr\{Q(y) \leqslant x \,|\, \text{data}\} = G(1 - y; b\overline{F}_0(x) + n - i, bF_0(x) + i).$$

It has a suitable density inside the data windows $(x_{(i)}, x_{(i+1)})$ and point masses

$$\begin{aligned}
\Delta H_{n,b}(x_{(i)}) &= G(1 - y; b\overline{F}_0(x_{(i)}) + n - i, bF_0(x_{(i)}) + i) \\
&\quad - G(1 - y; b\overline{F}_0(x_{(i)}) + n - i + 1, bF_0(x_{(i)}) + i - 1) \\
&= \text{const.} \, y^{bF_0(x_{(i)}) + i - 1}(1 - y)^{b\overline{F}_0(x_{(i)}) + n - i}
\end{aligned}$$

at the point $x_{(i)}$. In the case $b \to 0$, there is no posterior probability left between data points; the posterior of $Q(y)$ concentrates on the data points with probabilities

$$\Delta H_{n,0}(x_{(i)}) = p_{n,y}(x_{(i)}) = \binom{n-1}{i-1} y^{i-1}(1 - y)^{n-i} \qquad \text{for } i = 1, \ldots, n. \quad (4.3)$$

The Bayesian quantile estimator function is $\widehat{Q}_b(y) = \mathrm{E}\{Q(y) \,|\, \text{data}\}$. The non-informative limit is of particular interest here:

$$\widehat{Q}_0(y) = \sum_{i=1}^{n} \binom{n-1}{i-1} y^{i-1}(1 - y)^{n-1} x_{(i)} \qquad \text{for } y \in (0,1). \quad (4.4).$$

This estimator can be seen as a Bernshteĭn polynomial approximation to a version of the empirical quantile estimator. It has been worked with earlier by Cheng (1995) and others, but is here given additional interpretational weight as the non-informative limit of a natural non-parametric Bayesian estimator. Issues related to this are discussed further in forthcoming work with S. Petrone. They also exhibit the full posterior process $Q(\cdot)$, as opposed to concentrating on a single y at a time.

4.2   An automatic non-parametric density estimator

Note that $\widehat{Q}_b(y)$ and $\widehat{Q}_0(y)$ are smooth estimates of $F^{-1}(y)$, whereas the corresponding Bayes estimators $\widehat{F}_b(t)$ and $\widehat{F}_0(t)$ for F under quadratic loss have jumps at the data points. One may take the derivative to obtain an estimate of the quantile density function $q(y) = 1/f(F^{-1}(y))$. This may be inverted to find a density estimate $\widehat{f}_b(x) = 1/\widehat{q}_b(\widehat{F}_b(x))$. It requires numerically solving

$\widehat{Q}_b(\widehat{F}_b(x)) = x$ for $\widehat{F}_b(x)$, for each x. A particularly attractive automatic non-parametric density estimator emerges when $b \to 0$. The resulting $\widehat{f}_0(x)$ does not require a separate smoothing parameter. It is zero outside the data range $[x_{(1)}, x_{(n)}]$, with

$$\widehat{f}_0(x_{(1)}) = \frac{1}{(n-1)(x_{(2)} - x_{(1)})} \quad \text{and} \quad \widehat{f}_0(x_{(n)}) = \frac{1}{(n-1)(x_{(n)} - x_{(n-1)})},$$

and is positive and smooth inside. For large n it is approximately equal to a kernel-type density estimator with a Gaussian kernel and variable bandwidth proportional to $n^{-1/2}$.

4.3 Quantile pyramids

The following is an attempt to construct a prior process $Q(\cdot)$ directly on the set of quantile functions. Let us for convenience work on distributions on the unit interval $[0,1]$, so that $Q(0) = 0$ and $Q(1) = 1$. The starting point is a class of distributions that can be specified on arbitrary bounded intervals. Let $p_{m,[a,b]}$ denote a density concentrated at the sub-interval $[a,b]$, to be employed at level m in a growing pyramid, or tree. A simple possibility is to fix a density h on the broadest interval in question and then scale it to $h(x)/\int_a^b h(x)\,dx$ on the required sub-interval. To describe the intended prior quantile process, first draw the median $Q(\frac{1}{2})$ from the distribution $p_{1,[0,1]}$, say. Then draw the quartiles independently, say $Q(\frac{1}{4}) \sim p_{2,[0,Q(1/2)]}$ and $Q(\frac{3}{4}) \sim p_{2,[Q(1/2),1]}$. At stage three, one draws the four remaining octiles $Q(\frac{1}{8}), Q(\frac{3}{8}), Q(\frac{5}{8})$, and $Q(\frac{7}{8})$ independently from the appropriate $p_{3,\cdot}$ distributions on the appropriate intervals, and so on. At stage m new quantiles $Q(j/2^m)$ are generated for $j = 1, 3, \ldots, 2^m - 1$, conditional on the values already generated at levels 1 to $m-1$ above, and $Q(j/2^m)$ depends only upon its two parents $Q((j \pm 1)/2^m)$. In this way a 'quantile pyramid' or 'quantile tree' is grown. The construction resembles that of Pólya trees, see Ferguson (1974), Lavine (1992, 1994), and Walker *et al.* (1999), but is different in spirit and operation. With Pólya trees, the partitions are fixed (as dyadic intervals) but the probabilities are random (as Beta variables); this specifies a random distribution function F. Here we fix probabilities instead (in the natural dyadic fashion) and use random partitions; the result is the quantile function Q.

The quantile pyramid may be stopped at some level m, after which linear interpolation defines the remaining parts of the distribution. It may also be allowed to go on indefinitely to define a full stochastic process Q on $(0,1)$ not determined from a finite number of parameters, thus constituting a genuine non-parametric prior quantile process. Existence of the process follows by the tightness of the sequence of finite approximations. The quantile–Dirichlet process touched on above can be shown to be a special case.

4.4 Posterior quantile pyramids

Assume data $x_1, \ldots, x_n$ have been observed. The challenge is to determine the behaviour of the posterior quantile process. One point of view is that a Q process determines an F for which general principles for finding the posterior of F apply; hence $Q = F^{-1}$ may be analysed too. This is often complicated, however, and the following two alternatives appear quite fruitful.

Assume for illustration of the first idea that there is a simultaneous density for the 15 sedecimiles $q_j = Q(j/16)$ of the form given above, and that Q otherwise is defined by linear interpolation. This means that its F is also linear over the 16 intervals defined by the 15 quantiles, that is the distribution has a constant density $F'(x) = \frac{1}{16}/(q_j - q_{j-1})$ for $x \in (q_{j-1}, q_j)$, for each of the 16 sub-intervals. This gives a likelihood for the data proportional to

$$L_{n,1}(q) = \prod_{j=1}^{16} \left(\frac{1}{q_j - q_{j-1}} \right)^{N_j(q)} \qquad \text{for } q_1 < \cdots < q_{15},$$

where $N_j(q)$ is the number of data points falling in the quantile-defined x-interval (q_{j_1}, q_j). This makes it possible to write down the posterior density of $(q_1, \ldots, q_{15})$. Algorithms of Metropolis–Hastings type may be used to simulate from this distribution; see Hjort and Walker (2002).

A second route is that offered by what may be termed the substitute likelihood. In the setting above, assume that a pyramid-type prior is given for the 15 quantiles $q_1, \ldots, q_{15}$, but we avoid any further specification of Q. The substitute likelihood for data, say $L_{n,2}(q)$, is the multinomial probability

$$\binom{n}{N_1(q), \ldots, N_{16}(q)} \left(\frac{1}{16} \right)^{N_1(q)} \cdots \left(\frac{1}{16} \right)^{N_{16}(q)} = \frac{n!}{N_1(q)! \cdots N_{16}(q)!} \left(\frac{1}{16} \right)^n.$$

With the substitute likelihood and a pyramid-type prior for the quantiles there is a convenient way of expressing the (substitute-based) posterior density, as shown in Hjort and Walker (2002). The point is to rearrange the multinomial terms to match the tree structure of the prior. Here it means that the 15 quantiles follow the same pyramidal dependence structure given the data as they did in the prior. This partial conjugacy type result has the practical advantage that one may deal with one quantile at a time, following the tree. First, one simulates a median from $p(q_8 \,|\, \text{data})$, then the two quartiles from respectively $p(q_4 \,|\, \text{data}, q_8)$ and $p(q_{12} \,|\, \text{data}, q_8)$, and so on. These individual simulation steps could use a Metropolis–Hastings type algorithm. Repeating the full process many times finally gives the end posterior distributions for the quantiles of interest. This rearrangement can actually also be carried out for the first linear interpolation likelihood, and indeed $L_{n,1}$ and $L_{n,2}$ can be shown to behave similarly for large n.

5   Bayesian density estimation and consistency issues

Non-parametric Bayesian density estimation means placing a prior distribution on the set of densities and then analysing aspects of the posterior distribution; an overview with many approaches not yet fully explored is in Hjort (1996). A technical point worth mentioning is that one may compute the posterior mean and variance functions via simulations from the prior alone, that is without having to assess all aspects of the posterior distribution of the density.

Assume a prior π is constructed for an unknown continuous density f. If the data really follow a density f_0, will the posterior distribution $\pi(f \,|\, \text{data})$ concentrate around f_0 as the sample size increases? This topic is currently a hectic one, and various authors have reached different, highly technical and perhaps rather harsh sets of conditions sufficient to ensure consistency; see Wasserman (1998), Barron *et al.* (1999), Ghosal *et al.* (1999), Ghosal and van der Vaart (2000), and Shen and Wasserman (2001).

An important result was established as early as Schwartz (1965). She showed that under a condition which we will denote (A), which is that π puts positive mass on all Kullback–Leibler neighbourhoods $\{f : \int f_0 \log(f_0/f) < \varepsilon\}$ around f_0, the posterior is at least weakly consistent. This means that for almost all sequences under f_0, $\pi(U \,|\, \text{data}) \to 1$ for all weak neighbourhoods $U = \{f : w(F_0, F) < \varepsilon\}$ around f_0; here w is any metric on the cumulatives F_0 and F equivalent to convergence in distribution. Strong Hellinger consistency demands more, namely that $\pi(U \,|\, \text{data})$ should a.s. go to 1 also for the potentially much more complicated neighbourhoods $U = \{f : H(f_0, f) < \varepsilon\}$, where $H(f_0, f)^2 = \int (f^{1/2} - f_0^{1/2})^2 \, \mathrm{d}x = 2 - 2 \int (f_0 f)^{1/2} \, \mathrm{d}x$. Consistency is a statement concerning the pair (π, f_0); one typically wishes conditions under which a prior π gives consistency for large sets of f_0. Conditions ensuring strong consistency given in the many recent papers on the subject typically take the form '(A) and (B)', where (A) is the minimum requirement given above and where (B) varies in content, sharpness, and context from one article to another.

Here we make two points. The first is in the technical tradition and holds that versions of condition (B) given in several recent articles have been too harsh. Walker and Hjort (2001) work with sequences of suitably modified posteriors and show that these are truly strongly consistent under condition (A) alone. The modification in question may be seen as having arisen either from a modification of the prior or from a robustification of the likelihood function. A corollary gives easy and weak conditions for the Bayes estimator (posterior mean) to be Hellinger consistent. This provides a circumventive way of establishing strong consistency for the sequence of real posteriors for many classes of prior distributions.

To give one illustration, consider a Pólya tree prior employing $\text{Beta}(a_k, a_k)$ variables at level k; an old result of Kraft (1964) guarantees that the randomly chosen F a.s. has a density f as long as $\sum_{k=1}^{\infty} 1/a_k^{1/2}$ converges. As long as this holds and the Kullback–Leibler divergence between f_0 and the prior predictive is finite, condition (A) holds. Condition (B) used by Barron *et al.* (1999) leads to the very strict criterion $a_k = 8^k$ (or even faster), which means Pólya trees where the Beta components become almost predetermined even for low k, i.e. trees with leaves that hardly move after three or four levels. With the Walker and Hjort

(2001) strategy, however, the condition $\sum_{k=1}^{\infty} 1/a_k^{1/2} < \infty$ alone is sufficient to secure Hellinger consistency of the predictive distribution; for example, it suffices to have a_k of the type $ck^{2+\varepsilon}$ for some positive ε.

The second point worth raising here is that for most statistical and decision-related applications one would be quite content with weak consistency, which is secured under the basic non-parametric prior condition (A) alone. One may argue that with weak consistency one learns the true cumulative for large n, and this suffices to learn also the derivative, even in the few and rather special situations where strong Hellinger consistency fails. Walker (2000) discusses similar points.

6 Lévy frailty processes and proportional hazards

The assumption of proportional hazards functions plays a major role in survival and event history analysis. Judged by the extremely wide application of methods based on proportional hazards, especially in terms of Cox models, it seems clear that one ought to understand better what this assumption really means. For instance, when assuming proportional hazards this is a statement about averages: on 'average' the hazard in a group is, say, twice the hazard in another group. However, each group will contain a wide variation in individual risk, and one may ask what proportional hazards means for this variation. This, of course, is a frailty point of view. Although frailty considerations often lead to the prediction of decreasing hazard ratios, this is not always so.

Aalen and Hjort (2002) present classes of frailty constructions that necessarily lead to proportional hazards. The approach taken in that article is not Bayesian *per se*, but the classes worked with rely on probability measures being constructed on large sets and having interpretations in Bayesian terms. One construction, complementing that of Aalen and Hjort, is as follows. Individuals are pictured as being continuously exposed to an unobserved cumulative damage-type process of the form

$$Z(t) = \sum_{j \leqslant M(t)} \theta G_j \qquad \text{for } t \geqslant 0. \tag{6.1}$$

Here $G_1, G_2, \ldots$ are taken to be i.i.d. non-negative variables, interpreted as adding over time to the hazard level of the individual, while $M(\cdot)$ is a Poisson process with cumulative rate $\Lambda(t) = \int_0^t \lambda(s)\,\mathrm{d}s$, that is its increments are independent and Poisson($\lambda(s)\,\mathrm{d}s$). The θ is an additional parameter acting multiplicatively on the G_js. There is a certain over-parameterization in that θ may be subsumed into the G_js in (6.1), but it is convenient for later purposes to keep it present. From a modelling perspective, one may work from different sets of assumptions about the G_j distribution, or the Poisson intensity $\lambda(t)$, or the θ factor, depending in suitable ways on covariate information.

The specific connection to the person's survival prospects is to model $S(t \,|\, H_t)$ $= \Pr\{T \geqslant t \,|\, H_t\}$, the survival distribution given the full history of what has happened to the person up to time $t-$, as

$$S(t \mid H_t) = \exp\{-Z(t)\} = \prod_{j \leqslant M(t)} \exp(-\theta G_j) = \prod_{j \leqslant M(t)} (1 - R_j)^\theta. \qquad (6.2)$$

Here $R_j = 1 - \exp(-G_j)$. Letting $L_0(u) = \mathrm{E}\exp(-uG_j)$ be the Laplace transform of the G_js, it follows that the unconditional survival function must take the form

$$S(t) = \mathrm{E}\exp\{-Z(t)\} = \mathrm{E}L_0(\theta)^{M(t)} = \exp[-\Lambda(t)\{1 - L_0(\theta)\}]. \qquad (6.3)$$

Note that even though the survival function is discontinuous given the jumps of the unobservable damage process, it becomes continuous marginally, with cumulative hazard rate $H(t) = \Lambda(t)\{1 - L_0(\theta)\}$ and hazard rate function

$$h(s) = \lambda(s)\{1 - L_0(\theta)\} = \lambda(s)\{1 - \mathrm{E}(1 - R)^\theta\}. \qquad (6.4)$$

One may now add aspects of observable covariate information on to the framework above. Consider individuals $i = 1, \ldots, n$ with covariate vectors $x_1, \ldots, x_n$. For individual i, eqn (6.2) translates as $S(t \mid x_i, Z_i) = \exp\{-Z_i(t)\}$, with a cumulative risk factor process $Z_i(t) = \sum_{j \leqslant M_i(t)} \theta_i G_{i,j}$. Again, x_i may or may not enter the parameters of M_i and θ_i, or the distribution of $G_{i,j}$. For a first illustration, assume that the Poisson process $M_i(\cdot)$ for individual i has intensity $\lambda_i(s) = \lambda_0(s) \exp(\beta^t x_i)$, as happens with standard Poisson regression modelling, and that both the θ_is as well as the risk multipliers $R_{i,j} = 1 - \exp(-G_{i,j})$ have the same distribution across individuals. Then (6.4) implies that individual i has a hazard rate function

$$h_i(s) = \lambda_0(s) \exp(\beta^t x_i) \mathrm{E}\{1 - \exp(-\theta G)\}.$$

In other words, the Cox regression structure has been derived from the frailty process model. For a second illustration with a less standard outcome, let the $\lambda_i(s)$ be as above, take the θ_is to be 1, and model the $R_{i,j}$ as arising from a Beta distribution with parameters $(c\mu(x_i), c - c\mu(x_i))$, say. This allows individuals with different covariates to have different expected levels for their risk multipliers. A reasonable model emerging from this would be

$$h_i(s) = \lambda_0(s) \exp(\beta^t x_i)\mu(x_i) = \lambda_0(s) \exp(\beta^t x_i)\frac{\exp(\gamma^t x_i)}{1 + \exp(\gamma^t x_i)},$$

for example, with additional γ parameters to model the $\mu(x_i)$. The point to note is that the $1 - L_0(\theta)$ term is always inside $(0, 1)$. A particular case with a reasonable biological interpretation is the one with a common Poisson rate but different impacts $R_j = 1 - \exp(-G_j)$ for different individuals, entailing a hazard rate structure of the form $h_i(s) = \lambda_0(s) \exp(\gamma^t x_i)/\{1 + \exp(\gamma^t x_i)\}$.

More general Lévy frailty processes may also be considered here in the place of (6.1), and different specializations lead to different hazard regression structures.

Such are developed and discussed in Gjessing *et al.* (2003). We should point out also that additive regression forms may be derived for the hazards under other assumptions for the Lévy processes. The theme here is obviously of a general nature. It concerns the study of biologically plausible background process models, not immediately or not necessarily with the aim of analysing direct data from them, but rather to deduce plausible functional forms of important statistical quantities. This theme is also visible in some of the work reported by S. Richardson (Chapter 8 in this volume). Such lines of research do have a strong tradition in statistics and probability theory, dating back more than a century, but have perhaps been underplayed in much of contemporary work.

7   Beta processes in a linear hazard model

The purpose of this section is to indicate how the beta process, introduced in Hjort (1984, 1990) as a non-parametric Bayesian tool for modelling cumulative hazard rates in event history analysis, can be used also in Aalen's additive hazard regression model.

Assume survival data exist in the form of triplets (t_i, x_i, δ_i) for n individuals, where the t_is are life-times, possibly censored, the x_is are covariates of dimension p, and the δ_is are indicators for non-censoring. In contrast to the multiplicative Cox regression model, Aalen's linear hazard regression model takes an additive form $h_i(s) = \alpha_0(s) + x_{i,1}\alpha_1(s) + \cdots + x_{i,p}\alpha_p(s)$, with a consequent expression for the cumulative hazard rate H_i and for survival distributions

$$S(t\,|\,x_i) = \exp\{-H_i(t)\} = \overline{G}_0(t)\overline{G}_1(t)^{x_{i,1}} \cdots \overline{G}_p(t)^{x_{i,p}} \qquad (7.1)$$

for an individual with covariate vector x_i, where $\overline{G}_j = 1 - G_j$ is the survival function having α_j as the hazard rate. This model is typically analysed non-parametrically, with emphasis on Aalen plots for the cumulative risk factor functions; see Aalen (1989). We will use Lévy processes for some of these components, and need a framework able to handle discrete cumulative hazard rates as a function of continuous time. The canonical model formulation is that the cumulative hazard $H(t\,|\,x_i)$ for an individual with covariate information x_i should have increments obeying

$$1 - \mathrm{d}H(s\,|\,x_i) = (1 - \mathrm{d}A_0(s))(1 - \mathrm{d}A_1(s))^{x_{i,1}} \cdots (1 - \mathrm{d}A_p(s))^{x_{i,p}}$$

for all s. This implies (7.1) again, with $\overline{G}_j(t) = \prod_{[0,t]}(1 - \mathrm{d}A_j(s))$; see, for example, Hjort (1990) for the product integral.

One may now attempt non-parametric Bayesian modelling of the A_j or G_j functions within this framework. In the general Aalen model these increments are allowed to be both positive and negative (as long as (7.1) behaves like a survival function), and a possibility is to use $A_j = B_j - C_j$, where B_j and C_j are independent Lévy processes with non-negative infinitesimal increments bounded by 1. Let us here focus on a separate sub-model, where the A_js are to have non-negative increments. This is a meaningful model when the $x_{i,j}$s represent risk levels associated with risk factors that a priori increase the hazard. This sub-model is particularly suited for the case where normal and healthy individuals follow a life-time distribution governed by A_0, corresponding to each $x_{i,j} = 0$, and where increased $x_{i,j}$s means increased hazard.

In such a situation, a natural prior takes the form of independent beta pro-cesses Beta($c_j, A_{0,j}$) for A_j; the increments $dA_j(s)$ are independent and approx-imately Beta distributed with mean $dA_{0,j}(s)$ and variance $dA_{0,j}(s)/\{1+c_j(s)\}$. One now needs to generalize the original main theorem about beta processes to arrive at the posterior distribution of $A_0, \ldots, A_p$ given a set of (y_i, x_i, δ_i) data. Such a result has been derived in an unpublished technical report from 1997. Essentially, the A_js still behave like beta processes (with updated parameters) between observed data points $t_{(1)} < \cdots < t_{(n)}$, and there are jumps $\Delta A_j(t_{(i)})$ at the data points with a certain non-standard distribution. Formulae for $\widehat{A}_j = \mathrm{E}(A_j \mid \text{data})$ have been obtained; likewise for $\widehat{S}(t \mid x) = \Pr\{T \geqslant t \mid x, \text{data}\}$ and for posterior variances and covariances. One may also simulate from the poste-rior to allow Bayesian inference for all parameters of interest. Such a programme has been carried out in Beck (2000).

8 Local Bayesian regression

In this section we study a class of Bayesian non- and semiparametric methods for estimating regression curves and surfaces. The main idea is to model the regression as locally linear, and then place suitable local priors on the local parameters.

The non-parametric regression problem concerns data $Y_i = m(x_i) + \varepsilon_i$ for $i = 1, \ldots, n$, where the ε_is are zero-mean i.i.d. with standard deviation σ, and where the $m(x)$ function is unknown. The favoured frequentist method is that of local polynomials, special cases of which are the 'local constant' and the 'local linear' methods; see Fan and Gijbels (1996) for an exposition. The local linear method minimizes for a given position x the function $\sum_{i=1}^{n} K_h(x_i - x)\{Y_i - (a + b(x_i - x))\}^2$ w.r.t. (a, b), and uses $\widetilde{m}(x) = \widetilde{a} = \widetilde{a}_x$ as an estimator. Here $K_h(u) = h^{-1}K(h^{-1}u)$ is a scaled version of a kernel function $K(u)$.

Bayesian non-parametric regression must involve prior modelling of the curve $m(x)$ and calculations related to its posterior. The spline smoothing apparatus may be phrased in such terms. Here we outline methods that become Bayesian generalizations of the successful local polynomial modelling strategy. For illus-tration of the general ideas we focus here on the 'local constant' model and method. The classical Nadaraya–Watson estimator is the minimizer $\widetilde{m}(x)$ of $\sum_{i=1}^{n} K_h(x_i - x)(Y_i - a)^2$, which is $\sum_{i=1}^{n} \overline{K}(h^{-1}(x_i - x))Y_i / \sum_{i=1}^{n} \overline{K}(h^{-1}(x_i - x))$, where we take $\overline{K}(u) = K(u)/K(0)$ to be a symmetric, unimodal kernel function, supported on $[-\frac{1}{2}, \frac{1}{2}]$. Now consider

$$L_n(x, a, \sigma) = \prod_{i \in N(x)} f(y_i \mid x_i, a, \sigma)^{\overline{K}(h^{-1}(x_i - x))}, \qquad (8.1)$$

where $Y_i \mid x_i \sim N(a, \sigma^2)$ and the product is over a local neigbourhood $N(x) = [x - \frac{1}{2}h, x + \frac{1}{2}h]$. The local likelihood view is to interpret $\overline{K}(h^{-1}(x_i - x))$ as the information weight carried by the data pair (x_i, y_i) for the local $a = a_x$ parameter. Maximization of (8.1) gives the local likelihood estimator, which is also the local constant estimator, and for this operation the level of $\overline{K}$ is immaterial; $\overline{K}$ and K give the same results. For the local Bayesian computation we insist on using $\overline{K}$, but with the maximum value 1 corresponding to having full information weight for the underlying model, here the $N(a, \sigma^2)$ model. The scaled kernel smooths the information value down to zero for data pairs outside the $x \pm \frac{1}{2}h$ window. Note that $L_n(x, a, \sigma)$ is the genuine likelihood for the model over this window when the kernel is uniform.

The local Bayesian computation starts out with a prior for the local parameter, say $a = a_x$, for which we take the prior $N(m_0(x), \sigma^2/w_0(x))$ with a suitable local precision function $w_0(x)$. This prior is then combined with the local likelihood $L_n(x, a, \sigma)$, which is proportional to $\sigma^{-s_0(x)} \exp\{-\frac{1}{2}Q(x, a)/\sigma^2\}$. Here $s_0(x) = \sum_{i \in N(x)} \overline{K}(h^{-1}(x_i - x))$, which may also be expressed as $nhf_n(x)/K(0)$ in terms of the kernel density estimator f_n based on K, while $Q(x, a) = Q_0(x) + s_0(x)\{a - \tilde{m}(x)\}^2$, in which $Q_0(x) = \sum_{i \in N(x)} \overline{K}(h^{-1}(x_i - x))\{y_i - \tilde{m}(x)\}^2$. The result is

$$m(x) \mid \text{local data}, \sigma \sim N\left(\hat{m}(x), \frac{\sigma^2}{w_0(x) + s_0(x)}\right), \qquad (8.2)$$

with local Bayes estimator

$$\hat{m}(x) = \frac{w_0(x)}{w_0(x) + s_0(x)} m_0(x) + \frac{s_0(x)}{w_0(x) + s_0(x)} \tilde{m}(x).$$

Note that the non-informative prior case $\sigma^2/w_0(x) = \infty$ yields the frequentist local linear estimator.

This is 'so far, so good', and suffices if one really can come up with a prior guess curve $m_0(x)$ and a strength of belief function $w_0(x)$. More realistically these are not fully specified a priori, and a more general local Bayesian regression programme would comprise the following steps. (a) Give a prior guess function $m_0(\cdot)$ and a prior for σ. (b) For each x, use the local prior $a_x \sim N(m_0(x), \sigma^2/w_0(x))$ for the local constant a_x. (c) Carry out the local Bayesian prior to posterior calculation, employing the local likelihood. This is the calculation carried out above, with the general result

$$\hat{m}(x) = E(a_x \mid \text{local data}) = \hat{m}(x \mid w_0(\cdot), m_0(\cdot)).$$

(d) Use empirical Bayes methods to estimate or fine-tune $w_0(x)$, given $m_0(\cdot)$. (e) Finally, use hierarchical Bayes methods, involving a background or first-stage prior on $m_0(\cdot) = m_0(\cdot, \xi)$, say, to arrive at

$$\hat{m}(x) = E\big[\hat{m}(x \mid \hat{w}_0(\cdot, \xi), m_0(\cdot, \xi)) \mid \text{all data}\big]$$

$$= \int \hat{m}(x \mid \hat{w}_0(\cdot, \xi), m_0(\cdot, \xi)) \, d\pi(\xi \mid \text{all data}).$$

This would typically be computed via simulations of the ξ_js from $\mathrm{d}\pi(\xi \,|\, \text{all data})$; for each of these one computes the precision function $\widehat{w}_0(x, \xi_j)$ using empirical Bayes methods, giving via (8.2) a full curve $\widehat{m}(x \,|\, \widehat{w}_0(\cdot, \xi_j), m_0(\cdot, \xi_j))$. In the end one averages these curves to display the curve estimate.

We note that the computation leading to (8.2) and the Bayes estimator, corresponding to steps (a), (b), and (c), requires only studying the situation at a single position x at a time, so to speak. Steps (d) and (e) really require full simultaneous aspects of the prior modelling of the curve, however. A full description of the local constant prior used to exemplify the general scheme here is that the curve is constant on each of many windows of length h, with a simultaneous multinormal prior for the levels at these windows. For the local linear version of the scheme, the prior model takes the view that the curve is approximately linear inside each of the many small windows, with a simultaneous multinormal prior for the collection of local levels and local slopes. Details, discussion, and generalizations of the various ingredients at work here can be found in Hjort (1998).

Observe that when the width of the local data window is large, these methods reduce to familiar fully parametric Bayesian methods, whereas they are essentially non-parametric when the width is small. The apparatus also encompasses the possibility of using non-informative reference priors for the local parameters, in which case estimators coincide with the by now classical local polynomial frequentist methods.

9 Random shapes with smoothed Dirichlets

Consider the class of closed curves in the plane which can be represented as $R(s)\,(\cos(2\pi s), \sin(2\pi s))$ for $0 \leqslant s \leqslant 1$, with $R(s)$ being some smooth positive function with $R(1) = R(0)$. Various stochastic process models for the radius function $R(s)$ give rise to different random shape models. Kent *et al.* (2000) in effect use such an approach, based on a circularly symmetric Gaussian process for $R(s)$, following up earlier work by Grenander and Miller (1994). This works, but is moderately unsatisfactory in that the paths of such a process can be below zero. This also leads to some interpretational and statistical problems with the Gaussian likelihood approach used in these papers.

A different approach that avoids some of these difficulties is to use smoothed gamma and Dirichlet processes for the random radius function. Let g_0 be a smooth density on $[0, 1]$, periodic in the sense that $g_0(0) = g_0(1)$, with cumulative distribution G_0. Consider a gamma process G with parameters (bG_0, b) on $[0, 1]$; in particular, $\mathrm{E}\,\mathrm{d}G(s) = g_0(s)\,\mathrm{d}s$ while $\mathrm{var}\,\mathrm{d}G(s) = g_0(s)\,\mathrm{d}s/b$. It is fruitful to consider the model smoothing locally over these gamma process increments. Consider therefore $R(s) = \int K_a(s - u)\,\mathrm{d}G(u)$, where $K(u)$ is a kernel probability density, taken to be continuous and symmetric on its support $[-\tfrac{1}{2}, \tfrac{1}{2}]$, and where $K_a(u) = a^{-1}K(a^{-1}u)$ for a bandwidth parameter a. The radius integral is taken to be modulo the circle around which it lives, that is clockwise modulo its parameter interval $[0, 1]$. For pure shape analysis it makes sense to strip away any information about the size of the objects studied. Such size normalization can be achieved in several ways, but the most natural strategy here is to normalize by the average radius length, or, in other words, to condition on the event $\int_0^1 R(s)\,\mathrm{d}s = G(1) = 1$. Therefore let $\overline{G}(\cdot) = G(\cdot)/G(1)$, which is a Dirichlet (b, G_0), and $\overline{R}(s) = \int K_a(s-u)\,\mathrm{d}\overline{G}(u)$. This smoothed Dirichlet process is guaranteed to have total average radius length 1. The distribution of a set of random radii is quite complicated, but in principle is determined via the Hilbert transform results mentioned in Section 3.

Various models of interest emerge via the use of different g_0 functions, perhaps parameterized to reflect wished-for aspects of the shapes. Kent *et al.* (2000) focus on 'featureless' objects. This translates into requirements of circular symmetry and independence of starting point and leads to choosing the uniform density for g_0. Thus, we have a two-parameter model for a random shape, centred at the unit circle. The parameter b has to do with the concentration of the gamma increments around their expected values, while the parameter a reflects the degree of smoothing of the independent Gamma increments. Parameters a and b need to be estimated from one or more observed shapes. Useful properties include the formula $\pi\{1 + (b+1)^{-1}\int(K_a-1)^2\,\mathrm{d}u\}$ for the mean of the random area of the curve and formulae for $\mathrm{cov}\{\overline{R}(s), \overline{R}(s+h)\}$. In ongoing work I have used empirical covariance functions and a certain maximum simulated likelihood strategy to determine parameter estimates.

10 Non-parametric envelopes around parametric models

Some non-parametric Bayesian constructions can be viewed as providing 'non-parametric envelopes' around traditional parametric models. In this light, traditional parametric inference is the limiting case of zero envelope width. The non-parametric Bayesian solutions may hence be seen as robustifications of such procedures, allowing for some amount of modelling error.

10.1 A semiparametric Bayes model

Consider a regression situation with $Y_i = x_i^t \beta + \sigma \varepsilon_i$ for $i = 1, \ldots, n$, where the ε_is come from a distribution G. Let us study the prior where (β, σ) has some prior density π and G independently comes from a $\mathrm{Dir}(b, G_0)$, where G_0 is the standard normal. A large value of b corresponds to G being very close to G_0 and hence to the traditional parametric setup. Seeing the data and knowing the parameters amounts to knowing the ε_is, so G given the data and (β, σ) is an updated Dirichlet with total parameter $bG_0 + \sum_{i=1}^n \delta(\sigma^{-1}(y_i - x_i^t \beta))$. One may show that the posterior of the parameters is $\pi(\beta, \sigma \mid \mathrm{data}) = c\,\pi(\beta, \sigma)L_n(\beta, \sigma)$, where L_n is the likelihood under the null model $G = G_0$, that is the posterior is the same as it would be under the null model. This assumes that the y_is are distinct. Inference for quantities that depend also on G is affected by the non-parametric part of the prior, however. In particular,

$$\widehat{G}(t) = \mathrm{E}\{G(t) \,|\, \text{data}\} = w_n G_0(t) + (1-w_n)n^{-1} \sum_{i=1}^{n} \Pr\{\sigma^{-1}(y_i - x_i^{\mathrm{t}}\beta) \leqslant t \,|\, \text{data}\},$$

where $w_n = b/(b+n)$. This may be seen to be the integral of a smooth function $\widehat{g}(t)$, which is a convex combination of the normal prior density $g_0(t)$ and a kernel-type density estimator $g_n(t)$ with a variable bandwidth approximately proportional to $n^{-1/2}$. Interestingly, a very similar story emerges with the more general process studied in Section 2.3. Essentially, the formula for $\widehat{G}$ holds but with a value of w_n being determined by the distribution H of the B_js.

10.2 Model fitting with control sets

The extra randomness around the normal model introduced by the Dirichlet in the semiparametric setup above did not influence the posterior distribution of (β, σ). Suppose now that G is taken to be a $\mathrm{Dir}(b, G_0)$, but conditioned to have $G(B_j) = z_j$ for each of the chosen sets $B_1, \ldots, B_k$ partitioning the sample space. With such a pinning down of the Dirichlet, the posterior becomes proportional to $\pi(\theta)L_n(\theta)M_n(\theta)$, where $M_n(\theta) = \prod_{j=1}^{k}(bz_j)^{N_j(\theta)}/(bz_j)^{[N_j(\theta)]}$, writing θ for (β, σ). Here $N_j(\theta)$ is the number of $r_i(\theta) = \sigma^{-1}(y_i - x_i^{\mathrm{t}}\beta)$ in B_j, and $x^{[m]} = x(x+1)\ldots(x+m-1)$. This leads to non-standard asymptotics for Bayes estimators, as M_n is of the same stochastic order as L_n; see Hjort (1986). The Bayes estimator balances two aims of equal importance: to be close to the maximum of the likelihood, and to come close to having a fraction of z_j residuals $r_i(\theta)$ in the set B_j for $j = 1, \ldots, k$. This apparatus may be used when one of the intentions of fitting a model is to predict frequencies for certain sets, and can be tailor-made to model-robust quantile regression, for example.

10.3 Randomness around a parametric survival data model

Assume survival data of the familiar type (t_i, δ_i) are available, where δ_i is the indicator for non-censoring, and let $\alpha_\theta(s)$ describe some parametric model for the hazard rate function. To create model uncertainty around it, let A be a $\mathrm{Beta}(c, A_0)$ process centred at the unit rate model; its cumulative hazard rate mean is $A_0(t) = t$ and its variance is $t/\{1+c(t)\}$. Now postulate that $1-\mathrm{d}A_\theta(s) = \{1 - \mathrm{d}A(s)\}^{\alpha_\theta(s)}$ for positive s and give θ a prior π. For large c this becomes ordinary parametric inference for the α_θ model, while for moderate or small c we have a semiparametric Bayesian model around the given parametric one. The survival function for given θ and A is $S_\theta(t) = \prod_{[0,t]}\{1 - \mathrm{d}A(s)\}^{\alpha_\theta(s)}$. Here one may show that the posterior density of θ becomes proportional to $\pi(\theta)L_n^*(\theta)$, where

$$L_n^*(\theta) = \prod_{i:\,\delta_i=1} \Big[\psi\big(c(t_i) + \alpha_\theta(t_i)Y(t_i)\big) - \psi\big(c(t_i) + \alpha_\theta(t_i)(Y(t_i) - 1)\big)\Big]$$

$$\times \exp\left[-\int_0^\infty \big\{\psi\big(c(s) + \alpha_\theta(s)Y(s)\big) - \psi\big(c(s)\big)\big\}\, c(s)\,\mathrm{d}A_0(s)\right],$$

in terms of $Y(s) = \sum_{i=1}^{n} I\{t_i \geqslant s\}$, and where ψ is the derivative of the logarithmic gamma function. When $c \to \infty$ this can be shown to become the familiar likelihood $L_n(\theta)$. For moderate and smaller values of c this leads to model-robust Bayesian parametric inference.

11   Concluding remarks

This article has hopefully helped to illustrate that the field of Bayesian non-parametrics is rich in challenges and possibilities. That its reach is expanding is witnessed for example by the breadth of the contributions in Walker *et al.* (1999). Also, several other chapters in this volume touch upon aspects of non-parametric Bayes in various ways. That its future looks bright is also helped by computational advances over the last decade.

Several of the non-parametric Bayesian stories told breifly here have interesting extensions to more general settings. In particular, many of the models, methods, and results surveyed above for the i.i.d. situation can be generalized to situations with covariate information. For example, forthcoming work with Petrone uses the quantile–Dirichlet process to develop Bayesian inference methods for quantile regression. Such methods have also been developed by Kottas and Gelfand (2001). The Bayesian modelling of local parameters used in Section 8 is also clearly of a general nature, and can be used for example to develop Bayesian Poisson regression methods.

One may also point to further challenges for the field. A theme of interest is to build models that take prior notions of shape into account, such as unimodality in density estimation; see Hansen and Lauritzen (2002) for an interesting construction. Another line of research is that exemplified in Section 10, i.e. enveloping frequently used parametric models in larger models via Bayesian modelling of uncertainty. This may lead to model-robust inference methods with clear interpretations. One example could be to build a time series model where the autocorrelation function is a non-parametrically modelled function centred at say the parametric $AR(p)$ structure, with an extra parameter to dictate the degree of closeness to this centre function. Similar attempts could be geared towards modelling the covariance function in geostatistical models. Yet further challenges include constructing and polishing Bayesian extensions of generalized linear models via modelling of the link functions, as exemplified for example in Gelfand and Mallick (1995) who used mixtures of betas to model the covariate link function for proportional hazards.

Acknowledgements

I have benefited on many levels from my involvement with the HSSS programme, also regarding stimulus for work reported in this article. I have been privileged to work on these themes with Benoît Beck, Arnoldo Frigessi, Håkon Gjessing, Andrea Ongaro, Sonia Petrone, Jean-Marie Rolin, Stephen Walker, and Odd Aalen. Thanks are also due to my fellow editors and to Natal'ya Tsilevich for particularly constructive comments on an earlier version of this article.

References

Aalen, O. O. (1989). A linear regression model for the analysis of life times. *Statistics in Medicine*, **8**, 907–25.

Aalen, O. O. and Hjort, N. L. (2002). Frailty models that yield proportional hazards. *Statistics and Probability Letters*, **58**, 335–42.

Barron, A., Schervish, M. J., and Wasserman, L. (1999). The consistency of distributions in nonparametric problems. *Annals of Statistics*, **27**, 536–61.

Beck, B. (2000). *Nonparametric Bayesian analysis for special patterns of incompleteness*. PhD thesis, Department of Statistics, Université Catholique de Louvain, Belgium.

Billingsley, P. (1995). *Probability and Measure* (3rd edn). Wiley, New York.

Cheng, C. (1995). The Bernstein polynomial estimator of a smooth quantile function. *Statistics and Probability Letters*, **24**, 321–30.

Cifarelli, D. M. and Regazzini, E. (1990). Distribution functions of means of a Dirichlet process. *Annals of Statistics*, **18**, 429–42; corrigendum, ibid. (1994) **22**, 1633–4.

Dey, D., Müller, P., and Sinha, D. (1998). *Practical Nonparametric and Semiparametric Bayesian Statistics*. Springer-Verlag, New York.

Diaconis, P. and Kemperman, J. (1996). Some new tools for Dirichlet priors. In *Bayesian Statistics 5* (eds J. M. Bernardo, J. O. Berger, A. P. Dawid, and A. F. M. Smith), pp. 97–106. Oxford University Press, UK.

Doksum, K. A. (1974). Tailfree and neutral random probabilities and their posterior distributions. *Annals of Probability*, **2**, 183–201.

Efron, B. (2003). Robbins, empirical Bayes, and microarrays. *Annals of Statistics*, to appear.

Escobar, M. D. and West, M. (1995). Bayesian density estimation and inference using mixtures. *Journal of the American Statistical Association*, **90**, 577–88.

Fan, J. and Gijbels, I. (1996). *Local Polynomial Modelling and its Applications*. Chapman & Hall, London.

Ferguson, T. S. (1973). A Bayesian analysis of some nonparametric problems. *Annals of Statistics*, **1**, 209–30.

Ferguson, T. S. (1974). Prior distributions on spaces of probability measures. *Annals of Statistics*, **2**, 615–29.

Gelfand, A. E. and Mallick, B. K. (1995). Bayesian analysis of proportional hazards models built from monotone functions. *Biometrics*, **51**, 843–52.

Ghosal, S. and van der Vaart, A. (2000). Rates of convergence for Bayes and maximum likelihood estimation for mixtures of normal densities. Research Report, Vrije Universiteit, Amsterdam.

Ghosal, S., Ghosh, J. K., and Ramamoorthi, R. V. (1999). Posterior consistency of Dirichlet mixtures in density estimation. *Annals of Statistics*, **27**, 143–58.

Gjessing, H. K., Aalen, O. O., and Hjort, N. L. (2003). Frailty models based on Lévy processes. *Advances in Applied Probability*, to appear.

Green, P. J. and Richardson, S. (2001). Modelling heterogeneity with and without the Dirichlet process. *Scandinavian Journal of Statistics*, **28**, 355–75.

Grenander, U. and Miller, M. I. (1994). Representations of knowledge in complex systems (with discussion). *Journal of the Royal Statistical Society*, B, **56**, 549–603.

Guglielmi, A. and Tweedie, R. L. (2000). MCMC estimation of the law of the mean of a Dirichlet process. Technical Report TR 00.15, CNR–IAMI, Milano, Italy.

Hansen, M. B. and Lauritzen, S. L. (2002). Non-parametric Bayes inference for concave distribution functions. *Statistica Neerlandica*, **56**, 110–27.

Hjort, N. L. (1984). Contribution to the discussion of Andersen and Borgan's 'Counting process models for life history data: a review'. *Scandinavian Journal of Statistics*, **12**, 141–50.

Hjort, N. L. (1986). Contribution to the discussion of Diaconis and Freedman's 'On the consistency of Bayes estimates'. *Annals of Statistics*, **14**, 49–55.

Hjort, N. L. (1990). Nonparametric Bayes estimators based on Beta processes in models for life history data. *Annals of Statistics*, **18**, 1259–94.

Hjort, N. L. (1996). Bayesian approaches to semiparametric density estimation (with discussion contributions). In *Bayesian Statistics 5* (eds J. M. Bernardo, J. O. Berger, A. P. Dawid, and A. F. M. Smith), pp. 223–53. Oxford University Press, UK.

Hjort, N. L. (1998). Local Bayesian regression. Statistical Research Report, Department of Mathematics, University of Oslo, Norway.

Hjort, N. L. (2000). Bayesian analysis for a generalised Dirichlet process prior. Statistical Research Report, Department of Statistics, University of Oslo, Norway.

Hjort, N. L. and Ongaro, A. (2002). On the distribution of random Dirichlet means. Statistical Research Report, Department of Mathematics, University of Oslo, Norway.

Hjort, N. L. and Walker, S. G. (2002). Nonparametric Bayesian quantile inference. Statistical Research Report, Department of Mathematics, University of Oslo, Norway.

Ishwaran, H. and Zarepour, M. (2000). Markov chain Monte Carlo in approximate Dirichlet and beta two-parameter process hierarchical models. *Biometrika*, **87**, 353–69.

Kent, J. T., Dryden, I., and Anderson, C. R. (2000). Using circulant symmetry to model featureless objects. *Biometrika*, **87**, 527–44.

Kerov, A. and Tsilevich, N. (1998). The Markov–Krein correspondence in several dimensions. PDMI preprint 1.

Kingman, J. F. C. (1975). Random discrete distributions. *Journal of the Royal Statistical Society*, B, **37**, 1–22.

Kottas, A. and Gelfand, A. (2001). Bayesian semiparametric median regression modeling. *Journal of the American Statistical Association*, **96**, 1458–68.

Kraft, C. H. (1964). A class of distribution function processes which have derivatives. *Journal of Applied Probability*, **1**, 385–8.

Lavine, M. (1992). Some aspects of Polya tree distributions for statistical modeling. *Annals of Statistics*, **20**, 1222–35.

Lavine, M. (1994). More aspects of Polya tree distributions for statistical modeling. *Annals of Statistics*, **22**, 1161–76.

Lo, A. Y. (1984). On a class of Bayesian nonparametric estimates: I, density estimates. *Annals of Statistics*, **12**, 351–7.

Markov, A. A. (1896). Nouvelles applications des fractions continues. *Mathematische Annalen*, **47**, 579–97.

Regazzini, E., Guglielmi, A., and di Nunno, G. (2000). Theory and numerical analysis for exact distributions of functionals of a Dirichlet process. Research report, Università di Pavia.

Rubin, D. B. (1981). The Bayesian bootstrap. *Annals of Statistics*, **9**, 130–4.

Schwartz, L. (1965). On Bayes procedures. *Zeitschrift für Wahrscheinlichkeitstheorie und Verwandte Gebiete*, **4**, 10–26.

Sethuraman, J. (1994). A constructive definition of Dirichlet priors. *Statistica Sinica*, **4**, 639–50.

Sethuraman, J. and Tiwari, R. (1982). Convergence of Dirichlet measures and the interpretation of their parameter. In *Proceedings of the Third Purdue Symposium on Statistical Decision Theory and Related Topics* (eds S. S. Gupta and J. Berger), pp. 305–15. Academic Press, New York.

Shen, X. and Wasserman, L. (2001). Rates of convergence of posterior distributions. *Annals of Statistics*, **29**, 687–714.

Tsilevich, N. V., Vershik, A., and Yor, M. (2000). Distinguished properties of the gamma process, and related topics. Prépublication du Laboratoire de Probabilités et Modèles Aléatoires de l'Université Paris VI, No. 575.

Walker, S. G. (2000). A note on consistency from a Bayesian perspective. Manuscript, Department of Mathematical Sciences, University of Bath.

Walker, S. G. and Hjort, N. L. (2001). On Bayesian consistency. *Journal of the Royal Statistical Society*, B, **63**, 811–21.

Walker, S. G., Damien, P., Laud, P. W., and Smith, A. F. M. (1999). Bayesian nonparametric inference for random distributions and related functions (with discussion). *Journal of the Royal Statistical Society*, B, **61**, 485–528.

Wasserman, L. (1998). Asymptotic properties of nonparametric Bayesian procedures. In *Practical Nonparametric and Semiparametric Bayesian Statistics*, Lecture Notes in Statistics, No. 133 (eds D. Dey, P. Müller, and D. Sinha), pp. 293–304. Springer-Verlag, New York.

15A
Asymptotics of non-parametric posteriors

Aad van der Vaart
Free University Amsterdam, The Netherlands

1   Introduction

Nils Hjort is to be congratulated on his interesting overview of issues in non-parametric Bayesian statistics. He discusses a wide range of topics. Perhaps it is somewhat biased toward his own work to be a true overview of the field.

In this discussion I raise a number of additional points in the form of questions that we consider of interest for further research, and complement Hjort's contribution by mentioning an additional result on consistency and rates of convergence of posterior distributions.

In general, my discussion addresses questions concerning the theoretical properties of posterior distributions, or Bayes estimators, rather than computational issues or the construction of particular priors. Here the typical question is whether the posterior distribution will approximate a given distribution if the observations are sampled from this distribution. If the the number of observations tends to infinity, then this can be investigated in an asymptotic setting.

2   Questions

2.1   Semiparametric versus non-parametric

The past 20 years have seen a considerable progress in the development of semiparametric methods. The Cox model is now only one amongst many examples of semiparametric models that are routinely applied, in particular in survival analysis or econometrics. Many semiparametric methods are based on the principle of maximum likelihood. In some cases, for instance with semiparametric mixture models, it is maximum likelihood as usual, in other cases this is empirical likelihood (e.g. the Cox model), sieved likelihood, or likelihood after estimating a nuisance parameter or taking out a projection on nuisance scores. This development has been fairly successful in establishing a theory for semiparametric estimation that parallels the Fisher–Cramér theory for maximum likelihood in smooth parametric models: the estimators are asymptotically normal and their asymptotic variance is a minimal inverse 'efficient information' matrix. This is true for selected parameters that are estimable by a $\sqrt{n}$-rate. For parameters for which the 'inverse' from the model to the parameter is non-differentiable, we do not find the same rates or normal limits, but maximum likelihood has still been shown to work well in many inverse problems. One example is the problem of estimating a distribution function based on current status data discussed in the next section.

As yet there is no parallel of these results for semiparametric Bayesian estimators, apart from a few isolated examples where posteriors can be computed explicitly. This raises the following questions:

- What sort of priors on semiparametric models yield posteriors that converge at a $\sqrt{n}$-rate and are asymptotically normal, in those cases where (modified) maximum likelihood is known to have these properties?

- Can Bayesian methods be made to work in inverse problems?

2.2   Really high-dimensional problems

The complete class theorem says that the collection of risk functions of Bayes estimators in a problem dominate (from below) the risk functions of all estimators. If this theorem is true, then we would be assured that in every situation there are priors such that the corresponding posteriors do behave well, as desired in the preceding subsection. Unfortunately, the complete class theorem is true only in restrictive circumstances. For instance, with compact parameter spaces and continuously parameterized models.

'Compact' means small and hence appears to contradict 'non-parametric' or 'semiparametric'. Taking limits of Bayes estimators in high dimensions may require very weak topologies, so that true Bayes estimators may be far from being a complete class. Another question is:

- What does the complete class theorem have to say about non-parametric or semiparametric Bayesian methods?

A somewhat different angle on this issue arises from a comparison between Bayes and maximum likelihood. The method of maximum likelihood has been criticized for being virtually useless for truly high-dimensional problems. For instance, consider estimating a survival distribution $F(t)$ at a fixed point t based on current status data. Thus we observe a sample from the distribution of $(C, 1\{T \leqslant C\}, Z)$ for T the unobserved survival time (with distribution function F), Z a vector of covariates, and C an 'observation time'. To identify the parameter F from the observed data, it is often assumed that the variables T and C are independent given the covariate vector Z. Under this condition and without further modelling one can construct reasonably good estimators for $F(t)$. Maximum likelihood would require that we specify the likelihood, and in particular the conditional distribution $F(t \,|\, z)$ of T given Z. If Z is high-dimensional (say five-dimensional), then applying a likelihood-based method would require handling the six-dimensional function $(t, z) \mapsto F(t \,|\, z)$. This is hard to imagine without some sort of modelling (e.g. of the Cox form), given the usual number of observations. On the other hand, reasonable estimators for the marginal quantity $F(t)$ can be obtained by setting up a clever estimating equation, side-stepping the need for a (correct) model for $F(t \,|\, z)$.

Now if the complete class theorem applies in some way, then this reasonable estimator would be (almost) Bayes. This would suggest that where likelihood encounters a problem by wanting to search over too large a parameter space, Bayes can avoid it by specifying an appropriate prior. But does such a prior exist and how can we find it?

2.3 Proper and fixed

True Bayesians appear to prefer priors that do not depend on the data, or even the amount of data. Furthermore, true Bayesian interpretation of a prior does require that a prior be proper. This imposes some restrictions already for Bayesian procedures for parametric models. Are these restrictions too severe for many non-parametric situations? The fact that supports of probability distributions on complete metric spaces are necessarily σ-compact, and compact means small, would suggest so. It appears that consistency issues have not been addressed for improper infinite-dimensional priors. Data-dependent priors have been considered in consistency studies, but will be more relevant within the context of rates and limit distributions.

2.4 Model selection

In the past ten years there has been much interest in model selection and adaptation, for instance within the context of wavelet 'smoothing', cross-validation, and penalization techniques. One is given a list of models. For each model there is a preferred procedure, but it is necessary to choose the 'right model' first. The obvious Bayesian approach is to attach a prior probability to each of the models, next to spreading prior mass over the models. Implementation of such procedures has been investigated from a computational point of view (e.g. jump Markov chain Monte Carlo (MCMC)), but an open question is:

- Can Bayesian model selection in non-parametric or semiparametric situations achieve equally good results as other methods? What are the correct prior probabilities?

3 Rates of convergence

In this section I give a very brief discussion of results on rates of convergence of posterior distributions, in particular relative to global metrics on densities. These results may be viewed as giving 'consistency with a rate' and hence are stronger than consistency. The results mentioned here give such rates under weaker conditions than mentioned by Hjort. The results below come from Ghosal *et al.* (2000).

We agree that using a global distance is not the best choice for all problems. Furthermore, a rate of convergence may depend on the metric used. For instance, given a prior on a partitioned parameter (θ, η), with θ being finite-dimensional and η being the 'non-parametric part', we would hope that the posterior for θ might have a rate of convergence of $\sqrt{n}$, whereas the rate for the full model is determined by the rate for the harder estimation problem, that of estimating η.

However, at this time there appear to be only isolated results on rates of convergence relative to other metrics. Such results appear to be restricted to cases where a posterior is given by an explicit formula, which is very restrictive on both prior and model. Much work remains to be done.

We consider the situation where the observations $X_1, \ldots, X_n$ are an i.i.d. sample from a density p_0 relative to a σ-finite dominating measure. As a metric d on the set of densities, we choose either the Hellinger or L_1 metric, or the L_2 metric if the densities are uniformly bounded. The 'packing numbers' $D(\epsilon, \mathcal{P}, d)$ of a set of densities are defined to be the maximum number of densities in $\mathcal{P}$ such that the distance between each pair is at least ϵ. Let $B(\epsilon)$ be the set of all densities such that both $P_0 \log(p_0/p) \leqslant \epsilon^2$ and $P_0(\log(p_0/p))^2 \leqslant \epsilon^2$. Let Π_n be a sequence of priors on the set of densities.

Theorem 3.1 *Suppose that for some sequence $\epsilon_n \downarrow 0$ with $n\epsilon_n^2 \to \infty$, a constant C, and sets $\mathcal{P}_n$ with $\Pi_n(p : p \notin \mathcal{P}_n) \leqslant e^{-(C+4)n\epsilon_n^2}$,*

$$\log D(\epsilon_n, \mathcal{P}_n, d) \leqslant n\epsilon_n^2, \tag{3.1}$$

$$\Pi_n(B(\epsilon_n)) \geqslant e^{-Cn\epsilon_n^2}. \tag{3.2}$$

Then the posterior probability $\Pi_n(p : d(p, p_0) > M\epsilon_n \mid X_1, \ldots, X_n)$ tends to zero in P_0^n-probability, for any sufficiently large $M > 0$.

The two displayed conditions of the theorems are related to conditions (B) and (A) of Hjort. Condition (3.2) requires a minimum amount of prior mass in the neighbourhood $B(\epsilon)$ of the true density. Condition (3.1) is a characterization of the size of the model, the 'model' being interpreted as the set $\mathcal{P}_n$ that supports the prior (apart from the exponentially small fraction $\exp\{-(C+4)n\epsilon_n^2\}$). Clearly, the bigger the model, the slower the posterior rate. Condition (3.1) has nothing to do with Bayes, but arises exactly in the same way in bounds on the risk of minimum contrast estimators, or clever special constructs (Wong and Shen 1995). It can be replaced by a condition asserting the existence of appropriate tests, and is directly linked to the conditions used for consistency by Schwartz (1965).

As stated, none of the conditions is necessary. In fact, the paper by Ghosal *et al.* (2000) already contains better ones. Hjort criticizes several recent articles for having used too 'harsh' assumptions; there is nothing 'harsh' about these conditions, however, in particular since condition (3.1) has nothing to do with Bayes.

15B
A predictive point of view on Bayesian non-parametrics

Sonia Petrone
Bocconi University, Milano, Italy

Professor Hjort touches upon many topics, and gives many ideas and suggestions to be developed. In my discussion, rather than making a long list of comments on the many aspects touched upon, I would like to start with an example, which raises some questions and, I think, underlines some open issues in Bayesian non-parametric inference. The problem is described by Coram and Diaconis (2001) and regards studying, by probabilistic techniques, the correspondence between the eigenvalues of random unitary matrices and the complex zeros of Riemann's zeta function. Coram and Diaconis analyse some aspects of this problem by frequentist statistical procedures. Instead, one might want to adopt a Bayesian (non-parametric) approach, but, as I will try to show, many difficulties and open problems arise.

1 An example

Evidence exists of a close connection between the zeta zeros and the eigenvalues of typical unitary matrices. The unitary group U_n is the group of the $n \times n$ complex matrices M such that $MM^* = I$. Let us sample matrices from the Haar measure on U_n. In the literature, there are results on the distribution of the n eigenvalues of a matrix M drawn from U_n and of some statistics related to them. Let us fix $n = 42$. Then, Coram and Diaconis consider the data set given by the $50\,000$ consecutive zeta zeros starting around the $(10^{20} + 271959460)$th zero. For comparison with the eigenvalues of random matrices, they put the zeros onto the unit circle following a 'wrapping' procedure. In this way the data are 1190 realizations of 42 points on a circle, and the problem is to test whether this data set could have come from the Haar measure. As a first test, Coram and Diaconis compare the distribution of the trace of a random matrix with the 'trace' based on the zeta zeros. Results in the literature show that the norm of the trace of a random matrix is remarkably well approximated by an exponential distribution. Let us compute an analogous 'norm squared trace' for each of the 1190 blocks of 42 zeta zeros in the circle, and denote it by W_i, $i = 1, \ldots, 1190$. The Ws would be distributed according to the exponential distribution with mean 1, to very good approximation, if they were actually formed from the trace of random matrices U_n. Therefore, we are faced with a goodness-of-fit problem. Coram and Diaconis consider some frequentist tests for the hypotheses of exponential distribution of the Ws. Could we instead approach the problem from a non-parametric Bayesian point of view? I will discuss some questions that arise (at least, in my mind).

2   A predictive approach

Perhaps one contribution that a Bayesian approach might provide is the emphasis on the predictive aspects of the problem; in this sense, the interest would be on 'predicting the next zero' of the zeta function and one might ask whether knowledge about unitary matrices can suggest a good predictive rule. For predictive purposes, we have to consider the data as dependent, and the simplest dependence assumption is exchangeability. Let us assume that W_i, $i = 1, \ldots, 1190$, are part of an infinite sequence of exchangeable random variables. Then, from the de Finetti representation theorem, exchangeability is equivalent to assuming that the data are conditionally i.i.d. according to a distribution function F, which has a prior probability law π. With no further assumption besides exchangeability, the prior is non-parametric, i.e. its support is the class of all the distribution functions on the sample space. The problem is then to choose a specific non-parametric prior.

2.1   Non-parametric priors for high-dimensional data

We can find several proposals of non-parametric priors in the literature, and some of them are presented in Sections 2 – 4 of Hjort's article (quantile processes in Section 4.3 seem related to a general construction by Dubins and Freedman (1966)). Many non-parametric priors (such as the Dirichlet process, or Pólya trees) are based on a partition of the sample space. However, in the problem of the zeta zeros, if we are interested in the n eigenvalues rather than in the W statistic, the sample space would have dimension 42, so partitioning it seems quite problematic. One of the challenges for research in Bayesian non-parametrics is to deal with high-dimensional data sets.

   An alternative way of constructing a non-parametric prior is based on mixtures of parametric distributions. Early proposals in this direction go back to the 1980s, but encountered computational difficulties which delayed their development. Still, the literature is mainly focused on applications of mixtures in density estimation or regression, and a careful study of a non-parametric 'mixture prior' seems missing. In particular, it would be interesting to explore mixture priors for high-dimensional data, such as the case of the zeta zeros.

2.2   Characterization of the prior via predictive assumptions

So, which prior is well suited for expressing the researcher's information about the distribution of the Ws, or, more generally, of the zeta zeros? It is important that the structure of the prior remains fairly simple, so that the quantities on which one has to express a probability law have a physical interpretation. Nevertheless, the Bayesian approach, requiring one to express a prior on a class of probability measures, still appears complicated, and many researchers feel uncomfortable with that; there must be a way to make it simpler. The predictive approach might be fruitful. The representation theorem establishes a correspondence between the prior and the probability law P of the sequence $\{W_k, k \geqslant 1\}$, and consequently the sequence of predictive distributions. Therefore, one might choose or even characterize the prior through predictive assumptions (see, for example, Cifarelli *et al.* (2000)). This is an interesting line of research, which however has been pursued mainly for univariate exchangeable sequences; it would be of interest to extend the study to high-dimensional data and to more complex dependence structures.

3   Goodness of fit

For its many properties, the Dirichlet process provides the central class of non-parametric priors. However, as a consequence of its discrete nature, it is not appropriate as a prior on the alternative hypothesis in a goodness-of-fit problem (Ferguson 1973). Therefore, different priors have been used, usually centred on the null model. Non-parametric envelopes around parametric models are discussed in Section 10 of Hjort's article, and are regarded as a way of robustifying parametric procedures. However, caution is needed. For example, the peculiar probability of ties among the data implied by the Dirichlet process prior can lead to unexpected consequences for the posterior, so that the latter is not 'robust' (see, for example, Petrone and Raftery (1997)). It would be interesting to study the probability law of ties for the generalizations of the Dirichlet process presented by Hjort in Section 2.

Bayesian procedures for goodness of fit are still to be fully developed. As already mentioned, the problem might be studied from a predictive point of view, so that a model would be evaluated for its predictive properties, according to a loss function. A posteriori, different (Bayesian) procedures might be scored according to the accuracy of the resulting forecasts, possibly in some cross-validation way.

In a parametric framework, testing a precise null hypothesis can give rise to the so-called Lindley paradox as the sample size increases. A similar problem might arise in a non-parametric context, too, when testing a precise hypothesis on the unknown distribution function.

4   Computational issues and asymptotic behaviour

Computational aspects remain an open issue, especially if great accuracy is required, for example for distinguishing between very close hypotheses. Hjort presents interesting results, particularly regarding random Dirichlet means. These methods could also be applied for studying the distribution of functionals of the kind $\int \phi(x; a, \theta) \, dG(\theta)$, where ϕ is a kernel probability density and G is a Dirichlet process. This possibility is also mentioned by Hjort in Section 9.

In the zeta zeros problem, the sample size is 1190. Given the large sample size, would Bayesian and frequentist answers agree? This question is related to the problem of consistency. The notion of consistency discussed in Section 5 of Hjort's article is that, as the sample size goes to infinity, the posterior piles up around the true distribution P_0. This should hold for any given P_0 (or at least, for any P_0 in a certain family). Diaconis and Freedman (1990) suggest that consistency might be reformulated as a finite sample result, without exceptional null sets or 'true values' of parameters. For multinomial probabilities, they develop an explicit inequality which shows that the posterior must concentrate near the observed frequencies. Therefore, the inequality replaces the asymptotics and eliminates the null sets; observed frequencies stand for the true parameters. In a somewhat similar direction, Berti and Rigo (1997) prove Glivenko–Cantelli like results for exchangeable sequences which guarantee that the distance between the predictive and the empirical distribution functions converges uniformly to zero as the sample size increases, almost surely with respect to the probability law P of the exchangeable sequence. Again, the predictive distribution is compared with the empirical distribution function, rather than with the true distribution.

Besides consistency, it would be interesting to study when we can obtain a non-parametric extension of the Bernstein–von Mises–Laplace theorem, regarding asymptotic normality of the posterior.

5   Problems of dependence

Let us consider once more the zeta zeros problem, and the data consisting of the traces $W_1, \ldots, W_N$. Coram and Diaconis find that the sequence $\{W_i\}$ shows a negative serial correlation. In a Bayesian framework, this means that the assumption of exchangeability is too restrictive. Unfortunately, most of the literature on Bayesian non-parametrics considers the case of exchangeable data. Extending the analysis to more complex dependence structures is one of the main open areas of research.

Hjort touches problems of regression, survival analysis, and inference of random shapes, yet many ideas are still to be developed. For the case of partially exchangeable data, Guglielmi and Melilli (2000) show that any prior can be approximated by mixtures of products of Dirichlet processes. Such mixtures have been used in regression problems, where the partition of the data in groups is induced by some covariates (see Muliere and Petrone (1993) and references therein). A general class of priors for dependent data are Dependent Dirichlet processes (MacEachern 2000). The mixture priors briefly discussed in Section 2.1 might be extended in several directions. In particular, multidimensional mixture priors are of interest for application to stochastic regression. Furthermore, the distribution of the mixture weights might depend on covariates, or include a temporal or spatial structure.

Additional references in discussion

Berti, P. and Rigo, P. (1997). A Glivenko–Cantelli theorem for exchangeable random variables. *Statistics and Probability Letters*, **32**, 385–91.

Cifarelli, D. M., Muliere, P., and Secchi, P. (2000). Urn schemes for constructing priors. Technical Report no. 429/P, Politecnico di Milano, Italy.

Coram, M. and Diaconis, P. (2001). New tests for the correspondence between unitary eigenvalues and the zeros of Riemann's zeta function. Technical Report, Department of Statistics, Stanford University.

Diaconis, P. and Freedman, D. (1990). On the uniform consistency of Bayes estimates for multinomial probabilities. *Annals of Statistics*, **18**, 1317–27.

Dubins, L. and Freedman, D. (1966). Random distribution functions. *Proceedings of the Fifth Berkeley Symposium in Mathematical Statistics and Probability*, Vol. 2, pp. 183–214. University of California Press.

Ghosal, S., Ghosh, J. K., and van der Vaart, A. W. (2000). Convergence rates of posterior distributions. *Annals of Statistics*, **28**, 500–31.

Guglielmi, A. and Melilli, E. (2000). Approximating de Finetti's measures for partially exchangeable sequences. *Statistics and Probability Letters*, **48**, 309–15.

MacEachern, S. N. (2000). Dependent Dirichlet Processes. Technical Report, Ohio State University.

Muliere, P. and Petrone, S. (1993). A Bayesian predictive approach to sequential search for an optimal dose: parametric and nonparametric models. *Journal of the Italian Statistical Society*, **3**, 349–64.

Petrone, S. and Raftery, A. E. (1997). A note on the Dirichlet process prior in Bayesian nonparametric inference with partial exchangeability. *Statistics and Probability Letters*, **36**, 69–83.

Wong, W. H. and Shen, X. (1995). Probability inequalities for likelihood ratios and convergence rates of sieve MLEs. *Annals of Statistics*, **23**, 339–62.

Author index

Subject index